Determination of Toxic Organic Chemicals In Natural Waters, Sediments and Soils

Determination and Analysis

T.R. Crompton
Retired Laboratory Manager, National Rivers Authority,
Anglesey, United Kingdom

ELSEVIER

ACADEMIC PRESS
An imprint of Elsevier

Academic Press is an imprint of Elsevier
125 London Wall, London EC2Y 5AS, United Kingdom
525 B Street, Suite 1650, San Diego, CA 92101, United States
50 Hampshire Street, 5th Floor, Cambridge, MA 02139, United States
The Boulevard, Langford Lane, Kidlington, Oxford OX5 1GB, United Kingdom

British Library Cataloguing-in-Publication Data
A catalogue record for this book is available from the British Library

Library of Congress Cataloging-in-Publication Data
A catalog record for this book is available from the Library of Congress

ISBN: 978-0-12-815856-2

For Information on all Academic Press publications
visit our website at https://www.elsevier.com/books-and-journals

Publisher: Candice Janco
Acquisition Editor: Marisa LaFleur
Editorial Project Manager: Jennifer Horigan
Production Project Manager: Nilesh Kumar Shah
Cover Designer: Miles Hitchen

Typeset by MPS Limited, Chennai, India

Determination of Toxic Organic Chemicals In Natural Waters, Sediments and Soils

Contents

Preface xv

1. Hydrocarbons in nonsaline waters 1

1.1 Aliphatic hydrocarbons 1
Head space analysis 3
Gas stripping methods 4
Gas chromatography 5
High-performance liquid chromatography 7
Infrared spectroscopy 7
Thin-layer chromatography 10
Fluorescence techniques 11
Paper chromatography 12
Sampling 12
Oil spillages 13
Metals in spillage oils 16
Continuous monitoring of oil slicks 16
1.2 Aromatic hydrocarbons 17
Gas chromatography 17
Column chromatography 18
Spectrophotometric method 18
Infrared spectrometry 18
Ultraviolet spectroscopy 19
Polycyclic aromatic hydrocarbons 21
Gas chromatography 22
High-performance liquid chromatography 24
Thin-layer chromatography 25
Fluorescence spectrometry 26
Miscellaneous 26
References 28

2. Oxygen-containing compounds in nonsaline waters 35

2.1 Carboxylic acids 35
Gas chromatography 36
High-performance liquid chromatography 37
Thin-layer chromatography 39
Spectrophotometric methods 39
Fluorescence spectrometry 40

	Potentiometry	41
	Polarography	41
	Miscellaneous	41
2.2	**Phenols**	42
	Gas chromatography	42
	High-performance liquid chromatography	44
	Thin-layer chromatography	45
	Raman spectroscopy	46
	Spectrophotometric method	47
	Atomic absorption spectroscopy	49
	Miscellaneous	49
	Preconcentration	49
2.3	**Phenolic acids**	50
2.4	**Methyl *tert*-butyl ether**	51
2.5	**Alcohols**	51
	Spectrophotometric methods	51
	Gas chromatography	52
2.6	**Glycols**	52
	Spectrometry	52
	Gas chromatography	52
	Thin-layer chromatography	53
2.7	**Dioxans**	53
	Gas chromatography	53
2.8	**Esters**	53
	Gas chromatography	53
	High-performance liquid chromatography	53
	Electrokinetic chromatography	54
2.9	**Aldehydes**	54
	Gas chromatography	54
	Thin-layer chromatography	55
	Spectrophotometric methods	55
	Miscellaneous	55
2.10	**Ketones**	56
2.11	**Carbohydrates**	57
	Gas chromatography	57
	Spectrophotometric method	57
	Miscellaneous	57
2.12	**Lactams**	58
	Thin-layer chromatography	58
	Polarography	58
	Enzymic assay	58
References		59

3. Halogen-containing compounds in nonsaline waters 65

3.1	**Saturated aliphatic chloro compounds**	66
	Gas chromatography	66
	High-performance liquid chromatography	70

	Thin-layer chromatography	71
	Headspace analysis	71
	Purge and trap analysis	72
	Negative-ion mass spectrometry	72
	Miscellaneous	73
3.2	**Unsaturated chloroaliphatic compounds**	74
	Gas chromatography–mass spectrometry	74
	Mass spectrometry	75
	Headspace analysis	75
	Purge and trap analysis	75
	High-performance liquid chromatography	75
3.3	**Haloforms**	75
	Gas chromatography	77
	Headspace analysis	79
	Purge and trap methods	79
	Resin adsorption–gas chromatography	80
	High-performance liquid chromatography	81
	Gel-permeation chromatography	81
	Spectrophotometric method	81
	Preconcentration	81
3.4	**Chloroaromatic compounds**	82
	Gas chromatography	82
	Liquid chromatography	82
	Miscellaneous	82
3.5	**Halocarboxylic acids**	82
	Gas chromatography	82
	Isotope dilution mass spectrometry	83
	Miscellaneous	83
3.6	**Polychlorodibenzo-*p*-dioxins and polychlorodibenzofurans**	83
	Gas chromatography–mass spectrometry	83
3.7	**Chlorophenols**	84
	Gas chromatography	84
	High-performance liquid chromatography	86
	Thin-layer chromatography	87
	Miscellaneous	88
	Preconcentration	88
3.8	**Polychlorobiphenyls**	88
	Gas chromatography	89
	Gas chromatography–mass spectrometry	89
	High-performance liquid chromatography	91
	Thin-layer chromatography	91
	Polarography	92
	Miscellaneous	92
3.9	**Polychloroterphenyls**	93
3.10	**Miscellaneous**	93
3.11	**Bromine-containing compounds**	94
References		95

4. **Nitrogen compounds in nonsaline waters** 103

4.1 **Aliphatic amines** 103
Gas chromatography 103
High-performance liquid chromatography 104
Miscellaneous 104
4.2 **Aromatic amines** 105
Gas chromatography 105
High-performance liquid chromatography 105
Miscellaneous 105
4.3 **Amino acids** 106
Gas chromatography 106
High-performance liquid chromatography 106
Miscellaneous 106
4.4 **Amides** 107
Gas chromatography 107
High-performance liquid chromatography 107
Polarography 107
4.5 **Nitrophenols** 107
Gas chromatography 107
Thin-layer chromatography 108
Column chromatography 108
Spectrophotometric method 108
Capillary electrophoresis 108
4.6 **Trinitrotoluene** 109
4.7 **Chloroaniline** 109
Liquid chromatography 109
4.8 **Hydrazines** 109
4.9 **Nitriles** 110
4.10 **Nitrosamines** 110
Gas chromatography 110
Nuclear magnetic resonance spectroscopy 111
4.11 **Ethylenediaminetetraacetic acid** 111
4.12 **Nitriloacetic acid** 111
Column chromatography 112
Polarography 112
Atomic absorption spectrometry 112
Gas chromatography 112
Polarography 113
Spectrophotometric method 114
4.13 **Miscellaneous nitrogen compounds** 115
References 116

5. **Phosphorus containing compounds in nonsaline waters** 119

5.1 **Alkyl and aryl phosphates** 119
5.2 **Adenosine triphosphate** 120

5.3 **Inositol triphosphate** 120
5.4 **Plytase-hydrolysable phosphate** 120
5.5 **Phosphine** 120
5.6 **Organophosphorus compounds** 121
Miscellaneous 121
Nuclear magnetic resonance spectroscopy 121
5.7 **Organophosphorus insecticides and pesticides** 121
References 121

6. **Sulphur-containing compounds in nonsaline waters** 123

6.1 **Mercaptans and disulphides** 123
Gas chromatography 123
Titration method 123
6.2 **Dimethyl sulphoxide** 123
6.3 **Alkylthiols** 124
6.4 **Ethylene thiourea** 124
6.5 **Thiobenzamide** 125
6.6 **Chlorobenzo sulphonic acid** 125
Pyrolysis-gas chromatography—mass spectrometry with single-ion monitoring 125
6.7 **Miscellaneous** 125
References 125

7. **Surface active agents in nonsaline waters** 127

7.1 **Nonionic surface active agents** 127
Gas chromatography 127
Column chromatography 128
Ion-exchange chromatography 129
Spectrophotometric methods 130
Atomic absorption spectrometry 132
Miscellaneous 133
7.2 **Anionic surface active agents** 133
Gas chromatography 133
High-performance liquid chromatography 134
Miscellaneous 135
7.3 **Cationic surface active agents** 136
High-performance liquid chromatography 136
Gas chromatography 136
Titration methods 136
Spectrophotometric methods 136
Miscellaneous 137
Titration methods 138
Spectrophotometric methods 138
References 139

8. **Volatile organic compounds in nonsaline waters** 141

References 143

9. **Multiorganic compounds in nonsaline waters** 145

9.1 **Preliminary extraction of organic compounds** 146
9.2 **Determination of organic compounds** 149
Gas chromatography 149
High-performance liquid chromatography 149
Infrared spectroscopy 150
Miscellaneous 154
References 155
Further reading 158

10. **Pesticides and herbicides in nonsaline waters** 161

10.1 **Organochlorine insecticides** 162
Gas chromatography 162
Mixtures of chlorinated insecticides and polychlorinated biphenyls 166
Gas chromatography–mass spectrometry 167
High-performance liquid chromatography 168
Thin-layer chromatography 169
Other techniques 169
Preconcentration 170
10.2 **Organophosphorus insecticide** 170
Extraction procedures 172
Gas chromatography 173
High-performance liquid chromatography 176
Thin-layer chromatography 177
Spectrometric methods 177
Electrochemical methods 180
Miscellaneous 181
10.3 **Urea herbicides** 182
Gas chromatography 183
High-performance liquid chromatography 184
10.4 **Sulphonylurea herbicides** 186
10.5 **Phenoxyacetic acid–type herbicides** 186
Thin-layer chromatography 189
Paper electrophoresis 189
Miscellaneous 190
10.6 **Triazine type** 191
Gas chromatography 191
High-performance liquid chromatography 193
Thin-layer chromatography 194
Miscellaneous 194

10.7 **Carbamate type** 195
Gas chromatography 195
High-performance liquid chromatography 196
Thin-layer chromatography 197
Enzymic assay 198
Miscellaneous 199
10.8 **Pyrethroids** 199
10.9 **Other insecticides** 201
10.10 **Miscellaneous herbicides** 202
Microextraction of herbicides 202
10.11 **Pesticide survey** 206
References 206
Further reading 217

11. **Miscellaneous organic compounds in nonsaline waters** 219
11.1 **Plant pigments** 219
High-performance liquid chromatography 219
Thin-layer chromatography 220
Spectrophotometric and spectrofluorimetric methods 220
11.2 **Humic and fulvic acid** 221
Polarography 223
Gel permeation chromatography 223
Ultraviolet spectroscopy 223
Fluorescence spectroscopy 224
11.3 **Geosmin** 225
11.4 **Mestranol and ethynyloestradiol** 226
11.5 **Cobalamin (vitamin B12)** 226
11.6 **Algal toxins and blooms** 226
11.7 **Microcystins** 226
11.8 **Anthropogenic, ostragenic and oestrogenic hormones** 227
11.9 **Antibiotics** 228
11.10 **Pharmaceuticals** 229
11.11 **Miscellaneous pollutants** 230
References 230

12. **Organometallic compounds in nonsaline waters** 235
12.1 **Organotin compounds** 235
Gas chromatography—mass spectrometry 235
Inductivity-coupled plasma mass spectrometry 238
Miscellaneous 238
12.2 **Organomercury compounds** 239
Gas chromatography 239
Atomic absorption spectrometry 241
Neutron activation analysis 243
Miscellaneous 243
Storage of mercury-containing samples 243

12.3 **Organolead compounds** 244
Gas chromatography 244
Polarography 245
Atomic absorption spectrometry 245
Preconcentration 246
12.4 **Organoarsenic compounds** 246
Polarography 246
Gas chromatography 247
Atomic absorption spectrometry 247
Ion-exchange chromatography 248
Miscellaneous 248
12.5 **Organoantimony compounds** 248
12.6 **Organogermanium compounds** 249
12.7 **Organocopper compounds** 249
12.8 **Organoselenium compounds** 249
12.9 **Organosilicon compounds** 249
References 250
Further reading 253

13. Organic compounds in aqueous precipitation 255

13.1 **Polycyclic aromatic hydrocarbons** 255
13.2 **Phenols** 255
13.3 **Carboxylic acids** 256
13.4 **Pesticides** 256
13.5 **Organomercury compounds** 256
13.6 **Organotin compounds** 256
13.7 **Organolead compounds** 257
References 257

14. Organic compounds in soil, solvent extraction 259

14.1 **Conventional solvent extraction** 259
14.2 **Conventional solvent extraction from soil packed cartridges** 259
14.3 **Pressurised liquid extraction** 259
14.4 **Microwave-assisted extraction** 259
14.5 **Subcritical water extraction** 263
14.6 **Solid-phase microextraction** 263
14.7 **Subcritical fluid extraction** 263
References 265
Further reading 268

15. Determination of noninsecticidal compounds in soil 269

15.1 **Hydrocarbons** 269
15.2 **Oxygen-containing compounds in soil** 273
Oxalates 273
Nonylphenols 275

Organic acids and ketones 276
Methoxy groups 277
15.3 **Halogen-containing compounds in soil** 277
Chlorinated organic compounds 277
Chlorinated aliphatic hydrocarbons 281
Chlorobenzoic acid 281
Perfluorooctane sulphonyl fluoride 282
15.4 **Nitrogen-containing compounds in soil** 283
Nitro compounds 283
Polycyclic aromatic nitrogen heterocyclic 283
Hydrazines 285
Growth regulators 285
Miscellaneous 285
15.5 **Sulphur and phosphorus–containing compounds in soil** 285
Sulphur compounds dimethyl disulphide 285
15.6 **Miscellaneous organic soil** 286
Humic and fulvic acid 286
15.7 **Volatile organic compounds in soil** 288
15.8 **Mixtures of organic compounds in soil** 290
References 293
Further reading 301

16. Determination of insecticides and herbicides in soil 303

16.1 **Determination of chlorine containing insecticides and herbicides in soil** 303
16.2 **Determination of triazine herbicides in soil** 307
16.3 **Determination of phenoxy acetic acid herbicides in soil** 307
16.4 **Determination of carbamate type of insecticides in soil** 307
16.5 **Substituted urea-type herbicides in soils** 307
16.6 **Determination of imidazolinone herbicides in soils** 307
16.7 **Determination of organophosphorus-type herbicides in soil** 316
16.8 **Miscellaneous insecticides in soil** 318
16.9 **Review of earlier work on the determination of miscellaneous insecticides in soil** 332
16.10 **Determination of fungicides** 332
References 332
Further reading 339

17. Determination of organometallic compounds in soils 341

17.1 **Organoarsenic compounds** 341
17.2 **Organolead compounds** 345
17.3 **Organotin compounds** 345
17.4 **Organomercury compounds** 345
References 345
Further reading 347

18. Determination of organic compounds in sediments 349

18.1 Determination of concentration of typical organic compounds in sediment 349
18.2 Determination of hydrocarbons in sediments 349
18.3 Oxygen-containing compounds 349
18.4 Chlorine-containing compounds 351
18.5 Nitrogen-containing compounds 357
18.6 Phosphorus-containing compounds 358
18.7 Sulphur-containing compounds 358
18.8 Insecticides and herbicides 359
18.9 Miscellaneous organic compounds 362
References 366
Further reading 370

19. Organometallic compounds in sediments 373

19.1 Organoarsenic compounds 374
19.2 Organolead compounds 374
19.3 Organotin compounds 376
19.4 Organomercury compounds 380
19.5 Organosilicon compounds 386
References 386

Index 389

Preface

This book is concerned with a discussion of methods currently available in the world literature up to 2015 for the determination of organic and organometallic compounds in natural waters in soils and sediments.

In the case of soils, the presence of deliberately added or adventitious organic compounds can cause contamination of the tissues of crops grown on the land or animals feeding on the land and, consequently, can cause adverse toxic effects on man, animals, birds and insects and have a profound effect on the ecosystem. Drainage of these substances from the soil can cause pollution of adjacent streams, rivers and eventually the oceans. Some of the substances included in this category are pesticides, herbicides, growth regulators, organic fertilisers, crop sprays and sheep dip.

The presence of organic compounds in river sediments is due, in part, to manmade pollution and monitoring the levels of these substances in the sediment and sediment cores provides an indication of the time dependence of their concentrations over large time spans.

Another consideration is that fish, particularly bottom feeders and crustacean pick up contaminants when sediments enter their gills and the contamination of these creatures has definite toxicological implications both for the creatures themselves, for man who eats them and, in the case of fish meal, for animals.

Sediments have the property of absorbing organic contaminants from water within their bulk (accumulation) and, indeed, it has been shown that the concentration, for example, of some types of insecticide in river sediments is 10,000 times greater than occurs in the surrounding water. The subsequent slow release of the substances will occur from sediment into the surrounding water and, consequently, will continue to cause contamination even after the source of pollution into the water has been stopped.

To date, insufficient attention has been given to the analysis of soils and sediments and one of the objects of this book is to draw the attention of analysts and others concerned to the methods available and their sensitivity and limitations.

Examination for organic substances combines all the exciting features of analytical chemistry. First, the analysis must be successful and in many cases, must be completed quickly. Often the nature of the substances to be analysed for is unknown, might occur at exceedingly low concentrations and might, indeed, be a complex mixture. To be successful in such an area requires analytical skills of a high order and the availability of sophisticated instruments.

The work has been written with the interests of the following groups of people in mind: management and scientists in all aspects of the water industry, river management, fishery industries, sewage effluent treatment and disposal, land drainage and water supply; also management and scientist in all branches of industry. It will also be of interest to agricultural chemists, agriculturalists concerned with the ways in which organic chemicals used in crop or soil treatment permeate through the ecosystem, the biologists and scientists involved in fish, plant, insects and plant life, and also to the medical profession, toxicologists and public health workers and public analysts. Other groups or workers to whom the work will be of interest include oceanographers, environmentalists and, not least, members of the public who are concerned with the protection of our environment.

Finally, it is hoped that the work will act as a spur to students of all subjects mentioned and assist them in the challenge that awaits them in ensuring that the pollution of the environment is controlled so as to ensure that we are left with a worthwhile environment to protect.

Chapter 1

Hydrocarbons in nonsaline waters

Chapter Outline

1.1 **Aliphatic hydrocarbons** 1
Head space analysis 3
Gas stripping methods 4
Gas chromatography 5
High-performance liquid chromatography 7
Infrared spectroscopy 7
Thin-layer chromatography 10
Fluorescence techniques 11
Paper chromatography 12
Sampling 12
Oil spillages 13
Metals in spillage oils 16
Continuous monitoring of oil slicks 16
1.2 **Aromatic hydrocarbons** 17
Gas chromatography 17
Column chromatography 18
Spectrophotometric method 18
Infrared spectrometry 18
Ultraviolet spectroscopy 19
Polycyclic aromatic hydrocarbons 21
Gas chromatography 22
High-performance liquid chromatography 24
Thin-layer chromatography 25
Fluorescence spectrometry 26
Miscellaneous 26
References 28

1.1 Aliphatic hydrocarbons

The identification procedure for oils in natural water samples can be divided into three stages:

1. Isolation of the hydrocarbon components from the pollutant sample.
2. Identification of the same in terms of the petroleum product, for example crude oil, petroleum and gas oil.
3. Identification of the specific source of pollution, such as an individual tanker, tank truck, factory or domestic fuel tank.

Stage (2), a general classification of the oil is often satisfactorily achieved by gas chromatographic techniques possibly coupled with mass spectrometry or infrared spectroscopy applied to a sample of the oil pollutant. Stage (3), the true identification, invariably requires samples from potential sources for comparison with the pollutant.

Determination of Toxic Organic Chemicals In Natural Waters, Sediments and Soils.
DOI: https://doi.org/10.1016/B978-0-12-815856-2.00001-1

This is often attempted again using gas chromatography, by comparison of the resulting chromatograms, but in a less satisfactory and confident manner. Generally, when the comparisons of chromatograms are reasonably similar, the perpetrator of the pollution accepts liability in the case of accumulated circumstantial and scientific evidence and introduces the recommended remedial measures.

Existing gas chromatographic techniques can, in the majority of cases, classify petrol, paraffin, light fuel oils, intermediate fuel oils and, with less ease, lubricating, transformer and cutting oils. Higher boiling products with little volatility are not amenable to conventional gas chromatographic techniques, and recourse has to be taken to other techniques such as the use of capillary columns or non-gas liquid chromatography (GLC) techniques.

Techniques other than gas chromatography or, more commonly, combinations of techniques have been used to characterise oil spills. These include analytical determination such as the infrared spectra, asphaltene and paraffin contents, that provide a general classification of the pollutants (crude oils, fuel oils, oil sludges, etc.) and others, such as the Ni/V ratios, sulphur content and chromatographic profiles, that permit, by comparison with reference samples, their precise identification.

However, another approach involving only one analytical technique, but increasing the number of parameters considered, has been emphasised recently for analysis, that is infrared spectroscopy–gas chromatography. In these cases a multiparametric profile is used for identification, instead of a combination of different analytical determinations and pattern recognition techniques have, often, been applied to improve the diagnostic performance.

The main requirements that must fulfil these fingerprinting parameters besides their specificity are that they must remain unaltered during the weathering processes affecting the pollutant, namely by evaporation, solution, photooxidation and biodegradation. In consequences, both conditions, specificity and stability, need to be investigated in order to evaluate the reliability and the usefulness of any proposed method.

Vos et al. [1] have carried out a detailed study of the analysis of oil-contaminated groundwater to ascertain the rate of filtration of oil components and the effects of their biodegradation under conditions very close to those in a natural aquifer. Large-scale lysimeter experiments are reported in a sand-dune area where the groundwater level could be adjusted with an external overflow device. Details are given of hydrocarbon concentrations determined by adsorption onto Amberlite XAD-4 resins and investigations using chromatography, mass spectroscopy, high-resolution gas chromatography, infrared spectroscopy and ultraviolet spectroscopy.

Matsumoto and Hanya [2] compared the principal hydrocarbons from polluted river water and unpolluted surface water in Japan. The presence of aqualane, and unresolved mixture of hydrocarbons, and n-alkanes with an even number of carbon atoms was related to the occurrence of artificial hydrocarbons (fossil fuels and industrial products) while the occurrence of n-alkanes with an odd number of carbon atoms in unpolluted waters was due mainly to the presence of algae and higher aquatic plants.

Peitscher [3] detected and identified traces of oil on surface water. Samples of oil films on surface water were collected with a cloth made of polyester fabric. The cloth was fixed to telescopic rods, so that less accessible sites could be reached and it was kept on the oil interface for periods ranging from several minutes to 1 hour depending on the amount of oil. The adsorbed oil was extracted and analysed by infrared spectroscopy. Investigation by techniques including gas chromatography and mass spectroscopy facilitated identification of the sources of pollution, which could be confirmed by direct comparison of infrared spectra. Differences between spectra for five different types of oil were distinct for a film thickness of 0.2 mm but less distinct for a film thickness of 0.1 mm.

Various techniques for the determination of aliphatic hydrocarbons are now reviewed.

Head space analysis

Khazal et al. [4] and Drodz and Novak [5] examined and compared the methods of headspace gas and liquid extraction analysis, comparing the gas chromatography of samples of gaseous liquid-extract phases withdrawn from closed equilibrated systems and involving standard addition quantitation, for the determination of trace amounts of hydrocarbons in water. The liquid extraction method [6] is more accurate but it yields chromatograms with an interfering background due to the liquid extractant. The sensitivity of determination of volatile hydrocarbons in water is roughly the same for each method, and the concentration amenable to reliable determination amounting to tens of $\mu g\ L^{-1}$ on a packed column with a flame ionisation detector.

Drodz et al. [7] examined the reliability and reproducibility of qualitative and quantitative headspace analyses of parts per billion of various aliphatic and aromatic hydrocarbons in water using capillary column gas chromatography utilising a simple all-glass splitless sample injection system. They examined the suitability of the standard addition method for quantitative headspace gas analysis for concentrations in the condensed phase up the hundreds of parts per billion.

The headspace method of analysis is less accurate but more sensitive than methods based on liquid extraction. With this method an equilibration time of 10 minutes is adequate for equilibrium between the water sample and the headspace to be achieved.

Various other workers [8,9] have studied the application of headspace analysis to the determination of hydrocarbons in water. McAucliffe [10] determined dissolved individual hydrocarbons in 5 mL aqueous samples by injecting up to 5 mL of the headspace. For petroleum oils, which contain numerous hydrocarbons, very much larger aqueous samples are required. The percentage of hydrocarbons in the gaseous phase, after water containing the hydrocarbons in solution was equilibrated with an equal volume of gas, was found to be 96.7%–99.2% for most C_3-C_8 alkenes. In the case of benzene and toluene the values were 18%, 5% and 21.0% respectively, indicating that the lower aromatic hydrocarbons may be less amenable to the technique.

Gas stripping methods

Swinnerton and Linnenbom [11] were the first to examine the applicability of gas stripping methods to the determination of hydrocarbons in water. They determined C_1-C_6 hydrocarbons by stripping them from water with a stream of helium.

After gas stripping, the hydrocarbons can be passed directly to a gas chromatograph or, to increase sensitivity, trapped in a cold trap and then released into the gas chromatograph. Alternatively, the stripped hydrocarbons can be trapped in, for example active carbon then released into the gas chromatograph. This method offers the possibility of determining trace amounts of organic compounds in water even below the parts per trillion (ppt) level (1 part in 10^{12}, w/w), particularly for the most volatile compounds [12]. Grob and coworkers [13–16] reported an impressive improvement of the method by using a closed-loop system, provided with a small-volume effective charcoal filter, but several precautions are necessary when working at such low concentrations. The complication of the procedure and the sophisticated equipment required results, in view of the absolute amounts of pollutants involved, in overall were excellent results.

Kaiser [17] has described a sensitive degassing technique for trace hydrocarbons in which volatile hydrocarbons up to C_{12} are removed from aqueous solution at 20°C by a stream of dry nitrogen during 2–10 minutes, and passed into a gas chromatographic column cooled in liquid nitrogen. After the degassing period was completed, the column temperature was programmed at a rate of 7.5°C min^{-1} and the hydrocarbons eluted and detected in the usual manner. The detection limit achieved for individual hydrocarbons in water was 100 ppb (10^{-9} wt.%).

Polak and Lu [18] have described a gas stripping method for the determination of the total amount of volatile but slightly soluble organic materials dissolved in water from oil and oil products. Helium is bubbled through a sample of the aqueous liquid, and the gas carries the organic vapours directly to a flame ionisation detector. The detector response plotted against time gives an exponential curve from which the amount of organic material is derived with the aid of an electrical digital integrator.

Colenutt and Thorburn [19] applied a gas stripping technique to various synthetic and actual samples of hydrocarbons in water. Synthetic solutions of

10 μg L^{-1} n-alkanes between n-octane to n-hexadecane prepared by adding acetone solutions of the hydrocarbon to distilled water were put through the procedure. Gas chromatograms were prepared of carbon disulphide extracts. If the solution was analysed almost immediately after preparation a value close to the normal 10 μg L^{-1} for each component was obtained. However, if the aqueous sample was left exposed in an open laboratory for any length of time, the concentration of the lower molecular weight compounds decreased. Thus the concentrations of the lower alkanes are somewhat suspect in that the evaporation effects prior to sampling are unknown.

These workers found up to 10 times greater concentrations by hydrocarbons in rain water to that found in river water.

Drodz et al. [20] used water–air systems with low μg^{-1} levels of benzene, toluene and n-dodecane to evaluate the analytical method of repetitive stripping and trapping of analytes. Closed-circuit and open arrangements were investigated to determine the reliability of the method. In a closed circuit, the stripping/trapping process was accomplished under a conservation or an equilibrium regime, whereas in an open arrangement, conservation or pseudo-equilibration models of trapping were possible. All the above were used for quantitative analysis. Conservation trapping gave better results when working in an open arrangement. Systematic negative errors of 20% and 40% were obtained for the higher aliphatic hydrocarbons and were attributed to varying matrix effects associated with the adsorption of analytes at the air–water interface.

Gas chromatography

Gas chromatography is limited to sufficiently volatile materials. Since petroleum products are readily classified as a type by this technique, the method is slightly specific, and although substances with similar retention could interfere, the chromatogram profile enables many significant interferences to be noticed and discounted.

Gas chromatography has been used to estimate concentrations of volatile petroleum material in groundwater [21]. One-gallon (4.55 L) samples of well water were solvent extracted. Iso-octane was employed when a low-boiling petroleum solvent was thought to be the pollutant. A quantitative determination of 20 mg L^{-1} was achieved using n-octane as an internal standard.

Direct injection of a petrol-in-water solution (10 mg L^{-1}) was found impractical due to background signal or unknown interference peaks [22]. McAucliffe [23] found various impurity peaks in the direct injection of aqueous solutions of hydrocarbons, which limited sensitivity to about 1 mg L^{-1} of individual hydrocarbons.

Two disadvantages of quantitative gas chromatographic analysis of low concentrations are that if different aqueous injection are employed, then background 'noise' considerably affects sensitivity, and if a solvent is used for extraction of the oil, impurities in the solvent can be very significant. Headspace analysis and degassing techniques avoid these disadvantages but are usually applicable only to the more volatile petroleum products. Desbaumes and Imhoff [24] have described a method for the determination of volatile hydrocarbons and their halogenated derivatives in water.

Bridie et al. [25] have studied the solvent extraction of hydrocarbons and their oxidative products from oxidised and nonoxidised kerosene–water mixtures using pentane chloroform and carbon tetrachloride. Extracts are treated with FloriSil to remove nonhydrocarbons before analysis by temperature programmed gas chromatograph. It was concluded that, although each of the solvents extracts the same amount of hydrocarbons, pentane extracts the smallest amount of nonhydrocarbons. FloraSil effectively removes nonhydrocarbons from pentane extracts, but also removes 10%–25% of aromatic hydrocarbons. However, as the other solvents are less susceptible than pentane to treatment with FloraSil, pentane is considered by these workers to be the most suitable solvent for use in determining oil in water.

Belkin and Habre [26] have described the measurement of low levels (10, 100, 500 $\mu g\ L^{-1}$) of gasoline in water using a stripping thermal desorption procedure. A multicomponent collection tube, containing glass beads, Tenax TA, Ambersorb XE-340 and charcoal was used in place of the more common Tenax collection tube. Gasoline recovery from spiked samples was 94%–104%. The procedure used small water samples (15 mL) and was relatively rapid as three samples could be sparged simultaneously and thermal desorption gas chromatography time was less than 30 minutes.

Roberts and Thomas [27] have developed a capillary gas chromatographic method for distinguishing between jet fuel (JP-4) and diesel fuel in groundwaters at the $\mu g\ L^{-1}$ level.

Uhler et al. [28] applied a modified EPA 8200 gas chromatography–mass spectrometry procedure to the molecular fingerprinting of gasolines. The need to recognise and distinguish different types of gasolines that may be present at a contaminated site and the need to determine the relative proportion of inputs from different sources are often critical components of environmental 'forensic' investigations. Historically, identification and differentiation of automotive gasolines (particularly when weathered) has been hampered by analytical limitations of existing methods, notably US EPA Method 8260. In this Uhler et al. [28] describe a modified EPA Method 8260 that is suitable for environmental investigations involving gasolines (and other light petroleum products). In the modified EPA 8260 Method, 109 analytes that can occur in automotive gasoline are quantified in nonaqueous liquid samples, water and soil matrices. The accuracy and precision of the method is demonstrated through comparative analysis using several NIST SRM gasoline standards and replicate analyses.

High-performance liquid chromatography

Schönmann and Kern [29] have used online determination of parts per million analyses of aliphatic hydrocarbons and polycyclic aromatic hydrocarbons in water by high-performance liquid chromatography. They used an online trace enrichment technique allowing direct high-performance liquid chromatographic analysis of aqueous samples containing very low concentrations of polycyclic aromatic hydrocarbons.

This trace enrichment method is based on the affinity of nonpolar pollutants for reversed-phase chromatography supports. When aqueous samples are passed through a reversed-phase column these compounds and any other nonpolar organic compounds present in the sample are immobilised at the head of the column. When detectable quantities of pollutants have been accumulated on the column, they can be analysed by introducing a mobile phase of the desired eluent strength.

Bundt et al. [30] separated low-boiling petroleum hydrocarbons from the aliphatic mono-, di- and polyaromatics by column chromatography. This separation was simplified by removal of nonvolatile polar components using a silica gel–aluminium oxide column.

Infrared spectroscopy

Infrared spectroscopy is generally accepted as an excellent technique for determining aliphatic and aromatic petroleum products of the order of 1 mg or less. Invariably, adsorption intensities of C–H vibrations in aliphatic hydrocarbons are measured and related to the quantity of the oil present. However, Jeltes and Hepple [31] point out that when in very low concentrations, the substances dealt with are in true solution and are aromatic hydrocarbons, infrared spectroscopy is not suitable. In contrast to gas chromatography, the techniques are applicable to nonvolatile petroleum products.

Numerous workers [32–38] have applied infrared spectroscopy to the determination of hydrocarbons in a solvent extract (usually carbon tetrachloride) of water in amounts down to 1 ng L^{-1}.

Should any organic matter such as fatty acids, glycerides, chemical and biochemical oxidation products of petroleum oils be coextracted into the organic solvent, it can seriously interfere with the determination. To overcome this problem, a prior separation stage has been introduced, involving percolation of the carbon tetrachloride extract through a bed of alumina or FloraSil [39–42]. This method is preferred by Hughes et al. [43] who used a modified impeller and a sample bottle immersed in a 30k Hz ultrasonic cleaning bath for dispersion of the carbon tetrachloride solvent.

Polar materials, such as carboxylic acids, esters, ketones, phenols and amines, are strongly absorbed, whereas weak polar hydrocarbons are eluted preferentially and examined by infrared spectrometry [41,44,45].

Gruenfeld [46] compared the relative extraction efficiencies of carbon tetrachloride and trichlorotrifluoroethane in the extraction of oils from water. Although the two solvents were almost equally effective in extracting the oil, trichlorotrifluoroethane is recommended because of its lower toxicity.

Mallevialle [47] carried out a systematic study of the factors governing the determination of hydrocarbons in water by extraction with carbon tetrachloride, followed by FloraSil chromatography and measurement of infrared adsorption. This method was unsuitable for aromatic hydrocarbons.

Mallevialle [47] standardised on extracting 1 L of water, adjusted to pH 3.0 with 10 mL of carbon tetrachloride. The water was stirred mechanically for 30 minutes.

Efficiency ranged from 65% to 95% for the first extraction and from 95% to 100% for the second; these variations can be explained by the fact that hydrocarbons are in the state of pseudosolutions or microemulsions according to their nature and to the presence of surface-active elements.

The peaks of high intensity due to CH and CH_2 chains and can be used to measure hydrocarbon content. Measurements were made in Infrasil cells with 1, 10, 20 and 50 mm optical path. Mallevialle [47] used a reference mixture as follows: 37.5% trimethylpentane; 37.5% cetane; 25% benzene.

The peaks over the range 3000–3150 cm^{-1} are due to the vibrations of the CH groups, particularly of the aromatics. As the intensity of these peaks is much weaker, the aromatics are normally measured by fluorescence or by UV absorption. The peaks at 2962 and 2872 ($\pm$ 10)cm^{-1} correspond to the CH_2.

Using this method Mallevialle [47] was able to measure as little as 0.1–0.2 mg L^{-1} by means of two extractions with 10 mL of carbon tetrachloride from 5 L of water (optical path: 10 mm) with an accuracy of $\pm$ 0.5 mg L^{-1}.

Powell et al. [48] have described a near-infrared method for the determination of total hydrogen bonded to carbon, which they consider should be applicable to the characterisation of oil in polluted river waters. The demonstration is based on the integrated absorption of the first overtone of the C–H stretching band at 1680–1785 nm, which is rectilinearly related to the concentration of C-bonded H for the six hydrocarbons studied (as 0.01 M solution in carbon tetrachloride).

Golden [49] described a procedure which separates hydrocarbons into three groups:

> C_5–C_{10} aliphatic and short-chain aromatic hydrocarbons, C_{10}–C_{32} aliphatic hydrocarbons and polycyclic hydrocarbons such as 3,4-benzopyrene. Nitrobenzene was used to extract the first group which are then identified by gas chromatography. 'Heavy' hydrocarbons are extracted with carbon tetrachloride, passed through FloraSil and analysed by infrared spectroscopy, gas chromatography and by weighing. Polycyclic hydrocarbons are

extracted with cyclohexane and determined by thin-layer chromatography and ultraviolet spectrography.

Standard methods based on this procedure have been published by many authorities and these include Stichling CONCAWE [50] the United States Environmental Protection Agency [51] and the American Petroleum Institute (API-733-58) [52]. Coles et al. [53] reviewed these methods and used the less toxic Freon 113 (1,1,2-trichloro-1,2,2-trifluorethane) as solvent. Whittle et al. [54] set out to find a method of calibration which does not require a standard oil. They examined three of the four methods of calculation described in the literature.

1. Simard et al. [32] using the sum of the absorbances at 2925, 2860 and 2969 cm^{-1} which they attributed to CH_2 CH_3 and CH groups respectively. This method gave some compensation for variations on oil composition compared to an absorbance reading at a single wavelength.
2. Coles [53], also using the sum of three absorbances, but they used 2930, 2960 and 3030 cm^{-1}, attributing the 3030 cm^{-1} peak to CH aromatic groups, the 2960 cm^{-1} peak to CH_3 and the 2930 cm^{-1} peak to CH_2.
3. API-733-58 [52], using the sum of 2930 and 3030 cm^{-1} peaks and the authors [48,51], who use only a single absorbance reading at 2930 cm^{-1}.

These methods of calculation described by Whittle et al. [54] are based on measurements made on individual standard solutions of pure compounds, for example n-hexadecane and toluene, in carbon tetrachloride, and take into account the variations in composition of different oils.

Whittle et al. [54] selected a method based on the use of n-hexadecane, pristone and toluene for the studies on linearity. This method gave the least bias.

Hellman [55] investigated the possibilities and limits of infrared spectrometry for the determination of mineral and fuel oil in surface waters. The method is applicable to concentrations down to about 0.01–02 mg L^{-1}. It was used to detect traces of fuel or mineral oil, also to determine the dispersion of oil and its emulsions in water.

Geyer et al. [56] tested the feasibility of infrared methods in the case of water containing either toluene, trichloroethylene or methylene chloride. The characteristic absorption bands of these substances in a carbon tetrachloride extract following clean-up on an alumina column were illustrated in the presence or absence of mineral oil. Toluene and oil could only be distinguished with difficulty, while trichloroethylene did not interfere with the quantitative determination of mineral oil; quantitative separation only was possible in the presence of methylene chloride.

Peitscher [3] detected and identified traces of oil on surface water. Samples of oil films on surface water were collected with a cloth made of polyester fabric. The cloth was fixed to telescopic rods so that less accessible sites could be reached and it was kept on the oil interface for periods ranging from several minutes to 1 hour depending on the amount of oil. The absorbed oil was extracted and analysed by infrared spectrometry. Investigation by techniques including gas chromatography and mass spectrometry facilitated identification of the source of pollution, which could be confirmed by direct comparison of infrared spectra. Differences between spectra for five different types of oil were distinct for a film thickness of 0.2 mm but less distinct for a film thickness of 0.1 mm.

Shtivel et al. [57] determined petroleum products in water by infrared spectroscopy in the 3.57–3.12 μm range. The concentrations of aliphatic and aromatic hydrocarbons in a carbon tetrachloride extract could be determined separately after separation of the polar compounds on aluminium oxide. A drawback of the method was that it did not take into account the presence of CH groups and quaternary carbon atoms.

Ramsey and Wei [58] used supercritical fluid extraction coupled with fixed wavelength infrared detection to carry out oil in water analysis.

Thin-layer chromatography

The great advantage of thin-layer chromatography is its simplicity and low cost. Hydrocarbon components of petroleum products can be separated very efficiently from polar contaminants on the thin-layer chromatography plate. In all quantitative methods, a sample oil is compared to a range of known concentrations of an identical or similar oil or hydrocarbon mixture. Channel thin-layer chromatography has also been used by Berthod [59].

During development, the samples are confined to narrow bands (2 mm × 50 mm), formed by removal of parallel thin layers, and thus the amount of material in the spot can be calculated, after calibration, from the length of the spot. The technique was applied to trace analysis of hydrocarbons in water.

A micromolecular thin-layer chromatographic technique was developed by Koppe and Muhle [60] for the detection and determination of dissolved hydrocarbons in natural waters. Silica gel plates were employed and developed in a horizontal position. Aqueous samples were extracted with carbon tetrachloride and the extracted organic layer chromatographed using carbon tetrachloride as the initial developing solvent. Circular patterns were produced by spraying with 0.5% phosphomolybdic acid in butanol solution, the nonpolar components of the oil forming blue-grey peripheries, while weakly polar components produced violet-grey colours in the solvent field. Ethanol was applied as a second developing solvent, which caused the nonvolatile hydrocarbon components to concentrate on the periphery. Perhydropyrene was found to be a suitable comparison standard. Under the specified conditions, the lower detection limit is claimed to be polycyclic aromatic hydrocarbons, but for containing a considerable quantity of saturated hydrocarbons, the limit is 0.1 mg L^{-1}. A photobromination process was described for the determination of more volatile hydrocarbons down to n-hexane, which are not determined by the above procedure.

Hunter [61] discussed the quantitation of environmental hydrocarbons using thin-layer chromatography and compared the relative effectiveness of gravimetric and densitometric evaluations of the developed plate. The method involves the use of silica gel adsorbent, which is capable of separating samples into saturates (alkanes) and unsaturates (mainly aromatics) at sensitivities to 0.5 μg. Comparison of the data obtained by gravimetric and densitometric methods indicates that densitometry is less accurate owing to variable hydrocarbon response, and it is concluded that the need for repeated checks and calibration make this method only marginally more convenient than the more accurate gravimetric procedure.

Various other workers have discussed the application of thin-layer chromatography for the determination of aliphatic hydrocarbons in nonsaline waters [50–54,59–79].

Fluorescence techniques

Petroleum products contain many fluorescing components, for example aromatic hydrocarbons polycyclic aromatic hydrocarbons and various heterocyclic compounds. The development of improved techniques and instrumentation has led to the use of this technique for the identification of crude and residual oil pollutants in a marine environment and of motor and related products [80–83]. When applicable, fluorescence techniques are extremely sensitive. The predominant fluorescent substances in petroleum products are polycyclic aromatic hydrocarbons and heterocyclic compounds. Therefore fluorescence should be dependent on the type of oil being examined. This has been found to be an important factor [84], and the variations in fluorescence intensities of up to eight times have been found within a small number of lubricating oils examined. Therefore most fluorescent techniques for determining oil in water with useful accuracy demand a sample of the polluting oil, so that the fluorescence characteristics of the oil can be evaluated. Standard hydrocarbons do not show any significant fluorescence. The rivers, lakes and seawater in the vicinity of estuaries; the background fluorescence of naturally occurring organic matter; and sewage effluent can be very high, and cause serious interference, while in groundwater and in distant seawater is it usually less significant.

Various workers [84–92] have discussed methods for evaluating fluorescence in petroleum polluted natural water samples and petroleum spillages.

Paper chromatography

Sinel'nikov [63] determined bitumoids in open reservoir water using paper chromatography. The water sample is extracted repeatedly with chloroform at pH 7.0 and then at pH 3.0. The combined extracts are evaporated at 40°C to a small volume and the bitumenoids are concentrated in the zone of capillary rise on a strip of chromatographic paper and are separated by dipping the strip into 5 mL of 70% ethanol and allowing a chromatogram to develop for 12 hours. After drying, the separate bitumoid fractions are cut out of the paper and extracted with 10 mL of chloroform and the fluorescence of the extract measured.

Sampling

Gibb and Barcelona [64] pointed out that samples taken for quality control of groundwater may be adulterated by material used in the construction of monitoring wells as well as the containers used to collect the samples. Similar problems also occur when sampling from a distribution system. Physiochemical criteria in relation to such sampling are outlined. Criteria for construction of groundwater quality monitoring wells are outlined (the number, their location, diameter, type of cement) together with the types of sampling devices.

Pankow et al. [65] have described a syringe and cartridge method for downhole sampling of trace organics in groundwater. A sampling device is described, which is lowered down piezometers with a tube and consists of a small cylindrical cartridge of sorbent material attached to a syringe and operated from the surface. Sample analysis is performed using gas chromatography–mass spectrometry. Field tests conducted at an inactive landfill site show three of the 26 compounds identified to be landfill related. The advantages of this technique are minimisation of potential for volatilisation losses: avoidance of contamination or adsorption losses and convenience and high sensitivity.

Kola et al. [93] evaluated the application of ultraviolet–persulphate oxidation in a total organic carbon analysis for the determination of oil contamination from forestry in groundwater.

Different chain oils (tall, rapeseed and mineral oils) were used as model compounds to evaluate and optimise the applicability of a UV–persulphate TOC-analyser for quantitative determination of forestry oils and to follow the progress of their biodegradability. It is shown that the potassium persulphate–UV oxidation method is not sufficient to oxidise chain oils completely. There were differences in oxidation efficiency between different oils, changing from approximately 46% measured for tall oil to about 25% observed for rapeseed chain oil. The addition of Triton X-100 surfactant up to 2% (w/w) was observed to increase the oxidation efficiency, for example

75% for tall oil. The observations can be explained by assuming that in the presence of surfactant the emulsions are more homogenous and stable. Optimisation using two-level, full-factorial design (temperature of the oxidation chamber and the amount of persulphate) was studied. The results show that the UV–persulphate oxidation TOC-analyser is not a suitable method to monitor biodegradability of chain oils.

Oil spillages

A general classification of oil, for example crude oil, petroleum, gas oil, is often satisfactorily achieved by gas chromatographic techniques possibly coupled with mass spectrometry or infrared spectroscopy applied to a sample of the oil pollutant. The true identification invariably requires samples from potential sources for comparison with the pollutant.

Capillary columns provide greater resolution and therefore more detail for comparison between a polluting oil and suspect sample. The enormous separation power available has been demonstrated in their application to petroleum analysis. Gouw et al. [66] describes a versatile 10 mm × 0.25 mm capillary column coated with CV-101 and its application to the separation of hydrocarbon mixtures in the C_4–C_{58} n-alkane range [67].

Jeltes and Veldink [22] found that the polar liquid phase was more suitable for studying the major components of petrol gas oil and diesel oil forming true solutions in water. With such a phase, saturated hydrocarbons tended to elute before aromatic hydrocarbons, which were found to be to principal components in true solution. Invariably, dual-packed columns have been employed, and one of the earliest articles devoted to the identification of petroleum products is that of Lively [21], who use dual 1.2 m × 6 mm columns packed with 20% SE-30 as the liquid phase and a Chromosorb solid support. Most subsequent workers in this field have employed the same or a similar liquid phase. Liquid phases of similar properties that have been employed are 5% and 10% OV-1, 20% SE-52, 5% E301 [94], and 22% 301 [95], 10% and 20% Apiezon L. These liquids are essentially nonpolar substances, but more polar phases 5% and 10% polyethylene glycol 1500 have been used for investigating concentrations of soluble components in diesel and gas oil and in petrol.

Detailed salient points involved in classifying and identifying oils have been compiled [94]. Most distillate products, for example petrol, paraffin, fuel oils and diesel oil, and even very similar products such as turbojet fuel and kerosene, can be classified in the absence of excessive weathering effects. Various chromatograms have been published: kerosene, turbojet fuel, steam-cracked naphtha; lubricating oil–gas oil-weathered paraffin [96]; white spirit, turpentine substitute, paraffin, 30s fuel oil [97]; standard gasoline; petrol; diesel field and gas oil, various gasoline, diesel fuels and standard gasoline; petrol diesel fuel and gas oil, various gasolines, diesel fuels and aviation fuels [94].

A very practical description of the application of gas chromatography to quantitative analysis of petroleum products in aqueous and soil samples is contained in the report by CONCAWE [98]. Well-detailed procedures are given for three hydrocarbon mixtures with different boiling point ranges.

Jeltes [99] has described a gas chromatographic method for the determination of mineral oil–water in which a true solution of diesel fuel and gas oil in water were analysed by isothermal gas chromatography at 110°C with polyoxyethylene glycol 1550 as stationery phase and Chromosorb W as support. For two-phase oil–water systems, the oil was extracted with carbon tetrachloride and the extracts were analysed by temperature programmed gas chromatography with SE-30 as stationery phase by a method similar to that of Benyon [98]. Components of mineral oils, that is water-soluble types boiling below 300°C, could be identified and determined, and it is possible sometimes to establish the origin of the oil.

An interesting gas chromatographic technique for identifying petroleum products, including lubricating oils, is that of Dewitt Johnson and Fuller [100]. A gas chromatograph column effluent was split in order that it could be simultaneously sensed by a double-headed flame photometric detector, specific for both sulphur and phosphorus compounds, and by dual-flame ionisation detectors for carbon compounds in general. Since most petroleum products are claimed to contain sulphur and phosphorus compounds, three chromatographic traces were obtained for the products examined.

McAucliffe [101] reported that the sampling errors caused in the injection of water–hydrocarbon mixture onto a gas chromatographic column can be overcome by the addition of acetone for solubilisation only when the hydrocarbon chain length is less than 12.

Adlard and Matthews [102] applied the flame photometric sulphur detector to pollution identification. A sample of the oil pollutant was submitted to gas chromatography on a stainless steel column 1 m $\times$ 3 mm packed with 3% of OV-1 on AW-DMCS Chromosorb G (85–100 mesh). Helium was used as carried gas (35 mL min^{-1}). The column effluent was split between a flame ionisation and a flame photometric detector. Adlard and Matthews [102] claim that the origin of oil pollutants can be deduced from the two chromatograms. The method can also be used to measure the degree of weathering of oil samples.

Lur'e et al. [103] have described a method which involves extraction of 500 mL sample with hexane, concentration of the dried extract to 1 mL and gas chromatography of an aliquot of this solution. For lower levels of hydrocarbons, 2 L of sample is allowed to percolate through a column of activated carbon; the carbon is dried in air, then the hydrocarbons are extracted with chloroform and the extract is concentrated for gas chromatography.

Coles [104] found that in the identification of kerosene and aviation fuels, the usual packed columns provided sufficient detail only up to C_{13} n-alkane, a range readily altered by evaporated effects. A suitable capillary column was developed which revealed extra detail of minor components above C_{13} n-alkane and consisted of a 45 m × 0.25 mm stainless steel column coated with OV-101. This gave satisfactory resolution and stability in the 50°C–310°C temperature range.

Jeltes and Van Tonkelaar [105] compared gas chromatographic and infrared methods for the determination of dissolved mineral oil in water. They saturated various petroleum fractions with water by shaking for 4 minutes. Then, after 2 days, the clear aqueous phases were extracted with carbon tetrachloride. The polar compounds were removed from the carbon tetrachloride extracts by shaking with FloriSil and decanting after the FloraSil had settled. These extracts were analysed by infrared and gas chromatographic methods, both before and after treatment with FloraSil, and also after the two-phase systems had been exposed for 8 weeks to light and air.

Jeltes and Van Tonkelaar [106] investigated problems of oil pollution, the nature of the contaminants and the chemical methods used for their detection. In particular, the use of gas chromatography to obtain 'fingerprint' chromatograms of oil pollutants in water, and of infrared spectrophotometry to determine the oil contents of soils and sediments is discussed.

Kawahara et al. [107] have discussed the characterisation and identification of spilled residual fuel oils on surface water using gas chromatography and infrared spectrophotometry. The oily material was collected by surface skimming and extraction with dichloromethane, and the extract was evaporated.

Lysyj and Newton [108,109] have described a multicomponent pattern recognition and differentiation method for the analysis of oils in nonsaline waters. The method is based on that described earlier by Lysyj and Newton [109] which depends on the thermal fragmentation of organic molecules followed by gas chromatography. Dried algae and outboard motor oil were analysed and a specific pattern or numerical 'fingerprint' was obtained for each by polygraphic means. The algal pattern comprised of three specific peaks and seven peaks common to those of oil, whereas the oil pattern comprised two specific peaks and seven peaks common to those of the algae.

Garra and Muth [110] characterised crude, semirefined and refined oils by gas chromatography. Separation followed by dual-response detection (flame ionisation for hydrocarbons and flame photometric detection for sulphur-containing compounds) was used as a basis for identifying oil samples. By examination of chromatograms, it was shown that refinery oils can be artificially weathered so that the source of oil spills can be determined.

McAucliffe [23] found various impurity peaks in the direct injection of aqueous solutions by hydrocarbons, which limited sensitivity to approximately 1 m L^{-1} of individual hydrocarbons. He employed a 3.66 m × 6 mm column containing 25% SE-30 on firebrick and precolumn containing Ascarite to absorb the water and improve the sensitivity.

Bridie et al. [25] have studied the solvent extraction of hydrocarbons and their oxidative products from oxidised and nonoxidised kerosene–water mixtures, using pentane, chloroform and carbon tetrachloride. Extracts are treated with FloraSil to remove nonhydrocarbons before analysis by temperature programmed gas chromatography. From the results reported, it is concluded that, although each of the solvents extracts the same amount of hydrocarbons, pentane extracts the smallest amount of nonhydrocarbons. FloraSil effectively removes nonhydrocarbons from pentane extracts, but also removes 10%–25% or aromatic hydrocarbons. However, as the other solvents are less susceptible than pentane to treatment with FloraSil, pentane is considered by those workers to be the most suitable solvent for use in determining oil in water.

Metals in spillage oils

Determination of metals can assist in the characterisation of crude oil spillages. Distilled products should not contain these elements. In particular, the ratio of vanadium to nickel in oil can be characteristic of its source of origin. The ratios of trace elements, particularly V/Ni, have been shown to be unaffected by prolonged weathering effects.

Various techniques have been employed for the determination of metals in oils. These include atomic absorption spectrometry [111], neutron activation analysis [112–117], X-ray fluorescence spectrometry [118,119], gamma ray spectrometry [120] and oscillographic polarography [121].

Continuous monitoring of oil slicks

Although several methods have been developed for continuously monitoring oil or organic materials in water by techniques such as ultraviolet spectroscopy [122] and infrared analysis [123–128], only a very limited number of published works deal with the detection of surface slicks. This is probably due to the inherent difficulties involved in their design and economic utilisation.

Mattson [129] and others [130–132] have carried out investigations on the application of infrared spectroradiometry to the problem of remote detection and identification of oil slicks without physical contact with the oil or water. The spectroradiometer incorporates a circular variable interference filter as a monochromator and a small computer for instrumental control and data reduction. In the preliminary experiments, films of styrene and oleic acid were studied, and spectra between 8.55 and 12.5 μm obtained. Radiation was reflected from the oil film gathered by a Cassegrainian collector and focused on the circular variable filter wheel of a spectroradiometer, which in these particular experiments allowed scanning between 8.5 and 12.5 μm. Thin films of approximately 10 μm thickness tended to produce weakly characteristic styrene reflection spectra, but films of approximately 300 μm produced very definite styrene reflection spectra.

1.2 Aromatic hydrocarbons

Aromatics can occur in natural waters due to two causes: either the presence of aromatic substances or the occurrence of mineral oils, which although predominantly aliphatic do contain some aromatic constituents and indeed, some polycyclic aromatic hydrocarbons. As spectroscopic methods particularly infrared and fluorescence spectrometry are particularly suitable for determining aromatics, these have become the method of choice.

Various procedures for the determination of aromatic hydrocarbons in nonsaline waters are now discussed.

Gas chromatography

Wasik and Tsang [133,134] have described a method for the determination of traces of arene contaminants using isotope source dilution gas chromatography. They used perdeuterated benzene as the isotope source and analysed solutions containing 10–20 $\mu g\ L^{-1}$ of benzene and toluene. This method is most effective when the isotope can be completely separated from the parent compound and other contaminant, otherwise the isotope ratio must be determined by mass spectrometry.

Gas chromatography has found some applications in the determination of simple aromatics in water. Melkanovitskaya [135] has described a method for determining C_6–C_8 aromatics in subterranean waters. In this method the sample (25–50 mL) is adjusted to pH 8.0–9.0 and extracted for 3 minutes with 0.5 or 1.0 mL of nitrobenzene; the extract is washed with 0.3 mL of 5% hydrochloric acid of 5% sodium hydroxide solution and with 0.3 mL of water adjusted to pH 7.0. The purified extract is subject to gas chromatography at 85°C on a column (1 m × 4 mm) packed with 15% of polyoxyethylene glycol 2000 on Celite 545 960–80 mesh and operated with nitrogen (10 $mL\ min^{-1}$) as carrier gas, decane as internal standard and flame ionisation detection.

Senn and Johnson [136] used capillary column gas chromatography for the determination of aromatic hydrocarbons in natural waters.

Stolyarov [137] used headspace gas chromatography to determine microgram per litre quantities or aromatic hydrocarbons in waste water. Urano et al. [138] developed a steam carrier gas chromatographic method for determining aromatics in refinery wastes.

Mashayeteki et al. [139] have reported a method for the determination of biphenyl and biphenyl oxide in water using dispersive liquid–liquid microextraction followed by gas chromatography.

Semenov et al. [71] determined small amounts of petroleum products in chloroform extracts of natural water by extracting the sample (200–500 mL) followed by thin-layer chromatography on alumina. He developed the chromatogram with light petroleum–carbon tetrachloride–acetic acid (35:15:1) and examined the plate in UV radiation; the petroleum products exhibited three zones (pale blue, yellow and brown). Each zone is then extracted with chloroform, the fluorescence of the extracts measured and the results referred to a calibration graph. The sensitivity is 0.1 mg L^{-1}. The infrared and fluorescence spectra of the zone obtained with various petroleum products are discussed.

Column chromatography

Götz [140] has pointed out that the determination of hydrocarbons in water and sludge requires a preliminary treatment for removal of fats and fatty acids which interfere with the estimation. The efficiency of column chromatography for this purpose using Kieselgel, FloraSil and aluminium oxide as adsorbents was tested by these workers.

Louch et al. [141] have studied the dynamics of the extraction of benzene, toluene and *p*-xylene from water using liquid-coated fused silica films.

Spectrophotometric method

Dudova and Diterikhas [142] analysed down to 0.2 mg L^{-1} aromatics in underground water by a photometric procedure. The sample (100 mL) is shaken vigorously for 3 minutes with 10 mL of benzene-free hexane; the extract is then filtered through paper and3 mL of nitrating mixture added (10 g of dry NH_4NO_3 in 100 mL of concentrated H_2SO_4), with vigorous stirring. The lower layer is poured into a test tube heated in a boiling water bath for 30 minutes then neutralised with concentrated aqueous ammonia and extracted with 10 mL of ethyl ether. To determine benzene 1.0 mL of the extract is diluted with acetone to 5 and 5 mL of 20% aqueous sodium hydroxide added, and after 30 minutes is measured photometrically (green filter). To determine toluene and other aromatic hydrocarbons, ethanol, instead of acetone, is used, only one drop of 5% sodium hydroxide solution is added and the colour measured without delay.

Infrared spectrometry

Geyer et al. [56] tested the feasibility of infrared methods in the case of water containing toluene, trichloroethylene or methylene chloride. The characteristic adsorption bands of these substances in a carbon tetrachloride extract, following clean-up on an alumina column, were illustrated in the presence or absence of mineral oil. Toluene and oil could only be distinguished with difficulty, while trichloroethylene did not interfere with the quantitative determination of mineral oil; qualitative separation only was possible in the presence of methylene chloride.

Shtivel et al. [57] determined petroleum products in water by infrared spectroscopy in the 3.57–3.12 μm range. The concentrations of the aliphatic aromatic hydrocarbons in a carbon tetrachloride extract could be determined separately after separation of the polar compounds on aluminium oxide. A drawback of the method was that it did not take into account the presence of CH groups and quaternary carbon atoms.

Ducreux et al. [143] compared infrared spectroscopic, ultraviolet spectroscopic and gas chromatographic methods for the determination of aromatic hydrocarbons in water. Above 50 $\mu g\,L^{-1}$ hydrocarbons preconcentration by liquid–liquid extraction followed by gas chromatography is recommended. Below this level, a more effective preconcentration step is needed. Capture on a synthetic resin such as Amberlite XAD-2 is recommended, followed by ultraviolet spectrofluorimetry.

The Standing Committee of Analysis [144] compared oils and grease determinations in water obtained by solvent extraction–infrared spectroscopy and by solvent extraction gravimetric methods. In the case of the gravimetric method there was a minimal boiling point limitation on the materials determined. Oils and fats of natural origin are retained on the FloraSil column employed in the initial separation and may be determined separately by elution with a suitable solvent. The advantages and limitations of the CH_2 index are considered. A number of examples are presented for the use of the chromatographic techniques for obtaining organic profiles in conjunction with a variety of extraction methods. It was shown that infrared techniques can be used to give rapid and early indications of petroleum-based pollution.

Ultraviolet spectroscopy

Ultraviolet spectroscopy is a sensitive technique for quantitatively determining hydrocarbons and various heterocyclic compounds in water or organic solvents. Saturated hydrocarbons such as paraffins generally do not have any significant absorption. The ultraviolet spectroscopic technique is very often neither accurate nor specific, since it has been demonstrated that although the general shapes of spectra of oils are similar, the intensity of the maximum absorption varies according to the type of oil [143,145–156]. If the polluting oil is known, and a sample is available for calibration, then an accurate sensitive method can sometimes be developed. Unfortunately, the background absorption of river organics is often high. A river sample was found to produce seven times more absorption than water containing

100 mg L^{-1} benzene, while phenol and pyridine gave four and five times more absorption respectively [148]. A separation procedure based on the passage of an extract through a column of absorbent, for example silica gel, alumina or FloraSil, would elute saturated hydrocarbons almost quantitatively. However, the absorbing portion of petroleum oils, that is aromatic and polycyclic aromatic hydrocarbons, and hetrocylcic compounds, would be difficult to separate positively in a quantitative manner. This probably also applied to fluorescence. However, in the cases of groundwater, seawater distant from estuaries, and possibly very clean rivers, which can all have a much lower background absorption, the technique is more applicable.

Levy [157] has shown that for different oils the ratio of peak heights at 228 and 256 nm due to absorption by aromatics in the ultraviolet region is constant for a particular oil but varies with different oils. This ratio was independent of the concentration of oil over the range 8–7 mg L^{-1} (ratio A_{228}/A_{256} between 1.62 and 1.67). The peak height, at either 228 m or 256 nm, can be used for quantitative analysis. The relationship between concentration of oil and peak height in the concentration range 8–7 mg L^{-1} Levy [157] found that the ratio A_{228}/A_{256} for crude residual fuel oil ranged from 1.23 to 2.11. The ratio from a particular oil was fairly constant for more than 1 year.

Reitzel and Bjerg [156] determined toluene, ethylbenzene and xylene and other degradation products in groundwater by solid-phase extraction and in-vial derivatisation.

Benzylsuccinic acid and methylbenzylsuccinic acids are unambiguous indicators of anaerobic toluene and ethylbenzene/xylene degradation, and so the determination of these compounds in landfill leachates contaminated groundwater is highly relevant. Samples were diluted to <0.8 mS cm^{-1}, in order to reduce their ionic strength, and subsequently extracted through strong anion exchange dicks, followed by simultaneous in-vial elution and methylation. A detection limit of 0.1 μg L^{-1} was obtained for 100 mL sample. Using this method 19.3 μg L^{-1} of benzylsuccinic acid was measured in a landfill leachate, and low μg L^{-1} levels of the methylbenzene succinic acid were measured in gasoline-contaminated groundwater. The results were compared with the findings of benzylsuccinic acid at 16 other contaminated sites, and benzylsuccinic acid as indicators of biodegradation were evaluated. The estimation of biodegradation rates based on parent hydrocarbons and benzylsuccinic acid concentration or ratios is questionable. However, the degradation products serve as good qualitative in situ indicators for anaerobic biodegradation in contaminated groundwater.

Franke et al. [158] have described an isomer specific analysis of di-iso-propyl-napthylene in natural waters.

A polymeric nanofilm coated optical sensor has been used for speciation of aromatic by hydrocarbons [159].

Mishayekhi et al. [160] have developed a rapid sensitive method for the determination of biphenyl oxide in water samples using dispersive liquid–liquid microextraction followed by gas chromatography. This method involves the use of an appropriate mixture of extraction solvent (8.0 μL tetrachloroethylene) and dispenser solvent (1.0 mL acetonitrile) for the formation of cloudy solution in 5.0 mL aqueous sample containing biphenyl and biphenyl oxide. After extraction, phase separation was performed by centrifugation and biphenyl and biphenyl oxide in sedimented phase (5.0–0.3 μL) were determined by a gas chromatography-flame ionisation system. Type of extraction and disperser solvents and their volumes' salt effect on the extraction recovery of biphenyl and biphenyl oxide from aqueous salutation were investigated. Under the optimum conditions and without salt addition, the enriching factors for biphenyl and biphenyl oxide were 819 and 786, while the extraction recovery was 81.9% and 78.5%, respectively. The linear range was 0.125–100 $\mu g\ L^{-1}$ for both analytes. The relative standard deviation (RSD, $n = 4$) for 5.0 $\mu g\ L^{-1}$ of analytes were 8.4% and 6.7% for biphenyl and biphenyl oxide, respectively. The relative recoveries of biphenyl and biphenyl oxide from sea and river water at spiking level of 5.0 $\mu g\ L^{-1}$ were between 85.0% and 100%.

Polycyclic aromatic hydrocarbons

Many polycyclic hydrocarbons in trace quantities have been shown to be directly carcinogenic to mammals. They are adsorbed on to particular material which may be present in water samples and are also water soluble to some extent, so that their occurrence in the environment has caused widespread concern. At least 100 compounds of this type have been detected and characterised in environmental samples. The basic molecular structure consists of benzene rings either fused together or bridged by methylene side chains. Alkyl substituents also occur.

The compounds can be produced by biochemical degradation of other organic compounds under suitable conditions. They may occur in the environment from the combustion of material such as wood or leaves. Other sources or aromatic material from which polycyclic aromatic hydrocarbons may be derived include crude oil which can contain 20% by weight of dicyclic and higher polycyclic oxidation hydrocarbons and high-grade petrol, the aromatic content of which is over 50%. Unsaturated fatty acids, terpenoids and steroids may also be potential polycyclic oxidation hydrocarbons precursors. Some of the polycyclic aromatic hydrocarbons thus formed are relatively stable to further biodegradation. The significance of these compounds in the environment has been studied, but analytical difficulties preclude the examination of large numbers of samples necessary in a detailed study.

Four techniques have been used for the determination of polyaromatic hydrocarbons in water. These are gas chromatography or gas chromatography combined with mass spectrometry, fluorescence spectrometry, high-performance liquid chromatography and thin-layer chromatography. The gas-liquid chromatographic–mass spectrometric approach has the greater potential for obtaining a complete analysis of volatile environmental polycyclic aromatic hydrocarbons.

The World Health Organization recommends the two-dimensional thin-layer chromatographic method of Borneff and Kunte [161]. This allowed the determination of nanogram quantities, but the number of compounds which are analysed is limited. Six compounds are determined. The procedure is lengthy and repeated handling of the standards increase working hazards.

The data obtained using liquid chromatographic techniques [162–168] indicate that a less effective separation of the more volatile polycyclic aromatic hydrocarbons occurs when comparisons are made with the data using gas chromatography–mass spectrometric techniques. Although it might be pointed out that in the form of high-performance liquid chromatography continued improvements are being made in this technique.

Other techniques which have been applied to the determination of polycyclic aromatic hydrocarbon (PAHs) include X-ray, excided optical luminescence [169] and gel permeation chromatography [163], isotope dilution gas chromatography and collection on open-prepolyurethane columns [170,171].

Gas chromatography

Chatot et al. [172] coupled thin-layer chromatography with gas chromatography to determine polycyclic aromatic hydrocarbons. The hydrocarbons were separated by two-dimensional thin-layer chromatography on prewashed aluminacellelose acetate (21), with pentane as solvent in the first direction and ethanol–toluene–water (17:4:4) as solvent in the second direction. Zones fluorescent under ultraviolet radiation were extracted with benzene, the extracts were evaporated and a solution of each residue in benzene (20 μL) was injected on to a stainless steel column (2 m × 1/8 in (3 mm)) packed with 4.5% of SE-52 on Chromosorb G (DMCS). The carrier gas was nitrogen (60 mL min^{-1}) the column was temperature programmed from 100°C to 290°C per minute, and detection was by flame ionisation.

A gas chromatographic method using a short packed column has been described by Frycka [173] for separating critical pairs of polycyclic aromatic hydrocarbons for example benzo(a)pyrene and benzo(e)pyrene in the course of a few minutes.

Dunn and Stich [174] have applied flame ionisation gas chromatography to the determination of anthracene, phenanthrene, fluoranthene, pyrene, chrysene, benz(a)anthracene and triphenylene in solvent extracts of mussels and creosoted wood. Harrison et al. [175] studied the factors governing the extraction and gas chromatographic analysis of polycyclic aromatic hydrocarbons in water. Factors such as initial concentration, presence of suspended solids and prolonged storage of the samples considerably affected extraction efficiencies. It is recommended that water samples should be collected directly into the extraction vessel and that analysis should be carried out as soon as possible after extraction.

Hellmann [176] have also reported on the gas chromatography of polycyclic aromatic hydrocarbons.

Caddy and Meek [177] combined gas chromatography with the fluorescence detection technique for the determination down to 10 mg L^{-1} of polycyclic aromatic hydrocarbons in natural waters. The water sample was adjusted to pH2.5 then stirred with Celite. The Celite was then removed and extracted with cyclohexane. The cyclohexane was evaporated to a small bulk and then injected on to the gas chromatographic column. Linear concentration-peak area calibration curves were obtained using the technique for several hydrocarbons in the 0–3 μg range.

Various other workers [178–181] have investigated the application of combinations of gas chromatography with a fluorescence detector to the determination of nanogram quantities of polycyclic aromatic hydrocarbons.

Acheson et al. [182] studied the effect of suspended soils, in the form of purified Fuller's earth, on the extraction of two PAH's pyrene and benzoic perylene from water. Both the variation in extraction efficiency with the level of suspended solids and the variation with polycyclic aromatic hydrocarbons concentration at a fixed level of suspended solids were investigated. Extraction efficiency decreased from 80% ± 2% of the zero suspended contents level 50%–80% of the 200 mg L^{-1}. Continuous dichloromethane extraction for 36–72 hours was compared with the above procedures. The results showed that extractions from pure water were very similar to those achieved using the Ultra-Turrax, but in the presence of suspended solids the efficiency of extraction was reduced substantially. Again, no clear explanation of the effect of solids can be put forward.

Natusch and Tomkins [183] described a method for the isolation of polycyclic aromatic hydrocarbons by solvent extraction with dimethyl sulphoxide. Partition coefficients for a variety of aliphatic and aromatic hydrocarbons and their derivatives were determined to establish the types of compound which can be separated and to examine the interactions in the polycyclic aromatic hydrocarbons–dimethyl sulphoxide system.

The composition of the fractions extracted into dimethyl sulphoxide was determined by gas chromatography.

Basu and Saxena [184] investigated the isolation of benzo(a)pyrene fluoranthene, benzo(j)fluoranthene, benzo(k)fluoranthene, indeno(1,2,3-cd)pyrene and benzo(ghi)perylene from natural water. A FloraSil column clean-up procedure is described which is capable of removing impurities introduced from water and polyurethane-foam plugs to the extent necessary for their interference-free detection. This isolation step is followed by gas chromatography using a flame ionisation detection operated in a different mode.

Matsumoto and Hanya [185] used gas chromatography–mass spectrometry to identify and estimate nanogram per litre amounts of squalane in river water samples and sediments taken in the Tokyo area. Water samples were extracted three times with ethyl acetate and then chromatographed on a silica gel column. The column volumes of hexane eluate were analysed using gas chromatography–mass spectrometry. Squalane was identified in all samples of river waters at concentrations of 0.46–1.7 $\mu g\ L^{-1}$.

Constable et al. [186] compared solvent reduction methods for the concentration of polycyclic aromatic hydrocarbons in natural waters. The performance of volume reduction methods was compared using a seven-component standard polycyclic aromatic hydrocarbons mixture containing naphthalene, fluorine, pyrene, chrysene, 7,12-dimenthylbenz(a)anthracene, benzo(a)pyrene and perylene. Several solvents including methylene chloride, n-pentane, ethyl ether, acetone and acetonitrile were used. Volume reduction methods evaluated included rotary evaporation, nitrogen evaporation and four different distillation columns. Analyses were performed by gas chromatography. In general, the use of more volatile solvents, especially ethyl ether, gave higher recoveries.

High-performance liquid chromatography

Various other workers [184–196] have studied the application of high-performance liquid chromatography to the determination of polycyclic aromatic hydrocarbons in water samples. Hagenmaier et al. [195] used a reversed-phase high-performance liquid chromatography procedure for the determination of trace amounts of polycyclic aromatic hydrocarbons in water. Different column packing materials were tested, in conjunction with nonpolar stationery phases of various polarities, for separation efficiency, detection limits, and long-term stability. The method was suitable for concentrations as low as 2 $ng\ L^{-1}$ sample. Compounds studied included fluoranthene, benzofluoranthene isomers, benzopyrene and perylene derivatives.

Saxena et al. [197] have also studied the application and limitations of high-efficiency glass columns to the analysis of polycyclic aromatic hydrocarbons.

O'Donnell [198] screened Irish rivers for fluoranthene and benzopyrene using high-performance liquid chromatography on a reversed-phase C column. The solvents were methanol–water (1:1) and pure methanol. Detection was possible at concentrations of 0.018 $mg\ L^{-1}$. Levels averaged less than one-sixth of the WHO and EEC recommended tolerance limits for both compounds.

Workers at Perkin-Elmer Ltd [199] have studied the high-speed separation of polycyclic aromatic hydrocarbons using C_{18} − bonded phase packings (5 μm particles) using both isocratic and gradient elution. The analysis of several polycyclic aromatic hydrocarbons standards was performed using the 5 μm bonded phase column with gradient elution from 60% to 100% acetonitrile in 5 minutes at a flow rate of 4 mL min^{-1}. Although not completely resolving the components, this method offers a rapid analysis which is adequate for many of these compounds.

Schönmann and Kern [29] have used online enrichment for microgram per litre analysis of polycyclic aromatic hydrocarbons in water by high-performance liquid chromatography. The trace enrichment method is based on the affinity of nonpolar pollutants for reversed-phase chromatography supports.

When aqueous samples are passed through reversed-phase columns, these compounds and any other nonpolar organic compounds present in the sample are immobilised at the head of the column. When detectable quantities of pollutants have been accumulated on the column, they can be analysed by introducing a mobile phase of the desired eluent strength.

Chen et al. [200] applied a stochastic resonance algorithm to the quantitative analysis for weak liquid chromatographic signals of pyrene in water samples.

Thin-layer chromatography

Early work on the application of thin-layer chromatography to the separation and determination of polycyclic aromatic hydrocarbons was carried out by several workers [163,201−204].

Schossner et al. [205] developed a unidimensional thin-layer chromatographic separation technique with fluorimetric detection for the determination of polycyclic aromatic hydrocarbons in natural water. Separation is carried out on precoated plates after previous extraction with cyclohexane and clean-up in aluminium oxide or Sephadex LH 20 column. Amounts of the order of 1−10 ng per sample could be quantitatively estimated by fluorescence spectrometry with excitation wavelengths of 366 or 436 nm.

Haenni et al. [206] have described the use of dimethyl sulphoxide as a solvent suitable for extraction of polycyclic aromatic hydrocarbons from aliphatic solvents. It was incorporated in the procedure of Hoffmann and Wynder [207] in place of nitromethane, and the extracted polycyclic aromatic hydrocarbons were isolated by subsequent dilution with water and back-extraction into cyclohexane which was then evaporated. This procedure has a much improved efficiency of 84%−110%.

Thin-layer chromatography was carried out using the procedure of Borneff and Kunte [161] with slight modifications. The first development with iso-octane was omitted, and the plates were developed for a fixed distance of solvent front advance rather than a fixed time. After development, the plate was sprayed with a 2% w/v solution of tetrachlorophthalic anhydrides in acetone–chlorobenzene (10:1) and dried prior to measurement of fluorescence by irradiation with a UV lamp (360 nm) and visual comparison with standard chromatograms.

Fluorescence spectrometry

Early work on the determination of polycyclic aromatic hydrocarbons in water using fluorescence spectrometry was carried out by Scholz and Altmann [208] who devised a method for the determination of down to 0.1 ng^{-1} of benzo(a)pyrene in water. In this method the sample is extracted with cyclohexane, and the concentrated extract is subject to thin-layer chromatography on silica gel H impregnated with polyoxyethylene glycol 1000 using benzene–hexane (1:3) as solvent. The benzo(a)pyrene is located in 365 nm radiation by reference to standards and is extracted from the adsorbent with cyclohexane. The solution is evaporated, and a solution of the residue in dioxan is subject to fluorimetry at 429 nm (excitation at 365 nm). Down to 0.1 ng of benzo(a)pyrene per litre of sample could be determined with a relative error in the range 1–10 ng L^{-1} of $\pm$ 15%.

Schwarz and Wasik [209] carried out fluorescent measurements of benzene, naphthalene, anthracene, pyrene, fluoranthene and benzo(a)-pyrene in natural water. They reported fluorescence spectra, quantum yields and concentration dependencies for the aromatic hydrocarbons in water to ascertain the applicability of measuring polycyclic aromatic hydrocarbons in aqueous systems by spectrofluorescence. The detection limits are as follows: naphthalene, 0.3 μg L^{-1}, anthracene 0.03 μg L^{-1}, pyrene, 0.15 μg L^{-1}, fluoranthene, 0.17 μg L^{-1} and benzo(a)pyrene, 0.10 μg L^{-1}.

Further work on the fluorimetric determination of benzo(a)pyrene in water was conducted by Monarca et al. [210]. These workers utilised two-temperature spectrofluorimetry using Shpol'skii effect. This technique is capable of high selectivity and sensitivity in the determination of polycyclic aromatic hydrocarbons compounds at 77K and can permit this determination in mixtures without lengthy initial chromatographic separation procedures [211–213]. The aim in this work was to develop a method sufficiently sensitive to determine benzo(a) pyrene at the 0.3 ng L^{-1} level. The Environmental Protection Agency has been bound, under a Consent Decree, to set maximum concentration limits in effluent waters for a group of 'unambiguous priority pollutants'. Included on this list are several polycyclic aromatic hydrocarbons [214].

Miscellaneous

Afghan et al. [215] present results from studies on the feasibility of using polyurethane foams to extract and preconcentrate polycyclic aromatic hydrocarbons from water prior to their determination and on factors affecting adsorption. Preliminary filtration of the sample to remove particular matter is recommended. Polyurethane foam is particularly useful when large samples are routinely analysed in that it offers high flow rates, good capacities and low costs compared to solvent extraction. Polycyclic aromatic hydrocarbons can be extracted at the part per trillion (pg mL^{-1}) level from natural waters.

Hess et al. [216] used in situ precipitation with magnesium hydroxide to selectively preconcentrate aromatic hydrocarbons and polychlorinated biphenyls. Magnesium sulphate and ammonium hydroxide were added to a sample aqueous solution (1 L) spiked with polychlorinated biphenyls and polycyclic aromatic hydrocarbons. Centrifugation produced a precipitate which was then dissolved in ammonium chloride and sulphuric acid. The solution was extracted three times with methylene chloride and the combined extracts analysed by gas chromatography with flame ionisation detection. This method of in situ magnesium hydroxide precipitation was selective for polycyclic and polychlorinated biphenyls of high molecular weight. The method discriminated against acidic molecules (phenols), basic molecules (amines) and their neutral molecules such as phthalate esters.

Ghaoqui [217] has compared different adsorbents for the isolation of polycyclic aromatic hydrocarbons from waters.

Thomas and Womat [218] investigated the polycyclic aromatic hydrocarbon products resulting from the pyrolysis of catechol.

Polycyclic aromatic hydrocarbons of the $C_{24}H_{14}$ isomer class, some of which are potent mutagens and carcinogens, are produced during the burning of solid fuels and consequently can occur in water courses. For a clearer understanding of the formation of polycyclic aromatic hydrocarbons, pyrolysis experiments, Thomas and Womat [218] performed pyrolysis experiments in an isothermal laminar-flow reactor with the model fuel catechol (*ortho*-dihydroxybenzene) – a phenol-type compound representative of structural entities in complex solid fuels like coal, wood and biomass. The catechol pyrolysis experiments are conducted at 1000°C and at a residence time of 0.3 seconds. The pyrolysis products were analysed by high-pressure liquid chromatography with ultraviolet–visible absorbance detection and mass spectrometric detection. Product analysis reveals that the $C_{24}H_{14}$ polycyclic aromatic hydrocarbons products of catechol pyrolysis belong to three structural classes: perylene benzologues, fluoranthene, benzologues and pyrene benzologues. The $12C_{24}H_{14}$ polycyclic aromatic hydrocarbons identified in this study are: benzo[*b*]perylene, naphtho[1,2,-*b*] Fluoranthene, naphthol [1,2-*k*]fluoranthene, dibenzo [*b,k*]fluoranthene, naphtho[2,2-*b*]fluoranthene, naphtho[2,3-*k*]fluoranthene, naphthol[1,2,-*e*]pyrene, naphtho[2,3,-*e*]pyrene, naphtho[1,2,-*a*]pyrene, dibenzo[*a,e*]pyrene, dibenzo[*e,l*]pyrene, and dibenzo [*a,h*]pyrene. In addition to these identifications of naphtho[2,1-*a*]pyrene, naphthol[2,3,-*a*]pyrene and dibenzo[*a,i*]pyrene among the products of catechol pyrolysis bring the total number of $C_{24}H_{14}$ polycyclic aromatic hydrocarbons identified as products of catechol to 15. Of these, 12 have been reported to be mutagens and 6 have been reported to be carcinogens. The UV spectra establishing the identities of the 15 $C_{24}H_{14}$ catechol pyrolysis products are presented in this paper.

Lichens have been used as bio monitors for the determination of polycyclic aromatic hydrocarbons in environmental waters [219]. Sixteen polycyclic aromatic hydrocarbons were studied in 11 locations along the valley of Caracas (Venezuela). The results of this work indicate that 14 of the 16 analysed polycyclic aromatic hydrocarbons were highly accumulated into the lichen thalli of *Pyxine Coralligera* Malme. Polycyclic aromatic hydrocarbons levels in the samples revealed that the several volatile polycyclic aromatic hydrocarbons (naphthalene, acenaphtylene, acenaphtylene, acenaphtene and fluoranthene) have the highest levels in the majority of the studied locations. The fluoranthene/pyrene and phenanthrene/anthracene ratios suggested that the major sources of polycyclic aromatic hydrocarbons are anthropogenic, mainly associated with gasoline and diesel combustion (pyrolytic) and unburnt oil derivatives (petrogenic). The total polycyclic aromatic hydrocarbons concentrations obtained in this study were in the range of 0.24–9.08 μg/g, similar to those reported to other workers in European and Asian cities.

In this method the lichen sample was extracted with cyclohexane–chloromethane (4:1 v/v) the solvent removed from the extract by evaporation. After clean-up on a column of silica, the final determinations were carried out by high-performance liquid chromatography.

References

[1] L.L. Vos, A.L. Bridie, S. Herzberyg, H_2O 10 (1977) 277.
[2] G. Matsumoto, T. Hanya, Water Res. 15 (1981) 217.
[3] G.W. Peitscher, Fresenius Zeitung Für Analytsche Chimie 327 (1987) 86.
[4] W.J. Khazal, J. Vejrosta, J. Novak, J. Chromatogr. 157 (1978) 125.
[5] J. Drodz, J. Novak, J. Chromatogr. 152 (1978) 55.
[6] J. Drodz, J. Novak, J. Chromatogr. 136 (1977) 37.
[7] J. Drodz, J. Novak, J. Riiks, J. Chromatogr. 158 (1978) 471.
[8] C. McAucliffe, Chem. Technol. 1 (1971) 46.
[9] L.E. Kaiser, B.G. Oliver, Anal. Chem. 48 (1976) 2207.
[10] C. McAucliffe, J. Chem. Geol. 4 (1969) 225.
[11] J.W. Swinnerton, V.J. Linnenbom, J. Gas Chromatogr. 5 (1967) 570.
[12] J. Novák, J. Zlutický, V. Kubelka, J. Mostecký, J. Chromatogr. 76 (1973) 45.
[13] K. Grob, J. Chromatogr. 84 (1973) 255.
[14] K. Grob, F. Zurcher, J. Chromatogr. 117 (1976) 285.
[15] K. Grob, G. Grob, J. Chromatogr. 117 (1974) 303.
[16] K. Grob, G. Grob, K. Grob, J. Chromatogr. 106 (1975) 303.
[17] R. Kaiser, J. Chromatogr. Sci. 9 (1971) 227.
[18] J. Polak, B. Lu, Anal. Chim. Acta 63 (1974) 231.
[19] B.A. Colenutt, S. Thorburn, Int. J. Environ. Stud. 15 (1980) 25.
[20] J. Drodz, Z. Vodakova, J. Novák, J. Chromatogr. 354 (1986) 47.
[21] L. Lively, Engineering Bulletin Ext. Serv, Purdue University, 1965, 118, 657.

[22] R. Jeltes, R. Veldink, J. Chromatogr. 27 (1967) 242.
[23] C. McAucliffe, J. Phys. Chem. 70 (1966) 1267.
[24] E. Desbaumes, C. Imhoff, Water Res. 6 (1972) 885.
[25] A.I. Bridie, J. Bos, L.S.J. Herzberg, J. St. Petrol 59 (1973) 263.
[26] F. Belkin, M.A. Habre, Bull. Environ. Contam. Toxicol. 90 (1988) 244.
[27] A.J. Roberts, T.C. Thomas, Environ. Toxicol. Chem. 5 (1986) 3.
[28] R.M. Uhler, E.M. Healey, K.J. McCarthy, S.P. Uhler, S.A. Strout, Int. J. Environ. Anal. Chem. 83 (2003) 1.
[29] M. Schönmann, H. Kern, Varian Instrument News 15 (1981) 6.
[30] J. Bundt, W. Herbel, H. Steinhart, Fresenius Z. Analytica Chimie 328 (1987) 480.
[31] R. Jeltes, in: P. Hepple (Ed.), Water Pollution by Oil, London Institute of Petroleum, 1971, p. 43.
[32] R.C. Simard, W. Hasecawa, W. Boudaruk, C.E. Hedington, Anal. Chem. 23 (1951) 1384.
[33] C. Ruebelt, Helgoländer Wiss Meeresunters 16 (1967) 306.
[34] W. Fastabend, Chemi-Ingr.-Tech. 37 (1965) 728.
[35] C. Ruebelt, Gas-u. Wass. Fach 108 (1967) 893.
[36] C.Z. Ruebelt, Anal. Chem. 221 (1966) 299.
[37] V.M. Osipov, Khimiya Tekhnol. Topl. Masel 16 (1971) 52.
[38] M.T. Galubeva, Lab. Delo 11 (1966) 665.
[39] C.G. Lindgreen, J. Am. Water Works Assoc. 49 (1957) 55.
[40] F.J. Ludzack, C.E. Whitefield, Anal. Chem. 28 (1956) 157.
[41] L.R. Benyon, Methods for the Analysis of Oil in Water and Soil, CONCAWE, The Hague, Strching, 1968. 40 pp.
[42] R. Suzuki, N. Yamaguchi, N. Matsumoto, Japan Anal. 23 (1974) 1296.
[43] D.R. Hughes, R.S. Belcher, E.J. O'Brien, Bull. Environ. Contamin. Toxicol. 10 (1973) 170.
[44] L. Boisselet, Prox. 5th World Petroleum Congress, New York, 1959, pp. 165–175.
[45] H.Z. Hellmann, Anal. Chem. 269 (1973) 245.
[46] M. Gruenfeld, Environ. Sci. Technol. 1 (1973) 636.
[47] J. Mallevialle, Water Res. 8 (1974) 1071.
[48] D.L. Powell, E.J. Darland, T.R. Williams, Anal. Lett. 4 (1971) 479.
[49] J. Golden, Tech. Sci. Municipales 71 (1976) 17.
[50] Stichling, CONCAWE, The Hague Report, 1972, 9/12.
[51] US EPA Manual of Methods for Chemical Analysis of Water and Wastes, 1974, p. 232, Storet No. 00560.
[52] American Petroleum Institute, Manual of Disposal of Refinery Waste, vol. IV, Method, 1958, pp. 733–758.
[53] G.P. Coles, R.M. Dill, D.L. Shull, Am. Chem. Soc. Div. Petrol. Chem. Preparation 20 (1975) 641.
[54] P.J. Whittle, W.A. McCrum, M.W. Horne, Analyst 105 (1980) 679.
[55] H. Hellman, Deutsch Gewasser, Mitt, 1969, pp. 13, 19.
[56] D. Geyer, P. Martin, P. Adrian, Gus u Wasserfach Wasser Abwasser 119 (1978) 72.
[57] N.K. Shtivel, L.I. Gipokova, Z.S. Smirnova, Soviet J. Water Chem. Technol. 7 (1985) 66.
[58] E.D. Ramsey, G. Wei, Int. J. Environ. Anal. Chem. 88 (2008) 957.
[59] I.Z.Z. Berthold, Anal. Chem. 240 (1968) 320.
[60] P. Koppe, L.A. Muhle, Vom Wasser 35 (1968) 42.
[61] D. Hunter, Environ. Sci. Technol. 9 (1975) 241.
[62] D.L. Liu, Water Sewage Works 125 (1978) 40.

[63] V.W. Sinel'nikov, Gidrokhim Mater 50 (1969) 127 (Ref. Zhur Khim. 19GD, 1969 (24) Abstract No. 249. 264).
[64] J.P. Gibb, M.J. Barcelona, J. Am. Water Works Assoc. 78 (1986) 60.
[65] J.F. Pankow, L.M. Isabelle, J.P. Hewertson, J.A. Cherry, Groundwater 22 (1984) 230.
[66] T.H. Gouw, Anal. Chem. 42 (1970) 1394.
[67] F.K. Kawahara, Laboratory Guide for the Identification of Petroleum Products, 1014 Broadway, Cincinnati, OH, US. Department of Interior, Federal Water Administration, Analytical Quality Control Laboratory, 1969, 41 pp.
[68] Trent River Authority, Private Communication, Trent River Authority, Meadow Lane, Nottingham.
[69] S.J. Ramsdale, R.E. Wilkinson, J. Inst. Petrol. (London) 54 (1968) 326.
[70] F.C.A. Killer, R. Amos, J. Inst. Petrol. (London) 52 (1966) 315.
[71] A.D. Semenov, A.G. Stradomanskaya, L.F. Zurbina, Gidrokhim Mater 53 (1971) 51 (Ref. Khim 19GD, 1971).
[72] H. Kreiger, Gas u Wasserfach 104 (1963) 695.
[73] G. Lambert, Bull Cert, Belge Stud Docm Eaux, 1966, 20, 271.
[74] K. Triems, Chem. Techn. Berl 20 (1968) 596.
[75] K. Triems, O. Heinze, Anal. Abstracts 16 (1969) 2556.
[76] J.W. Farrington, J.M. Teal, J.G. Quinn, T. Wade, K. Burnes, Bull. Environ. Contam. Toxicol. 10 (1973) 129.
[77] W.A. Sauer, G.E. Fitzgerald, Environ. Sci. Technol. 10 (1976) 893.
[78] P.J. Matthews, Water Poll. Control 67 (1968) 588.
[79] P.J. Matthews, J. Appl. Chem (London) 20 (1970) 87.
[80] J.B.F. Lloyd, J. Forens. Soc. Soc. 11 (1971) 83.
[81] J.B.F. Lloyd, J. Forens. Sci. Soc. 11 (1971) 153.
[82] J.B.F. Lloyd, J. Forens Sci. Soc. 11 (1971) 235.
[83] C.A. Parker, W.J. Barnes, Analyst 85 (1960) 3.
[84] K. Bauer, H. Driescher, Fortschr. Wasserchem Ihrer Grezeb, 1968, 10, 31.
[85] F. Danyl, B. Nietsch, Mikrochemie Mikrochem, Acta 39 (1952) 333.
[86] B. Nietsch, Angew. Chem. 66 (1954) 571.
[87] B. Nietsch, Gass. Wass. Warme 10 (1956) 66.
[88] B. Nietsch, Mikrochim. Acta (1956) 171.
[89] E.M. Levy, Water Res. 5 (1971) 723.
[90] V.E. Sinel'nikov, Kazan. Med. Zh. 3 (1968) 83.
[91] Environmental Protection Agency Industrial Environmental Research Laboratory Cincinnati, OH, 1979, Report No. 32428, p. 256.
[92] P. John, E.C. McQuat, L. Sontar, Analyst 107 (1982) 221.
[93] H. Kola, T. Kuokhanen, I. Valinaki, P. Peramake, B. Lauhanen, Int. J. Environ. Anal. Chem. 83 (2003) 157.
[94] M.E. Le Pera, No. AD 646−382 Identification Characterisation of Petroleum Fuels Using Temperature Programmed Gas Liquid Chromatography, Va National Technical Information Service Department, Springfield, VA, 1966.
[95] M. Blumer, Marine Biol. 51 (1970) 95.
[96] D.F. Duckworth, in: P. Heppler (Ed.), Aspect of Petroleum Pollutant Analysis in Water Pollution and Oil, Institute of Petroleum, London, 1971, p. 165.
[97] R. Ellerker, Water Poll. Control 67 (1968) 542.
[98] L.R. Benyon, 40 pp Method for the Analysis of Oil in Water and Soil, CONCAWE, The Hague, Striching, 1968.

[99] R. Jeltes, Anal. Abstract 15 (1968) 3633.
[100] W.W. Dewitt Johnson, E.D. Fuller, Proceedings 13th Conference, Great Lakes, Res., 1970, p. 128.
[101] C. McAucliffe, Chem. Geol. 4 (1969) 255.
[102] E.R. Adlard, P.H.D. Matthews, Nat. Phys. Sci. 233 (1971) 83.
[103] Y.U.Y.U. Lur'e, V.A. Panova, Z.V. Nickolaeva, Gidrokhim Water, 1971, 55, 108 Ref Zhur Khim 199D, Abstracts, No. 16G 258.
[104] R.D. Coles, Nature 233 (1971) 546.
[105] R. Jeltes, W.A.M. Van Tonkelaar, Water Res. 6 (1972) 271.
[106] R. Jeltes, W.A.M. Van Tonkelaar, H_2O 5 (1972) 288.
[107] F.K. Kawahara, J. Chromatogr. Sci. 10 (1972) 629.
[108] J. Lysyj, P.R. Newton, Anal. Chem. 44 (1972) 2385.
[109] J. Lysyj, P.R. Newton, Anal. Abstract 22 (1972) 459.
[110] M.E. Garra, J. Muth, Sci. Technol. 8 (1974) 249.
[111] R.A. Stuart, R.D. Branch, Instrument News, vol. 21, Perkin-Elmer Corporation, Norwalk, CT, 1970, p. 2.
[112] D.W. Bryan, V.P. Guinn, Virginia National Information Service, USA, EC Report No GA 9889, 1970, p. 134.
[113] V.P. Guinn, S.C. Bellanca, Washington USA P O National Bureau of Standards Special Publications, 1967, No. 312. p. 83.
[114] H.R. Leukens, D. Bryan, N.A. Hiatt, H.L. Schlesinger, Report Atomic Energy Commission US, 1971, GULF-RT-A-10684, p. 134.
[115] H.R. Leukens, Report Atomic Energy Commission, GULF-RT-A-10973, p. 74.
[116] J.P. Flaherty, H.B. Eldridge, Appl. Spectrosc. 24 (1970) 534.
[117] K.S. Scott, B.M. Bean, Patent No 3, 574 550, Washington Patent Office.
[118] R. Luis, Chim. Ingr.-Tech. 40 (1968) 538.
[119] H. Kubo, R. Brenthal, R. Wildman, Anal. Chem. 50 (1978) 899.
[120] H.B. Eldridge, J.P. Flaherty, Ind. Eng. Chem. Pro. Res. Div. 9 (1970) 422.
[121] G.K. Budnikov, E.P. Medyantseva, Zhur, Analit. Khim. 28 (1973) 301.
[122] H.C. Braemer, Proc. Am. Chem. Soci. Div. Wat. Waste Chem. (1966).
[123] G.R.E.C. Gregory, P.L. Palmer, Apparatus for Detecting Oil in Water, Brit Patent 1, (London Patent Office)1, 221, 066.
[124] G. Muskewity, Apparatus for the Continuous Determination of Oils in Liquid and Gaseous Media, Ger. Pat. (Berlin Patent Office), 1 810, 604.
[125] F. Grabb, Industusabwasser 45 (1968).
[126] Emeschregenossenschaft, Essen. Eff. Wat. Treat., 1970, 10, 667 Environmental Science and Technology, 1971, 5, 356.
[127] A.D. Goolsby, Environ. Sci. Technol. 5 (1971) 356.
[128] A. Vasicek, Optics of Thin Films, North-Holland, Amsterdam, 1960. 26, 41.
[129] J.S. Mattson, Environ. Sci. Technol. 5 (1971) 415.
[130] H.W. Fust, R.E. Kreider, K.W. Gardiner, Seventeenth Annual ISA Analysis Instrument Symposium, Houston, Texas, April, Analysis Instrumentation 9, Publ. Instrument Society of America, Pittsburgh, May 1971, Paper 45 Automated Instrument Approach for Oil in Water.
[131] M. Ahmadijian, C.W. Brown, Environ. Sci. Technol. 7 (1973) 452.
[132] A.I. McMullen, J.F. Monk, M.J. Stuart, Int. Lab. 60–61 (1975) 5447. Jan–Feb.
[133] S.P. Wasik, W.W. Tsang, Anal. Chem. 42 (1970) 1649.
[134] S.P. Wasik, W. Tsang, Anal. Abs. 21 (1971) 1854.

[135] G.C. Melkanovitskaya, Gidrokhim, Mater 53 (1972) 153. Zhur. Khim, 1972, 199D, Abstract No. 129265.
[136] R.B. Senn, M.S. Johnson, Ground Water Monit. Rev. 7 (1987) 58.
[137] B.V. Stolyarov, Chromatographia 14 (1981) 699.
[138] K. Urano, H. Meada, K. Qgura, H. Wada, Water Res. 16 (1982) 323.
[139] H. Ali Mashayeteki, A. Pabroamonok, M. Saher-Tehrani, S. Wagu-Husain, Int. J. Environ. Anal. Chem. 91 (2011) 516.
[140] R. Götz, Facum Stadt Hygiene 29 (1978) 10.
[141] D. Louch, S. Motlagh, S. Palissgn, Anal. Chem. 64 (1992) 1187.
[142] M. Ya Dudova, O.D. Diterikhas, Gidrohim. Mater. 50 (1969) 115.
[143] J. Ducreux, R. Boulet, N. Petroff, J.C. Roussel, Int. J. Environ. Anal. Chem. 12 (1982) 195.
[144] Standing Committee of Analysts. H.M. Stationery Office, London, Methods for the Examination of Waters and Associated Materials, 1984, p. 19 (P-22B:C ENV) 1984, The Determination of Hydrocarbon Oils in Waters by Solvent Extraction.
[145] L.A. Imanvilov, Transp. Khranenie Nefteprod, Uglevovdorodn Syr'ya 4 (1986) 7.
[146] R.S. Saltzman, Anal. Instrum 6 (1968) 79.
[147] P.A. Obnorlenskii, Khimiya Teknol, Topl, Masel 15 (1970) 54.
[148] K.N. Nadzhafova, Russian, Trudy, Vses. Nauchno0issled, Inst., Vodosnabzh., Kanaliz, Gidtotekhn, Soorunzhenii, Inzk, Gidrogeol 23 (1970) 1007.
[149] K.V. Minasyan, Russian, Prom. Arm. 11 (1969) 50.
[150] V.M. Ospimovpp, T.D. Belovapp, Russian, Khimiya Tekhnol, Topl, Masel. 13 (1968) 56.
[151] L.A. Skotnikovapp, Russian, Izv. Vyssh. Ucheb. Zaved., Neft. Gaz 12 (1969) 111.
[152] R.E. Polinskayapp, Russian, Proekt. Issled. Rab., Neftedob, Prom, Giprovostokneft 3 (1967) 106.
[153] Yu. Yu. Lurje, In Russian, Nauch, Soobshch, Veses. Nauchno-issled, Inst. Vodosnabzh, Kanaliz, Gidrotekhm, Sooruzhenii, Inzh. Gidrogeol, Ochistka, Prom. Stoch Vod. Moscow (1963), 34.
[154] O. Harva, A. Somersalo, Acta Chem. Fenn. 31 (1958) 384.
[155] Z. Przybylski, Chemia Analit 8 (1963) 601.
[156] K.L. Reitzel, P.L. Bjerg, Int. J. Environ. Anal. Chem. 85 (2005) 1075.
[157] E.M. Levy, Water Res. 6 (1972) 57.
[158] S. Franke, J. Grunenberg, J. Schwarzbauer, Int. J. Environ. Anal. Chem. 87 (2007) 437.
[159] L.I.B. Silva, A.M. Costa, A.C. Freitas, T.A.P. Santos, A.C. Duarte, Int. J. Environ. Anal. Chem. 89 (2009) 183.
[160] H.A. Mishayekhi, P. Abromand-Azar, M. Saber-Terrani, S. Wagif-Husain, J. Environ. Anal. Chem. 91 (2011) 516.
[161] J. Borneff, H. Kunte, Arch. Hyg. Bakt 153 (1968) 202.
[162] R.E. Jentoff, T.H. Gouw, Anal. Chem. 40 (1968) 1787.
[163] J.F. McKay, D.R. Latham, Anal. Chem. 45 (1973) 1050.
[164] C.G. Vaughan, B.B. Wheals, H.J. Whitehouse, J. Chromatogr. 78 (1973) 203.
[165] C. Melchiorri, L. Chiacchiarini, A. Grilla, S.U.N. D'Area, Ann. Ig. Mier 24 (1973) 279.
[166] S. Monarcha, R. Conti, G. Scassellato Sforzolini, A. Savino, Iq. Mod 69 (1976) 331.
[167] J. Borneff, in: I.H. Suffet (Ed.), Fate of Pollutants in the Air and Water Environments, vol. 2, John Wiley, Chichester, 1977.
[168] T. Outkiewicz, S. Ryborz, J. Maslowski, J. Environ. Prot. Eng. 4 (1978) 263.
[169] C.S. Woo, A.P. D'Silva, D.A. Fassell, Anal. Chem. 52 (1980) 159.
[170] A.D. Thrusten, R.W. Knight, Environ. Sci. Technol. 5 (1971) 64.
[171] J.D. Navra'til, R.E. Sievers, W.F. Walton, Anal. Chem. 49 (1977) 2260.

[172] G. Chatot, W. Jequier, M. Jay, R. Fontages, P. Obaton, J. Chromatogr. 45 (1969) 415.
[173] J.J. Frycka, Chromatography 65 (1972) 432.
[174] B.P. Dunn, H.F. Stich, J. Fish. Res. Bd Can 33 (1976) 2040.
[175] R.M. Harrison, R. Perry, R.A. Wellings, Wter Res. 10 (1976) 207.
[176] H.Z. Hellman, Anal. Chem. 275 (1975) 109.
[177] D.E. Caddy, D.M. Meek, Technical Report TR36, Water Research Centre, Stevenage, Herts, 1976.
[178] M.C. Bowman, M. Benoza, Anal. Chem. 40 (1968) 535.
[179] H.P. Burchfield, R.J. Wheeler, J.B. Bernos, Anal. Chem. 43 (1971) 1976.
[180] D.J. Freed, L.R. Faulkner, Anal. Chem. 44 (1971) 1194.
[181] J.W. Robinson, J.P. Goodbread, Anal. Chim. Acta 66 (1973) 239.
[182] M.A. Acheson, R.M. Harrison, R. Perry, R. Wellings, Water Res. 10 (1976) 207.
[183] D.F.S. Natusch, B.A. Tomkins, Anal. Chem. 50 (1978) 1429.
[184] D.K. Basu, J. Saxena, J. Environ. Sci. Technol. 12 (1978) 791.
[185] G. Matsumoto, T. Hanya, J. Chromatogr. 194 (1980) 199.
[186] D.J.T. Constable, S.R. Smith, J. Tanka, J. Environ. Sci. Technol. 18 (1984) 975.
[187] W.M. Lewis, Water Treat. Exam 24 (1975) 243.
[188] B. Crathone, M. Fielding, Pract. Anal. Div. Chem. Soc. 11 (1978) 155.
[189] R.I. Crane, B. Crathorne, N. Fielding, Hydrocarbons and halogenated hydrocarbons in aquatic environments, in: B.K. Afghan (Ed.), Environmental Science Research Series, vol. 16, New York, Plenum.
[190] M. Dong, D.C. Locke, E. Ferrant, Anal. Chem. 48 (1976) 368.
[191] A.M. Krstulovic, D.M. Rosie, P.R. Brown, Anal. Chem. 48 (1976) 1383.
[192] E.P. Lankmayr, K. Muller, J. Chromatogr. 170 (1979) 139.
[193] T.J. Nielson, J. Chromatogr. 170 (1979) 147.
[194] N.T. Crosby, D.C. Hunt, L.A. Philip, I. Patel, Analyst 106 (1981) 135.
[195] H. Hagenmaier, R. Feirabend, W.W. Jager, Zeitschrift Für Wasser und Abwasser Forschung 10 (1971) 99.
[196] J.J. Black, P.P. Dymerski, W.F. Zapisek, Bull. Environ. Contamin. Toxicol. 22 (1979) 278.
[197] J. Saxena, D.K. Basu, J. Kozuchowski, Method Development and Monitoring of Polynuclear Aromatic Hydrocarbons in Selected US Water, Health Effects Research Laboratory, TR-563, No. 24, Cincinnati, OH.
[198] M. O'Donnell, J. Environ. Sci. 1 (1980) 77.
[199] Perkin-Elmer Ltd, Anal. Bull. 2/2 (1) (1981) 4.
[200] M.A. Cy Chen Ma, B. Zhao, S.F. Xie, B.R. Xian, Int. J. Environ. Anal. Chem. 91 (2011) 112.
[201] C.A. Gilchrist, A. Lynes, G. Steel, B.T. Whitham, Analyst 97 (1972) 880.
[202] M.I. Stepanova, R.I. Li'ina, Z.H. Shaposhnikov Zhur, Analyt. Khim 27 (1972) 1201.
[203] K. Nowacka-Barezyk, J. Adamiak-Ziemba, A.Z. Janina, Chem. Anglit 18 (1973) 223.
[204] R.I. Crane, B. Crathorn, M. Fielding, LR 4407 Determination of Polycyclic Aromatic Hydrocarbons in Waters, Water Research Centre, 1980.
[205] H. Schossner, W. Falkenberg, H.Z. Althaus, Wasser Abwasser Forsch 16 (1983) 132.
[206] E.O. Haenni, J.W. Howard, E.L. Joe, J. Assoc. Off. Agric. Chem. 45 (1962) 67.
[207] D. Hoffman, O. Wynder, Anal. Chem. 32 (1961) 295.
[208] L. Scholz, H.J. Altmann, Zeitschrift für Analytische Chimie 240 (1968) 81.
[209] F.P. Schwarz, S.P. Wasik, Anal. Chem. 48 (1976) 525.
[210] S. Monarca, B.S. Causey, F.G. Kirkbright, Water Res. 13 (1979) 503.
[211] L.V.S. Hood, J.D. Winefordner, Anal. Chim. Acta 42 (1968) 199.

[212] D. Lavalette, B. Muel, M. Habert-Habert, L. René, R. Latarjet, J. Chem. Phys. 65 (1968) 2141.
[213] T. Ya Gaevaya, A. Ya Khesina, Anal. Chem. USSR 29 (1974) 1913.
[214] Sampling and Analysis Procedures for Screening of Industrial Effluents for Priority Polyllutants. US Environmental Protection Agency, Environmental Monitoring and Support Laboratory, Cincinnati, OH, April 1977.
[215] B.K. Afghan, R.J. Wilkinson, A. Chow, T.W. Findley, H.D. Gessler, K.S. Srikameswaren, Water Res. 18 (1984) 9.
[216] G.G. Hess, D.E. McKenzie, B.M. Hughes, J. Chromatogr. 366 (1986) 197.
[217] L. Ghaoqui, J. Chromatogr. 399 (1987) 69.
[218] S. Thomas, M.J. Womat, Int. J. Environ. Anal. Chem. 88 (2008) 825.
[219] R. Fernandez, F. Gallarraga, S. Benzo, G. Marques, A.J. Fernandez, M.G. Regniz, et al., Int. J. Environ. Anal. Chem. 91 (2011) 230.

Chapter 2

Oxygen-containing compounds in nonsaline waters

Chapter Outline

2.1 Carboxylic acids **35**
Gas chromatography 36
High-performance liquid chromatography 37
Thin-layer chromatography 39
Spectrophotometric methods 39
Fluorescence spectrometry 40
Potentiometry 41
Polarography 41
Miscellaneous 41
2.2 Phenols **42**
Gas chromatography 42
High-performance liquid chromatography 44
Thin-layer chromatography 45
Raman spectroscopy 46
Spectrophotometric method 47
Atomic absorption spectroscopy 49
Miscellaneous 49
Preconcentration 49
2.3 Phenolic acids **50**
2.4 Methyl *tert*-butyl ether **51**
2.5 Alcohols **51**
Spectrophotometric methods 51
Gas chromatography 52
2.6 Glycols **52**
Spectrometry 52
Gas chromatography 52
Thin-layer chromatography 53
2.7 Dioxans **53**
Gas chromatography 53
2.8 Esters **53**
Gas chromatography 53
High-performance liquid chromatography 53
Electrokinetic chromatography 54
2.9 Aldehydes **54**
Gas chromatography 54
Thin-layer chromatography 55
Spectrophotometric methods 55
Miscellaneous 55
2.10 Ketones **56**
2.11 Carbohydrates **57**
Gas chromatography 57
Spectrophotometric method 57
Miscellaneous 57
2.12 Lactams **58**
Thin-layer chromatography 58
Polarography 58
Enzymic assay 58
References **59**

2.1 Carboxylic acids

There are two sources of fatty acids in river waters: man-made contamination and naturally occurring. Regarding the latter the determination of free and bound fatty acids present in aquatic systems is important; first because fatty acids are sufficiently diverse in structure that they can be used to

Determination of Toxic Organic Chemicals In Natural Waters, Sediments and Soils.
DOI: https://doi.org/10.1016/B978-0-12-815856-2.00002-3

determine the source and cycling of organic carbon and second because fatty acids in vivo function primarily as structural components of membranes and energy storage products. In the reduced state the amount and type of fatty acids may be indicative of the trophic status of the ecosystem at the time of fatty acid formation. Unsaturated, short chain ($\leq C_{20}$) and microbial fatty acid are indicative of productive systems, whereas long chain dominates in oligotrophic systems. Fatty acids have been shown to comprise 5%–10% of the weight of humic–fulvic acid structure. As an integral portion of the structure of these refractory materials, fatty acids can be used to determine the source of organic carbon and the physical/chemical characteristics of these materials. The behaviour of humic/fulvic acids may determine the transport of toxic trace metals and anthropogenic organics.

Gas chromatography

Kawahara [1,2] carried out the micro determination of pentafluorobenzyl ester derivatives of benzoic, oleic, linoleic and undecanoic acids by electron-capture gas chromatography (GC). The method was applied to chloroform extracts of natural water.

Khomenkô and Goncharova [3] separated and concentrated the following acids dissolved in natural water: oxalic, malonic, succinic, glutaric, adipic, fumaric, malic, tartaric, citric, pyruvic and gallic. Ethyl ether–butanol (1:1) and isobutyl alcohol and ethyl acetate were used as solvents to prepare extracts for GC. The preferred method was extraction (3 × 5 minutes) and pH 2.0 into an equal volume of butanol.

Van Huyssteen [4] completely separated normal and iso acids on glass columns (2.13 m × 3 mm i.d.) packed with 3% of FFAP (a reaction product of polyethylene glycol 20,000 and 2-nitrophthalic acid developed by Varian Aerograph) on Chromosorb 101 (180–100 mesh) at 180°C in an instrument equipped with a dual-flame ionisation detector (FID); nitrogen was used as carried gas (77 mL min^{-1}).

Bethge and Lindstroem [5] first removed metal cations from a 10 mL sample of water by elution with water from a Dowex to W-X8 ion-exchange column and the eluate was titrated to pH 8.0 with standard tetrabutylammonium hydroxide. A calculated amount (as determined from the titration) of hexanoic acid was added as internal standard; the solution was concentrated to a syrup, the syrup was dissolved in acetone and α-bromotoluene was added in slight excess. After 2 hours to ensure complete reaction, 1 μL of the acetone solution was injected into a stainless steel column (2 m × 2 mm) packed with 3% of butane-1-4-diol succinate with nitrogen (30 mL min^{-1}), as a carrier gas and an FID. The column was kept at 120°C for 17 minutes, then temperature programmed to 150°C min^{-1}. Down to 50 μm concentration of the acids could be determined by this procedure.

Richard and coworkers [6,7] employed macroreticular XAD-4 resin aminated with trimethylamine for the concentration, isolation and determination of acidic material from aqueous solutions. Acidic material is separated from other organic material by passing the water sample through a resin column in hydroxide form; other organic compounds are removed with methanol and diethyl ether. The acids are eluted with diethyl ether saturated with hydrogen chloride gas. After concentration the eluate is treated with diazomethane and the esters formed are separated by GC.

Richard et al. [7] concentrated and determined organic acids in river water samples. Concentration was achieved by anion-exchange chromatography with subsequent elution, esterification to methyl derivatives and GC. The process described was shown to work well for a variety of complex aqueous samples and is simple and reasonably quick.

Vairavamurthy et al. [8] determined acrylic acid by electron-capture capillary GC after extraction with tri-*n*-phosphine oxide and derivatisation with pentafluorobenzyl bromide.

Mebran et al. [9] determined acetic acid by direct aqueous injection GC. Fatoki and Vernon [10] used capillary GC of chloroform extracts of natural water to determine free fatty acids.

Olsen et al. [11] used a combination of isotope dilution GC and Fourier transformed infrared spectroscopy to analyse methyl esters of several carboxylic acids including malonic, hexabenzene carboxylic, octanoic and octanedioic acids.

High-performance liquid chromatography

Application of high-performance liquid chromatography to the resolution of complex mixtures of fatty acids in water [12,13] had provided an alternative to the high-temperature separation obtained by GC. Both techniques have similar limits of detection but lack the ability to analyse environmental samples directly. Analysis requires that the fatty acids be separated from the organic and inorganic matrices, followed by concentration.

Typically, these processes can be accomplished simultaneously by the appropriate choice of methods. Initial isolation of the fatty acids is based on relative solubility of the material of interest in an organic phase compared to the aqueous phase. Secondary separation is determined by the functional group content and affinity for a solid support.

Hullett and Eisenreich [14] used high-performance liquid chromatography for the determination of free and bound fatty acids in river water samples. The technique involves sequential liquid–liquid extraction of the water sample by 0.1 m hydrochloric acid, benzene–methanol (7:3) and hexane–ether (1:1). The resultant extract was concentrated and the fatty acids were separated on FloraSil using an ether–methanol 1:1 and 1:3 elution. Final determination of individual fatty acids was accomplished by forming the chromatophoric phenacyl ester and separating by high-performance liquid chromatography. Bound fatty acids were released by base saponification or acid hydrolysis of a water sample from which the fatty acids had been removed by solvent extraction.

Kieber et al. [15] determined formate in nonsaline waters by a coupled enzymatic/high-performance liquid chromatographic technique. The precision is approximately $\pm 5\%$ relative standard deviation. Intercalibration with an anion chromatographic technique showed an agreement of 98%. Down to 0.5 μmol L^{-1} absolute of formate could be determined. Kieber et al. [15] found 0.2–0.8 mol L^{-1} formate in seawater and 0.4–10 μmol in rainwater [16].

The procedure involves precolumn oxidation of formate with formate dehydrogenase which is accompanied by a corresponding reduction of β-nicotinamideadenine dinucleotide [β-$(NAD)^{+}$] to reduce β-nicotinamide dinucleotide (i.e. β-NADH). The latter is quantified by high-performance liquid chromatography.

Pontin Gautier et al. [17] have developed a method based on the relation between the vapour pressure of an organic micropollutant and its concentration.

The sample is injected into a stream of nitrogen passing through a series of saturators separated by electrically operated valves. A mixture of pollutants is successively separated into its constituent parts; a single compound may be analysed in a single saturator. On leaving the apparatus the compound is recorded by an FID. The method has been used for determination of concentrations of phthalic esters at concentrations of 20 mg L^{-1} and has achieved a high degree of selectivity for undivided esters.

Nimura et al. [18] used 1-pyrenyl diazomethane as a labelling reagent for the liquid chromatographic determination of down to 20–30 fmol carboxylic acids including palmitate, 1-pyrenyl palmitate, lactate, propionate and formate. The fluorescent reagent is excited at 340 nm and emits at 395 nm.

Kishida et al. [19] carried out a specific determination of ascorbate. The ascorbate is derivatised to dehydroascorbic acid bis-(dinitrophenyl) hydrazone (osazone) using 2,4-dinitrophenylhydrazine as the reagent. The osazone is then determined by high-performance liquid chromatography.

Brown [20] has described a technique for the routine determination of acrylic acid monomer in natural and polluted waters which used high-performance liquid chromatography for separation from interferences and using ultraviolet detection for quantification.

Okada [21] has described a redox suppression for the ion exclusion chromatography of carboxylic acids with conductometric detection. The reaction between hydriodic acid (the eluent) and hydrogen peroxide (the precolumn reagent) is used as the redox suppressor for ion exclusion chromatography of carboxylic acids. The suppressor is useful with highly acidic eluents and reduces background conductance more effectively than a conventional ion exchange suppressor. Saari Nordhaus and Anderson [22] studied the chromatographic separation on a mixed-mode stationary phase of mixtures of carboxylic acids and inorganic anions and showed that by careful selection of eluent pH and ionic strength simultaneous determinations could be carried out of acetic, lactic, propionic, butyric and isobutyric acids, and also of chloride, bromide nitrate, phosphate, selenite and sulphate.

Hirajama and Kuwamoto [23] carried out a numerical analysis of the elution behaviour of substituted benzoate anions in ion chromatography. Anions studied included benzoic, chlorobenzoic, various anisates, dimethylbenzoic, salycyclic, hydroxybenzoic, resorcylic, protocatechuic and gallic acids.

Berglund et al. [24] used two-dimensional conductometric detection with sequential suppressed and single-column detection in the ion chromatography of glyoxylate and malonate, also fluoride, borate, chloride, arsenate and nitrate in natural waters.

Thin-layer chromatography

Khomenkô and coworkers [25,26] used thin-layer chromatography for determining nonvolatile organic acids dissolved in nonsaline water. The organic acids are extracted from the water and concentrated, then separated on a silica gel column into four groups which are concentrated to 0.1–0.2 mL. Thin-layer chromatography is carried out in layers of silica gel KSK previously air-dried for 10 minutes and activated for 30 minutes at 105°C. The acids in the first, second and third fractions are developed in butanol–benzene–acetic acid (10:20:3, 15:85:2 and 15:35:8, respectively) and in the fourth fraction in ethyl acetate–water–formic acid (9:1:1). After drying the chromatograms for 1.5 hours at 120°C, the organic acids are detected by spraying with 0.4% solution of bromocresol green in 20% ethanolic alkali, and the spot areas are measured for a semiquantitative determination of the acids.

Spectrophotometric methods

Stradomskaya and Goncharova [27] describe a spectrophotometric method for the determination of lactic acid in amounts down to 5 μg in natural water. The method is based on oxidation of lactic acid to acetaldehyde and reaction of the latter in 1-naphthol to form a coloured product. Amino acids, aldehydes (other the acetaldehyde) and organic acids do not interfere.

Goncharova and Khomenkô [28] determined benzoic acid in ether extracts of natural and contaminated waters. The benzoic acid was back-extracted from the ether into 01 M aqueous hydrochloric acid and evaluated spectrophotometrically at 230 nm. Beer's law is obeyed with 200–140 μg of benzoic acid in 20 mL. Phenol, benzene and carboxylic acids (succinic, citric, lactic, oxalic, malic, glutaric, propionic, acetic and formic) do not interfere. The presence of humic and fulvic acids has little effect since they are not extracted into ether.

Smotrakov et al. [29,30] determined malic acid spectrophotometrically in natural water. Malic acid is oxidised to carbonyl compounds with potassium permanganate in sulphuric acid medium, and then the carbonyl compounds are reacted with 2,4-dinitrophenylhydrazine. The 2,4-dinitrophenylhydrazones are then dissolved in potassium hydroxide in 50% ethyl alcohol and evaluated at 554 nm.

Hatsue and Masayuki [31] determined phenylglyoxylic and mandelic acids spectrophotometrically as follows: the sample (0.5 mL) is acidified with 0.05 mL of 0.1 M hydrochloric acid and shaken with 5 mL of ethyl ether for 10 minutes. A portion of the ether layer is evaporated to dryness on a water bath at 70°C. A volume of 4 mL concentrated sulphuric acid–40% aqueous formaldehyde (100:1) reagent is added to the residue, and after 15–16 minutes the extinctions at 350–450 nm are measured. The extinction due to phenylglyoxylic acid has a maximum at 350 nm and that due to mandelic acid a maximum at 450 nm, and the extinctions are proportioned to the amounts of each acid and are additive when mixtures of the acids are analysed.

Krotova [32] determined methacrylic acid spectrophotometrically. A portion of the effluent sample was extracted with toluene or benzene to remove *N*-phenyl-2-naphthylamine, μα-dimethylbenzyl alcohol and acetophenone, and a 50 mL sample was adjusted to pH 8.0 (to prevent polymerisation) and subjected to distillation in a stream of nitrogen to remove benzene, toluene, butadiene, acetone or other volatile substances. The solution was diluted to 50 mL with water and adjusted to pH 4.0–5.0 with 0.1 M hydrochloric acid. The extinction is measured at 208 nm.

Jordan [33] described a procedure for the determination of traces of formic acid and formaldehyde in river water. Formic acid plus formaldehyde are determined after reducing the acid with nascent hydrogen; formaldehyde alone is determined without reduction. The chromatographic acid spectrophotometric method is carried out in 6–7 M sulphuric acid. The limit of determination is 50 μg L^{-1} without preconcentration and 50 ng L^{-1} with preconcentration by extraction with diethyl ether.

Fluorescence spectrometry

Bustin and West [34] determined traces of citrate in aqueous systems. The citrate is converted into ammonium 2,6-dihydroxyisonicotinate, and the resulting blue fluorescence gives a measure of the amount of citrate. The calibration graph is rectilinear for 0.01–10 μg mL^{-1}; the optimum working range is 0.1–10 μg mL^{-1}. There was no interference from species normally encountered in water samples.

Potentiometry

Hasson et al. [35] used a potentiometric gas sensing probe for the selective determination of acetate ion in natural waters.

Polarography

Kopanica et al. [36] described a polarographic determination of small quantities of lactic acid and pyruvic acid (α-ketopropionic acid) in polluted waters. For differential pulse polarographic determination of pyruvic acid, a damping solution of pH 4.0–5.0 is used with the potential of the polarography wave of −1.10 V and a peak width of 110 mV. Determination of lactic acid is performed similarly after oxidation by potassium permanganate to pyruvic acid. The method has high precision.

Miscellaneous

Escher and Schwarzenbach [37] evaluated the liposome–water partitioning of organic acids and bases and evaluated three methods for the experimental determination of this parameter. These methods are based on potentiometry and immobilisation of lipid material on a solid support. A method based on immobilised liposomes in the form of noncovalently coated large partides gave promising results in the determination of partitioning coefficient.

Erikssen and Ledin [38] analysed long-chain fatty acids in water with in-vial derivatisation.

The acids were purified and concentrated by solid-phase extraction on strong anion-exchange discs, followed by in-vial derivatised to their corresponding methyl ester and subsequently analysed by GC–MS (mass spectrometry). The method was able to quantify the acids at a concentration $<1\ \mu g\ L^{-1}$ with a recovery of 31%–97%. The treatment efficiency of a treatment plant was evaluated by comparing concentrations of fatty acids at the inlet and outlet. It was found that the treatability decreased with increasing chain length for the saturated acids (19%–100% degradation), whereas the corresponding monounsaturated acids were more easily degraded.

Richard and Fritz [6] employed macroreticular XAD-4 resin aminated with trimethylamine for the concentration, isolation and determination of acidic material from aqueous solutions. Acidic material is separated from other organic material by passing the water sample through a resin column in hydroxide form; other organic compounds are removed with methanol and diethyl ether. The acids are eluted with diethyl ether saturated with hydrogen chloride gas. After concentration the eluate is treated with diazomethane, and the esters formed are separated by GC.

Valto et al. [39] determined fatty acids in water using solid-phase extraction GC analysis. The suitability of liquid–liquid extraction and solid-phase extraction methods was studied for the rapid separation of resin and fatty acid fractions from waters. A novel procedure for online monitoring of selected individual acid components (limit of detection 11–18 μg L^{-1}) by combined solid-phase extraction combined with atmospheric pressure chemical ionisation–MS was also developed. The stability of this technique for quality control was tested. The method is suitable for the analysis of resins and fatty acids.

2.2 Phenols

Trace amounts (<1 mg L^{-1}) of phenolic compounds can have significant detrimental effects on water quality. Phenols are toxic to aquatic life [40] and mammals [41,42] and can impart objectionable tastes and odours to water and fish [40,43,44]. The US Environmental Protection Agency [45] recommends a maximum of 1 μg L^{-1} for total phenolic compounds in domestic water supplies.

Gas chromatography

One of the earliest GC methods for the determination of phenols in water involved absorption of phenols on carbon, extraction of the carbon with chloroform to remove the phenols, concentration of the phenols in the chloroform by evaporation and determination of phenols in the chloroform extract by GC [46]. Since then, various workers have studied the application of GC to the determination of traces of mono-dihydroxy phenols in natural and polluted waters, including phenol, cresols, xylenols and resorcinol [16,47–64]. Some GC methods are applied directly to the sample [65–67], while others use a prior concentration technique, frequently solvent extraction [68–73].

Semenchenko and Kaplin [51] acidified the sample with hydrochloric acid then saturated it with sodium chloride. The phenols were extracted with diethyl ether. The extract was shaken with aqueous sodium bicarbonate (5%) to remove any organic acids then the phenols extracted into aqueous sodium hydroxide and methylated with dimethyl sulphate prior to GC.

Eichelberger et al. [52] collected the phenols from water samples on columns of activated carbon. The phenols were stripped from the carbon with chloroform prior to GC of the cleaned-up extract. Clean-up was achieved by a treble extraction of the chloroform phase with aqueous sodium hydroxide. The aqueous extracts were combined, acidified with concentrated hydrochloric acid to pH 2.0 and extracted three times with ethyl ether, and the combined ethyl ether extracts were passed through a 10 cm FloraSil column topped with anhydrous sodium sulphate (2 cm). The phenols were eluted with ether, and the eluate was concentrated by evaporation and analysed by GC.

Cooper and Wheatstone [54] converted phenols to their pentafluorobenzyl derivatives prior to GC.

Murray [74] determined low concentrations of phenols, cresols and xylenols in chloroform extracts of water. The trimethylsilyl derivatives of the phenols were formed and analysis completed by GC. The method was rapid and required a minimum of sample manipulation. The lower limit of detection was 0.100 mg L^{-1}, 0.025 mg L^{-1} for xylenols. The percentage extraction from a 1 L sample was poor, namely phenol 20%, *m*-cresol, 30%, *p*-cresol, 40%, *o*-cresol, 48%, 3,4-xylenol, 8%, 2,6-xylenol, 83%, *o*-xylene, 95%. Phenol is the most hydrophilic compound, and its extraction was the lowest. To achieve a 90% recovery of phenol, more than 10 extractions would be required.

Voznakova and coworkers [75,76] departed from the usual approach to phenol analysis by GC made by earlier workers, that is direct injection GC or solvent extraction followed by GC. These workers described a method for determining phenol, *o*-cresol and 2,6-xylenol in water in which the phenols are sorbed on a macroporous polymer sorbent (Separon SE) and then undergo thermal desorption for GC analysis. This method is claimed to be quick and reproducible and to have a detection limit of between 1 and 10 μg L^{-1} depending on the phenol.

Several workers have discussed the determination of phenols in river waters [77,78]. Coleman et al. [77] described a field technique for monitoring *p*-cresol in streams and rivers.

Goldberg et al. [79] showed that in the solvent extraction of phenols from water samples it is not unusual to lose 50% or more of the extracted solute during evaporation.

Goldberg and coworkers [79,80] used solvent-heavier-than-water, two-cycle liquid extractors to concentrate phenols at the μg L^{-1} level from water. The solvent used was dichloromethane. The nonaqueous solution containing the extract was concentrated by Kuderna–Danish concentrators, and this was followed by GC. Overall concentration factors were around 1000 with efficiencies 23.1%–87.1%. Determination could be made with an accuracy of only 15%–20% because of solute losses during concentration. The range of concentration used was of the order of 1 μg L^{-1}.

The Department of Environment (United Kingdom) [81] has published details of five procedures for the determination of monohydric and some other phenols in water. The first two methods are for media in which the phenol concentration exceeds 100 μg L^{-1} and the remaining three methods for lower concentrations.

DiCorcia et al. [82] carried out a GC analysis of phenolic compounds in natural water using acid-washed graphitised carbon black. Except in the cases of 2,4-dinitrophenol and 4,6-dinitor-*o*-cresol, experiments show that GC, using acid-washed graphitised carbon black modified with trimesic acid and polyethylene glycol 20 M, provides an accurate method for the fractionation of the phenols.

Janda and Krijt [83] have described a method for the isolation of phenols from water by distillation–extraction. The water sample (150 mL) is acidified and strongly salted, then continuously steam-distilled with 2 mL diethyl ether for 1.5 hours. The extract is analysed by splitless glass chromatography. The recovery, for a concentration range of about 0.1–30 mg L^{-1} approaches 100%. The phenols could be isolated into diethyl ether from suspensions and from water with high suspended solids content. The detection limit, using splitless injection glass capillary columns, is approximately 10 μg L^{-1}.

Polar volatile analysis by GC has been used to determine phenols in water [84].

Matsumoto et al. [71] and others [68–72] described a GC–mass spectrometric identification of phenols and aromatic acids in river waters. The organics are separated from the water sample by extraction with ethyl acetate and then further separated by silica gel column chromatography prior to GC–MS.

High-performance liquid chromatography

Armentrout et al. [85] carried out trace determination of phenolic compounds in water at the ppm level by reversed-phase liquid chromatography using electrochemical detection with a carbon-polyethylene tubular anode.

Hashimoto et al. [86] carried out a high-performance liquid chromatography study on the radiolysis of phenol in aqueous solution. This showed that in the oxygenated solution, hydroquinone, pyrocatechol, hydroxyhydoquinone and trace amounts of resorcinol and phloroglucinol are produced. In the deaerated solution, hydroquinone, pyrocatechol and small amounts of resorcinol and hydroxyhydroquinone were produced. The decomposition rates of phenol, hydroquinone and pyrocatechol were approximately five times higher in the oxygenated that in the deaerated solution.

The application of high-performance liquid chromatography to the determination of phenols has been discussed by Sharp and Myer [87].

Ratanathanwongs and Crouch [88] have described an online postcolumn reaction based on air-segmented continuous flow for the determination of phenol in natural waters by high-performance liquid chromatography. The reaction used was the coupling of diazotised sulphanilic acid with the phenol to form high-coloured azo dyes. The detection limit for phenol was 17 μg L^{-1}, which represents a 16-fold improvement over the determination of phenol with ultraviolet detection.

Rennie [89] has discussed the application of automated high-performance liquid chromatography to monitoring phenol in the River Dee. This monitor used trace enrichment followed by reversed-phase chromatography and UV detection.

Li and Liu [90] used dispersive liquid–liquid microextraction in combination with high-performance liquid chromatography for the determination of bisphenol-A in natural waters.

In this microextraction method, several parameters, such as extraction solvent volume, sample volume, disperser solvent, ionic strength, pH and disperser volume, were optimised with the acid of interactive orthogonal array and a mixed-level experiment design. First, an orthogonal array design was used to screen the significant variable for the optimisation. Second, the significant factors were optimised by using a mixed-level experiment. Under the optimised extraction conditions (extraction solvent: ionic liquid, 60 μL; dispersive solvent: methanol, 0.4 mL; and pH 4.0), the performance of the established method was evaluated. The extraction recoveries were 20% and 50% under the two sets of conditions, respectively. The optimised dispersive liquid–liquid microextraction conditions of extraction volume 60 μL and sample volume 7 μL were chosen. In this method, it is evident that bisphenol-A was separated from some other endocrine disruption chemicals such as ethinylestradiol and estrone.

Thin-layer chromatography

Thieleman [91] has described a thin-layer chromatographic method for the identification of phenolic impurities in diethyl ether or active carbon extracts of water. The separation and identification of *o*, *m*- and *p*-cresols, phenol, 2-naphthol, and 2,5-, 3,4- and 3,5-xylenols, after coupling with Fast Red Salt AL (CI Azoic Diazo Component 36) was carried out on Kieselgel G impregnated with potassium carbonate with the use of one of three solvents. Polyhydric phenols were not identifiable by this method but were separated by thin-layer chromatography on Kieselgel G and D with dioxan–benzene–anhydrous acetic (25:9:4) or benzene–acetone (9:10) as solvent, and detection by spraying with diazotised sulphanilic acid or with molybdophosphoric acid.

Thieleman [92] also described a thin-layer method for the determination of phenols, cresols, xylenols and naphthols in active carbon extracts of natural and potable water. Although the direct separation of these phenols was not possible, products of their coupling with Fast Blue Sale BB (C.I. Azoic Diazo Component 20) could be separated. The *RF* values are reported for the dye complexes for phenol *o*-, *m*- and *p*-cresols, 1- and 2-naphthols, and 3,4-, 3,5- and 2,5-xylenols.

Thieleman [93] also achieved a thin-layer chromatographic separation of phenol, three xylenols and 1- and 2-naphthols after coupling water sample with diazodised Fast Blue Salt BB (C.I. Azoic Diaz Component 20). The separation of the coupled phenolic compounds was achieved with three solvent systems on plates coated with Kieselgel G-potassium carbonate (1:2). *RF* values are reported. The separation of phenol and the cresols by paper chromatography is also reported.

Edeline et al. [94] described a method for determining phenols in environmental samples based on separation on thin layers of activated carbon.

C'hmil [95] determined alkylchloro- and nitrophenols by thin-layer and gas-liquid chromatography. Separate determination of nitrophenols and nitrophenol compounds in water in the presence of each other by partition (reversed) thin-layer chromatography has been investigated. Procedures for analysing chlorophenols and nitrophenol compounds in water using a combination of gas-liquid and thin-layer chromatography (adsorption and partition) have been suggested.

Afghan et al. [96] mention the use of fluorescence methods for determining phenols in natural waters. This method uses *n*-butyl acetate or isoamyl extraction followed by back extraction with sodium hydroxide solution to extract the phenol from the water sample.

Raman spectroscopy

Various workers have examined the applicability of Raman spectroscopy [97–99] for determining the traces of phenols in natural waters. Marley et al. [100,101] studied *o*-chlorophenol, 2-4-dichlorophenol, 2,4,6-trichlorophenol, 2-chloro-5-methylphenol and 2-chloro-4-nitrophenol in water solution using the 514.5, 488.0 and 457.9 nm lines of an argon ion laser. The effects of source intensity, optical alignments and background fluctuations on quantitative results were examined. Peak measurement and cross-correlation were used in quantitative measurements. Accuracy was improved by using an internal standard with each measurement. Limits of detection ranged between 100 and 0.3 mg L^{-1} depending on the component. The occurrence of resonance enhancement was identified as the most important factor in controlling sensitivity.

Van Haverbeke and coworkers [102,103] used laser-excited resonance Raman spectroscopy to determine phenolic compounds at the 2–200 ppb level in distilled water, tap water and pond water. The phenols were first converted to derivatives by reaction with the diazonium salt of 4-nitroaniline. Haverbeke and Herman [103] used a closed-loop flow-through setup in their experiments. In this setup a large volume of the sample solution, usually 650 mL, continuously flowed through the laser beam. Since the individual dye molecules are only momentarily exposed to the laser beam, any decomposition occurring will be reduced to a minimum, thus assuring more accurate measurements.

There is a definite contribution from the fluorescence band of dye, on which the resonance Raman spectrum is superimposed. The individual resonance Raman bands are, however, clearly observed. The broad, weak band around 1640 cm^{-1} is due to deformation vibration of the water solvent molecules.

Spectrophotometric method

Cheesman and Wilson [97] investigated five chromogenic reagents for the spectrophotometric determination of down $0.1\ mg\ L^{-1}$ phenols in natural waters. The chromogenic agents are for use with unchlorinated water sample. This method is also recommended by the American Public Health Association [100,101,103,104]. This method is sensitive to para-substituted phenols [105,106].

The traditional method for phenol analysis is based on the oxidising coupling of 4-aminoantipyrine (4-APP) with phenol in alkaline solution. This method has been reevaluated by Katsaounos et al. [107]. These workers combined spectrophotometry with micellar-assisted preconcentration (cloud point extraction). The method explores the conventional reaction pathway while extinction is facilitated by surfactant-based precipitation during which the nonpolar derivative of 4-APP is entrapped in the micelles and concentrated into a surfactant-rich phase. This latter is resolubilised, and the complex is quantified spectrophotometrically in the presence of a surfactant. Compared to the traditional method, the modification proposed offers certain analytical advantages, normally lower detection limits and shorter time of analysis. The method was applied in various samples of different origin with satisfactory results.

Other workers have investigated the 4-APP (4-amino-phenazone) spectrophotometric methods for the determination of phenols in water [108–115]. In the procedure described by Stroehl [113] a 500 mL sample of water is adjusted to pH 2.0–3.0 with phosphoric acid and distilled down to 450 mL to separate the volatile phenols. The remaining nonvolatile components constitute only a small proportion of the phenol content, but may, if desired, be separated by liquid–liquid extraction. The distillate is treated with 4-APP in the presence of potassium ferricyanide to give an orange-red complex, which is measured photometrically against a standard.

The earliest references found to the use of 3-methyl-2-benzothiazolinone hydrazone as colorimetric reagent for phenols are those of Pays and Bourden [116] and Freistad and coworkers [117,118]. However, the first workers to apply this reagent to water samples were Freistad and coworkers [117,118] followed by Goulden et al. [119], Lei and Pang [105] and Gales [106]. Although the method had a limit of detection of approximately $10\ \mu g\ L^{-1}$ for phenol, this limit could possibly be improved if the coloured reaction products of the reagent and phenols could be concentrated by solvent extraction (as in the aminoantipyrine procedures). This reagent generally reacts with para-substituted phenols to a greater extent than aminoantipyrine.

Goulden et al. [119] and Lei and Pang et al. [105] described an automatic continuous flow apparatus for this spectrophotometric determination which is based on the reaction of the phenols with 3-methylbenzothiazolin-2-one hydrazone. Up to 80% of the sample flow is distilled and condensed, and after the colour-formation step, the dye is extracted into chloroform before measurement. At a concentration of 5 μg of phenol per litre, the coefficient of variation was 3.

Fuxinmel et al. [120] determined 4-nonylphenol and 4-tetroctylphenol in natural waters using a chromogenic agent 1-butyl-3-methyl-imidozolium hexafluorophosphate in combination with ionic liquid–based liquid–liquid microextraction.

Gales [106] investigated the use of 3-methyl-2-benzothiazoline hydrazone for the determination of down to 1 $\mu g\,L^{-1}$ of phenols in waters. Phenolic compounds were determined by manual and automated methods based on coupling with 3-methyl-2-benzothiazoline hydrazone in an acidic medium using ammonium cerium (IV) sulphate as an oxidant.

Bosch et al. [121] developed a procedure for the determination of traces of phenols in natural waters and wastewaters without the need for preliminary distillation. It involved reaction of the phenols with iodine monobromide, followed by extraction of the products with cyclohexane and spectrophotometric measurement. The method could also be used to determined *p*-substituted phenols and chlorophenols; the wavelengths at which several compounds show maximal absorption are listed.

Karenman et al. [122] have described a spectrophotometric method based on a diazotised sulphanilic acid or 4-nitroaniline procedure which is capable of determining down to 1 $\mu g\,L^{-1}$ phenols in natural waters.

Bajeva and Gunta [123] described an extractive photometric method for phenol determination in river water based on using potassium dichromate–potassium ferricyanide as oxidiser in a sodium carbonate buffer medium (pH 10.5–11.5). The resultant red dye was extracted in *n*-pentanol. The purple *n*-pentanol extracted exhibited an absorption maximum at 535 nm and was stable for 24 hours. Slight differences in pH produced considerable changes in colour development. The colour reaction was independent of time, over the range 2–60 minutes, and temperature, over the range 15°C–60°C. The system was free from common interferents including cresols. Beer's law was followed in the range 4–70 μg phenol. The molar absorptivity and Sandell's sensitivity were 30, 500 $L\,mol^{-1}\,cm^{-1}$ and 0.0029 $\mu g\,cm^{-2}$, respectively.

Hassan et al. [124] based a colorimetric method for detecting phenols in water on the reaction of nitrosophenol and resorcinol, with absorbance measured at 480 nm. Using the proposed method, percentage recoveries from natural water samples did not differ significantly from those obtained with the 4-APP method for phenol, *o*-cresol, *m*-cresol and *p*-substituted phenols (which did not react with 4-APP). The minimal detection limit was 4 $\mu g\,L^{-1}$, compared to 10 μg^{-1} for the 4-aminonantipyrine method.

Atomic absorption spectroscopy

Bo-Xing et al. [125] have described an indirect determination of trace phenol in water by atomic absorption spectroscopy. The method is based on the bromination of phenol to form tribromophenol. The excess bromine is reacted with iodine and the iodine formed extracted into cyclohexane and reduced back to iodine. The iodine is complexed with cadmium in sulphuric acid solution and the complex extracted into methyl isobutyl ketone. The extract is analysed for cadmium (and hence indirectly for phenol). The linear concentration range for determination of phenol is 0.6–90 μM.

Miscellaneous

Goldberg and Weiner [126] have described methods for the extraction and concentration of phenolic compounds from water and sediment.

A detailed study has been made [127] of the method of preparation and storage of stable standard solutions of phenol, *o*- and *m*-cresols and guaiacol for analysis. The usual standard solution of phenol in distilled water with a concentration of 100 mg L^{-1} must be used on the day of preparation as its concentration of phenol fell by 25%–28% within 30 days.

Normal and derivative flow injection–cyclic voltammetry have been used for the simultaneous determination of phenolic compounds in natural water at levels below 1 μg L^{-1} [128].

Molecular emission cavity analysis has also been applied to the determination of traces of phenols in water [129,130].

Zgola-Grzeskowiak and Grzeskowiak [131] have discussed the determination of alkyl phenol and their short-chained ethoxylates in water.

Martianov et al. [132] developed an enzyme-linked immunoassay for nonylphenol. Preconcentration of phenols in water has been achieved by absorption on 5-vinyl pyridine-divinylbenzene copolymer or silica gel–modified cellulose [133] or *n*-vinyl-2 pyrrolidene copolymer [133] or silica gel–modified cellulose [133] or cation-exchange resins [134–135].

Preconcentration

Preconcentration on 5-vinyl pyridine-divinylbenzene copolymer resin

Kawabata and Ohira [130] have studied the removal and recovery of phenols from aquatic samples using 5-vinyl pyridine-divinylbenzene copolymer as an adsorbent. Although the analytical implications of this work were not discussed, these clearly exist. Elution of the concentrated phenols from the resin column was accomplished by a treatment with acetone or methanol.

Preconcentration on silica gel–modified cellulose

Columns containing silica gel chemically linked to different groups have been compared for their suitability for preconcentration of selected phenols from dilute aqueous solution [134]. The chemically bonded phases comprised octyl, octadecyl, phenyl, diol and cyanide radicals of which the phenol-bonded silica gel gave the best recoveries following reversed-phase high-performance liquid chromatography analysis. Recoveries from *m*-cresol, *p*-chlorophenol, 2,5-dimethylphenol, 2,6-dichlorophenol and *o*-phenylphenol were all in excess of 90% when the water sample was first acidified to pH 2.0 and 30 g sodium chloride per 100 mL added before passage through the column. For phenol, however, a recovery of only 42% was achieved under these conditions.

Preconcentration on N-vinyl-2-pyrrolidone copolymer

Carpenter et al. [134] separated phenolic materials from aqueous solutions on cross-linked aqueous insoluble *N*-vinyl-2-pyrrolidone polymers. The pH for maximal binding of the phenolic compound to the resin was found to be dependent on the acidity of the phenolic compound. Binding to resin was particularly favourable for polyhydroxyl and extended aromatic compounds. Columns packed with this resin removed more than 95% of simple phenolic compounds from aqueous solution and quantitative recovery of the bound phenolic compound was possible by elution with 4 M urea solution.

Preconcentration on cation-exchange resins

Ambersorb XE-34 cation-exchange resin followed by desorption with diethyl ethers or dichloromethane has been used to preconcentrate phenols [135]. XAD-2 resin and pyridine desorbent have been used for the same purpose. Other methods of preconcentrating phenols include adsorption on active carbon and coprecipitation with ferric hydroxide [136–139].

2.3 Phenolic acids

Rump et al. [73] have described a cellulose this layer method for the detection of phenolic acids such as *m*-hydroxybenzoic, *m*-hydroxyphenylacetic acid and *m*-hydroxyphenyl propionic acid in water samples suspected to be contaminated with liquid manure. The phenolic acid is extracted with ethyl acetate from a volume of acidified sample equalling 1 mg of oxygen consumed (measured with potassium permanganate). The ethyl acetate is evaporated and the residue dissolved in ethanol. After spotting of a 1 μm aliquot on a cellulose plate, the chromatogram is developed by capillary ascent with the solvent *n*-propanol–*n*-butanol–25% NH_3–water (4:4:1:1 by vol.). The solvent front is allowed to advance 10 cm. The air-dried plate is sprayed with a diazotised *p*-nitroaniline reagent to make the phenolic acids visible.

2.4 Methyl *tert*-butyl ether

Purge-and-trap GC with flame ionisation detection [140] has been applied to the determination of methyl *tert*-butyl ether. The sample solutions were added with 10% (w/w) sodium sulphate and adjusted to pH 4.0 by acetic acid and sodium acetate buffer solution to improve the purge efficiency before the analysis. A CP-4010 purge-and-trap injector was used to purge methyl *tert*-butyl ether from water and cool it in the cold-trap kept at 75°C, then the cooled trap was flash heated to release the analytes onto a HP-1 capillary column. The detection was GC-FID. A good linear response was obtained and the detection limit was 0.1 $\mu g\ L^{-1}$. This method has been successfully applied to the determination of *tert*-methyl ether in river samples.

Aguera et al. [141] describe the application of solar photocatalysis by photo-Fenton reaction to the degradation of methyl *tert*-butyl ether in aqueous solutions at a pilot scale in a compound parabolic collector reactor. The results show a rapid and complete oxidation of methyl *tert*-butyl ether after only 35 minutes of treatment. However, 54% remains to be mineralised after that time, almost complete mineralisation (97%) being reached only after 155 minutes of treatment. This indicated the formation of a large number of transformation products more resistant to degradation than the parent compound. Identification and quantification of six transformation products (*tert*-butyl formate, *tert*-butyl alcohol, acetone, methyl acetate, isobutene and formaldehyde) were possible with the combined use of two complementary techniques: GC–MS, both coupled to a purge-and-trap system. Ionic chromatography was used for the identification of organic ions usually present in the last stages of the mineralisation process. Only formate ions were detected and these may be considered as the last step in the oxidation process previous to total mineralisation. A degradation pathway, fixing the most critical compounds in terms of kinetic degradation, is proposed.

2.5 Alcohols

Spectrophotometric methods

Romantsova [142] described a spectrophotometric determination of small amounts of cyclohexanol (and cyclohexanone) in river water samples and effluents. The sample (0.2 mL containing down to 10 $mg\ L^{-1}$ of cyclohexanol and 3 $mg\ L^{-1}$ of cyclohexanone) is mixed with 0.4 mL of a 5% solution of 4-dimethylaminobenzaldehyde in 83% sulphuric acid and 4.6 mL of 83% sulphuric acid and heated on a boiling water bath for 15 minutes. Extinctions are measured against a reagent blank at 435 nm (max. for cyclohexanone) and 520 nm (max. for cyclohexanol).

Igarashi [143] has described a spectrophotometric method for the determination of methanol using chromotropic acid. The test solution containing 1–15 μg of methanol is treated with 0.2 mL of 0.3% v/v propionaldehyde solution and 0.4 mL of potassium permanganate solution. After 5 minutes unconsumed permanganate is reduced with 0.22 mL of 20% sodium sulphite solution. Then 2% aqueous chromotropic acid and 75% v/v sulphuric acid are added; the mixture is heated at 80°C–85°C for 10 minutes and cooled to room temperature, and the extinction is measured at 575 nm against a reagent blank.

Gas chromatography

An automated GC method has been described for the determination of methanol and glycols [144].

Chromadistillation has been used to determine alcohols in water [145]. This method isolates and concentrates impurities from the sample. Experiments were carried out using methanol, *n*-butanol, isopentanol and *n*-pentanol. Alcohol concentration down to 0.1 mg L^{-1} could be determined.

2.6 Glycols

Spectrometry

Ethylene glycol and di- and triethylene glycols in amounts down to 0.5 mg L^{-1} have been determined by spectrophotometry [146]. A volume of 0.5 mL 2 M potassium permanganate is added to a 5 mL sample. After heating for 1 minute at 100°C, 0.07 M sodium arsenite (1 mL) and 1 mL of 2% 3-methylbenzothiazolin-2-one hydrazone hydrochloride are added and the solution again heated for 6 minutes at 100°C. After cooling, 1 mL of 2% ferric chloride 96H_2O–3% sulphuric acid is added and the solution diluted to 10 mL. After 20 minutes the extinction is measured at 630 nm against a reagent blank.

Gas chromatography

GC has been used [147] to determine parts per million of glycols in water.

Nevinnaya and Kovanov [148] have developed a GC method for the determination of mono-, di- and triethylene glycols. The glycols were extracted with mixtures of distilled water and organic solvents, such as dioxane or acetone, acetylated with acetic anhydride in the presence of boron trifluoride and subjected to GC on a nitrile rubber column using an FID. The actual concentration, the determined concentration (mg L^{-1}) and the deviance of the results at 95% confidence limit were, respectively, for monoethylene glycol, 100%, 96.7% and 5.3%; for diethylene glycol, 10%, 9.7%, 0.3% and for triethylene glycol, 40%, 38.2%, 1.6%.

Thin-layer chromatography

Tachsteinova and Kopanica [149] determined caprolactam in natural water by absorptive voltammetry after separation by thin-layer chromatography.

2.7 Dioxans

Gas chromatography

Eptein et al. [150] compared two methods for the determination of 1,4-dioxane in water. The first was a heated purge-and-trap GC–MS system in which salting out with sodium sulphate preceded the determination. The second method employed adsorption on coconut-shell charcoal and desorption with carbon disulphide/methanol, before desorbate analysis by GC with flame ionisation detection. The second technique was also successful with 2-butanone, 4-methyl-2-pentanone and butoxyethanol in water. Analysis of samples by both methods yielded comparable results. Recoveries ranged from 63% to 129% with an average recovery of 77.5% with a standard deviation of 19. The experimental detection limit was 1 μg L^{-1} in a 4 L sample and the determination limit was approximately 2 μg L^{-1}.

2.8 Esters

Gas chromatography

Thuren [151] determined phthalates in water using solvent extraction, cleanup with deactivated FloriSil and quantitative analysis by GC. The detection response was linear between 0.5 and 100 ng phthalate. Compounds mentioned are dimethyl phthalate, dibutyl phthalate, di-(2-ethylhexyl) phthalate and benzyl butyl phthalate.

Hites and coworkers [152–154] used computerised GC–MS, high-resolution MS, and high-pressure liquid chromatography to determine the plasticisers dibutyl phthalate-bis(2-2-butoxyethoxy) methane, bis-(2-ethyll-hexyl) adipate and various isomers of dioctyl phthalate and di-isodecyl phthalate and diethyl, dibutyl and bis-(2-ethyl)hecylphthalates in river water from 1 to 30 ppb.

High-performance liquid chromatography

Mori [155] has identified and determined very low levels of phthalates esters in river water using reversed-phase high-performance liquid chromatography using an ultraviolet detector. Phthalates were extracted with *n*-hexane and the uncleaned or concentrated extracts were injected into three chromatographic systems, these being cross-linked porous beads (Shodex

HP-225, Showa Penko Co.), porous polymer beads and polystyrene GPC gel. The eluents were, respectively, *n*-hexane (system A), methanol (system B) and chloroform (system C).

The presence of *n*-dibutyl phthalate and di-2-ethylhexyl phthalate was observed in a river water sample. The concentrations of phthalates in the extract were 450 ppb of *n*-dibutyl phthalate and 100 ppb of di-2-ethylhexyl phthalate and their concentrations in river water were 45 and 100 ppb, respectively.

Schouten and coworkers [156,157] and others [17] used high-performance liquid chromatography to determine very low levels of di-ethylhexl phthalate and di-*n*-butyl phthalate in Dutch river waters and compared results with those obtained by gas-liquid chromatography. Good agreement was obtained between the two techniques, although high-performance liquid chromatography was shown to be the less time-consuming technique. These two esters account for approximately 95% of the total phthalate production in Western Europe, used mainly in plastic production.

Electrokinetic chromatography

Takeda et al. [158] studied the migration behaviour of phthalate esters in micellar electrokinetic chromatography with and without the addition of butyl alcohol. This procedure was applied to the determination of phthalate esters in environmental waters.

Barroso et al. [159] performed on electrochemical study of the determination of butylate in natural waters.

Reid et al. [160] investigated possible sources of phthalate contamination in the environmental analytical laboratory.

Richard and Fritz [6] employed macroreticular XAD-4 resin aminated with trimethylamine for the preconcentration, isolation and determination of acidic material from aqueous solutions. Acidic material is separated from other organic material by passing the water sample through a resin column in hydroxide form; other organic compounds are removed with methanol and diethyl ether. The acids are eluted with diethyl ether saturated with hydrogen chloride gas. After concentration the eluate is treated with diazomethane, and the esters formed are separated by GC.

2.9 Aldehydes

Gas chromatography

A GC method has been proposed [161–164] for the determination of formaldehyde and acetaldehyde in water. One method is based on the reactions between the aldehydes and cyanide ion to form cyanohydrins. The cyanohydrins are analysed by gas–solid chromatography and selectively detected with a nitrogen–phosphorus detector. Concentrations of formaldehyde and acetaldehyde in the ranges 0.5–10 mg L^{-1}, respectively, were determined with relative standard deviations of the order of 1.5% and 1.0%, respectively. The method is specific for formaldehyde and acetaldehyde. Optimal conditions were 40 μg L^{-1} cyanide and pH less than 3.

Nishikawa et al. [162] give details of a procedure for the specific determination of acrolein, based on bromination of acrolein-*o*-methyloxime, followed by GC with electron-capture detection. The method has a detection limit of 0.4 μg L^{-1} and a relative standard deviation of 4.5%.

Thin-layer chromatography

Maktaz et al. [163] determined low concentrations of glutamic aldehyde in chloroform extracts of natural waters in the form of hydrazone and separated it from accompanying aldehydes and ketones by thin-layer chromatography. The mean relative error when determining 0.1–0.4 mg L^{-1} of glutaric aldehyde in natural waters was 11.3%.

Spectrophotometric methods

Manual and automated spectrophotometric methods have been described for the determination of free and bound formaldehyde in water [144]. The term 'combined formaldehyde' (i.e. total minus free) is used to represent these forms of formaldehyde including hexamine, which produce free formaldehyde under certain conditions, principally either distillation or hydrolysis. The limit of application of the method varies from 2.0 to 25.0 mg L^{-1} for formaldehyde.

Miscellaneous

Song and Hay [165] have described a rapid sensitive chemiluminescence flow sensor for the determination of formaldehyde. Formaldehyde was sensed by measuring the decrement of chemiluminescence intensity, which was observed linearly over the logarithm of formaldehyde concentration in the range of 5.0–1000.0 ng L^{-1}, and the limit of detection was 1.8 ng mL^{-1}. At a flow rate of 2.0 mL min^{-1}, including sampling and washing, analysis could be performed in 0.5 minutes with a standard deviation of less than 3.0%. The flow sensor offered reagentless procedures and remarkable stability in the determination of formaldehyde and could be easily reused. The proposed flow microsensor was applied successfully in the determination of formaldehyde in artificial water samples and air.

Howe [164] has described a differential pulse polarographic method for the determination of formaldehyde in natural waters.

Jahangir and Samuelson [166] have discussed the preconcentration by sorption of cyclohexane derivatives with hydroxyl or carbonyl group separated by one or two methylene groups from aqueous media by ion exchange resins. They found that these compounds adsorbed more strongly than the corresponding aromatic compounds both on sulphonated styrene-divinylbenzene resins and on nonionic styrene-divinylbenzene resins. These observations and the lower temperature dependence observed for the cyclohexane derivatives indicate that the hydrophobic interactions have a marked influence on the absorption.

Zeolite 24 H-5 has been used for the preconcentration of low molecular weight aldehydes as their 2,4-dinitrophenyhydrazones in potable water [167]. In this method the 2,4-dinitrophenylhydrazones were then separated by liquid chromatography and determined by GC using flame ionisation detection. Results are presented for the separation and detection of a range of aldehydes and ketones in aqueous solution at concentrations of 1000 mg L^{-1}. Recoveries approximated to 100% for model compounds tested, with the exception of formaldehyde.

2.10 Ketones

Middleditch et al. [168] determined acetone, methyl ethyl ketone, methyl isobutyl ketone and tetrahydrofuran in natural water by direct aqueous injection GC.

Karyakin and Chirkova [169] described a method for determining down to 10 ppm acetone in water based on the photochemical reaction of acetone with fluorescein sodium. The latter is converted into a nonfluorescent compound. The acetone concentration is determined from a graph of the log of the fluorescence intensity at 515 nm of a solution of fluorescein sodium at approximately 20°C against the log of the acetone concentration after exposure for 10 minutes to UV radiation. Hydrocarbons and ethers do not interfere but organic acids do.

Vajta et al. [170,171] investigated the determination of down to 5 μg L^{-1} acetone in aqueous petroleum refinery effluents. The effluent (1 mL) containing acetone (or ethyl methyl ketone) is mixed with 1 mL of saturated methanolic 2,4-dinitrophenylhydrazine and one drop of concentrated hydrochloric acid. The mixture is maintained at 50°C for 30 minutes and after being set aside for 1 hour is made alkaline with 5 mL of a 10% solution of potassium hydroxide in aqueous methanol (1:4). After 10 minutes the extinction is measured at 490 and 540 nm. The concentration of acetone is determined by reference to a calibration graph.

2.11 Carbohydrates

Gas chromatography

Josefsson [172] and Stabel [173] determined monosaccharides in natural water by GC of the trimethyl derivatives.

Ochiai [174] used GC to determine dissolved carbohydrates in nonsaline water. The carbohydrates are first hydrolysed to alditol acetates of monosaccharides by refluxing for 7 hours with 1 M hydrochloric acid under nitrogen. The hydrolysate is then reduced with sodium hydroxide at 60°C. The gas chromatograph was equipped with an FID. A glass column (2 m × 3 mm) is packed with 5% OV-275 on Chromosorb W was employed at a nitrogen rate of 40 mL min^{-1}. A temperature-programmed analysis from 160 to 240°C min^{-1} is required for 40 minutes to elute the acetyl derivatives of the monosaccharides and the internal standard inositol.

Spectrophotometric method

Shaova and Kaplin [175] have described a spectrophotometric method for the determination of down to 0.1 mg L^{-1} of reducing sugars in natural water. A volume of 10 mL zinc sulphate $7H_2O$ (5%) to precipitate proteins, then 2.5 mL sodium hydroxide (5%) is added to a 200 mL portion of sample until zinc is completely precipitated. The precipitate is filtered and the filtrate passed through an ion-exchange column (KU-1 and AN-22 resins) discarding the first 100 mL of percolate.

Miscellaneous

Shaheen and Senn [176] have found that hydrolysis with sulphuric acid is more efficient than hydrolysis with IM hydrochloric acid for the hydrolysis of polysaccharides in lake water prior to their determination by the method based on spectrophotometry and capillary electrophoresis with laser induced fluorescence.

These methods were applied to quantify the vertical polysaccharide profile of the Lake Geneva (0–309 m). Maximum concentration of polysaccharide (1.02 ± 0.03 mg L^{-1}) was observed at the surface of the water. The concentration gradually decreased with a minimum concentration of polysaccharides at 290 μm (0.27 ± 0.04 mg L^{-1}). A fairly good correlation ($r^2 = 0.83$, $P < .001$, $N = 9$) between the total organic carbon and spectrophotometry saccharide adduct measured by fused silica capillary was observed. The vertical profile of the lake indicated glucose and fructose as dominant sugars with a minor contribution of xylose.

Chanuder and Filella [177] carried out a critical evaluation of a method based on 3-methyl-2-benzothiazolinone hydrochloride for the determination of total dissolved carbohydrates in freshwaters.

The objective of this study was to critically examine the application of this method to the analysis of carbohydrates in freshwater samples in order to understand what the method really measures. By simultaneously determining the total dissolved organic carbon and humic-type compounds in the same samples, it was shown to be possible to demonstrate that sometimes a significant part of the organic carbon remains undetected. This seems to indicate that a substantial amount of carbohydrate present in some natural waters is probably not 'seen' by this method.

2.12 Lactams

Thin-layer chromatography

Tachsteinova and Kopanica [149] determined caprolactam in natural water by absorptive voltammetry after separation by thin-layer chromatograph.

Polarography

Erimin and Kopylova [178] have described a polarographic method for the determination of ε-caprolactam in synthetic fibre production effluents. The sample (500 mL) is adjusted to pH 7.0 with sulphuric acid and shaken with light petroleum (3 × 35 mL), the combined extracts are washed with water (3 × 5 mL) and then discarded and the main aqueous phase plus washings is made 0.5 M in sulphuric acid and heated under reflux for 1 hour to hydrolyse ε-caprolactam to 6-aminohexanoic acid. The cooled solution is adjusted to pH 2.0 with 50 mL of 10.6 M sodium hydroxide and passed through a column of KU-1. The 6-aminohexanoic acid is eluted with 1 M sodium bicarbonate; the eluate is mixed with 0.5 mL of 40% aqueous formaldehyde and diluted to 25 mL with 1 M sodium bicarbonate and the polarogram of the deoxygenated solution is recorded from −90 to 1.4 V.

The mixture is heated at 100°C for 20 minutes and 0.5 mL cupric chlorides $2H_2O$ (2%) and 1 mL oxalic acid IN are added to the hot solution. After 1 minute the colour of the solution is compared with that of similarly prepared standards.

Enzymic assay

Cavari and Phelps [179] have described a sensitive enzymic assay for glucose in natural waters in amounts down to 0.1 mg L^{-1} without prior concentration or extraction. The method is based on reaction with adenosine 5′-triphosphate catalysed with hexokinase to form glucose-6-phosphate. Adenosine 5′-triphosphate consumed is measured by the luciferin–luciferase assay.

References

[1] F.K. Kawahara, Anal. Chem. 40 (1968) 2073.

[2] F.K. Kawahara, Anal. Abstr. 17 (1969) 1827.

[3] A.N. Khomenkô, I.A. Goncharova, Gidrokhim. Mater. 48 (1968) 77 (Ref. ZH. Khim., 199D, Abstract N. 13G 267 (13) 1969).

[4] J.J. Van Huyssteen, Water Res. 4 (1970) 645.

[5] P.O. Bethge, K. Lindstroem, Analyst 99 (1974) 137.

[6] J.J. Richard, J.S. Fritz, J. Chromatogr. Sci. 18 (1980) 35.

[7] J.J. Richard, C.D. Chriswell, J.S. Fritz, J. Chromatogr. 199 (1980) 143.

[8] A. Vairavamurthy, M.O. Andreae, J.M. Brooks, Anal. Chem. 58 (1986) 2684.

[9] M.F. Mebran, M. Galker, M. Meran, W.J. Cooper, J. High Resolut. Chromatogr. Chromatogr. Commun. 11 (1988) 610.

[10] O.S. Fatoki, F. Vernon, Water Res. 23 (1989) 123.

[11] E.S. Olsen, J.W. Diehl, M.L. Froelich, Anal. Chem. 60 (1988) 1920.

[12] R.F. Borch, Anal. Chem. 47 (1975) 2437.

[13] N.W. Hoffman, J.C. Liao, Anal. Chem. 48 (1976) 1104.

[14] D.A. Hullett, S.J. Eisenreich, Anal. Chem. 51 (1979) 1953.

[15] D.J. Kieber, G.M. Vaughan, K. Hopper, Anal. Chem. 60 (1988) 1654.

[16] J. Szewczyk, R. Desal, Koks Smola Gaz 13 (1968) 355.

[17] M. Pontin Gautier, P. Bonastr, Grenir, Environ. Technol. Lett. 1 (1980) 460.

[18] N. Nimura, T. Kinoshita, T. Yoshida, A. Uetake, C. Nakai, Anal. Chem. 60 (1988) 2067.

[19] E. Kishida, Y. Nishimoto, S. Koto, Anal. Chem. 64 (1992) 1505.

[20] L. Brown, Analyst 104 (1979) 1165.

[21] T. Okada, Anal. Chem. 60 (1998) 1666.

[22] R. Saari Nordhaus, J.M. Anderson, Anal. Chem. 64 (1992) 2283.

[23] N. Hirajama, T. Kuwamoto, Anal. Chem. 65 (1993) 141.

[24] I. Berglund, P.K. Dasgupta, T.L. Lopez, O. Nara, Anal. Chem. 65 (1993) 1192.

[25] A.N. Khomenkô, I.A. Goncharova, A.G. Stradomskaya, Anal. Abstr. 14 (1967) 7929.

[26] A.N. Khomenkô, Anal. Abstr. 14 (1967) 7927.

[27] A.G. Stradomskaya, I.A. Goncharova, Godrokhim. Mater. 48 (1968) 72.

[28] I.A. Goncharova, A.N. Khomenkô, Godrokhim. Mater. 47 (1968) 161 (Ref. ZH. Khim, 19D, Abstr. No. 9G275 (9) 1969).

[29] V.G. Smotrakov, A.G. Stradomskaya, I.A. Goncharova, Gidrokhim. Mater 49 (1969) 202.

[30] V.G. Smotrakov, A.G. Stradomskaya, Zh. Khim. 239 (1969) 17, (19D. Abstr. No. 17G).

[31] O. Hatsue, I. Masayuki, Br. J. Ind. Med. 27 (1970) 150.

[32] Z.A. Krotova, Zavad Lab. 40 (1974) 263.

[33] D.W. Jordan, Anal. Chim. Acta 113 (1980) 189.

[34] R.M. Bustin, P.W. West, Anal. Chem. 68 (1974) 317.

[35] S.S. Hasson, M.S. Ahmed, A. Mageed, Anal. Chem. 66 (1964) 492.

[36] M. Kopanica, M. Stara, J. Jenik, Vodni Hospod. Ser. B 33 (1983) 49.

[37] B.I. Escher, R.P. Schwarzenbach, Environ. Sci. Technol. 34 (2000) 3962.

[38] E. Erikssen, A. Ledin, Int. J. Environ. Anal. Chem. 83 (2003) 987.

[39] P. Valto, S. Knuutinen, R. Alen, Int. J. Environ. Anal. Chem. 87 (2007) 87.

[40] Environmental Protection Agency, Quality Criteria for Water, Superintendent of Documents, US Government Printing Office, Order No. 005-001-01049-4, Washington, DC, 1976.

[41] N.I. Sax, Dangerous Properties of Industrial Materials, second ed., Reinhold, New York, 1963.
[42] P.G. Stecher, The Merck Index, eighth ed., Merck, Rahway, NJ, 1968.
[43] R.A. Baker, J. Am. Water Works Assoc. 55 (1963) 913.
[44] National Academy of Science and National Academy of Engineering, Water Quality Criteria: A Report of the Committee on Water Quality Criteria, Environmental Studies Board, Superintendent of Documents, US Government Printing Office, Order No. 55501-00520, Washington, DC, 1972.
[45] US Environmental Protection Agency, Manual of Methods for Chemical Analysis of Water and Wastes, Office of Technology Transfer, Washington, DC, 2000.
[46] H.W. Breidenbach, T.J. Litchenberg, C.F. Heink, D.J. Smith, J.W. Eichelberger, US Department of Interior Publications, WP-22, 1966.
[47] V.Y. Kaplin, L.V. Semenchenko, N.G. Fesenko, Gidrokhim. Mater. (1966) 41.
[48] L.V. Semenchenko, V.T. Kaplin, Gidrokhim. Mater. 43 (1967) 42.
[49] V.T. Kaplin, L.V. Semenchenko, Gidrokhim. Mater. 46 (1968) 182.
[50] V.A. Panova, Ochistka Proizvid Stochnykh Vod 4 (1968) 184.
[51] L.L. Semenchenko, V.T. Kaplin, Zhur. Anat. Khim 23 (1968) 1257.
[52] J.W. Eichelberger, R.C. Dressman, A. Longbottom, J. Environ. Sci. Technol. 4 (1970) 576.
[53] F.K. Kawahara, Environ. Sci. Technol. 5 (1971) 235.
[54] R. Cooper, K.C. Wheatstone, Water Res. 7 (1973) 1375.
[55] A.D. Paklomova, V.L. Berendeeva, Ukr. Zh. 40 (1974) 1211.
[56] R.B. Baird, C.L. Kuo, J.S. Shapiro, Yankow Arch. Environ. Contam. Toxicol. 2 (1974) 165.
[57] O.A. Plechova, Y. Filippovy, I.M. Aremova, Teknol. Ochiski Prirod Stochnvod 200 (1977) 1.
[58] R.A. Baker, J. Am. Water Works Assoc. 58 (1966) 751.
[59] R.A. Baker, J. Am. Water Works Assoc. 1 (1967) 977.
[60] D.C. Chriswell, C. Chang, J.S. Fritz, Anal. Chem. 47 (1975) 1325.
[61] A. DiCorcia, J. Chromatogr. 80 (1973) 69.
[62] S. Goren-Stul, H.F. W. Kleijn, A.E. Mostaert, Anal. Chim. Acta 34 (1966) 322.
[63] D.W. Grant, G.A. Vaughan, in: M. Swaay (Ed.), Gas Chromatography, Butterworth's, London, 1962, p. 205.
[64] V.J. Kusy, Chromatography 57 (1971) 132.
[65] F. Dietz, J. Traud, P. Koppe, Chromatographia 9 (1976) 380.
[66] D. Smith, J.J. Litchenberg, ASTM Spec. Techn. Publ 448 (1968) 78.
[67] R.A. Baker, Air Water Pollut. 10 (1966) 591.
[68] J.M. Derek, J. Fish. Res. Board Can. 32 (1975) 292.
[69] K. Nagasawa, H. Uchiyama, A. Ogamo, T. Shinozuka, J. Chromatogr. 144 (1977) 77.
[70] J. Zerbe, J. Chem. Anal. 22 (1977) 575.
[71] G. Matsumoto, R. Iashiwateri, T. Hanya, Water Res. 11 (1977) 693.
[72] A.P. Meijers, R.C. Van der Leer, Water Res. 10 (1976) 597.
[73] O. Rump, Water Res. 8 (1974) 889.
[74] D.A. Murray, J. Fish. Res. Board Can. 32 (1975) 292.
[75] Z. Voznakova, M.J. Popl, J. Chromatogr. Sci. 17 (1979) 682.
[76] Z. Voznakova, M.J. Popl, M. Barker, J. Chromatogr. Sci. 15 (1978) 123.
[77] R.A. Coleman, R.D. Edstron, M.A. Unger, R. Huggett, J. Anal. Chem. 542 (1982) 631.
[78] P.J. Rennie, Analyst 107 (1982) 327.

[79] M.C. Goldberg, L. Long, M. Sinclair, Anal. Chem. 45 (1973) 89.
[80] M.C. Goldberg, E.R. Weiner, Anal. Chim. Acta 115 (1980) 373.
[81] Department of the Environment/National Water Council Standing Committee of Analysts, Methods for the Examination of Waters and Associated Materials, Phenols in Waters and Effluent by Gas-Liquid Chromatography, 4-Aminoantipyrine or 3-Methyl-2-Benzothiazolinone Hydrazone, H.M. Stationery Office, London, 1981, p. 39 (RP 22B: CENV).
[82] R. DiCorcia, E. Sauperi, Sebastiani, C. Severini, Chromatographia 14 (1981) 86.
[83] V. Janda, K.J. Krijt, J. Chromatogr. 283 (1984) 309.
[84] T. Ramstad, T.J. Nestrick, Water Res. 15 (1981) 375.
[85] D.N. Armentrout, L.D. McLean, M.W. Long, Anal. Chem. 51 (1979) 1039.
[86] S. Hashimoto, T. Miyata, M. Washino, W. Kawakami, Environ. Sci. Technol. 13 (1979) 71.
[87] R.E. Sharp, G.X. Meyer, Anal. Chem. 54 (1982) 1164.
[88] S.K. Ratanathanwongs, S.R. Crouch, Anal. Chim. Acta 192 (1987) 277.
[89] P.S. Rennie, Anal. Proc., Lond. 13 (1987) 1396.
[90] Y. Li, J. Liu, Int. J. Environ. Anal. Chem. 90 (2010) 880.
[91] H.Z. Thieleman, Chem. Lpz. 9 (1969) 189.
[92] H.Z. Thieleman, Pharmazie 25 (1970) 365.
[93] H.Z. Thieleman, Anal. Chem. 253 (1972) 38.
[94] F. Edeline, R. Deswaed, G. Lambert, Trib. CEBEDEAU 31 (1978) 137.
[95] V.O. C'hmil, Zhur Analicleskei Khimil 36 (1981) 279.
[96] B.K. Afghan, D.E. Belliveau, R.H. Larose, J.F. Ryan, Anal. Chim. Acta 71 (1974) 355.
[97] R.V. Cheesman, A.L. Wilson, The Absorbtiometric Determination of Phenols in Water Report, TP 84, The Water Research Association, Marlow, Buckinghamshire, SL7 2HD, UK, 1972.
[98] American Public Health Association, Standard Methods for the Examination of Water and Wastewater Including Bottom Sediments and Sludges, twelfth ed., American Public Health Association, New York, 1965, pp. 517–520.
[99] American Public Health Association, Standard Methods for the Examination of Water and Wastewater, thirteenth ed., Method 222 through 222E, American Water Works Association, and Water Pollution Control Federation, New York, 1971.
[100] N.A. Marley, C.J. Mann, T.J. Vickers, Appl. Spectrosc. 38 (1984) 540.
[101] N.A. Marley, C.J. Mann, T.J. Vickers, Appl. Spectrosc. 39 (1975) 628.
[102] L. Van Haverbeke, J.F. Janssens, H.A. Herman, Int. J. Environ. Anal. Chem. 10 (1981) 205.
[103] L. Van Haverbeke, M.A. Herman, Anal. Chem. 51 (1979) 932.
[104] Z.C. Kalsaaonos, E.K. Paleologos, D.L. Guokas, M.I. Karayannis, Int. J. Environ. Anal. Chem. 83 (2003) 507.
[105] C.F. Lei, S.W. Pang, Huaxue Tongbae (Chem. Bull.) 19 (1974) 1.
[106] M.E. Gales, Analyst 100 (1975) 841.
[107] C.Z. Katsaounos, E.K. Paleologos, D.L. Giokas, M.I. Karayannis, Int. J. Environ. Anal. Chem. 83 (2003) 507.
[108] American Society for Testing Materials, Manual on Industrial Water and Industrial Wastewater, Part 23, Method D-1783-70, American Society for Testing Materials, Philadelphia, PA, 1973.
[109] S.D. Faust, E.W. Mikulewicz, Water Res. 1 (1967) 405.
[110] E.F. Molder Jr, N. Jacobs, Anal. Chem. 29 (1957) 1369.

[111] J.E. Fountaine, P.B. Josipura, P.N. Keliher, J.D. Johnson, Anal. Chem. 46 (1974) 62.
[112] R.C. Whitlock, S. Sigga, J. Suida, Anal. Chem. 44 (1972) 532.
[113] G.W. Stroehl, Staedtehygiene 19 (1968) 142.
[114] J.A. Vinson, Environ. Lett. 5 (1973) 199.
[115] A. Malz, Foeder. Eur. Gerammerschutz 11 (1964) 19.
[116] M. Pays, R. Bourden, Anis Pharamir 26 (1981) 681.
[117] H.O. Freistad, D.E. Ott, F.A. Gunthers, Anal. Chem. 41 (1969) 1756.
[118] H.O. Freistad, Anal. Chem. 41 (1969) 1753.
[119] P.D. Goulden, P. Brooksbank, M.P. Day, Anal. Chem. 45 (1973) 2430.
[120] Fuxinmel, D. Shugui, Z.L. Yu, Int. J. Environ. Anal. Chem. 86 (2006) 985.
[121] F. Bosch, G. Font, J. Mones, Analyst 112 (1987) 1335.
[122] Y.I. Karenman, N.N. Selmonshchuk, S.G. Kharitonova, Sov. J. Water Chem. Technol. 3 (1981) 24.
[123] A.K. Bajeva, U.K. Gunta, Asian Environ. 7 (1985) 36.
[124] S.M. Hassan, E.B. Salem, A.W. El-Salam, Anal. Lett. 20 (1987) 677.
[125] X. Bo-Xing, X. Tongming, S. Ming-Neng, F. Yu-Zhu, Talanta 32 (1985) 215.
[126] M.C. Goldberg, E.R. Weiner, Anal. Chim. Acta 112 (1980) 373.
[127] I.U.A. Belen'Kaya, E.F. Koritshaya, V.A. Sapozhnikov, S.A. Andronati, J. Water Chem. Technol. 5 (1983) 57.
[128] F. Conete, M. Fios, M.D. Castrom, M. Valcarcel, Anal. Chim. Acta 214 (1988) 375.
[129] O. Osi-Banio, S.O. Aiava, Anal. Chim. Acta 120 (1980) 371.
[130] N. Kawabata, K. Ohira, Environ. Sci. Technol. 13 (1979) 1396.
[131] A. Zgola-Grzeskowiak, T. Grzeskowiak, Int. J. Environ. Anal. Chem. 91 (2011) 576.
[132] A.A. Martianov, A.V. Zherdev, S.E. Eremin, B.B. Dzantiev, Int. J. Environ. Anal. Chem. 84 (2004) 965.
[133] B. Rossner, G. Schwedt, Fresenius Z. Anal. Chem. 315 (1983) 610.
[134] A. Carpenter, S. Siggia, S. Carter, Anal. Chem. 48 (1976) 225.
[135] C. Cavelier, M. Gilber, L. Vivien, P. Hamblin, Rev. Franc. Sci. L'eau 3 (1984) 19.
[136] O. Osi-Banio, S.O. Alava, Anal. Chim. Acta 211 (1988) 375.
[137] J.W. Eichelberger, R.C. Pressman, A.J. Long, J. Environ. Sci. Technol. 4 (1970) 576.
[138] P.M. Williams, A. Zirino, Nature 304 (1964) 462.
[139] M. Tasumota, W.T. Williams, J.M. Prescott, J. Mar. Res. 19 (1961) 89.
[140] C. Wei, L. Jiemin, J. Guibin, Y. Ziwei, Int. J. Environ. Anal. Chem. 83 (2009) 285.
[141] A. Aguera, M. Mezaua, D. Hermando, S. Malato, J. Carras, A. Fernandez-Alba, Int. J. Environ. Anal. Chem. 84 (2004) 149.
[142] G.I. Romantsova, Zavod. Lab. 36 (1970) 289.
[143] S.I. Igarashi, Jpn. Anal. 22 (1973) 444.
[144] Department of the Environment, National Water Council, Standing Technical Committee of Analysts, Methods for the Examination of Water and Associated Materials, Formaldehyde Methanol and Related Compounds in Raw and Potable Waters, 1982, Tentative Methods, HM Stationery Office, London, 1983.
[145] S.K. Yanovkii, D.N. Alksnis, A.A. Zhukhovitskii, B.P. Okhotnikov, J. Anal. Chem., USSR 37 (1982) 106.
[146] W.H. Evans, A. Dennis, Analyst 98 (1973) 782.
[147] A. Di Corcia, R. Samperi, Anal. Chem. 98 (1973) 782.
[148] I.V. Nevinnaya, V.I. Kovanov, J. Water Chem. Technol. 6 (1984) 54.
[149] A. Tachsteinova, M. Kopanica, Anal. Chim. Acta 199 (1987) 77.
[150] P.S. Eptein, T. Mauer, M. Wagner, S. Chase, B. Giles, Anal. Chem. 59 (1987) 1087.

[151] A. Thuren, Bull. Environ. Contam. Toxicol. 36 (1986) 33.
[152] R.A. Hites, J. Chromatogr. Sci. 11 (1973) 570.
[153] R.A. Hites, O. Biemann, Science, N.Y. 158 (1972) 178.
[154] R.A. Hites, Environ. Health Perspect. 3 (1973) 17.
[155] S. Mori, J. Chromatogr. 129 (1976) 53.
[156] M.J. Schouten, J.M. Copius Peereboom, U.A.T. Brinkman, Int. J. Environ. Anal. Chem. 7 (1979) 13.
[157] M.J. Schouten, J.M. Copius Peereboom, U.A.T. Brinkman, H.E. Schwart, C.J.M. Anzion, H.P.M. Van Vleit, Int. J. Environ. Anal. Chem. 6 (1979) 133.
[158] S. Takeda, S. Wakida, M. Yamari, A. Kawashara, K. Higashi, Anal. Chem. 65 (1993) 2489.
[159] M.F. Barroso, M.J. Ramalhose, C. Delern-Matos, M. Carmo, V.F. Vaz, G.F. Sales, Int. J. Environ. Anal. Chem. 88 (2008) 1049.
[160] A.M. Reid, C.A. Braughan, A.M. Fogortis, J.J. Roche, Int. J. Environ. Anal. Chem. 87 (2007) 125.
[161] C. Importa, G. Norta, C. Ferreti, M. Paple, J. Chromatogr. 285 (1984) 385.
[162] H. Nishikawa, T. Havakawa, T. Sakai, Analyst 112 (1987) 45.
[163] E.D. Maktaz, L.E. Botvinova, A.D. Kruchinia, Sov. J. Water Chem. Technol. 6 (1984) 59.
[164] L.H. Howe, Anal. Chem. 48 (1976) 2167.
[165] Z. Song, S. Hay, J. Int. Environ. Anal. Chem. 83 (2003) 807.
[166] J.M. Jahangir, O. Samuelson, Anal. Chim. Acta 100 (1978) 53.
[167] O.I. Ogawa, J.S. Fritz, J. Chromatogr. 329 (1985) 81.
[168] B.S. Middleditch, N.J. Sung, A. Alekis, C. Settemlike, Chromatography 23 (1987) 73.
[169] A.V. Karyakin, T.S. Chirkova, Zh. Priki. Spertrock. 13 (1970) 468 (Ref. Zhur Khim. 19GD, Abstract No. 55G338, 1971, 5).
[170] A.V. Vajta, G. Paimai, E. Verms, I. Szebenyi, Period. Polytech. Chem. Eng. 15 (1974) 263.
[171] A.V. Vajta, G. Paimai, W. Verms, I. Szebenyi, Anal. Abstr. 16 (1969) 1948.
[172] B.O. Josefsson, Anal. Chim. Acta 52 (1970) 65.
[173] V.H.H. Stabel, Arch. Hydrobiol. 80 (1977) 216.
[174] M. Ochiai, J. Chromatogr. 194 (1980) 224.
[175] L.G. Shaova, V.T. Kaplin, Girokem Matter 53 (1970) 42.
[176] R. Shaheen, J.P. Senn, Int. J. Environ. Anal. Chem. 85 (2005) 177.
[177] V. Chanuder, M. Filella, Int. J. Environ. Anal. Chem. 86 (2006) 693.
[178] U. Erimin, G.A. Kopylova, Zavod Lab. 88 (1973) 1065.
[179] B.Z. Cavari, G. Phelps, Appl. Environ. Microbiol. 33 (1977) 1237.

Chapter 3

Halogen-containing compounds in nonsaline waters

Chapter Outline

3.1 Saturated aliphatic chloro compounds **66**
Gas chromatography 66
High-performance liquid chromatography 70
Thin-layer chromatography 71
Headspace analysis 71
Purge and trap analysis 72
Negative-ion mass spectrometry 72
Miscellaneous 73
3.2 Unsaturated chloroaliphatic compounds **74**
Gas chromatography–mass spectrometry 74
Mass spectrometry 75
Headspace analysis 75
Purge and trap analysis 75
High-performance liquid chromatography 75
3.3 Haloforms **75**
Gas chromatography 77
Headspace analysis 79
Purge and trap methods 79
Resin adsorption–gas chromatography 80
High-performance liquid chromatography 81
Gel-permeation chromatography 81
Spectrophotometric method 81
Preconcentration 81
3.4 Chloroaromatic compounds **82**
Gas chromatography 82
Liquid chromatography 82
Miscellaneous 82
3.5 Halocarboxylic acids **82**
Gas chromatography 82
Isotope dilution mass spectrometry 83
Miscellaneous 83
3.6 Polychlorodibenzo-*p*-dioxins and polychlorodibenzofurans **83**
Gas chromatography–mass spectrometry 83
3.7 Chlorophenols **84**
Gas chromatography 84
High-performance liquid chromatography 86
Thin-layer chromatography 87
Miscellaneous 88
Preconcentration 88
3.8 Polychlorobiphenyls **88**
Gas chromatography 89
Gas chromatography–mass spectrometry 89
High-performance liquid chromatography 91
Thin-layer chromatography 91
Polarography 92
Miscellaneous 92
3.9 Polychloroterphenyls **93**
3.10 Miscellaneous **93**
3.11 Bromine-containing compounds **94**
References **95**

Determination of Toxic Organic Chemicals In Natural Waters, Sediments and Soils.
DOI: https://doi.org/10.1016/B978-0-12-815856-2.00003-5

As seen in Table 3.1, halogen-containing compounds are found extensively in the aqueous ecosystem including rivers, lakes and surface waters.

It is seen that up to 13.4 μg L^{-1} total haloforms have been found in river waters (up to 70.5 μg L^{-1} in lakes) and up to 4.3 μg L^{-1} polycyclic aromatic hydrocarbon in rivers. At these maximum levels, there is a cause for ecological concern.

The various types of halogen compounds are discussed below.

3.1 Saturated aliphatic chloro compounds

Gas chromatography

Murray and Riley [1,2] described gas-chromatographic methods for the determination of trichloroethylene, tetrachloroethylene, chloroform and carbon tetrachloride in natural waters. These substances were separated and determined on a glass column (4 m × 4 mm) packed with 3% SE-52 on Chromosorb W (AW DMCS) (80–100 mesh) and operated at 35°C, with argon (30 mL min^{-1}) as carrier gas. An electron-capture detector was used, with argon–methane (9:1) as quench gas. Chlorinated hydrocarbons were stripped from water samples by passage of nitrogen and removed from solid samples by heating in a stream of nitrogen. In each case the compounds were transferred from the nitrogen to the carrier gas by trapping a copper column (30 cm × 6 mm) packed with Chromosorb W coated with 3% SE-52

TABLE 3.1 Summary of organochlorine compounds in natural waters (rivers, lakes and surface waters) (concentration μg L^{-1}).

Haloforms	Rivers	Lakes
$CHCl_3$	0.02–0.75	54.6–59.1
$BrCl_2CH$	<0.1–7.6	–
Br_2ClCH	<0.1–4.66	–
Br_3CH	<0.1–0.51	–
CCl_4	0.02–0.12	11.8–14.3
$CH_2CH\ CH_2Cl$	0.05–0.09	7.8–11.4
$Cl_2CH\ CH_2Cl$	–	8–20
Total haloforms	0.92–13.4	62.4–70.5
Total polycyclic aromatic hydrocarbons	<0.1–4.3	–

and cooled at −78°C and subsequently sweeping it on to the gas-chromatographic column with the stream of argon. A limitation of this procedure is that compounds which boil considerably above 100°C could not be determined [3].

A different approach was to pass the water through a bed of activated carbon, which was subsequently extracted in a Soxhlet unit; the extract was evaporated and analysed, but the results were quantitatively uncertain [4]. A method has been published by which the water sample was codistilled with cyclohexane, and the organic phase was then injected into an electron-capture detector gas chromatograph [5]. Extraction with *n*-pentane followed by gas chromatography (GC) has also been used, [6] but although the extraction was easy and effective, the chromatographic conditions described were time-consuming and unsuitable for compounds heavier than perchloroethylene.

Chlorinated normal paraffins up to the C_{30} carbon number range are of low volatility and relatively thermally unstable, producing hydrogen chloride on decomposition; hence direct GC is not attractive.

Zitko [7] has devised a method based on column chromatography followed by microcoulometric detection. The procedure is not specific.

Zitko [8] have also described a confirmatory method in which the chloroparaffins are reduced to normal hydrocarbons which are then analysed by GC. Both methods lack sufficient sensitivity for trace (sub-ppm) analysis and the confirmatory method may be difficult to apply. Freidman and Lombardo [9] have described a gas-chromatographic method applicable to chloroparaffins that are slightly volatile.

This method is based on microcoulometric detection and photochemical elimination of chlorinated aromatic compounds that otherwise interfere.

Deetman et al. [10] and Guann and Wong [11] have devised an electron-capture gas-chromatographic technique applicable to water, for the determination of down to 1 ng L^{-1} of 1,1-trichloroethane, trichloroethylene, perchloroethylene, 1,1,1,2-tetrachloroethane, 1,1,2,2-tetrachloroethane, pentachloroethane, hexachloroethane, pentachlorobutadiene, hexachlorobutadiene, chloroform and carbon tetrachloride. These workers used extraction of the water samples with *n*-pentane followed by analysis on an Apiezon-L column. This was satisfactory as a means of isolating the chlorinated compounds from the samples. Recoveries of 95% were obtained in a single extraction.

Hagenmaier et al. [12] have described a method for the quantitative gas-chromatographic determination of volatile halogenated hydrocarbons in lake water samples. Sample enrichment was affected by liquid−liquid extraction with pentane, followed by separation on a capillary gas−liquid-chromatographic column with electron-capture detection. A 1:25 pentane/water ratio was employed in conjunction with a standard solution of a reference compound (1-bromobutane) for estimating extraction efficiency. The detection limit using split injection was approximately 0.005 μg L^{-1} and could be increased to less than 1 ng L^{-1} by on-column injection. The method was applied to samples of water taken from Lake Constance. The water contained appreciable amounts of trichloroethylene (8–20 μg L^{-1}) and tetrachloroethylene (2–5 μg L^{-1}).

Chiba and Haraguchi [13] determined halogenated organic compounds in natural water by GC atmospheric pressure–induced–helium microwave–induced plasma emission spectrometry using a heated discharge tube for pyrolysis.

Blanchard and Hardy [14] have described a method for determining benzene, toluene, ethylbenzene, dichloromethane, chloroform and carbon tetrachloride in natural water. Samples were collected using a silicone polycarbonate membrane and adsorbed on to activated carbon. The volatile components were desorbed with carbon disulphide and their levels determined by GC. Linearity of response and response time of the membrane were evaluated. The response time was less than 5 minutes in all cases.

Burgasser and Calaruotolo [15] have described a gas-chromatographic method for determining semi- or nonvolatile chlorinated organics such as hexachlorobutadiene in solvent extracts of water in amounts down to 0.1 μg L^{-1} in natural waters.

Direct injection GC [16] coupled with a mass-spectrometric detector [17] and headspace analysis with a photoionisation detector [18] have all been used to determine vinyl chloride in natural water. In the direct gas-chromatographic method [16], vinyl chloride, arenes and other volatile halogen compounds are separated from the water sample by stripping them in a closed system. The stripped compounds were absorbed on Porapak N in a glass tube within the closed system and eluted with methanol. They were then separated by GC on a 3 m Chromosorb 102 column.

Malaiyandi and Jenkins [19] have discussed some of the difficulties associated with the gas-chromatographic determination of ultratrace levels of hexachlorohexane isomers in environmental samples. A supposedly operator-independent method was tested for the recovery of hexachlorocyclohexane isomers from water. This recovery method, designed to reduce losses during extraction and evaporation, was tested on samples containing 30–350 ng L^{-1} of alpha-, beta-, gamma- and delta-hexachlorohexane isomers employing a Kuderna–Danish evaporator and modified Snyder column. Mean recoveries of alpha-, beta- and gamma-hexachlorocyclohexane isomer were better than 90%. However, recovery of the delta hexachlorocyclohexane isomer was 63%–85%. This was attributed to degradation at the injection port (250°C). Significant differences in operator variability were observed.

Oliver and Bothen [20] discuss a capillary gas-chromatographic procedure which can be used, after sample preconcentration, to identify and quantify 12 chlorinated benzenes in water samples. Preconcentration factors of 1000 with a small column of macroreticular resin and 2500 with liquid–liquid extraction using pentane were achieved. The pentane extraction technique is preferred because of its simplicity. Well-coated open-tubular glass capillary columns were used for GC with a nickel-60 electron-capture detector. Using both preconcentration techniques, more than 80% of all the chlorobenzenes were recovered. Detection limits in water for the pentane extraction technique vary from about 1 ng L^{-1} for dichlorobenzene to approximately 0.01 mg L^{-1} for hexachlorobenzene. Italia and Uden [21] used electron-capture GC to compare the volatile halogenated compounds produced during the chlorination of humic acid.

Prath and Pauliszyn [22] studied the gas extraction of volatile organic species, for example 1,1,1-trichloroethane from water using a hollow film membrane.

Li et al. [23] and others [24–26] have described a simple, rapid and inexpensive procedure for extraction and analysis of volatile halocarbons in water samples using a headspace single-drop microextraction technique and GC with microcell electron detector (GC-μECD). Single-drop microextraction is a technique in which the analyte is partitioned between the aqueous phase and a very small volume of organic solvent.

It is very important in this technique to select an appropriate extraction solvent for this method. Three basic requirements must be met: a high boiling point and a low volatility, and a good extraction efficiency for the target compounds should be high, so as not to interfere with the analysis of the target compounds in the chromatography. In accordance with the above requirements, ethylene glycol, 1-octanol and hexadecane were selected as extraction solvents and tested for the suitability. Each solvent was evaluated using an enrichment factor in the extraction of a 40-mL water sample with a 5-minute extraction time at 25°C in the stirred solution with a 1.0-μL organic drop. Each enrichment factor was calculated as the ratio of analyte concentration in both the aqueous solution and the solvent drop (Table 3.2). Of the three solvents examined, nonpolar hexadecane has the best chromatographic behaviour, but both 1-octanol and ethylene glycol had relatively better enrichment factors than hexadecane.

Headspace volume is an important factor that affects the analytical precision, repeatability and accuracy. In this study, headspace volumes of 25, 40 and 55 mL were tested in a 65-mL vial with 5 minutes of extraction at 25°C in the stirred solution with 1.0 μL of 1-octanol. The lower the ratio of the gas phase to liquid phase, the larger the area of the target chromatographic peak. Although the sensitivity of the determination is better with a smaller headspace, it is difficult to manipulate the experiment if the headspace is too small. A liquid phase-gas phase ratio of about 1:2 is suitable, and so the 25-mL headspace volume or 40-mL liquid volume is appropriate.

Operation parameters, such as extraction solvent, headspace volume, organic drop volume, sale concentration, temperature and sampling time, were studied and optimised. Extraction of 10 volatile halocarbon compounds

TABLE 3.2 Enrichment factors for different extraction solvents.

Compounds	1-Octanol	Ethylene glycol	Hexadecane
Chloroform	53	40	27
1,1,1-Trichloroethane	133	119	120
Carbon tetrachloride	209	186	203
Trichloroethylene	174	164	168
Dichlorobromomethane	80	58	37
1,1,2-Trichloroethane	50	47	30
Chlorodibromomethane	172	105	67
Tetrachloroethane	645	583	798
Bromoform	218	201	136
1,1,2,2-Tetrachloroethane	212	205	102

was achieved using the optimised method. Calibration curves of 10 target compounds yielded good linearity in the respective range of concentration ($R^2 \geq 0.9968$, chlorodibromomethane in the concentration range of 0.05–50 μg L^{-1}).

The limits of detection were found between 0.002 (tetrachloroethene) and 0.374 μg L^{-1} (1,1,2-tricholorethane) and relative standard deviations (RSDs) ranged between 4.3 (chloroform) and 9.7% (1,1,2,2-tetrachloroethane). Spiked recoveries of ground water agreed well with the known values between 118% (20.0 μg L^{-1} of 1,1,2-trichloroethane) and 82.61% (10.0 μg L^{-1} of tetrachloroethane), demonstrating that the headspace single-drop microextraction procedure combined GC-μECD was a useful and reliable technique for the rapid determination of volatile halocarbon compounds in water samples.

Single-drop microextraction is a type of technique in which the analyte is partitioned between the aqueous phase and a very small volume of organic solvents.

High-performance liquid chromatography

Kummert et al. [27] described a method for the trace determination (down to 0.06 μmol) of tetrachloroethylene in natural waters using direct aqueous injection – high-pressure liquid chromatography, and Stozek and Beumer [28] also used the technique to determine chlorinated solvents in water.

Studies have been carried out of the liquid-chromatographic approaches used on a Pye moving wire transport system with a ^{63}Ni electron-capture detector. However, studies of this technique were discontinued because of the poor sensitivity to chloro-*n*-paraffins with low chlorine contents. The second technique involving GC with a Coulson conductivity detector led to decomposition of the chloro-*n*-paraffins when they were volatised. The peaks from the decomposition products were broad and irreproducible.

Thin-layer chromatography

Hollies et al. [29] found that chloro-*n*-paraffins could be chromatographed on a silica gel plate from which an image of the chromatogram could be printed on an aluminium oxide plate by heating the two face to face, so that the high sensitivity of detection on aluminium oxide would be utilised. This detection procedure, followed by suitable preliminary cleanup and separation steps, is the means of this basis of this method.

Headspace analysis

Hellman [30] determined chloroform, carbon tetrachloride, trichloroethylene and perchloroethylene in natural waters. Studies of the operating variables are described including the effects of filling volume, bath temperature and duration of heating in the thermoblock. The method gave satisfactory and reproducible results, with detection limits of 0.05 $\mu g\ L^{-1}$.

Mehran et al. [31] evaluated various gas-chromatographic methods employing direct headspace and water injection into fused silica capillary columns for their ability to separate a model system composed of deionised water and trace amounts of 16 halocarbons. For headspace injection, both separation and detection sensitivity were affected by the lengths of solute bands and enhanced by focussing.

Kirshen [32] analysed 17 halogenated organic compounds by a purge and trap method.

Kaiser and Oliver [33] have determined volatile halogenated hydrocarbons at the 0.1–10 $\mu g\ L^{-1}$ level in water by headspace and GC. Hrivnak et al. [34] determined chlorinated C_1–C_4 hydrocarbons in water using capillary GC. For the isolation of chlorinated hydrocarbons (*n*-butyl chloride, di-, tri- and tetrachloromethane, 1,2-dichloroethane, 1,2-dichloropropane and trichloroethylene), a stripping technique was used. The hydrocarbons were analysed in a capillary stainless steel column at 80°C. Using electron capture, it is possible to determine down to 0.1 $\mu g\ L^{-1}$ of these substances [34].

Hellman [35] has applied the headspace technique to the determination of tetrachloromethane and trichloroethylene in river water. McLary and Barker [36] used headspace analysis to monitor levels of trichloromethane, 1:1:1 trichloroethylene and tetrachloroethylene in ground water samples.

Biebier et al. [37] compared headspace and solvent extraction methods for the determination of halogenated hydrocarbons in natural waters. Yurteri et al. [38] studied the effects of salts, surfactants and humic material in clean and polluted water on the Henry's law constants governing headspace analysis.

Purge and trap analysis

Lopez-Avila et al. [39] determined dichloromethane and aromatic compounds using photoionisation and Hall electrolytic conductivity GC in which the gas chromatograph is connected in series to a purge and trap analyser.

Cochran [40] used a Naflon tube drier in a purge–whole column cryotrapping method for selectively removing water from an analyte-containing purge stream during the analysis of volatile chlorinated and fluorinated hydrocarbons in natural water.

Chichester-Constable et al. [41] have developed an improved sparger for a purge and trap concentrator for the analysis of halocarbons in the 1–500 $\mu g\ L^{-1}$ range.

Mosesman et al. [42] have considered factors influencing the analysis of volatile pollutants by wide-bore capillary chromatography and a purge and trap system using five volatile gas mixtures: dichlorodifluoromethane, chloromethane, vinyl chloride, bromomethane and chloromethane. The factors studied were initial column temperature (10°C) carrier gas (helium), flow rate, speed with which pollutants were desorbed from the trap and the type of detector used (flame ionisation detector or electrolytic conductivity detector). After optimising the analytical conditions for the gases, a mixture of 36 volatile pollutants was analysed. Resolution due to peak broadening was also done for the most volatile compounds. The remaining 30 compounds were refocused at the column inlet, producing sharp, well-resolved peaks. A carrier gas flow rate of 10 $mL\ min^{-1}$ and an initial column temperature of 10°C were the optimal conditions for purge and trap analysis of volatile priority pollutants from a VOCOL wide-bore capillary column.

Negative-ion mass spectrometry

Rusina et al. [43] compounded various quantification methods for the analysis of short-chain polychlorinated alkanes using GC electron-capture negative ionisation low-resolution mass spectrometry. The method based on visual comparison of congener group patterns of external standards used for quantification was very sensitive to the choice of the quantification standard. Two other methods used the relation of the response factors with the chlorine content of short-chain chlorinated paraffin mixtures for quantification. Results from the three methods above deviated from normal values less than 20%. This was ~50% when individual polychlorinated alkane standards were applied for quantification of short-chain chlorinated paraffins. The deviation is probably caused by the fact that only C_{10} carbon chain length standards with five to nine chlorine atoms could be used. However, quantification using individual polychlorinated alkane standards is a promising method, provided more standards will become commercially available. The clear advantage of this method is that the standards are defined, which makes quantification comparable between different laboratories. The method gave results that agreed with the median values with $\pm 40\%$.

Miscellaneous

Simonov et al. [44] have described an ultraviolet spectrophotometric method for determining down to 1 ppm of tetrachloroethylene, hexachloropropene, hexachlorobutadiene and hexachlorocyclopentadiene from their extinction at 202, 240, 255 and 335 nm, respectively.

Dilling and Kalios [45] have studied the evaporation rates in aqueous solution of various polychlorinated compounds such as methylene dichloride, chloroform, 1,1,1-trichloroethane, trichloroethylene and tetrachloroethylene. The compounds were studied at concentrations of 1 ppm in water. All the compounds examined had evaporated by 50% in less than 30 minutes and by 90% in less than 90 minutes when stirred in an open container at 25°C. The addition of salt, clay, limestone, sand, peat moss and kerosene in the water has relatively little effect on the rates of disappearance. These workers conclude that low molecular weight–chlorinated hydrocarbons would not persist in agitated natural water bodies owing to evaporation.

Friedman et al. [46] studied the recovery of several volatile organic chloro compounds from simulated water samples. Solutions of volatile organic compounds in organic free water in 3% methanol were submitted to two US Geological Survey laboratories for volatile organic analysis by gas-chromatographic separation and mass-spectrometric detection. After 3 days, the analytical recovery of dilute concentrations of bromoform, dichloromethane, ethylbenzene and 1,1,2,2-tetrachloroethane was not statistically different from the recovery of these compounds from methanol solutions which had been kept 100 times more concentrated until immediately prior to analysis. There was no significant difference between values reported by the two laboratories despite altitude difference of 1.6 km and the use of different instruments. Recovery efficiency was more than 80% in more than half the determinations. The recoveries of bromomethane and vinyl chloride were hindered by addition of 2% methanol to the storage containers. Recovery of 2-chlorovinylether from the 2% methanol was greater than from distilled water. However, recoveries from both decreased after 5 days. Recovery of dichloropropene from distilled water decreased after 11 days. There was no significant decrease in the recovery of bromomethane, chlorobenzene, chloroethane, dichlorobromomethane, ethylbenzene and vinyl chloride after 34 days.

Hellman [47] studied the behaviour of volatile chlorinated hydrocarbons in flowing waters in a series of model experiments designed to establish the relative importance of physical and biochemical processes in the degradation of these substances. The effects of turbulence, temperature, radiation, adsorption and remobilisation were investigated, together with biochemical degradation over periods of approximately 9 days, and correspond to the transport time in the Rhine between Basle and the Dutch/German border. Release to the atmosphere was the principal dissipation route. Monitoring results for chlorinated hydrocarbons in Rhine water samples obtained under widely differing flow conditions are also discussed. For trichloromethane, the mass flow appeared to be almost independent for the hydrographic conditions, whereas for other compounds, the mass flow appeared to increase as discharge increased. Possible explanations for such apparently anomalous behaviour are discussed.

Burguera and Burguera [48] have used the emissions from iridium chloride, iridium bromide and iridium iodide, generated at an iridium-lined MECA cavity to determine organic halogen compounds separated on a gas-chromatographic column. Emissions are measured, respectively, at 360, 376, and 410 nm. Linear calibration ranges are 5–60, 10–150 and 5–500 mg for chloro-, bromo- and iodo-alkanes, respectively.

Martinsen et al. [49] preconcentrated organochlorine compounds on activated carbon and XAD-4 resin prior to determination by neutron activation analysis, thin-layer and gel-permeation chromatography.

Chloromethanes have been preconcentrated on XAD-4 macroreticular resin [50].

3.2 Unsaturated chloroaliphatic compounds

Alberti and Jonke [51] describe a gas-chromatographic method for the determination of vinyl chloride in surface water using a flame ionisation detection on a Poropack Q and Chromosorb 101 column. The detection limit is 0.3 mg L^{-1}, and samples of wastewaters from vinyl chloride or PVC factories can be injected direct into the gas chromatograph, while water samples with lower concentrations require preliminary enrichment for which a gradient-tube is described.

In the direct gas-chromatographic method [16], vinyl chloride arenes and other volatile halogen compounds were separated from the water sample by stripping them in a closed system. The stripped compounds were absorbed on Poropak N in a glass tube within the closed system and eluted with methanol. They were then separated by GC on a 3 m Chromosorb 102 column.

Gas chromatography–mass spectrometry

Fujii [17] has combined mass spectrometry with GC for the direct determination of submicrogram per litre amounts of vinyl chloride in potable and river waters. The method is based on mass fragmentography followed by chromatography–mass spectrometry by simultaneously recording *m/e* 62 and 64.

Mass spectrometry

Rivera et al. [52] have described a direct mass-spectrometric method for determining volatile chlorinated hydrocarbons, including vinyl chloride in water. They give details of a highly sensitive technique for the determination of chlorinated hydrocarbons.

The charcoal filter was quantitatively transferred to the previously cooled direct inlet probe of a MS-902S AEI high-resolution mass spectrometer that could be temperature programmed from −150°C to +350°C. Vinyl chloride desorption took place in the range of −30°C to +100°C. In this temperature interval and with the use of the peak matching technique (resolution 1000) the signals at *m/e* 62 and 64 corresponding to the molecular ions $C_2H_3\,^{35}Cl$ and $C_2H_3\,^{37}Cl$ were recorded.

Quantitative measurements of vinyl chloride in water were made by interpolating measured curve areas on a linear plot obtained by running standard water samples in the 0.05–10.0 ppb range. Standard deviation was 9%. Quantitation was difficult below 0.05 ppt.

Headspace analysis

Headspace analysis with a photoionisation detector [18] has been used to determine vinyl chloride in a natural water.

Purge and trap analysis

Bellar et al. [53] described a purge and trap method for determining vinyl chloride at the microgram per litre level in water. An inert gas was bubbled through the sample to transfer vinyl chloride to the gas phase, and the vinyl chloride was then concentrated on silica gel or Carbosieve B under noncryogenic conditions and determined by GC with a halogen-specific detector. Gas-chromatographic–mass-spectrometric methods were used to provide confirmatory identification of vinyl chloride. Data obtained by Bellar et al. [53] showed that a quantitative recovery of vinyl chloride is obtained on silica gel.

High-performance liquid chromatography

Kummert et al. [27] have described a method for the trace determination (down to 0.06 μmol) of tetrachloroethylene in natural waters using direct aqueous injection–high-performance liquid chromatography.

3.3 Haloforms

An epidemiological study of Mississippi River water [54] is of some importance as it suggests a relationship between cancer in certain communities and

TABLE 3.3 Haloform concentrations (μg L^{-1}) in potable water.

Haloforms	Chlorinated	Chlorinated and treated with carbon
$CHCl_3$	0.57–182	0.5–4.06
$BrCl_2CH$	0.2–74.7	0.08
Br_2ClCH	<0.01–36.8	0.16–15.0
Br_3CH	<0.1–8.0	0.13–0.37
CCl_4	0.017–1.2	0.02–0.15
CH_2ClCH_2Cl	0.06–24.5	0.06–24.5
Cl_2CCCl_2	0.008–0.	–
CH_3CCl_3	0.06–0.5	–
$CHClCCl_2$	0.015–0.9	–
CH_2Cl_2	0.3	–
$CH_2ClCHClCH_3$	0.8	–
Total haloforms	2.4–259	0.95–44.2

the haloform level in Mississippi-derived water supply [55]. Some typical levels of are haloforms illustrated in Table 3.3.

There have been numerous reports on the presence of chlorinated and brominated haloforms in river and polluted water [56].

Luong et al. [57] have drawn attention to the role of bromide in water supplies in the formation of brominated trihalomethanes with reference to its interaction during the chlorinated process with humic material present in natural waters. Changes during water treatment were examined and subsequent trihalomethane formation on chlorination of those waters evaluated. For bromide levels in lowland waters of up to 120 μg L^{-1}, brominated trihalomethanes were shown to account for up to 54% of the total trihalomethanes formed on treatment.

Chloroform apparently results from reaction between hypochlorite and any of several types of organic precursors in the chlorinated raw water. The brominated and mixed brominated/chlorinated trihalomethanes are presumed to be formed from the reaction of hypobromite and hypochlorite with the same precursors; the hypobromite is formed from the oxidation of bromide by hypochlorite. If iodide salts are also present in the water being chlorinated, an analogous reaction with hypochlorite results in trihalomethanes containing iodine. Bunn et al. [58] detected all 10 possible mixed and single halogen-containing trihalomethanes of chlorine, bromine and iodide when salts containing fluoride, bromide and iodide were added to a river water sample before chlorination (no fluorinated trihalomethanes were detected). Glaze et al. [59] identified seven trihalomethanes containing chloride bromine, and iodine in potable water samples. Also found in potable water, but not necessarily formed during the chlorinated process, are compounds such as methylene chlorine, dichlorobenzene, hexachlorobutadiene [60], tetrachloroethylene [61], trichloroethylene [60,61], carbon tetrachloride [62], 1,2-dichloroethane [61,62] and ethylene dibromide [63].

The formation of volatile organohalogen compounds by the chlorination of waters containing organic contaminants [64] has received wide attention [65–67].

As a consequence of the concern regarding possible adverse effects of minute quantities of trihalomethanes in potable water, in 1978 the US Environmental Protection Agency [68,69] drew up an amendment to US National Interim Primary Drinking Water Regulations designed to protect the public from exposure to undesirable amounts of trihalomethanes (including chloroform) in potable water. A maximum contaminant level of $100\,\mu g\,L^{-1}$ has been prescribed for total trihalomethanes applicable initially to community water supplies for populations as part of the treatment process [70].

Various methods of determining haloforms in environmental waters are now discussed.

Gas chromatography

Blanchard and Hardy [71] have described a method for determining benzene, toluene, ethylbenzene, dichloromethane, chloroform and carbon tetrachloride in natural waters. Samples were collected using a silicone polycarbonate membrane and adsorbed on to activated carbon. The volatile components were desorbed with carbon disulphide and their levels determined by GC. Linearity of response and response time of the membrane were evaluated. The response time was less than 5 minutes in all cases.

Eklund et al. [72,73] have developed a method for the determination of down to $1\,\mu g\,L^{-1}$ volatile organohalides in waters which combines the resolving power of the glass capillary column with the sensitivity of the electron-capture detector. The eluate from the column was mixed with purge gas of the detector to minimise band broadening due to dead volumes. This and low column bleeding give enhanced sensitivity. Ten different organohalides were quantified.

Comba and Kaiser [75] analysed samples of ground water for volatile contaminants. The headspace collection technique, a vacuum distillation with cryogenic trapping of the distillate, was used. The analysis used capillary GC with electron-capture detection. Recovery values are tabulated and comparisons made with detection limits obtained using other analytical techniques. The headspace technique is simple, inexpensive, with good sensitivity and is also applicable to the analyses of haloforms and associated halomethanes and haloethanes. Detection limits of 1.0 ng L^{-1} or better were achieved for carbon tetrachloride.

Renberg [50] has reported a resin adsorption method for the determination of trihalomethanes and chloromethanes and dichloromethane in water. In this method, halogenated hydrocarbons are determined by adsorption on to XAD-4 polystyrene resin and eluted with ethanol. The extract was analysed by GC and is sufficiently enriched in hydrocarbon to be suitable for other chemical analysis or biological tests.

Quimby et al. [76] used GC with microwave emission detection to identify the aqueous chlorination and bromination products of fulvic and humic acid in water.

Kroner [77] has described a microprocessor-controlled gas chromatograph–monitoring system for determining purgeable carbon tetrachloride, haloforms and halomethane at a concentration of 0.1 μg L^{-1} in river water. Chloroform was found in 70% of the sample analysed; other chlorinated solvents found regularly were trichloroethane, trichloroethylene, tetrachloroethylene, methylene chloride and carbon tetrachloride.

Libbey [63] used Amberlite XAD-4 resin to extract ethylene dibromide from water for determination by GC and electron-capture detection. Samples spiked with 1 μg^{-1} ethylene dibromide exhibited a mean of 0.98 μg L^{-1} and a standard deviation of 0.105 μg L^{-1}.

Various other workers have discussed resin adsorption methods for the determination of trihaloforms in water [78–84]. Kissinger and Fritz [85] have described a method based on adsorption of trihaloforms on acetylated XAD-2 resin, followed by elution with pyridine and then GC.

Arbon and Grimsrud [86] carried out a selective detection of microgram amounts of halogenated aliphatic hydrocarbons (e.g. methyl iodide, ethyl iodide and brominated compounds) in environmental natural water, using an electron-capture detector with negative-ion hydration and photo detachment.

Bognar et al. [87] carried out a selective detection of brominated hydrocarbons ($CF_2Br_2CH_2Br_2$) in the presence of $CFCl_3$, $CHCl_3$ and CCl_4 using a photo detachment–modulated electron-capture detection. This detection responds to any halogen compounds present in the water sample.

Headspace analysis

Dynamic headspace techniques for the determination of trihalomethanes have been studied by Symons et al. [88], Keith [78], and other workers from the Health and Welfare Department, Canada [89] and static headspace techniques have been studied by Keith [78], Bush and Narang [90], Morris and Johnson [91]. Gomella and Belle [92] and Friant [93] have described a direct head gas analysis procedure for the isolation of chloroform from aqueous environmental samples. The technique included GC and mass spectrometry. This worker carried out fundamental studies of partitioning of organic compounds between the aqueous and vapour phases and systematically examined effects of variations in operating parameters. Possible errors and limitations of this method are discussed.

Varma et al. [94] carried out a comparative study of the determination of trihalomethanes in water. They compared the results of chloroform extraction using six liquid–liquid extraction solvents (pentane, methylcyclohexane, isooctane, hexane, *n*-heptane and *n*-nonane) with the vapour space extraction method. The vapour space method yielded the poorer results.

Headspace GC has been used to determine 0.12 ppb–1 ppm chloroform carbon tetrachloride, trichloroethylene and tetrachloroethylene in natural and industrial waters [95]. These workers carried out a systematic study of the parameters affecting the accuracy and precision of results obtained by this technique.

Dietz and Singley [95] also showed the equilibration of the aqueous sample with the headspace with no agitation is very low. Even after 2 hours, equilibration is far from complete when vials are not shaken. One minute of agitation is sufficient for sample phase equilibration. Equilibrium is completely attained by rapid hand agitation which is comparable to a long-term agitation on a mechanical shaker.

Purge and trap methods

Analytical methods for the determination of trace amounts of volatile relatively insoluble organic compounds in water generally require a preconcentration step. The target species can be enriched either by concentrating the gases in the headspace on adsorption traps [60,96–98] or by purging the aqueous phase with a stream of gas followed by trapping [99–106].

In one gas purging method the volatile compounds are extracted from the water sample by passing through the water sample and collecting the volatile compounds on a small adsorption column. The compounds are introduced into chromatograph by heating the adsorption column. The determination is carried out by temperature-programmed GC using a halogen-specific detector flame ionisation or mass spectrometry [64,102–106]. This technique is quite sensitive ($>0.5\ \mu g\ L^{-1}$).

Bellar and Lichtenberg [102] used a purge and trap method in conjunction with temperature-programmed GC to resolve 23 volatile organohalides, all of which have been identified at one time or another in various water samples. This method entails pumping of the water sample with an inert gas and collection of the purged trihalomethane on an adsorbent (e.g. Tenax GC), followed by the thermal desorption.

Dressman et al. [107] compared determination of haloforms by methods involving extraction with methylcyclohexane, isooctane and pentane with results obtained by a purge and trap method involving purging the sample with nitrogen and adsorption on a Tenax-GC trap, followed by thermal desorption from the trap and GC.

For comparison, the average percentage recovery of the trihalomethanes obtained by the purge and trap method in the 1–200 mg L^{-1} range was 75%–95%, much the same as in the liquid–liquid extraction methods.

The US Environmental Protection Agency has published two methods for trihalomethane analysis [90,108,109]: methods 501.1 [109] and 501.2. Method 501.1 is a procedure for the analysis by the purge and trap technique; method 501.2 is a liquid–liquid extraction technique and also discusses the determination of trihalomethane precursors.

Resin adsorption–gas chromatography

Two approaches to this analysis are possible. In the first one the water [110] sample is purged with purified nitrogen or helium and the purged trichloroforms are trapped on a column of macroreticular resins such as Amberlite XAD-2 or XAD-4. The trapped trihaloforms are then desorbed from the column with a small volume of a polar solvent prior to GC. In the second approach the water sample is contacted directly with the resin and then the trihaloforms desorbed as before. Alternatively, the resin can be injected into the gas chromatograph injection port [111]. Both methods provide a very useful built-in concentration factor which improves method sensitivity.

Brass [112] has compared purge and trap and solvent extraction methods for the determination of halomethanes. He claims similar results are obtained by both methods.

Kirschen [113] has investigated the Environmental Protection Agency standard purge trap method 601 [114] for determining trihalomethane and other halogenated volatiles in water.

A purging method has been described for the specific determination of low levels of methyl bromide fumigant in water [115]. The analysis was performed with a Hewlett Packard Purgatrator mounted on a PTC (packed trap capillary) module with two electron-capture detectors. The module was used in a Hewlett Packard Model 433 gas chromatograph.

Pierce et al. [116] describe the use of the purge and trap method in conjunction with an electrolytic conductivity detector for the determination of volatile halogenated organics including the trihalomethanes.

Ammonia has been shown to be a major interferent in the gas-chromatographic analysis of trihalomethanes and other organohalogens in water using the purge and trap technique [117]. The ammonia must be removed prior to analysis.

Leppine and Archambault [118] have carried out determination of trihalomethanes ($CHCl_3$, $BrCl_2$, CH_2, Br_2ClCH_2, $CHBr_3$) at the 10–50 $\mu g\ L^{-1}$ level in natural waters utilising a purge and trap analysis and a gas chromatograph equipped with an electron-capture detector.

High-performance liquid chromatography

The high-performance liquid chromatography of down to 1 $\mu g\ L^{-1}$ of haloforms has been discussed [119].

Gel-permeation chromatography

Schnoor et al. [74] have determined the apparent molecular weight range of trihalomethane precursor compounds in the Iowa River and a reservoir near Iowa City. Soluble organics were size fractionated by gel-permeation chromatography and the fractions were chlorinated and analysed for trihalomethane yields by electron-capture GC. Of the trihalomethanes formed, 75% were derived from organics of molecular weight less than 3000 and 20% from those of molecular weight less than 1000.

Spectrophotometric method

Okumura et al. [120] give details of a procedure for the spectrophotometric determination of trace amounts of chloroform in water; it is based on the fact that chloroform produces a blue fluorophore when heated with nicotinamide and concentrated sodium hydroxide solution.

Preconcentration

Oake and Anderson [121] studied the adsorption of the trihalomethanes from water on to carbon.

Haloforms in potable water have been preconcentrated on a mini column of active carbon [122].

3.4 Chloroaromatic compounds

Gas chromatography

Kapila and Aue [123] have studied the determination of hexachlorobenzene using electron-capture GC. The technique involves routing of gas-chromatographic peaks of chlorinated hydrocarbons to a built-in flow-through reactor where they are partially dechlorinated in carrier gas doped with hydrogen, over a nickel catalyst and the reaction products flow on to a second column for separation.

Oliver and Bothen [20] discuss a capillary gas-chromatographic procedure which can be used, after sample preconcentration, to identify and quantify 12 chlorinated benzenes in pentane extracts of water samples. Preconcentration factors of 1000 with a small column macroreticular resin and 2500 with liquid–liquid extraction using pentane were achieved. Detection limits in water for the pentane extraction technique vary from about 1 ng L^{-1} for dichlorobenzene to approximately 0.01 mg L^{-1} for hexachlorobenzene. Italia and Uden [122] used electron-capture GC to compare the volatile halogenated compounds formed in the chlorination of humic acid.

Ryan et al. [124] used a helium discharge detector for the quantitation of volatile haloaromatics in natural waters. Down to 20 pg of *p*-dibromobenzene could be determined.

Liquid chromatography

Di Corcia and Samperi [125] determined down to 0.1 μg L^{-1} of chloroaniline in environmental waters. Acetonitrile was used as extractant with two extractors in tandem.

Miscellaneous

Stetter and Cao [126] used a gas sensor and a permeation apparatus for the determination of 1–5 mg L^{-1} of chlorinated hydrocarbons in water.

3.5 Halocarboxylic acids

Gas chromatography

Ozawa and Tsukioka [127] have described a gas-chromatographic method for determining down to 0.6 μg L^{-1} of sodium monofluoroacetate in natural waters.

Isotope dilution mass spectrometry

This technique has been used to determine trichloroacetic acid in natural waters [128,129].

Miscellaneous

Trifluoroacetic acid is a significant breakdown product of chlorofluorocarbons present in the atmosphere and water and is therefore a marker for the presence of these substances in the ecosystem.

Zehavi and Seiber [130] have discussed a headspace gas-chromatographic method for the determination of microgram per litre levels of trifluoroacetic acid in fog and rain water and surface waters. The described method determines trace levels of trifluoroacetic acid, and the atmospheric breakdown product of several of the hydrofluorocarbon and hydrochlorofluorocarbon replacements for the chlorofluorocarbon refrigerants in water and air. Trifluoroacetic acid is derivatised to the volatile methyl trifluoroacetate and determined by automated headspace GC with electron-capture detection or manual headspace GC using GC–mass spectrometry in the selected ion monitoring mode. The method is based on the reaction of an aqueous sample containing trifluoroacetic acid with dimethyl sulphate in concentrated sulphuric acid in a sealed headspace vial under conditions favouring distribution of methyl trifluoroacetate to the vapour phase. Water samples are prepared by evaporative concentration during which trifluoroacetic acid is retained as the anion, followed by extraction with diethyl ether of the acidified sample and then back-extraction of trifluoroacetic acid (as the anion) in aqueous bicarbonate solution. The extraction step is required for samples with a relatively high background of other salts and organic materials. Air samples are collected in sodium bicarbonate-glycerine-coated glass denuder tubes and prepared by rinsing the denuder contents with water to form an aqueous sample for derivatisation and analysis. Recoveries of trifluoroacetic acid from spiked water, with and without evaporated concentration, and from spiked air were quantitative, with estimated detection limits of 10 ng mL^{-1} (not concentrated) and 25 pg mL^{-1} (concentrated 250 mL:1 mL) for water. Several environmental air, fog water, rain water and surface water samples were successfully analysed; many showed the presence of trifluoroacetic acid.

3.6 Polychlorodibenzo-*p*-dioxins and polychlorodibenzofurans

Gas chromatography–mass spectrometry

Yasuhara and Itoh [131] have described a gas-chromatographic method combined with mass-spectrometric detection for the identification and determination of 2,3,7,8-tetrachlorodibenzo-*p*-dioxin in river water.

Taguchi et al. [132] applied high-resolution mass spectrometry coupled with selective ion monitoring (HRMS-SIM) to the determination of polychlorodibenzo-*p*-dioxins and tetra-, octa-, chloro-, and polychlorodibenzofurans in natural waters.

Laramee et al. [133] applied negative-ion mass spectrometry to the determination of polychlorodibenzo-*p*-dioxins and found a correlation between observed mass fragmentations and calculated internal energies. Fifteen congeners and isomers were examined.

Schimmel et al. [134] studied the molar response of polychlorodibenzo-*p*-dioxins and polychlorodibenzofurans and polychlorobiphenyls using a mass-spectrometric detector.

Plomley et al. [135] studied the mass spectrometry of polychlorodibenzo-*p*-dioxins and polychlorodibenzofurans in a quadrupole ion trap and compared results obtained in single-frequency modulation and multifrequency resonant excitation modes. These workers concluded that small-bandwidth multifrequency irradiation represents the best mode of resonant excitation for the mass spectrometry−mass spectrometry analysis of multiple target analytes such as polychlorodibenzo-*p*-dioxins and polychlorodibenzofurans.

3.7 Chlorophenols

Gas chromatography

Korhonen and Knuutinen [136] and others [127,137−139] studied the GC of chlorinated phenols and catechols using a nonpolar SE-30 quartz capillary column. A wide range of degrees of chlorination were analysed with flame ionisation detection. The elution order of the compounds was confirmed and the relative retention times presented. While all the compounds could be separated without derivatisation, simultaneous determination resulted in overlapping peaks. Retention times tended to be shorter than those for corresponding acetate esters and a stronger ortho-effect observed in the catechols.

Optimal conditions for the detection of trace quantities of chloro-, nitro- and methyl phenols in 0.4−1.0 mL water samples have been described [140]. Phenols were extracted with toluene, derivatised with heptafluorobutyric anhydride and separated and quantified by electron-capture GC or flame ionisation GC. Detection limits for phenol, chlorophenols, *p*-cresol, dimethyl and nitrophenols were 0.01−0.02 and 1.6 $\mu g\ L^{-1}$ for 2,4-dinitrophenol.

Huang and Zhu [141] reported that a capillary column coated with SE-36 gave better separations of chlorophenols than a similar column coated with QV-1. These workers extracted trace levels by hexachlorophenol from environmental waters. Following extraction and methylation, the extracts were analysed by GC with electron-capture detection. Down to 0.005 $\mu g\ L^{-1}$ of hexachlorophenol could be determined.

Malissa et al. [142] described methods for the extraction and enrichment of chlorophenols from natural water for their determination by capillary GC-Fourier-transform infrared spectroscopy.

Derivatisation was used to improve the chromatographic detection of phenols. This process reduced the differentiating power inherent in spectrometric GC detectors by introducing functional groups common to all environmental samples. The separation and identification of phenol and 19 chlorinated phenols from spiked river water without derivatisation is reported. The method used capillary GC and Fourier-transform infrared spectroscopy, with an SE54 fused silica WCOT column. An on-column injection technique with a thermodesorption and cold trap injector with solvent removal was used. Mean recovery of chlorophenols from river water samples with concentrations of 10 $\mu g\ L^{-1}$ was in the 90% range.

Renberg [143,144] and others [145] determined down to 100 $\mu g\ L^{-1}$ pentachlorophenol in water by an electron-capture gas-chromatographic method. Hexane (0.5 mL) and 40 μL of an acetylation reagent (2 mL pyridine plus 0.8 mL acetic anhydride stored in the cold) are added to the combined aqueous extracts and shaken for 1 minute. The hexane phase was analysed by GC on a glass column (1 m × 1.5 mm) packed with 5% of QF-1 on Varaport 300 (100–120 mesh) operated at 105°C with nitrogen as carrier gas (25 mL min^{-1}).

Renberg [144] has used an ion exchange technique for the determination of chlorophenols and phenoxyacetic acid herbicides in water. The water sample was mixed with Sephadex QAE A-25 anion exchanger and the absorbed materials were eluted with a suitable solvent. The chlorinated phenols were converted into their methyl ethers and the chlorinated phenoxy acids into their methyl or 2-chloroethyl esters for GC.

Renberg [144] recommends the use of γ-BHC (lindane), DDE, or DDT as a gas-chromatographic internal standard in the determination of tri-, tetra- and pentachlorophenols. The relative retention times of the derivatives corresponding to these internal standards are shown in Table 3.4. The detection

TABLE 3.4 Halogen-containing compounds in nonsaline waters.

	OV-17	QF + SF 96
Γ-BHC	7 min at 160°C	9 min at 150°C
Methyl ethers of		
2,4,6-Trichlorphenol	0.096	0.14
2,3,4,6-Tetrachlorphenol	0.25	0.32
Pentachlorophenol	0.64	0.70

limits for the different substances for 1 L of a water sample are 0.0001–0.1 ppb.

Chriswell and Cheng [145] have shown that chlorophenols and alkylphenols in the parts per billion to parts per million range in nonsaline waters can be determined by sorption on macroporous anion exchange resin, elution with acetone and measurement by GC.

Lee et al. [146] have described a method for the quantification and isomer-specific determination of pentachlorophenol and 19 other chlorophenols. Acetic anhydride and pentafluorobenzoyl chloride were used to derivatives in-situ chlorophenols in water samples. The derivatives were extracted from water and determined by GLC in amounts down to 0.1 ppb with a recovery of 80%–95%.

High-performance liquid chromatography

The best method for determining pentachlorophenol is conversion into the methyl ether followed by analysis using GC with an electron-capture detector, or GC coupled with mass spectrometry [147]. Both of these methods require an extensive amount of pretreatment.

Ervin and McGinnis [148] developed a high-performance liquid-chromatographic method for determining low concentrations of pentachlorophenol in water and also chlorinated impurities that occur in technical grade materials such as 2,3,4,6-tetrachlorophenol, mono-, di- and trichlorophenols, octa-, hepta- and hexachlorodibenzo-*p*-dioxins, and a variety of other polychlorinated aromatic compounds.

The method involves chloroform extraction of acidified water samples and rotary evaporated without heat. After redissolving in chloroform, the samples were analysed directly by high-performance liquid chromatography on a microparticulate silica gel column. The minimum detectable concentration is 1 ppm (without sample concentration) and the coefficient of variation is 1%–2%.

Sara Fraz-Yazdi et al. [149] determined chlorophenols in environmental water samples using directly suspended droplet liquid–liquid phase microextraction prior to high-performance liquid chromatography ion chromatography [150].

Silgouer et al. [151] have described an automated liquid-chromatographic–mass-spectrometric method for the determination of methyl-, nitro- and chlorophenols in river water samples. This method for determination of 19 priority phenols in water samples has been developed using on-line liquid–solid extraction followed by liquid chromatography–mass spectrometry with atmosphere pressure chemical ionisation and ion spray interfaces in the negative ionisation mode. Sixteen phenols were determined by liquid chromatography–atmosphere pressure chemical ionisation–mass spectrometry with high sensitivity.

Limits of detection ranging from 5 to 0.1 ng L^{-1} were found when 50–100 mL of river water was processed in full-scan and time-scheduled selective ion monitoring modes, respectively.

Suspended droplet liquid–liquid phase microextraction was used in this method for the determination of three chlorophenols in environmental water samples. The analytes (2-chlorophenol, 3-chlorophenol and 4-chlorophenol) were extracted from 4.5 mL acidic donor phase (pH 2, P1) into an organic phase, 350 μL of benzene/1-octane (90:10 v/v, P2) and then were back-extracted into a 7 μL droplet of a basic (pH 3) aqueous solution (acceptor phase, P3). In this method, contrary to the ordinary single-drop liquid-phase microextraction technique, an aqueous large droplet is freely suspended on the surface of the organic solvent, without using a microsyringe as supporting device. This aqueous microdroplet is delivered at the top-centre position of an immiscible organic solvent which is laid over the aqueous donor sample solution while the solution is being agitated. Then the acceptor phase containing chlorophenols was withdrawn back into a HPLC microsyringe and neutralised by adding 7 μL HCl 0.1 M. The total amount was eventually injected into the HPLC system with UV detection at 225 nm for further analysis. Parameters such as the organic solvent, phase volumes, extraction and back-extraction times, stirring rate of pH values were optimised. The calibration graphs are linear in the range of 10–200 μg L^{-1} with $r \geq 0.9973$. The enrichment factors were ranged from 115 to 170, and the limit of detection ($n = 7$) varied from 5 to 10 μg L^{-1}. The RSDs ($n = 5$) were 6.8 to 7.4 at $S/N = 3$. All experiments were carried out at room temperature (22°C ± 0.5°C).

Brandt and Kettrup [152] determined organic group parameters AOCl, AOBr and AOS in water by ion chromatography detection. Model compounds studied included 4-chlorophenol, 4-bromophenol and thiobenzamide.

Thin-layer chromatography

Various workers [153–157] have used thin-layer chromatography for determining pentachlorophenol and other chlorophenols in water samples. Thielemann and Luther [156] separated the chlorophenols or Kieselgel G plates with benzene as solvent. The spots were located by spraying with diazotised sulphanilic acid solution or with a mixture (1:1) of 15% ferric chloride solution and 1% $K_3Fe(CN)_6$ solution.

Ch'mil [157] determined alkyl-, chloro- and nitrophenols by thin-layer and gas–liquid chromatography. Separate determination of nitrophenols and nitrophenol compounds in water in the presence of each other by partition-reversed thin-layer chromatography has been investigated. Procedures for analysing chlorophenols and nitrophenol compounds in water using combination of gas–liquid and thin-layer chromatography (adsorption and partition) have been suggested.

Miscellaneous

Carr et al. [158] give details of a simple spectrophotometric procedure for the determination of pentachlorophenol in water. Although this method is not as sensitive or specific as GC, the number of potential interfering compounds is limited, and it is useful for routine monitoring.

Pentachlorophenol has been determined in water in amounts down to 0.3 mg L^{-1} by differential pulse polarography [159]. Boyle et al. [160] have studied the degradation of pentachlorophenol in a simulated lentic environment.

Ingram et al. [161] used a mass-spectrometric isotope dilution technique to determine approximately 0.2 μg pentachlorophenol in water with a RSD of 8%.

Preconcentration

Lipidex 500 [162], (1-4-demethacryloxy methyl) naphthalenedivinyl benzene copolymer [163], activated carbon [164], XAD-4 resin [164] and C_{18} silica [165] have all been studied as solid adsorbents for the preconcentration of chlorophenols prior to analysis by gas-chromatographic [162,165] and quartz spring-balance techniques [164].

3.8 Polychlorobiphenyls

Polychlorobiphenyls have been prepared industrially since 1929 by chlorination of biphenyl with anhydrous chlorine using either iron filings or ion (III) chloride as a catalyst. The product obtained is a complicated mixture of several PCBs. In the United Kingdom, PCBs are marketed as Arochlors by Monsanto. All Arochlors are characterised by a four-digit number; the first two digits represent the type of molecule (e.g. 12 represents biphenyl, 54 terphenyl and 25 and 44 are mixtures of biphenyls and terphenyl); the last two digits give the percentage by mass of chlorine (e.g. Arochlor 1260 is a 12-carbon system with 60% m/m of chlorine). PCBs are sold under a variety of trade names of which Arochlor is one. The following is a list of the principal trade names used for PCB-based dielectric fluids which are usually classified as Askarels: Arochlor (United Kingdom, United States), Pyoclor (United Kingdom); Inerteen (SA); Pyanol (France), Clophen (Germany), Apirolio (Italy); Kaneclor (Japan); Solvol (Russia). Other names were used for PCB products intended for different applications but are no longer in current use. These include Santotherm FR (United Kingdom, prior to 1972 for heat transfer); Therminol FR (United States, prior to 1972 for heat transfer); Pydraul (ISA, prior to 1972 for hydraulic applications); Phenoclor (France) and Fenclor (Italy). The trade names Santotherm, Therminol and Pydraul are still in use, but they are now used to refer to nonchlorinated products.

Gas chromatography

Webb and McCall [166] isolated 27 isomers found in Arochlors 1221, 1242 and 1254 by preparative GC and identified them by comparison of their retention data and infrared spectra with those of known synthetic compounds. Analytical GC was carried out on a support coated open-tubular column [100 ft. × 0.2 in. (30.5 m × 5 mm)] of SE-30 at 190°C with helium as carrier gas (6.5 mL min^{-1}) and flame ionisation detection.

Bauer [167] has described a gas-chromatographic procedure for the determination of PCBs in water. The PCBs were extracted from water with hexane, and the extract was dried and concentrated before gas-chromatographic analysis.

Berg et al. [168] separated PCBs from chlorinated insecticides on an activated carbon column prior to derivatisation and gas-chromatographic separation on a capillary column. Separation is based on the observation that PCBs adsorbed on activated charcoal cannot be removed quantitatively with hot chloroform but can be removed with cold benzene. Insecticides of the DDT group and a variety of others (e.g. γ-BHC, Aldrin, Dieldrin, Endrin and Heptachlor and its epoxide) can be eluted from the charcoal with acetone-ethyl ether (1:3). Typical recoveries from a mixture of *p,p′*-DDE (1,1,-dichloro-2,2-bis(4-chlorophenyl)ethylene), *p,p′*-TDE and *o,p′*-DDT and a PCB (Arochlor 1254) by successive elution with 90 mL of 1:3 acetone-ether and 60 mL of benzene were 91%, 92%, 92%, 94% and 90%, respectively. Identification and determination of the PCBs was affected by catalytic dechlorination to bicyclohexyl or perchlorination to decachlorobiphenyl, followed by GC.

Beezhold and Stout [169,170] studied the effect of using mixed standards on the determination of PCBs. Mixtures of Arochlors 1254 and 1260 were used as comparison standards and gas chromatograms of these mixtures were compared with those obtained from a hexane extract of the sample after clean-up on a Florasil column.

Shulte and Ackerz [171] gas chromatographed PCBs with glass capillary column at temperatures up to 320°C.

Dunn et al. [172] propose a technique for the quantitative determination of constituents in complex mixtures characterised by GC data using partial least squares in latent variables. The technique was applied to the gas chromatograms of Arochlor 1242, 2148, 1252 and 1260. Two SIMCA pattern recognition analyses carried out classification of an unknown sample of a specific Arochlor and a calibration in which the relative amounts of specific Arochlors were estimated in a classified sample.

Gas chromatography–mass spectrometry

The earliest reference to the application of this technique occurs in 1971 when Stalling and Huckins [173] discussed GC–mass spectrometry characterisation of Arochlors and ^{36}Cl labelling of Arochlors 1248 and 1254. These workers mention the separation of PCBs from polyterphenyls.

Ahnoff and Josefsson [174] carried out confirmation studies on PCBs from river waters using mass fragmentography, a form of GC–mass spectrometry in which the mass spectrometer is focused at one or more individual mass numbers while the components emerge from the column. This technique is a more specific and sensitive technique than electron-capture GC for detecting and identifying traces of PCBs in river waters. The specificity is further increased by using the known intensity ratio of the isotope peaks of chlorine-containing species. Webb and McCall [175] determined the weight of PCBs producing each peak in chromatograms of several Arochlors by GC–mass spectrometry and the use of an electrolytic conductivity detector. Electron-capture detector response factors were derived, and these were applied in association with rules for the division of chromatograms to the quantitative analyses of environmental samples containing one or more Arochlors. Eichelberger et al. [176] applied GC–mass spectrometry with computer-controlled repetitive acquisition from selected specific ions to the analyses of PCBs in lake sediments. There is a substantial gain in sensitivity without loss of qualitative information contained in the complete mass system. This technique provides a basis for a sensitive qualitative and quantitative (from ion-abundance chromatograms obtained from subset scanning) analysis for PCBs. Karlruber et al. [177] and Ahnoff and Josefsson [178] have discussed the application of mass fragmentography to the detection of traces of PCBs and herbicides in water.

Voyksner et al. [179] compared GC–high-resolution mass spectrometry and mass spectrometry–mass spectrometry for the detection of PCBs and tetrachlorodibenzofuran in water.

Pellazari et al. [180] have reviewed advances in the detection, identification and quantification of PCBs. Areas covered included high-resolution GC using capillary columns and detection systems (electron capture, negative-ion chemical ionisation mass spectrometry), selected ion monitoring and pulsed positive-ion/negative-ion chemical ionisation mass spectrometry. The availability of chlorobiphenyls isomer standards is discussed.

Tanabe et al. [181] have described a method for the determination of specific tetra-, penta- and hexachlorinated PCB isomers related to the highly toxic 2,3,7,8-tetrachlorodibenzodioxin. Stages consist of alkaline extraction, clean-up and fractionation on activated carbon followed by high-resolution GC with electron-capture detection and mass-spectrometric identification of individual isomers.

Sample composition (SSC) is a technique which allows the reduction of the number of analytical determinations to be carried out in screening campaigns down to the very number of the original sample specimens while providing information particularised to the original sample specimens instead of average information. Quintana et al. [182] applied the technique in environmental screening studies. Technical mixtures of polychlorinated biphenyls (Arochlors) were chosen as model contaminants in water samples to show the use and potential of the technique in this field. EPA 1668/1668a protocols were used for sample treatment. Gas chromatography hyphenated to tandem mass spectrometry was used for the analysis of the samples. Two types of sample composition design matrices (a conventional Plackett–Burman screening matrix and a supernatural matrix) were used and compared in the study. A total of 22 sample specimens were considered. Four of these sample specimens were contaminated at levels between 200 and 750 ng L^{-1} (total PCB concentration). A total of 24 experiments are needed to process these 22 sample specimens when applying the conventional Plackett–Burman matrix. Comparatively, only analytical determinations are needed when using the supersaturated matrix. Both types of matrices allow clear identification of the contaminated sample specimens and produced satisfactory estimations of their concentration levels. Special emphasis was put in investigating and demonstrating the robustness of the SSC technique.

High-performance liquid chromatography

Albro and Fishbein [183] determined the retention indices on six stationary phases for some mono-, di-, tri-, tetra- and hexachlorobiphenyls. They confirmed the additivity of the half-retention values on predicting retention indices.

Kaminsky and Fasko [184] investigated the potential of reversed-phase liquid chromatography to the analysis of PCB mixture in environmental samples. They used mixtures of water and acetonitrile as the mobile phase to achieve analysis of 49 different PCBs and of samples of Arochlor 1221, 1016, 1254 and 1260.

Various workers have discussed the high-performance liquid chromatography of PCBs [185–187]. Petrick et al. [187] used high-performance liquid chromatography to remove aliphatic compounds and polycyclic aromatic hydrocarbons from an *n*-hexane extract of a natural water sample prior to its analysis by GC for PCB congeners.

Thin-layer chromatography

de Vos and Peet [188] applied mixtures of PCBs in hexane solution (20 μL) to a 0.25 mm layer of Kieselgel G previously saturated with light petroleum containing 8% of liquid paraffin and dried. The chromatograms were developed (× 3) with acetonitrile–methanol–water (20:9:1) previously saturated with liquid paraffin. After drying in air, the plates were sprayed with 0.85% ethanolic silver nitrate and exposed to UV radiation for detecting the spots (black and a white background).

Sackmauer et al. [189] determined PCBs in water by thin-layer chromatography on silica gel plates impregnated with 8% paraffin oil. A mixture of acetonitrile, acetone, methanol and water (20:9:20:1) was used as a mobile phase. For detection, a solution of silver nitrate and 2-phenoxyethanol was used, followed by irradiation with UV light. The detection sensitivity for Arochlor 1243 is 0.5–1.0 μg.

Polarography

PCBs chlorinated insecticides and polychlorinated naphthalenes and benzenes have been identified voltammetrically [190] using three electrode potentiostatic control circuitry with interruptible linear voltage sweep control and a normal voltage scan rate of 42 mV s^{-1}.

Neeley [191] determined PCBs in water in Lake Michigan.

Kopanica [192] applied polarographic and voltammetric techniques for polarography analysis of trichlorobiphenyl in natural waters using fast scan different pulse with a hanging mercury drop electrode.

Miscellaneous

Aga et al. [193] determined Arochlor and its sulphonic metabolite in natural waters using solid-phase extraction and enzyme-linked immunosorbent assay.

Polyurethane foam has been used as a liquid–liquid partitioning filter for the concentration of PCBs from water samples [194,195].

Ahnoff and Josefsson [196] tested various clean-up procedures for PCB analysis on river water extracts including treatment with sulphuric acid and activated Raney nickel, with a Florasil column and activated Raney nickel, and with a Florasil column and potassium hydroxide. The Florasil column procedure is more effective in removing contaminants than the sulphuric acid procedure. However, neither procedure could remove sulphur, which could be removed by the activated Raney nickel or potassium hydroxide procedures.

The US Environmental Protection Agency [197] and the Inland Waters Directorate, Canada [198] have described a method for the determination of PCBs in water [198].

Preconcentration of PCBs from water has been achieved by solvent extraction ion exchange resins [83,198–201], polyurethane plugs [194,202], charcoal [203], Chromasorb [202], macroreticular resins [199,201,204,205] cellulose triacetate [203] and covalently bonded silica [206].

Bayer and Buffle [207] studied the dynamic aspects including diffusion in sorption of hollow-fibre liquid-phase microextraction of PCBs in water.

The hollow-fibre liquid-phase microextraction was conducted under non-equilibrium conditions, to investigate the bioavailability of PCBs in natural waters. The study was conducted for 12 PCB congeners (log K_{Ow} ranging from 5.2 to 8.2) in the nanogram per litre range. This appeared as a major challenge since aqueous solutions in this concentration range tended to evolve rapidly due to adsorption of PCBs on glass walls. The average aqueous diffusion layer was measured to be 43 ± 2 μm at 500 rpm. Aqueous diffusion coefficients of PCBs estimated from experimental data were found to be about two times lower than those predicted by the Hayduk–Laurie equation, possibly due to the underestimation of the molar volume of the PCBs aggregation of PCBs in the aqueous phase or a decrease of the actual aqueous concentration during the time of extraction. The presence of humic acid in the solution decreased, as expected, the mass transfer of PCBs to the fibre, but the flux was not linked either to the total or to the free PCB concentration. This suggests a semilabile behaviour for the humic acid PCB complex, which was confirmed by the effect of stirring speed on the amount of PCBs extracted in the presence and in the absence of humic acid. These observations suggest that diffusion in solution is not only one of the limiting processes for the extraction of PCBs but also supports the need for more experimental data to understand in detail the mechanism of extraction of hydrophobic compounds, and their bioavailability in the presence of aquatic complexants.

Schneider et al. [208] also used solid-phase microextraction to rapidly measure PCBs in natural waters.

Zorita and Mathiasson [209] have discussed the determination of particle bound PCB congeners in natural waters.

Brinkmann et al. [206,210] used silica gel columns to separate lower chlorinated PCBs but was less efficient in separating higher PCBs.

3.9 Polychloroterphenyls

The occurrence of these substances has been reported in river water [211]. The samples were extracted with hexane. The extracts were cleaned by column chromatography, on either alumina or Florasil. The analysis of polychlorinated terphenyls was carried out with a combination of mass spectrometer and a gas chromatograph in the mass fragmentography mode.

Approximately 0.1 ppm of polychlorinated terphenyl was found in river water, the corresponding figure for PCBs was 0.1 ppb.

3.10 Miscellaneous

Other chlorine-containing compounds that have been determined in water include: Macromolecular chloro lingo sulphonic acid [212], chloroaniside and chloromethyl anisoles [213] and chlorinated carboxylic acids [214].

Lu Shi et al. [215] developed a practical analytical method based on solid-phase extraction and reversed-phase liquid chromatography with fluorescent detection for the analysis of light fluoroquinolones in water at the trace level. Different solid-phase extraction materials, pH conditions and eluents were modified to find effective solid-phase conditions. Aqueous samples (pH 2–3) were extracted using Aopel MEP cartridges which were subsequently eluted with 6 mL of 2% formic acid in methanol. The aqueous extracts were analysed by gradient elution liquid chromatography with fluorescent detection whose initial mobile phase was composed of acrylonitrile and 10 mmol L^{-1} tetrabutylammonium bromide (4:96:vlv) at pH 3. The limit of detection was in the range of 0.32–2.12 ng L^{-1}.

3.11 Bromine-containing compounds

Koida et al. [216] determined ethylene bromide and 1,2-dibromo-3-chloropropane in amounts down to 0.1 μL^{-1} of ground water by extraction of the sample with hexane containing 2,3-dibromo-1-propene as internal stand and followed by GC–mass spectrometry.

Leri and May [217] have described a method for the absolute determination of organochlorine and organobromine compounds in environmental water samples.

This method is based on X-ray absorption near-edge spectroscopy (XANES). C1 XANES spectra reflect contributions from all chlorine species present in a sample, providing a definitive measure of total Cl concentration in chemically heterogenous samples. Spectral features near the chlorine k-absorption edge provide detailed information about the bonding state of chlorine, whereas the absolute fluorescence intensity of the spectra is directly proportioned to total chlorine concentration, allowing for simultaneous determination of chlorine speciation and concentration in natural water samples. Absolute chlorine concentrations are obtained from C1 1s XANES spectra using a series of chlorine standards in a matrix of uniform density. With the high sensitivity of synchrotron-based X-ray absorption spectroscopy, it is possible to monitor chloride concentration which can be measured down to the 5–10 ppm range in solid and liquid samples. Referencing the characteristic near-edge features of chlorine in various model compounds, it is possible to distinguish between inorganic chloride (Cl_{inong}) and organochlorine (Cl_{org}), as well as between aliphatic Cl_{org} and aromatic Cl_{org}, with uncertainties in the range of 6%. In addition, total organic and inorganic bromine concentrations can be quantified using a combination of Br 1s XANES and X-ray fluorescence (XRF) spectroscopy. Bromine concentration is detected down to 1 ppm by XRF, and bromine 1s XANES spectra allow quantification of the Br_{inorg} and Br_{org} fractions. These procedures provide nondestructive, element-specific techniques for quantification of chlorine and bromine concentrations that preclude extensive sample preparation.

Hulteroth et al. [218] described a method based on negative-ion electrospray ionisation mass spectrometry with induced in-source fragmentation for the detection of unknown bromine compounds in water.

Gas chromatography–quadrupole mass spectrometry and isotope dilution have been used in the optimisation of analytical methods for the determination of octa-, nona- and decabrominated diphenyl ethers [219].

Wei et al. [220] have discussed the use of programmable temperature vaporisation and large column injection in the gas-chromatographic determination of polybrominated diphenyl ethers.

Polybrominated biphenyls have been detected in environmental samples by a procedure involving ultraviolet irradiation followed by GC of the photodegradation products [211,220–223].

This method was optimised using GC ionisation–quadrupole mass spectrometer. Several polybrominated diphenyl ethers and three $^{13}C_{12}^{-}$ labelled congeners in biological and environmental samples were analysed. In the optimised, instrumental conditions, abundance and repeatability improved with increase in temperature of the ion source. The instrumental detection limits (IDLs) for BDE-196, BDE-197, BDE-206, BDE-207 and BDE-209 were 0.1, 0.1, 0.2, 0.3 and 0.6 pg, respectively.

References

[1] A.J. Murray, J.P. Riley, Anal. Chim. Acta 65 (1973) 261.
[2] A.J. Murray, J.P. Riley, Nature 242 (1973) 37.
[3] J. Novak, J. Zluticky, V. Kubulka, J. Mostecky, J. Chromatogr. 76 (1973) 45.
[4] R.D. Kleopfer, B.T. Fairless, Environ. Sci. Technol. 6 (1972) 1036.
[5] S. Jensen, A. Jernelov, R. Large, K.H. Palmork, Food and Agriculture Organisation of the United Nationals Report No. FIR: MP 170/E-88, 1972.
[6] F. Dietz, J. Traud, Vom Water 41 (1973) 137.
[7] V.J. Zitko, Chromatography 81 (1973) 152.
[8] V.J. Zitko, J. Assoc. Off. Anal. Chem. 57 (1974) 1253.
[9] D. Freidman, P.J. Lombardo, J. Assoc. Off. Anal. Chem. 58 (1975) 703.
[10] A.D. Deetman, P. Demeulemeester, M. Garcia, G. Hauck, J.I. Hollies, D. Krockenberger, et al., Anal. Chim. Acta 82 (1976) 1.
[11] C.S. Guann, M.K. Wong, J. Chromatogr. 72 (1972) 283.
[12] H. Hagenmaier, G. Werner, W.Z. Jager, Wasser Abwasser Forsch. 15 (1982) 195.
[13] K. Chiba, H. Haraguchi, Anal. Chem. 55 (1983) 1504.
[14] R.D. Blanchard, J.K. Hardy, Anal. Chem. 56 (1984) 1621.
[15] A.J. Burgasser, J.F. Calaruotolo, Anal. Chem. 49 (1977) 1508.
[16] R.S. Narang, B. Bush, Anal. Chem. 52 (1980) 2076.
[17] T. Fujii, Anal. Chem. 49 (1977) 1985.
[18] V.B. Stein, R.S. Narang, Bull. Environ. Contam. Toxicol. 27 (1981) 583.
[19] M. Malaiyandi, E. Jenkins, P. Lee, M.J. Bowron, Environ. Sci. Health A20 (1985) 219.
[20] B.G. Oliver, K.D. Bothen, Anal. Chem. 52 (1980) 2066.
[21] M.P. Italia, P.C.J. Uden, Chromatography 449 (1988) 326.
[22] K.F. Prath, J. Pauliszyn, J. Anal. Chem. 64 (1992) 2101.
[23] X. Li, X. Xu, X. Wang, L. Ma, Int. J. Environ. Anal. Chem. 84 (2004) 633.
[24] Y. He, H.K. Lee, Anal. Chem. 69 (1997) 4634.
[25] M.A. Jeannot, F.E. Cantwell, Anal. Chem. 68 (1996) 2336.

[26] V. He, H.K. Lee, Anal. Chem. 69 (1997) 4635.
[27] R. Kummert, E. Molnar-Kbuices, W. Giger, Anal. Chem. 50 (1978) 1637.
[28] A. Stozek, W. Beumer, K. Pondenz, Abwasser 26 (1979) 632.
[29] J.J. Hollies, P.F. Pinnington, A.J. Handley, M.K. Baldwin, O. Bennett, Anal. Chim. Acta 111 (1979) 201.
[30] Z.Z. Hellman, Wasser Abwasser Forsch. 18 (1985) 92.
[31] M. Mehran, M. Cooper, J. Jennings, J. Chromatogr. Sci. 24 (1986) 142.
[32] N. Kirshen, Am. Lab. 16 (1984) 60.
[33] K.L.E. Kaiser, B.G. Oliver, Anal. Chem. 48 (1976) 2207.
[34] F. Hrivnak, P. Siskupic, J. Hassler, Vodni Hospodarstivi Ser. B 28 (1978) 195.
[35] H.Z. Hellman, Z. Wasser Abwasser Forsch. 21 (1988) 67.
[36] T.A. McLary, J.F. Barker, Groundwater Minot. Rev. 7 (1988) 63.
[37] V. Biebier, W. Funk, G. Von Mareard, D.Z. Rinne, Z. Wasser Abwasser Forsch. 20 (1987) 22.
[38] C. Yurteri, D.F. Ryan, J.J. Callow, M.D. Gurol, J. Water Pollut. Control Fed. 59 (1987) 950.
[39] V. Lopez-Avila, N. Heath, A.J. Hu, J. Chromatogr. Sci. 25 (1987) 351.
[40] J.W. Cochran, J. High Resolut. Chromatogr. 10 (1987) 573.
[41] D.J. Chichester-Constable, M.E. Barbeau, S.L. Liu, S.R. Smith, J.D. Stuart, Anal. Lett., London 20 (1987) 403.
[42] N.H. Mosesman, L.M. Sidisky, S.D. Corman, J. Chromatogr. Sci. 25 (1987) 351.
[43] T.P. Rusina, P. Kopytar, J. De Boer, Int. J. Environ. Anal. Chem. 91 (2011) 319.
[44] V.D. Siminov, L.N. Popova, T.M. Shamsutdinov, Dokl. Neftehim. Sekt. Basgkirk. Resp. Provl. Vses. Khim. Obshch, Mendeleeva 6 (1971) 357.
[45] W.L. Dilling, G.J. Kalios, Environ. Sci. Technol. 9 (1975) 813.
[46] L.C. Friedman, S. Schorodee, M.G. Brooks, Environ. Sci. Technol. 20 (1986) 826.
[47] F. Hellman, Z. Wasser Abwasser Forsch. 18 (1985) 210.
[48] L. Burguera, J.L. Burguera, Anal. Chim. Acta 53 (1983) 153.
[49] K. Martinsen, A. Kringstad, G.E. Carlberg, Water Sci. Technol. 20 (1988) 13.
[50] L. Renberg, Anal. Chem. 50 (1978) 1836.
[51] S. Alberti, F. Jonke, Z. Wasser Abwasser 8 (1975) 140.
[52] J. Rivera, M.R. Cuberes, J. Albaiges, J. Bull. Environ. Contam. Toxicol. 18 (1977) 624.
[53] T.A. Bellar, J.J. Litchenberg, J.W. Eichelberger, Environ. Sci. Technol. 10 (1976) 926.
[54] EPA, Analytical Report – New Orleans Area Water Supply Study, EPA 906/10-74-002, Surveillance and Analysis Division, EUEPA Region, Dallas Texas, Mimeo, 1974.
[55] Environmental Defence Fund, The Implications of Cancer Causing Substances in Mississippi River Water, Environmental Defence Fund, Washington, DC, 1974.
[56] W.H. Glase, R. Rawley, J. Am. Water Works Assoc. 71 (1979) 509.
[57] T. Luong, C.J. Peters, R.J. Young, R. Perry, Environ. Technol. Lett. 1 (1980) 299.
[58] W.W. Bunn, B.B. Haas, E.R. Deane, R.D. Kleopfer, Environ. Lett. 10 (1975) 205.
[59] W.H. Glaze, J.E. Henderson, G.G. Smith, Private Communication, 1972.
[60] J.J. Rook, Water Treat. Exam. 21 (1972) 259.
[61] B.K. Dowty, D.R. Carlisle, J.J. Laseter, Environ. Sci. Technol. 9 (1975) 762.
[62] USEPA, New Orleans Area Water Supply Study Draft Analytical Report, Lower Mississippi River Facility, USEPA, 1974.
[63] A.J. Libbey, Analyst 111 (1986) 1221.
[64] A.A. Stevens, G.L. Slocum, D.R. Sieger, G.G. Robeck, Conference Environmental Impact of Water Chlorination, Oak Ridge National Laboratory, Oak Ridge, TN, 1975.

[65] M. Fielding, K. McLaughlin, C. Steel, Water Research Centre Enquiry Report ER 532, August Water Research Centre, Stevenage Laboratory, Ebdon Way Stevenage, Herts, UK, 1977.

[66] J.M. Symona, T.A. Bellar, J.K. Carswell, J. Demarco, K.I. Kropp, G.G. Roebeck, et al., National Organics Reconnaissance Survey for Halogenated Organics in Drinking Water, EPA, Cincinnati, OH, 1975.

[67] A.A. Nicholson, O. Mersz, Bull. Environ. Contam. Toxicol. 14 (1975) 453.

[68] M.W. Tabor, Environ. Sci. Technol. 17 (1983) 324.

[69] US Environmental Protection Agency, US Environmental Protection Agency Federal Register No. 28, 43, 1978.

[70] USEP, USEP Control of Organic Chemical Contamination in Drinking Water, Federation Register, 44, 5755, 1978.

[71] R.O. Blanchard, J. Hardy, Anal. Chem. 56 (1984) 57.

[72] G. Eklund, B. Josefsson, C.J. Roos, High Resolut. Chromatogr. Commun. 1 (1978) 34.

[73] G. Eklund, B. Josefsson, C.J. Roos, J. Chromatogr. 142 (1977) 575.

[74] J.L. Schnoor, J.L. Nitzschke, R.D. Hureas, J.N. Veenstra, Environ. Sci. Technol. 113 (1979) 1134.

[75] M.E. Comba, K.L.E. Kaiser, Int. J. Environ. Anal. Chem. 16 (1983) 17.

[76] B.R. Quimby, M.F. Delaney, D.C. Uden, R.M. Barnes, Anal. Chem. 52 (1980) 259.

[77] R.C. Kroner, Pub. Works 111 (1980) 81.

[78] L.H. Keith, Identification and Analysis of Organic Pollutants in Water, Ann Arbor Science Publishers, Ann Arbor, MI, 1976.

[79] N.V.J. Brodtmann, Am. Water Works Assoc. 67 (1975) 558.

[80] J.A. Coburn, I.A. Valdmanis, A.S.Y. Chau, The Extraction of Organochlorine Pesticides and P.C.B.s From Natural Waters with XAD-2, Water Quality Branch, Canada Centre of Inland Waters, Burlington, ON, Canada.

[81] P.R. Musty, G. Nickless, J. Chromatogr. 89 (1974) 185.

[82] G.A. Junk, J.J. Rickard, M.D. Grieser, D. Witlak, J.L. Witlak, M.D. Arguello, et al., J. Chromatogr. 99 (1974) 745.

[83] J.A. Coburn, I.A. Valdmanis, A.S.U. Chau, J. Assoc. Off. Anal. Chem. 60 (1977) 224.

[84] G.A. Junk, D.C. Chriswell, R.C. Chiang, L.D. Kissinger, J.J. Richard, J.S. Fritz, et al., Fresenius Z. Anal. Chem. 283 (1976) 331.

[85] L.D. Kissinger, J.S. Fritz, J. Am. Water Assoc. 68 (1976) 435.

[86] R.E. Arbon, E.P. Grimsrud, Anal. Chem. 62 (1990) 1762.

[87] J.A. Bognar, W.B. Knighton, E.P. Grimsrud, Anal. Chem. 64 (1992) 2451.

[88] J.M. Symons, T.A. Bell, J.K. Carswell, J. DeMarco, K.L. Kropp, G.G. Robeck, et al., J. Am. Water Works Assoc. 67 (1975) 643.

[89] National Survey for Halomethanes in Drinking Water Health and Welfare Canada, 77-EHD-9, 1977.

[90] B. Bush, R.S. Narang, Bull. Environ. Sci. Technol. 13 (1977) 436.

[91] R.L. Morris, L.G. Johnson, J. Am. Water Works Assoc. 68 (1976) 492.

[92] C. Gomella, J.P. Belle, Tech. Sci. Munic. 73 (1978) 125.

[93] S.L. Friant, (Thesis), Drexel University Microfilms Ltd, London, 468, pr 29162, 1972.

[94] M.M. Varma, M.R. Siddique, K.T. Doty, H. Machis, J. Am. Water Works Assoc. 71 (1979) 389.

[95] E.A. Dietz, K.F. Singley, Anal. Chem. 51 (1979) 1809.

[96] J.J. Rook, Water Treat. Exam. 23 (1974) 234.

[97] J.E. Lovelock, Nature, London 256 (1975) 193.

[98] J.E. Lovelock, R.J. Maggo, R.J. Wade, Nature, London 241 (1973) 194.
[99] A. Zlatkis, A. Lichtenstein, A. Tuchbee, Chromatographia 8 (1973) 67.
[100] J.P. Mieure, M.M.J. Dietrich, J. Chromatogr. Sci. 11 (1973) 559.
[101] B. Dowty, J.L. Laseter, Anal. Lett. 8 (1975) 25.
[102] T.A. Bellar, J.J. Lichtenberg, J. Am. Water Works Assoc. 66 (1974) 739.
[103] K. Grob, J. Chromatogr. 84 (1973) 255.
[104] K. Grob, G. Grob, J. Chromatogr. 90 (1974) 303.
[105] K. Grob, K. Grob, G. Grob, J. Chromatogr. 106 (1975) 299.
[106] W. Bertsch, E. Anderson, G. Holzer, J. Chromatogr. 112 (1975) 701.
[107] R.C. Dressman, A.A. Stevens, J. Fair, B.J. Smith, Am. Water Works Assoc. 71 (1979) 392.
[108] US Environmental Protection Agency, Part III. Appendix C, Analysis of Trihalomethanes in Drinking Water, Federal Register, 44, No 231 to 68672, 1979.
[109] US Environmental Protection Agency, Method 501. 1. The Analysis of Trihalomethanes in Finished Waters by the Purge and Trap Method, US Environmental Protection Agency, EMSL, Cincinnati, OH 45268, 1979.
[110] Anonymous, Chem. Eng. News 54 (1976) 35.
[111] V.V. Ligon, R.I. Johnson, Anal. Chem. 48 (1976) 481.
[112] H.J. Brass, J. Am. Lab. 12 (1980) 23.
[113] N.A. Kirschen, Varian Instrum. Appl. 15 (1981) 2.
[114] US Environmental Protection Agency, Method 60. US Environmental Protection Agency, Federal Register 69464, 44, No. 223, 1979.
[115] Packard Tech. News 1 (No. 7258) (February 1981).
[116] C.T. Pierce, R.J. Grochoweski, R.R. Kongovi, K. Narangajvanaca, G.L. Brock, Am. Lab. April (1981) 34.
[117] Department of the Environmental/National Water Council Standing Committee of Analysts. H.M. Stationery Office, London, Methods for the Examination of Waters and Associated Materials Chloro and Bromo-Tri Halogenated Methanes in Waters, 1981.
[118] L. Leppine, J.F. Archambault, Anal. Chem. 64 (1992) 810.
[119] V. Janda, V. Hospodorstvi, Ser. B 31 (1981) 137.
[120] K. Okumura, Kawada, T. Lino, Analyst 107 (1982) 1498.
[121] R.J. Oake, I.M. Anderson, The Determination of Carbon Absorbable Organohalide in Water, Water Research Centre, Medmenham, 1984.
[122] R.C. Dressman, A.A. Stevens, J. Am. Water Works Assoc. 75 (1983) 431.
[123] S. Kapila, W.A. Aue, J. Chromatogr. 15 (1977) 569.
[124] D.A. Ryan, S.M. Argentine, G.W. Rice, Anal. Chem. 62 (1990) 853.
[125] A. Di Corcia, R. Samperi, Anal. Chem. 62 (1990) 1490.
[126] J.R. Stetter, Z. Cao, Anal. Chem. 62 (1990) 1490.
[127] H. Ozawa, T. Tsukioka, Anal. Chem. 59 (1987) 2914.
[128] D.L. Norwood, R.F. Christman, J.D. Johnson, J.R. Hass, J. Am. Water Works Assoc. 78 (1986) 175.
[129] J.W. Miller, P.C. Uden, R.M. Barnes, Anal. Chem. 54 (1982) 485.
[130] D. Zehavi, J.N. Seiber, Anal. Chem. 68 (1996) 3450.
[131] A. Yasuhara, M.M. Itoh, Environ. Sci. Technol. 21 (1987) 971.
[132] V.V. Taguchi, E.J. Reiner, D.T. Wong, D. Meresz, B. Hassas, Anal. Chem. 60 (1988) 1429.
[133] J.A. Laramee, B.C. Arbogast, M.L. Dienzer, Anal. Chem. 60 (1988) 1937.
[134] H. Schimmel, B. Schmid, R. Backer, K. Ballschmitter, Anal. Chem. 65 (1993) 640.

[135] J.B. Plomley, R.E. March, R.S. Mercer, Anal. Chem. 68 (1996) 2345.

[136] I.O.O. Korhonen, J. Knuutinen, Chromatographia 17 (1983) 154.

[137] D.R. Thielen, G. Olsen, Anal. Chem. 60 (1988) 1332.

[138] K. Kimata, K. Hosoya, T. Araki, N. Tanaka, E.R. Barnhard, L.R. Alexzander, et al., Anal. Chem. 65 (1993) 2502.

[139] E.B. Gonzalez, R.A. Baumann, G. Gooljer, W.H. Velthorst, R.U. Frei, Chemosphere 16 (1987) 1123.

[140] G. Bengtsson, J. Chromatogr. Sci. 23 (1985) 397.

[141] G. Huang, R. Zhu, Sepu 5 (1987) 113.

[142] H. Malissa, G. Sziogyenyn, K. Winsauer, Fresenius Z. Anal. Chem. 17 (1985) 321.

[143] L. Renberg, Water Res. 4 (1970) 533.

[144] L. Renberg, Anal. Chem. 46 (1974) 459.

[145] C. Chriswell, R. Cheng, Anal. Chem. 47 (1975) 1325.

[146] M.B. Lee, Y. Stokker, A.S. Chan, Anal. Chem. 70 (1987) 1003.

[147] H.J. Hoben, S.A. Ching, L.J. Casarette, R.A. Young, Bull. Environ. Contam. Toxicol. 15 (1976) 78.

[148] H.E. Ervin, G.D. McGinnis, J. Chromatogr. 190 (1980) 203.

[149] A. Sara Fraz-Yazdi, F. Mofazzebi, A. Eslhagi, Environ. Anal. Chem. 90 (2012) 1108.

[150] C. De Ruiter, J.F. Bohle, G. De Jong, U.A.T. Brinkman, R.W. Frei, Anal. Chem. 60 (1988) 666.

[151] D.P. Silgouer, M. Grasserbauer, D. Barcelo, Anal. Chem. 69 (1997) 2756.

[152] G. Brandt, A. Kettrup, Fresenius Z. Anal. Chem. 127 (1987) 213.

[153] B.G. Henshaw, J.W.W. Morgan, N.J. Williams, J. Chromatogr. 110 (1975) 37.

[154] M./G. Zigler, W.F. Phillips, Environ. Sci. Technol. 1 (1967) 65.

[155] A. Wolkoff, R.H. Larose, J. Chromatogr. 99 (1974) 731.

[156] H. Thielemann, M. Luther, Pharmazie 25 (1970) 367.

[157] Ch'mil, Ananliticheskoi Khimn 36 (1983) 729.

[158] R.C. Carr, P. Thomas, J.M. Neff, Bull. Environ. Contam. Technol. 28 (1982) 477.

[159] A.L. Wade, F.M. Hankridge, H.P. Williams, Anal. Chim. Acta 105 (1979) 91.

[160] P. Boyle, E.F. Robinson-Wilson, J.D. Petty, W. Weber, Bull. Environ. Contam. Technol. 24 (1980) 177.

[161] L.I. Ingram, G.D. McGinnis, S.V. Porikl, Anal. Chem. 51 (1979) 1077.

[162] K. Noren, J. Sjovall, J. Chromatogr. 414 (1987) 55.

[163] J. Gaivdzik, B. Gaivdzik, A. Czerwinska-Billi, Chromatographia 25 (1988) 504.

[164] K.F. Noll, A. Gounaris, Water Res. 22 (1988) 815.

[165] S. Fingler, V. Drevenkar, Z. Vasitic, Microchim. Acta No. 4/6 (1987) 163.

[166] R.G. Webb, A.C. McCall, J. Anal. Chem. 55 (1972) 746.

[167] U. Bauer, G.-U. Wasserfacht, Wasser Abwasser 113 (1972) 58.

[168] O.W. Berg, P.L. Diolady, G.A. Rees, Bull. Environ. Contam. Toxicol. 7 (1972) 338.

[169] F.L. Beezhold, V.F. Stout, Bull. Environ. Contam. Toxicol. 10 (1973) 10.

[170] Chromopak News, June 1979.

[171] E. Shulte, L. Ackerz, Anal. Chem. 268 (1974) 260.

[172] W.J. Dunn, D.L. Stalling, T.R. Schwartz, J.W. Hogan, J.D. Johonsson, S. Wold, Anal. Chem. 56 (1984) 1308.

[173] D.L. Stalling, J.N. Huckins, J. Assoc. Off. Anal. Chem. 54 (1972) 801.

[174] M. Ahnoff, B. Josefsson, Anal. Lett. 6 (1973) 1083.

[175] R.G. Webb, A.C. McCall, J. Chromatogr. Science 11 (1973) 366.

[176] J.W. Eichelberger, L.E. Harris, W.J. Budde, Anal. Chem. 46 (1974) 227.

[177] B.A. Karlruber, W.D. Hormann, K.A. Ramsteinen, Anal. Chem. 47 (1975) 2453.
[178] M. Ahnoff, B. Josefsson, Anal. Lett. 6 (1973) 1036.
[179] R.D. Voyksner, J.R. Hass, G.W. Sovocaol, Anal. Chem. 55 (1983) 744.
[180] E.E. Pellazari, M.A. Moseley, S.D. Cooper, J. Chromatogr. 334 (1985) 277.
[181] S. Tanabe, N. Kannan, T. Wakimoto, R. Tatsukawa, Int. J. Environ. Anal. Chem. 29 (1987) 199.
[182] J.B. Quintana, E. Martinez, A.M. Carro, R.A. Lorenzo, R. Cela, Int. J. Environ. Anal. Chem. 83 (2003) 269.
[183] P.W. Albro, O. Fishbein, J. Chromatogr. 69 (1972) 273.
[184] L.S. Kaminsky, M.J. Fasko, J. Chromatogr. 155 (1978) 363.
[185] L.L. Needham, S.L. Surek, S.L. Head, V.W. Burse, J.A. Liddle, Anal. Chem. 52 (1980) 2227.
[186] M.N. Dong, J.L. DiCesare, J. Chromatogr. Sci. 20 (1982) 517.
[187] G. Petrick, D.E. Schulz, J.C. Duinker, J. Chromatogr. 435 (1988) 241.
[188] R.H. de Vos, E.W. Peet, Bull. Environ. Contam. Toxicol. 6 (1971) 164.
[189] O.M. Sackmauer, O. Pal'usova, A. Szokolay, Water Res. 11 (1977) 551.
[190] S.O. Farwell, F.A. Beland, R.D. Geer, Bull. Environ. Contam. Technol. 10 (1973) 157.
[191] W.B. Neeley, Sci. Total Environ. 7 (1977) 117.
[192] M. Kopanica, Sci. Total Environ. 37 (1984) 83.
[193] D.S. Aga, E.M. Thurman, M.L. Dowes, Anal. Chem. 66 (1994) 1495.
[194] H.D. Gesser, A. Chow, F.C. Davis, J.F. Uthe, J. Reinke, Anal. Lett. 4 (1971) 883.
[195] J.F. Uthe, J. Reinke, H. Gesser, Environ. Lett. 3 (1972) 117.
[196] M. Ahnoff, B. Josefsson, Bull. Environ. Contam. Toxicol. 13 (1975) 159.
[197] US Environmental Protection Agency, US National Information Service, Springfield, VA Report No PB 279547, 30381, 1977.
[198] Inland Waters Directorate, Analytical methods manual, in: Part 2, Organic Constituents, Water Quality Branch, Ottawa, ON, 1974.
[199] C.D. Chriswell, R.I. Ericson, G.A. Junk, K.W. Lee, J.S. Fritz, H.J. Svec, J. Am. Water Works Assoc. 69 (1977) 669.
[200] P.R. Musty, G. Nickless, J. Chromatogr. 100 (1974) 83.
[201] P.R. Musty, G. Nickless, J. Chromatogr. 120 (1976) 369.
[202] J.W. Bedford, Bull. Environ. Contam. Technol. 12 (1974) 662.
[203] J.P. Thome, Y. Vandaele, Int. J. Environ. Anal. Chem. 29 (1987) 95.
[204] J. Lawrence, H.M. Todine, Environ. Sci. Technol. 10 (1976) 381.
[205] R.F.C. Mantoura, C.A. Llewellyn, Anal. Chim. Acta 151 (1983) 297.
[206] A. Brinkmann, J.W.F.L. Seetz, H.G.M. Remer, J. Chromatogr. 116 (1976) 353.
[207] S. Bayer, J. Buffle, Anal. Chem. 89 (2009) 277.
[208] A.R. Schneider, A. Paolichi, J.E. Baker, Int. J. Environ. Anal. Chem. 86 (2006) 789.
[209] S. Zorita, L. Mathiasson, Int. J. Environ. Anal. Chem. 85 (2005) 531.
[210] A.T. Brinkmann, A. DeKok, De Vries, H.G.M. Remer, J. Chromatogr. 128 (1976) 101.
[211] J. Freudenthal, P.A. Greve, Bull. Environ. Contam. Toxicol. 10 (1973) 108.
[212] W.H.G.M. Van Loon, J.S. Bron, B. De Groot, Anal. Chem. 65 (1993) 1726.
[213] H.P. Lee, J. Assoc. Anal. Chem. 71 (1988) 803.
[214] S. Onodera, T. Udagawa, M. Tabata, S. Ishkura, S. Suzuki, J. Chromatogr. 287 (1984) 176.
[215] X. Lu Shi, Y. Zhou, G. Lhany, R.Y. Gu, Surampalli, T.C. Zhang, Int. J. Environ. Anal. Chem. 90 (2010) 1085.

[216] K. Koida, Y. Tokumori, J. Saimatsu, K. Sel, Y. Kuhno, A. Oka, Horoshima-Shi Elsei, Kenyusho Nenpo 5 (1988) 34.
[217] A.C. Leri, M.B. May, Anal. Chem. 78 (2006) 5711.
[218] A. Hulteroth, A. Patschew, M. Jekel, Int. J. Environ. Anal. Chem. 87 (2007) 415.
[219] A. Eguchi, T. Isobe, K. Ramu, S. Tonable, Int. J. Environ. Anal. Chem. 91 (2011) 348.
[220] H. Wei, P.S. Dassanayeike, A. Li, Int. J. Environ. Anal. Chem. 9 (2010) 7.
[221] K.A. Banks, D.D. Bills, J. Chromatogr. 33 (1968) 450.
[222] W.G. Potter, LIB 1609, FDA, Minneapolis District, MN, 1969.
[223] D.R. Erney, J. Am. Oil Colour Assoc. 58 (1975) 1202.

Chapter 4

Nitrogen compounds in nonsaline waters

Chapter Outline

4.1 Aliphatic amines **103**
Gas chromatography 103
High-performance liquid chromatography 104
Miscellaneous 104
4.2 Aromatic amines **105**
Gas chromatography 105
High-performance liquid chromatography 105
Miscellaneous 105
4.3 Amino acids **106**
Gas chromatography 106
High-performance liquid chromatography 106
Miscellaneous 106
4.4 Amides **107**
Gas chromatography 107
High-performance liquid chromatography 107
Polarography 107
4.5 Nitrophenols **107**
Gas chromatography 107
Thin-layer chromatography 108
Column chromatography 108
Spectrophotometric method 108
Capillary electrophoresis 108
4.6 Trinitrotoluene **109**
4.7 Chloroaniline **109**
Liquid chromatography 109
4.8 Hydrazines **109**
4.9 Nitriles **110**
4.10 Nitrosamines **110**
Gas chromatography 110
Nuclear magnetic resonance spectroscopy 111
4.11 Ethylenediaminetetraacetic acid **111**
4.12 Nitriloacetic acid **111**
Column chromatography 112
Polarography 112
Atomic absorption spectrometry 112
Gas chromatography 112
Polarography 113
Spectrophotometric method 114
4.13 Miscellaneous nitrogen compounds **115**
References **116**

4.1 Aliphatic amines

Gas chromatography

Hermanson et al. [1] used aluminium column packed with 80–100 mesh Chromosorb W supporting 8.9% of amine 220 at 95°C with nitrogen as

Determination of Toxic Organic Chemicals In Natural Waters, Sediments and Soils.
DOI: https://doi.org/10.1016/B978-0-12-815856-2.00004-7

carrier and flame ionisation detection. A rectilinear response was obtained between peak area and amount of propylamine, dipropylamine and propanol; between 0.2 and 2.0 μg.

Gas flame ionisation chromatography has been used to determine dimethylamine [2,3], dimethylformamide [4], propylamine [3], diispropylamine [1] in river water. To separate C, C-, mono-, di- and trialkylamines, Onuska et al. [3] adjusted the pH of the sample between 5 and 8. A 1 μL aliquot of the filtrate was injected into a stainless steel column (185 cm × 2 mm i.d.) packed with 28% of Pennwalt 223 and 4% of potassium hydroxide on Gas-Chrom R (80–100 mesh) and maintained at 134°C. A dual-flamed ionisation detector was used and the carrier gas was helium (flow rate 52.2 mL min^{-1}). The detector response was rectilinear between 10 ng and at least 100 μg of diemthylamine, and the reproducibility was good. The column could be regenerated by increasing the column temperature to more than 180°C.

Kimoto et al. [5] used gas chromatography (GC) to determine simple aliphatic amines in dichloromethane extracts of water samples. They used a gas chromatograph equipped with a nitrogen–phosphorous detector and a 1.8 m × 2 mm i.d. glass column packed with Carbopack B-Carbowax 20-potassium hydroxide. The flow rates were helium 20 mL min^{-1}, hydrogen 3.5 mL min^{-1}. The injector and detector temperatures were 200°C and 250°C respectively. The column was programmed from 70°C to 150°C at 4°C min^{-1} and held at the upper temperature for 4 additional minutes. Then 2 μL samples were injected.

Avery and Junk [6] converted trace levels of primary aliphatic amines in aqueous solutions to imine derivatives using pentafluorobenzaldehyde. The derivatives were extracted into hexane and measured by combined high-resolution GC–mass spectrometry (MS) using multiple ion monitoring. The method was successfully applied to the analysis of primary amines in river water down to a detection limit of 10 μg L^{-1} for 0.5 mL samples.

High-performance liquid chromatography

Nishikawa [7] described a liquid-chromatographic procedure for the determination of aliphatic diamines as their acetylacetonates in natural water with a detection limit of 1.4–20 μg L^{-1}.

Miscellaneous

Hexane 1,6-diamine has been determined in amounts down to 0.5 μg L^{-1} in water by thin-layer chromatography and paper chromatography [8].

Florence and Farner [9] determined parts per billion (10^9) amines in natural water using a spectrophotometric procedure.

4.2 Aromatic amines

Gas chromatography

Wegman and DeKorte [10] have described a method for the determination of aniline toluidines, monochloroanilines and their derivatives in surface water. The aromatic amines in an isooctane extract of the water sample were brominated and determined by GC with a capillary column and an electron-capture detector. The results of analysis of Rhine water samples in the Netherlands were presented. The detection limits were in the 5–15 ng L^{-1} range.

Bao et al. [11] have reported a capillary column GC–ECD method for the determination of aromatic amines in water in amounts down to 2–50 ppt. In this method the amines were converted to the corresponding amides by reaction with pentafluorobenzyl chloride in 1% aqueous sodium bicarbonate followed by toluene extraction.

High-performance liquid chromatography

Zhou et al. [12] developed a rapid and sensitive method for the determination of aromatic amines such as *o*-nitroaniline, alpha-naphtylamine, *o*-chloroaniline in environmental water samples based on dispersive liquid phase microextraction coupled with high-performance liquid chromatography (HPLC). Excellent results were obtained when chlorobenzene and acetonitrile were employed as the extraction solvent and dispersive solvent, respectively. Some other parameters that would have important effect on the preconcentration of aromatic amines were investigated in detail. Under the optimal conditions the method had an excellent linear relationship between the peak area and the concentration for the above compounds in the concentration range of 1–50 μg L^{-1}. The limits of detection and precisions of the method were in the range of 0.1–0.7 μg L^{-1} and 6.3%–9.7%, respectively. The method has been validated with two real water samples, and the results showed that excellent spiked recoveries in the range of 92.8%–111.5% were achieved.

Benzidine has been determined in natural water using column chromatography [13].

Miscellaneous

Aniline derivatives at the microgram level have been determined in natural waters by a spectrophotometric procedure involving diazotization with sulphuric acid and naphthol [14].

Nielen et al. [15] used a strongly acidic cation-exchange resin for on-line preconcentration of polar anilines in water. The method could be automated and a detection limit for the nine anilines could be examined corresponding to 0.02–0.3 μg L^{-1}.

Stuber and Henloar [16] assessed the aqueous elution, selective concentration approach for isolating aromatic bases from water, and the factors controlling the concentration process identified using polymethylacrylic ester as an absorbent. The degree of concentration attainable depends on the ratio of the capacity of the natural form of the amine to that of the ionised form. Capacity factors of ionic forms of amines on polymethylacrylic ester resins are 20–250 times lower than those of the neutral forms and increase with the hydrophobicity of the amine.

4.3 Amino acids

Gas chromatography

Gardner and Lee [17] described an early gas-chromatographic method for the estimation of dissolved free and dissolved total free and combined amino acids in lake water. The amino acids were first concentrated by ion- and ligand-exchange chromatography. *N*-Trifluoroacetyl methyl esters were prepared and determined on glass columns (4 m × 3 mm o.d.) packed with 0.7% of XE-60, 0.5% of OV-101 and 0.2% of QF-1 on Diatoport S, with flame ionisation detection. Internal standards are added before derivatisation. Recoveries were from 36% to 97% for the free acids and 55%–93% for the hydrolysed samples. Recoveries were poor for phenylalanine and lysine.

High-performance liquid chromatography

Dai and Helz [18] used an amperometric detector employing a carbon paste electrode to determine aminopolycarboxylic acid in water samples by liquid-chromatographic separation on a reverse-phase column with aqueous trichloroacetic and as mobile phase. The amino carboxylic acids were directly oxidised at the detector electrode without involving intermediate species. Amino acids, citric acid and fulvic acid did not interfere.

Miscellaneous

Yu and Dovich [19] used capillary zone electrophoresis with thermo-optical absorbance detection to determine sub $\mu g\,L^{-1}$ concentration of 18 amino acids.

Gardner [20] isolated free amino acids at the 20 nmol L^{-1} level from as little as 5 mL of sample, by cation-exchange, and measured concentrations on a sensitive amino acid analyser equipped with a fluorimetric detector.

4.4 Amides

Gas chromatography

Arkell and Croll [21] and Croll and Simpkins [22] determined acrylamide monomer in amounts down to 0.1 μg L^{-1} by a gas-chromatographic procedure.

Hashimoto [23] used GC to determine acrylamide monomer in water. This method utilised the response of the brominated form of acrylamide (α,β-dibromopropionamide) to electron-capture detection. The acrylamide is brominated by an ionic reaction using bromine in the presence of potassium bromide and hydrobromic acid. The α,β-dibromopropionamide thus formed is extracted from water using ethyl acetate and the concentrate is then analysed by electron-capture gas–liquid chromatography. The disadvantage of this method, however, is that electron-capture detectors are prone to contamination.

High-performance liquid chromatography

Earlier high-performance liquid-chromatographic methods [24,25] for the determination of acrylamide monomer had detection limits of approximately 0.1 μg L^{-1}. Brown and Rhead [26] improved the sensitivity of HPLC to 0.2 μg L^{-1}. The procedure consists of bromination, extraction of the α,β-dibromopropionamide with ethyl acetate and quantification using HPLC with ultraviolet detection. Samples tested include river, sea and estuarine waters. The levels of inorganic ultraviolet absorbing impurities found in water samples did not interfere in this procedure. The solvent extraction procedure lowered interferences in all samples tested without removal of acrylamide or excessive use of solvent. The experimental yields of a α,β-dibromopropionamide encountered gave a mean of 70.13% ± 8.52% (95 confidence level) for acrylamide-spiked waters, over the concentration range of 0.2–8.0 μg L^{-1} of acrylamide monomer.

Polarography

This technique has been used to determine acrylamide in natural waters [27].

4.5 Nitrophenols

Gas chromatography

Cranmer [4] has described a gas-chromatographic method for the determination of down to 0.05 ppm 4-nitrophenol in human urine and this procedure would, no doubt, be applicable to water samples.

Bengtsson [28] also described optional conditions for the determination of trace amounts of nitrophenols (0.01–0.2 μg L^{-1}) and dinitrophenols

(1.6 μg L^{-1}). The phenols are toluene extracted from the water sample derivatised with heptafluorobutyric anhydride and separated and quantified by electron-capture GC or flame ionisation GC.

Thin-layer chromatography

Chambou and Chambon [29] have described a thin-layer chromatographic method for the determination of down to 5 μg L^{-1} of 4,6-dinitro-*o*-cresol in diethyl ether extracts of water samples. The nitrophenol spots are located by the bright yellow colour formed with ammonia, then removed from the plate and extracted with 50% acetic acid (0.05 mL)–methanol (5 mL); the extinction of the extract is measured at 370 nm [30].

Column chromatography

This technique coupled with a spectrophotometric detection has been used to determine mixtures of nitrophenols in natural waters [31].

Spectrophotometric method

Hicks and Riley [32] have presented a method for determining the natural levels of nucleic acids in lake waters, which involves preconcentration by adsorption.

Capillary electrophoresis

Hortstroffe et al. [33] developed a capillary electrophoresis system, which was coupled to sequential injection analysis for analysis of nitrophenols.

For characterisation this system was successfully applied to the separation and quantification of nitrophenols. A blue LED was used as light source, and hydrodynamic injection was carried out by using a pressure-stable solenoid valve and an inflatable pressure reservoir. A good reproducibility of migration time (0.5%) and peak heights (5%) were obtained. The calibration by using peak heights was found to be linear up to 776 μmol L^{-1}. The system was robust and reliable for autonomous analysis. All maintenance requirements including the conditioning of the capillary and flushing of both buffer reservoirs were carried out automatically. Instrumentation aspects of the capillary electrophoresis part are compared with hyphenated flow systems showing maximal operation versatility. Instrumental control and data evaluation were carried out using an auto analyser software package.

4.6 Trinitrotoluene

Spectrophotometric methods have been described for the determination of trinitrotoluene [34] and nitrobenzene [35].

Shankaran et al. [36] have developed molecular recognition electrodes for surface plasma resonance detection of trinitrotoluene in water.

Tall [37] performed electro analysis of some nitro compounds using bulk bismuth electrodes.

Lx Tian Fu et al. [38] determined trinitrotoluene in water by GC with electron-capture detection aminophenols.

A gas-chromatographic procedure has been described for the determination of down to 0.1 mol of aminophenols in natural waters [39].

4.7 Chloroaniline

Liquid chromatography

Di Corcia and Samperi [40] have described a method for the determination of down to 0.1 $\mu g\ L^{-1}$ of chloroaniline in environmental waters. Extraction of the sample with acetonitrile with two traps in tandem is followed by liquid-chromatographic analysis.

4.8 Hydrazines

Kinetic methods based on fluorimetry [41] and lectin glycoenzyme multilayer film–modified biosensor [403] have been used to determine hydrazine [41] and phenyl hydrazine respectively in environmental waters [42].

This approach to the analytical extraction of the transformation products of 1,1-dimethylhydrazine from soils based on the use of subcritical acetonitrile ensures the good recoveries of mobile analyte species (70%–112%) with a minimum expenditure time (30 minutes). It significantly exceeds the previously used Soxhlet extraction and ultrasonic extraction in effectiveness, rapidity and possibly of the simultaneous extraction of a wide range of compounds from soils with high concentrations of organic substances.

A combination of pressurised extraction with the subsequent analysis of the acetonitrile extracts detailed by GC–MS/MS does not require additional stages of sample preparation and ensures the rapid simultaneous determination of formaldehyde, acetaldehyde and 2-furaldehyde, dimethylhydrazones and also 1,1,4,4-tetramethyl-2-tetrazene, *N,N*-dimethylformamide, *n*-nitrosodimethylamine, 1-methyl-1*H*-1,2,4-triazole and 1-formyl-2,2-dimethylhydrazine in complex matrixes, such as peaty soils. The LODs of analytes for method developed lie in the range of 1.8–15 $\mu g\ kg^{-1}$, which is 1–2 orders of magnitude lower as compared to currently used approaches without additional extract preconcentration.

The application of this method to the analysis of soil samples permitted the detection of the main product of 1,1-dimethylhydrazine transformation in the peaty soil as 1-methyl-1*H*-124-triazole.

4.9 Nitriles

GC has been used to determine acrylonitrile at the microgram per litre level in natural waters [43].

Stefanescu and Ursu [44] determined acrylonitrile and acetonitrile in waters by spectrophotometric and titrimetric procedures after separation from the sample by azeotropic distillation. Down to 2 mg of each substance per litre can be determined.

Ghersin et al. [45] compared colorimetric and titrimetric procedures for the determination of acrylonitrile in waters. A titration based on the addition of sodium sulphide to the acrylonitrile followed by titration of liberated sodium hydroxide gave a sensitivity of 10 mg L^{-1} acrylonitrile.

The titrimetric method involving the use of mercaptoacetic acid has a sensitivity of 2 mg acrylonitrile L^{-1}, or down to 0.4 mg L^{-1}.

4.10 Nitrosamines

Gas chromatography

Nikaido et al. [46] give details of procedures for the recovery of low levels of dialkylnitrosamines from nonsaline waters, including lake water before subsequent detection by GC. The recovery technique involves the addition of potassium carbonate to the sample and concentration of the nitroso compounds on Amberlite XAD-2 resin. More than 90% recoveries were obtained for dimethylnitrosamine and dimethylnitrosamine. The dimethylnitrosamine recoveries obtained by this procedure were 97% or higher and were particularly good for nitrosamine levels of about 10 part per 10^9.

Tompkins and Griest et al. [47,48] have described a gas-chromatographic method for the determination of *N*-nitrosodimethylamine in contaminated waters. This procedure utilises a solid-phase extraction procedure, which extracts *N*-nitrosodimethylamine at the nanogram per litre level from aqueous samples using C_{18} (reversed phase) membrane extraction disc layers over a carbon-based extraction disc. The reversed-phase disc removes nonpolar water insoluble neutrals and is set aside; the carbon-based disc is extracted with a small volume of dichloromethane. *N*-Nitrosodimethylamine is quantified in the organic extract using a gas chromatograph equipped with both a short-path thermal desorber and a chemiluminescence nitrogen detector. The detection limit for the procedure is 3 ng L^{-1} *N*-nitrosodimethylamine with a recovery of about 75%.

Mills and Alexander [49] have described the factors affecting the formation of dimethylnitrosamine in samples of water.

Nuclear magnetic resonance spectroscopy

Fulton et al. [50] determined down to 510 μg L^{-1} of *N*-nitrosodimethylamine in natural waters using 500 MHz proton nuclear magnetic resonance spectroscopy.

4.11 Ethylenediaminetetraacetic acid

Ethylenediaminetetraacetic acid (EDTA) salts are present in low concentrations in detergent preparations and some food products. They are not biodegradable and might have an effect in mobilising trace metals in river waters, that is in reducing their tendency to be removed from solution by adsorption and precipitation reactions and possibly causing desorption from contaminated river sediments. Hence, there is an interest in determining EDTA and its salts in river waters, sewage and sewage effluents.

Gardiner [51] described a gas-chromatographic method for the determination of EDTA in aqueous environmental samples. The separation of the major peaks is increased by preparing the ethyl derivatives of the sample compounds, 1,6-hexanediaminetetraacetic acid being used as internal standard. The lower limit of detection of the method is approximately 15 μg L^{-1} with 25 mL samples.

Lockhart and Blakely [52] used GC to estimate ferric EDTA and its degradation products. GC analysis of the *N*-trifluoroacetyl-*n*-butyl esters of FE (III)–EDTA and photolysis products was performed on a 4 ft. × 1/8 in. i.d. glass column of 3% QF-1 on 100/120 mesh Gas Chromo-Q using a Hewlett Packard Model 7610A gas chromatograph with flame ionisation detectors, with nitrogen as the carrier gas. Oven temperature was programmed from 110 to 260°C min^{-1} starting at 2 minutes postinjection and holding for 2 minutes at 260°C. Injection temperature was 250°C and detector temperature was 300°C.

The *N*-trifluoroacetyl-*n*-butyl esters separated on the gas-chromatographic column were identified by fragmentation pattern analysis on a Varian 1800 gas chromatograph coupled via a molecular separator to an AEI-MS-30 double beam mass spectrometer.

4.12 Nitriloacetic acid

Zoccolilo and Ronchetti [53] have described a method for determining nitriloacetic acid in water, which involves extraction, conversion to the trimethyl ester and determination by capillary chromatography.

Column chromatography

Various workers [54–57] have pointed out that only high-performance liquid-chromatographic methods have sufficient sensitivity for the determination of EDTA.

Nowack et al. [58] have described a procedure for the determination of dissolved EDTA in rivers and waste water treatment plants. The procedure can be made between Fe(III)–EDTA and all other species. Nickel–EDTA can be detected semiquantitatively. After complexation with Fe(III)–EDTA, the EDTA is detected by reversed-phase ion-pair liquid chromatography as the Fe(III)–EDTA complex at a wavelength of 258 nm. The behaviour of a number of metal–EDTA complexes during analysis was checked. Fe(III)–EDTA was found to be the main species (60%–70%); Ni–EDTA was less than 10% in most samples.

Polarography

Dietz et al. [59] used polarography to determine nitriloacetic acid, EDTA and other complexing agents in surface and ground waters using bismuth complexes at pH 2. Concentrations of nitriloacetic acid and EDTA in the range of 0.1–3 $mg\ L^{-1}$ could be determined selectively using cathode ray, impulse or modified alternating voltage polarography.

Fayyad et al. [60] described an indirect potentiometric stripping analysis method for determining EDTA as its 1:1 bismuth complex in natural waters.

Atomic absorption spectrometry

Kunkel and Manahan [61] have described an atomic absorption method for determining strong heavy-metal chelating agents, such as EDTA and nitriloacetic acid in natural and waste waters. The method involved solubilisation of cupric ions (added as 0.5 M $CuCO_4$) by the chelating agents at pH 10 in the boiling solution, filtration of the cool mixture and then determination of cupric ions in the filtrate. The concentration of total strong heavy-metal chelating agents is proportional to the amount (in mg) of copper chelated in a standard volume of sample.

Gas chromatography

Gallasi et al. [62] Q5 have described a procedure for the determination of down to 0.7 $ng\ L^{-1}$ of nitriloacetic acid in surface waters. This method involved concentration on an ion-exchange column. Recovery of the analyte from the exchange column with formic acid followed esterification and chromatographic analysis of the butyl esters.

Chau and Fox [63] concentrated nitriloacetic acid in lake water samples by passing them down a Dowex 1 column (formate form) and elution with 2.5 and 8 M formic acid. The nitriloacetic acid is then esterified, with heptadecanoic acid added as internal standard, by heating for 1 hour at 100°C in a sealed ampoule with propanol saturated with hydrogen chloride. The propyl esters are analysed on a stainless steel column (1.8 m × 6 mm) packed with 3% of OV-1 on Chromosorb WHP (80–100 mesh). Temperature programmed from 180 to 225°C min^{-1}.

The limit of detection is 0.01 μg L^{-1}, the standard deviation was 1.33 μg L^{-1} and the coefficient variation was ±6.3%.

Warren and Malek [64] determined nitriloacetic acid and related aminopolycarboxylic acids (iminodiacetic acid, glycine and sarcosine) in inland waters and sewage effluents by converting to the butyl *N*-trifluoracetyl esters followed by chromatography on duel glass U-shaped columns (1.9 m × 2 mm) packed with 0.65% of ethanediol adipate on acid-washed Chromosorb W. Aue et al. [65] claimed a limit of detection of 1 ppb nitriloacetic acid for a 50 mL water sample. The nitriloacetic acid was converted to its butyl ester prior to GC using a nitrogen-specific detector, which gave a linear response over the range 1–1000 ng injected of the tri-*n*-butyl ester of nitriloacetic acid.

Murray et al. [66] and Schaffner and Giger [67] and Stolzberg and Hume [68] have also investigated the application of GC to the determination of traces of nitriloacetic acid in natural waters.

Polarography

Voulgarpoulos et al. [69] carried out indirect determination of nitriloacetic acid and EDTA in natural waters by differential pulse anodic stripping voltammetry. Both substances are determined by complex formation with excess trivalent bismuth ions at pH 2 followed by a back-titration of the excess using a hanging mercury drop electrode at a potential of −0.15 V. The peak current due to uncomplexed bismuth from the Bi–NTA complex can be recorded independently of the accounted for by EDTA. The lower limit of determination is 9.2 μg L^{-1} nitriloacetic acid for the deposition of 2 minutes. Interference due to ferric ions may be eliminated by the addition of ascorbic acid as a reducing agent, but copper ions, if present in excess of 40 μg L^{-1} must first be removed by preelectrolysis with a mercury pool electrode.

Wernet and Wahl [70] removed interfering cations and heavy metals that form complexes with nitriloacetic acid from surface water and effluent samples by equilibrating at pH 3 with Dowex 50W-8X resin (sodium form) prior to polarography in the presence of ammoniacal cadmium buffer solution at pH 8.

Haberman [71] determined down to 0.02 ppm of nitriloacetic acid in water and sewage by first passing the sample, adjusted to pH 3, through a cation-exchange resin After adjustment to pH 7.0, the percolate is then passed through an anion exchange to absorb the nitriloacetic, which is subsequently eluted with sodium chloride in 0.1 M acetic acid buffer of pH 4.7. Nitriloacetic acid is then determined by adding a known amount (in excess) of trivalent indium to the solution and measuring the height of the wave due to reduction of the In^{3+}-nitriloacetic acid complex at −0.79 V versus the SCE. An isotope dilution technique usually ^{14}C-labelled nitriloacetic acid is used to correct for incomplete recovery.

Low concentrations (1−10 ppm) of nitriloacetic acid have been determined in lake water by a method [72], which involved conversion to its 1:1 cadmium complex. The resulting solution (adjusted to pH 9) is subjected to polarography between −0.15 and 1.2 V (vs the SCE) in 0.1 M potassium nitrate, as supporting electrolyte; Cd^{2+} gives a wave at −0.060 V, and the Cd−nitriloacetic acid complex gives a wave of 0.97 V. To determine nitriloacetic in submicromolar concentrations it is advantageous to use a pH of 8.5 to avoid precipitation of cadmium hydroxide.

Afghan et al. [73] developed an automated method for the determination of nitriloacetic acid in natural water samples. The method is based on the formation of the bismuth nitrilotriacetic acid complex at pH 2 followed by determination by twin cell oscillographic DC polarography. As little as 10 $\mu g\ L^{-1}$ of nitrilotriacetic acid can be determined with no preconcentration of the sample being required. The coefficient of variation for 100 $mg\ L^{-1}$ was 1.3%.

Haring and Van Delft [74] studied the application of derivative pulse polarography at a hanging mercury drop electrode to the determination of nitriloacetic acid in water.

Wernet and Wahl [70] removed interfering cations and heavy metals that form complexes with nitriloacetic acid from the surface water and effluent samples by equilibrating at pH 3 with Dowex 50W-X8 resin (sodium form) prior to polarography in the presence of ammoniacal cadmium buffer solution at pH 8.

Afghan and Goulden [75] used potential sweep chronoamperometry of the nitriloacetic acid-lead complex to determine down to 10 $\mu g\ L^{-1}$ of nitriloacetic acid in water.

To determine total nitriloacetic acid the sample is acidified to pH 1 to release the nitriloacetic acid from heavy metal complexes, and these metals are masked by adding EDTA before bringing the pH back to 8 to form the lead complex. Large amounts of bis-(2-aminoethocyl)-ethane-*N,N,N′N′*-tetraacetic acid interferes in this method.

Spectrophotometric method

Coombs et al. [76] analysed mixtures of aminopolycarboxylic acids at the ppb level by chemical kinetics. The method is based on the reaction of the nickel complexes of these acids with cyanide ion and the large differences in reaction rates for the formation of $Ni(CN^{2-})_4$. The reaction is monitored spectrophotometrically (at 267, 285 or 310 nm) with use of a stopped-flow system, and the results are calculated from a computer program that provides on-line data acquisition and performs a regressive differential kinetic analysis for the two components. The acids can be determined singly or up to three in admixture. The error is within $\pm 5\%-10\%$ for mixtures of ligands at the micromolar level, and the sensitivity is 0.04 μm (8 parts per 10^9) in water.

4.13 Miscellaneous nitrogen compounds

Other organic compounds that have been determined in environmental waters include: pyridine bases [77], monoalkyl quaternary compounds [78], *N*-chloroamine [79], hydrazines [80], urea [81], azarenes [82], 1-methyl-2-pyrrolidine [83], diazo compounds [84], organic nitrogen [85], heterocyclic nitrogen compounds [86], peroxyacetyl nitrate [87] and sulphonated melamine-formaldehyde condensates [88].

The latter method Rojan et al. [88] developed a separation detection procedure using reversed-phase HPLC coupled with ultraviolet and fluorescence detection for the determination of sulphonated melamine-formaldehyde condensates and liquid sulphonates in environmental water samples with a detection limit of 0.05–0.3 μg. The extraction/enrichment procedure was based on solid-phase microextraction using polystyrene–divinylbenzene eriochrome P resin as sorbent. Approximately 70% recovery was achieved from aqueous samples at starting concentration in the 10 $\mu g\ L^{-1}$ range.

A structure confirmation of styrene–divinyl benzene was established by HPLC–MS with electrospray interference by using collision-induced ion fragmentation (in source CID mode). This procedure was applied to the determination of sulphonated styrene–melamine-formaldehyde condensates and sulphonic acids in ground waters.

Svabensky et al. [86] have carried out a detailed study of the application of HPLC built on octadecyl stationery phase to the determination of polycyclic aromatic nitrogen hetrocyclic in water.

The retention of a mixture of these compounds was studied under different chromatographic conditions. A mixture of phosphate buffer/acetonitrile was used as mobile phase in isocratic and gradient modes. The effect of different pH mobile phase in the range from 2.5 to 6.5 was investigated to describe retention changes of the heterocycles as a function of their acid/base properties. Different concentrations of phosphate buffer as a component of the mobile phase were used to study the effect of ionic strength. Very good reverse phase-HPLC separation of 24 nitro hetrocyclics obtained by two detection techniques, namely 16 polycyclic automatic hydrocarbons were ultraviolet-diode array detection and fluorescence detector are compared.

References

[1] H.P. Hermanson, K. Helrich, W.F. Carey, Anal. Lett., London 1 (1968) 941.
[2] J. Metera, V. Rabhl, J. Mostecky, Water Res. 10 (1976) 137.
[3] F.I. Onuska, Water Res. 7 (1973) 835.
[4] M. Cranmer, Bull. Environ. Contam. Toxicol. 5 (1970) 329.
[5] W.I. Kimoto, C.J. Dooley, J. Canne, W. Fiddler, Water Res. 14 (1980) 869.
[6] M.J. Avery, G.A. Junk, Anal. Chem. 57 (1985) 790.
[7] K. Nishikawa, J. Chromatogr. 392 (1987) 319.
[8] F.G. Dyat-Lovitskaya, L.W. Botvinova, Khim. Volkna. 1 (1970) 64. Ref. Z. Khim. 19GD (13) Abst. No 13G308.
[9] T.M. Florence, Y.J. Farner, Anal. Chim. Acta 63 (1973) 255.
[10] R.C.C. Wegman, G.O.A. DeKorte, Int. J. Environ. Anal. Chem. 9 (1981) 1.
[11] Z. Bao, U. Zhao, Huan Jing Huayine 6 (1987) 67.
[12] Q. Zhou, L. Pang, G. Yie, J. Xiao, P. Li, H. Bai, et al., J. Int. Environ. Anal. Chem. 90 (2010) 1099.
[13] T. Wanatabe, A. Hongu, K. Honda, M. Nakazato, M. Kanno, S. Saitoh, Anal. Chem. 56 (1984) 251.
[14] M.A. El-Dib, J. Assoc. Anal. Chem. 54 (1971) 1383.
[15] M.W.F. Nielen, R.W. Frei, U.A.T. Brinkman, J. Chromatogr. 317 (1984) 557.
[16] H. Stuber, J.A. Henloar, Anal. Chem. 55 (1983) 111.
[17] W.G. Gardner, G.F. Lee, Environ. Sci. Technol. 7 (1973) 719.
[18] J. Dai, G.G. Helz, Anal. Chem. 9 (1988) 44.
[19] M. Yu, N. Dovich, Anal. Chem. 61 (1989) 87.
[20] W.G. Gardner, Mar. Chem. 6 (1978) 15.
[21] G.H. Arkell, B.I. Croll, The Determination of Acrylamide in Water, The Water Research Association Medmenham, Harlow, UK, 1972, Report TP78.
[22] B.T. Croll, G.M. Simpkins, Analyst 97 (1972) 281.
[23] A. Hashimoto, Analyst 101 (1976) 932.
[24] E.R. Husser, R.H. Stehl, D.R. Price, R.A. DeLap, Anal. Chem. 49 (1977) 154.
[25] F.J. Ludwig, M.F. Besand, Anal. Chem. 50 (1978) 185.
[26] L. Brown, M. Rhead, Analyst 104 (1979) 391.
[27] B.T. Croll, G.M. Arkell, R.P. Hodge, Water Res. 8 (1974) 989.
[28] G. Bengtsson, J. Chromatogr. Sci. 23 (1985) 397.
[29] P. Chambou, R. Chambon, J. Chromatogr. 87 (1973) 287.
[30] C.U. Makarova, Cig. Sanit, Zh. Khim. 5 (1971) 61.
[31] J.C. Hoffsammer, D.J. Glover, C.V. Hazzard, J. Chromatogr. 195 (1980) 435.
[32] E. Hicks, J.P. Riley, Anal. Chim. Acta 116 (1980) 137.
[33] B. Hortstroffe, O. Elsholz, V.C. Martin, Int. J. Environ. Anal. Chem. 87 (2007) 797.
[34] C.A. Heller, G.R. Greni, E.D. Ericson, Anal. Chem. 54 (1982) 286.
[35] K.K. Verma, D. Gupta, Analyst 199 (1987) 33.
[36] D.R. Shankaran, T. Kawaguchi, S.J. Kim, K. Matsumoto, K. Tako, N. Miura, Int. J. Environ. Anal. Chem. 87 (2002) 771.
[37] O.E. Tall, D. Deli, A. Jattrezik-Renault, O. Vittori, Int. J. Environ. Anal. Chem. 90 (2012) 40.
[38] R.N. Lx Tian Fu, Int. Jahrstug Freunhofer Inst. Tech. Exploivist, Propellant Explos. Chem. Phys. Method (1979) 63/1–63/9.

[39] L.T. Coulte, E.E. Hargechinger, F.M. Paento, G.B. Baker, J. Chromatogr. Sci. 19 (1981) 151.
[40] A. Di Corcia, R. Samperi, Anal. Chem. 62 (1990) 1490.
[41] J. Fan, J. Kong, S. Feng, F. Wang, P. Peng, Int. J. Environ. Anal. Chem. 86 (2006) 995.
[42] L. Tang, G.M. Zeng, Y.H. Yang, G.U. Shen, G.H. Huang, R.G. Niu, et al., Int. J. Environ. Anal. Chem. 85 (2005) 111.
[43] J.M. Walren, R.K. Bcasley, Anal. Chem. 56 (1984) 1963.
[44] T. Stefanescu, G. Ursu, Mater. Plate 10 (1973) 330.
[45] Z. Ghersin, H. Stitzi, R. Wanea, Rev. Chim. 20 (1969) 689.
[46] M.M. Nikaido, D.D. Raymond, A.J. Francis, M. Alexander, Water Res. 11 (1977) 1085.
[47] B.A. Tompkins, W.H. Griest, Anal. Chem. 68 (1996) 2533.
[48] B.A. Tompkins, W.H. Griest, C.W. Higgins, Anal. Chem. 67 (1995) 4387.
[49] A.I. Mills, M.J. Alexander, Environ. Qual. 5 (1976) 437.
[50] D.B. Fulton, B.G. Sayer, A.D. Bain, H.V. Malle, Anal. Chem. 64 (1992) 349.
[51] J. Gardiner, Analyst 102 (1977) 120.
[52] H.B. Lockhart, R.W. Blakely, Environ. Sci. Technol. 9 (1975) 1035.
[53] I. Zoccolilo, M. Ronchatti, Anal. Chem. 77 (1987) 735.
[54] D.L. Venezby, W.E. Rudzinski, Anal. Chem. 56 (1984) 315.
[55] P.J.M. Bergers, A.C. de Groat, Water Res. 28 (1994) 639.
[56] M. Sillanpaa, B. Kokkonen, M.L. Sihvonen, Anal. Chim. Acta 303 (1995) 187.
[57] L. Ye, C.A. Lucy, Anal. Chem. 67 (1995) 2534.
[58] A. Nowack, F.G. Kari, S.U. Hilger, L. Sigg, Anal. Chem. 68 (1996) 561.
[59] F.Z. Dietz, Wasser Abwasser Forsch 7 (1974) 74.
[60] M. Fayyad, M. Tutunji, Z. Taka, Anal. Lett., London 21 (1988) 1425.
[61] R. Kunkel, S.E. Manahan, Anal. Chem. 45 (1973) 1465.
[62] S. Gallasi, T. La Noca, A. De Paolis, Metodi Anal. Aegu 6 (1986) 23.
[63] A. Chau, M.E. Fox, J. Chromatogr. Sci. 9 (1971) 271.
[64] C.B. Warren, E. Malek, J. Chromatogr. 64 (1972) 219.
[65] W. Aue, C.R. Hastings, K.O. Gerhard, J. Chromatogr. 72 (1972) 259.
[66] D. Murray, D. Pavoledo, J. Fish, J. Res. Board Can. 28 (1971) 1043.
[67] C. Schaffner, W. Giger, J. Chromatogr. 312 (1984) 413.
[68] R.J. Stolzberg, D.N. Hume, Anal. Chem. 49 (1977) 374.
[69] A. Voulgarpoulos, P. Valenta, H.W. Nunberg, Fresenius Z. Anal. Chem. 367 (1984) 367.
[70] J. Wernet, K.Z. Wahl, Anal. Chem. 251 (1970) 373.
[71] J.P. Haberman, Anal. Chem. 43 (1971) 63.
[72] J. Asplund, E. Wanninen, Anal. Lett., London 4 (1971) 267.
[73] B.K. Afghan, P.D. Goulden, J.P. Ryan, Anal. Chem. 44 (1972) 354.
[74] J.A. Haring, W. Van Delft, Anal. Chem. 94 (1977) 201.
[75] B.K. Afghan, P.D. Goulden, Environ. Sci. Technol. 15 (1971) 60.
[76] L.O. Coombs, J. Vasiliades, P.N. Margeruno, Anal. Chem. 44 (1972) 2235.
[77] T. Tsukioka, T. Murakami, J. Chromatogr. 396 (1987) 319.
[78] V.T. Wee, J.H. Kennedy, Anal. Chem. 54 (1982) 1631.
[79] F.E. Skully, D.M. Oglesby, H.J. Buck, Anal. Chem. 56 (1984) 1449.
[80] W.D. Basson, J.F. Van Staden, Analyst 103 (1978) 998.
[81] R.T. Emmet, Anal. Chem. 41 (1969) 1648.
[82] T.R. Steinheimer, M.G. Ondrus, Anal. Chem. 58 (1986) 1839.
[83] V.F. Fedonina, N.V. Fedyainov, S.A. Agapova, Sov. J. Water Chem. Technol. 3 (1981) 57.

[84] A.P. Bruins, L.O.G. Weidaft, J.D. Henion, W.L. Budde, Anal. Chem. 59 (1987) 2647.

[85] M. Rogora, M. Minelella, A. Orru, G.A. Tartari, Int. J. Environ. Anal. Chem. 86 (2006) 1065.

[86] R. Svabensky, K. Koci, Z. Simek, Int. J. Environ. Anal. Chem. 87 (2007) 337.

[87] B. Wang, M. Shoo, J.M. Roberts, G. Yang, F. Yang, M.H. Limin, Int. J. Environ. Anal. Chem. 90 (2010) 518.

[88] G. Rojan, C. Carsel, F. Commarata, A. Marcomini, C. Crescenzi, Int. J. Environ. Anal. Chem. 83 (2003) 51.

Chapter 5

Phosphorus containing compounds in nonsaline waters

Chapter Outline

5.1 Alkyl and aryl phosphates 119
5.2 Adenosine triphosphate 120
5.3 Inositol triphosphate 120
5.4 Plytase-hydrolysable phosphate 120
5.5 Phosphine 120
5.6 Organophosphorus compounds 121
Miscellaneous 121
Nuclear magnetic resonance spectroscopy 121
5.7 Organophosphorus insecticides and pesticides 121
References 121

5.1 Alkyl and aryl phosphates

Murray and Fishers [1] have described a gas chromatographic method for the determination in water of triarylphosphate esters (1 mol S-140, tricresyl phosphate, cresol phosphate). These substances are used commercially as lubricant oil and plastic additives, hydraulic fluids and plasticisers. The method involves extraction from the samples, hydrolysis and measurement of the individual phenols by gas chromatography as the trimethylsilyl derivatives. The lower detection limit was about 3 ppm.

In this procedure, a weighed sample was placed in a 20 mL ampoule, and 10 mL of 5% potassium hydroxide in 95% methanol was added. The ampoule was sealed and autoclaved at 25 psi (172 kPa) for 90 minutes. The ampoule was opened, and the contents washed into 950 mL of distilled water, acidified with 6 M hydrochloric acid to pH1 − 2 and made up to 1000 mL after addition of *o*-xylene internal standard and the mixture was extracted with 50 mL of chloroform. The solvent layer was evaporated to 1–2 mL and treated with Tri-Sil concentrate to form the trimethylsilyl derivative. This was allowed to react overnight and analysed by gas chromatography the following day. The concentration of the individual phenols were calculated from calibration graphs, and the composition of the ester was determined.

Flame ionisation gas chromatography was used to determine the phosphate ester.

Determination of Toxic Organic Chemicals In Natural Waters, Sediments and Soils.
DOI: https://doi.org/10.1016/B978-0-12-815856-2.00005-9

Ishikawa et al. [2] have developed procedures for determining alkyl and aryl phosphates in river waters. The procedure involves extraction of the sample with dichloromethane followed by gas chromatography using a flame photometric detector and gas chromatography–mass spectrometry after clean-up through a Florisil column.

5.2 Adenosine triphosphate

The quantity of adenosine triphosphate in water can be used to estimate the living biomass or microorganisms in the water, making it an important assessment of water quality.

Andre et al. [3] discuss the determination of adenosine triphosphate by luciferin-luciferase assay. This method was applied to the determination of adenosine triphosphate in bacterial colonies filtered from samples of polluted water after incubation for different periods. The adenosine triphosphate was extracted from the residue in the filter and the amount compared with the biochemical oxygen demand of the filtered water. The oxygen uptake rate and the rate of formation of adenosine triphosphate were then plotted against time, the two curves being similar up to 3–4 days' incubation, after which adenosine triphosphate production declined markedly, although oxygen uptake continued to increase.

Luminescence analysis has been applied to the determination of adenosine triphosphate. Shoaf and Lium [4] have discussed sampling techniques and sample preservation. Quick freezing of the sample in the acetone dry ice bath maintains the adenosine triphosphate concentration in the sample at a constant level [5].

Tobin et al. [6] give details of two extraction procedures for the determination of adenosine triphosphate in environmental samples by luciferin-luciferase assay.

5.3 Inositol triphosphate

Kuenerowicz and Verstraate [7] have described methods for the determination of inositol phosphate.

5.4 Plytase-hydrolysable phosphate

Omaka et al. [8] have described a method based on enzymatic flow injection for the determination of Plytase-hydrolysable phosphorus in natural waters.

5.5 Phosphine

Geng et al. [9] have reported the presence of traces of phosphine in lake water.

5.6 Organophosphorus compounds

Miscellaneous

Gas chromatographic detection with supported copper–cuprous oxide island film

Organophosphorus compounds have been detected in environmental waters at the parts per million level using a supported copper–cuprous oxide island film gas chromatographic detector. Both alternating current [10] and direct current studies were used [11].

Conducted chemiresistant sensors for gas chromatographic detection

Grate et al. [12] have studied the role of selective sorption in chemiresistant sensors for the gas chromatographic detection of organophosphorus compounds.

Surface acoustic wave sensors for gas chromatographic detection

Grate et al. [13] have described a smart sensor system for detecting traces of organophosphorus compounds. Vapour detection of these compounds is achieved employing a temperature-controlled array of surface wave detectors, automated sampled preconcentrations and pattern recognition. Down to 0.01 mg m^{-3} of organophosphorus compounds can be detected on the detector surface.

Nuclear magnetic resonance spectroscopy

Verweij et al. [14] studied the hydrolysis of some methylphosphonites and methylphosphonates in the pH range 4.5–7.5 by phosphorus-31 nuclear magnetic resonance spectrometry. For methylphosphonites, a two-step reaction mechanism was observed in which formation of the corresponding methylphosphonates was followed by production of methylphosphinic acid. Implications of the results obtained on the applicability of the verification procedure for nerve gasses in surface waters are discussed.

5.7 Organophosphorus insecticides and pesticides

The determination of organophosphorus insecticides is discussed in Section 10.2.

References

[1] D.A. Murray, J. Fishers, Res. Board, Canada 32 (1975) 457.
[2] S. Ishikawa, M. Taketomi, R. Shino Hara, Water Res. 19 (1985) 119.
[3] M. Andre, P. Van Beneden, J. Bassleer, Tribune du Cebedeau 31 (1978) 251.
[4] W.T. Shoaf, B.W. Lium, J. Res. US Geol. Surv. 4 (1976) 241.

[5] L.H. Siegrist, Sch. Z. Hydrol. 34 (1976) 49.
[6] R.S. Tobin, J.F. Ryan, B.K. Afghan, Water Res. 12 (1978) 783.
[7] F. Kuenerowicz, W. Vertstraete, J. Chem. Technol. Biotechnol. 29 (1979) 707.
[8] N. Omaka, M. Keith-Roach, I.D. McKelvie, P.J. Worsfold, Int. J. Environ. Anal. Chem. 88 (2008) 91.
[9] J. Geng, X. Niu, X. Wang, M. Edwards, D. Glindermann, Int. J. Environ. Anal. Chem. 90 (2010) 737.
[10] E.S. Kolesar, R.M. Walser, Anal. Chem. 60 (1988) 1737.
[11] E.S. Kolesar, R.M. Walser, Anal. Chem. 60 (1988) 1731.
[12] J.W. Grate, M. Klusty, W.R. Borger, A.W. Snow, Anal. Chem. 62 (1990) 1927.
[13] J.W. Grate, S.L. Rose Pehrsson, D.L. Venezky, M. Klusty, H. Wohitjen, Anal. Chem. 65 (1993) 1868.
[14] A. Verweij, W.H. Dekker, H.C. Beck, H.L. Boter, Anal. Chim. Acta 151 (1983) 221.

Chapter 6

Sulphur-containing compounds in nonsaline waters

Chapter Outline

6.1 Mercaptans and disulphides 123
Gas chromatography 123
Titration method 123
6.2 Dimethyl sulphoxide 123
6.3 Alkylthiols 124
6.4 Ethylene thiourea 124
6.5 Thiobenzamide 125
6.6 Chlorobenzo sulphonic acid 125
Pyrolysis-gas chromatography–mass spectrometry with single-ion monitoring 125
6.7 Miscellaneous 125
References 125

6.1 Mercaptans and disulphides

Gas chromatography

Leek and Bagander [1] determined reduced sulphur compounds in natural waters by gas chromatography using a flame ionisation detector. Substances determined include methyl mercaptan, dimethyl sulphide, hydrogen sulphide and carbon disulphide. Detection limits ranged from 0.2 mg L^{-1} (carbon disulphide) to 0.6 ng L^{-1} (methyl mercaptan).

Titration method

Duane and Stock [2] have discussed a titration method for determining thiols.

6.2 Dimethyl sulphoxide

Simo et al. [3] have discussed the determination of nanomolar concentrations of dimethyl sulphoxide, along with dimethyl sulphide and dimethyl sulphoniopropionate, at nanomolar levels in nonsaline waters. After removal of dimethyl sulphide by purge and cryotrapping, dimethyl sulphoniopropionate is removed by the same method after alkaline hydrolysis, and dimethyl sulphoxide is reduced to dimethyl sulphide using a combination of sodium borohydride and hydrochloric acid. The dimethyl sulphide produced is

Determination of Toxic Organic Chemicals In Natural Waters, Sediments and Soils.
DOI: https://doi.org/10.1016/B978-0-12-815856-2.00006-0

stripped, cryotrapped and analysed by gas chromatography. Detection of 3 pmol of dimethyl sulphoxide was achieved, resulting in a detection limit of 0.05 nM for a 50 mL sample.

Andreae [4] has described a gas chromatographic method for the determination of nanogram quantities of dimethyl sulphoxide in nonsaline waters, seawater and phytoplankton culture waters. The method involves a chemical reduction to dimethyl sulphide, which is then determined by gas chromatographically using a flame photometric detector.

Andreae [4] investigated two different apparatus configurations. One consisted of a reaction-trapping apparatus connected by a six-way valve to a gas chromatograph—equipped ionisation detector; the other apparatus combined the trapping and separation functions in one column, which was attached to a flame with a flame photometric detector. The gas chromatographic flame ionisation detector system was identical to that described by Andreae [4] for analysis of methylarsenicals, with the exception that a reaction vessel which allowed the injection of solid sodium borohydride pellets was used. The flame photometric system is a modified version of a design by Braman et al. [5].

6.3 Alkylthiols

Salgado-Petinal et al. [6] have described a method for the determination of alkylthiols in water by in situ derivatisation then solid-phase microextraction followed by gas chromatography with an electron capture detector. The dinitrobenzylation reaction was selected because the high chemical stability of the corresponding thioethers formed provides a significant increase in the distribution coefficient between the solid-phase microextraction fibre and the aqueous phase, and a potential increase in the selectivity and sensitivity. Different derivatisation reaction conditions (i.e. pH, temperature, reaction time and derivatisating reagent concentration) were the main parameters affecting the solid-phase microextraction process, after coating selection, extraction time profile, extraction and desorption temperatures had been optimised. Salgado-Petinal et al. [6] used a method based on a simple 2,4-dinitropheynlation reaction at pH8—10, in 60 minutes at 75°C, coupled to direct solid-phase microextraction using PDMS-DVB fibres at 30°C for 45 minutes. The performance of the method provided good linearity and precise data, and the detection limits were in the low nanogram per litre level.

6.4 Ethylene thiourea

Vad De Poll et al. [7] have described a method based on gas chromatography with an alkali flame detector for the determination of ethylene thiourea in river waters which is based on extraction with methylene dichloride in the presence of sodium ascorbate.

6.5 Thiobenzamide

Brandt and Kettrup [8,9] showed it was possible to determine the parameters of adsorbable organic sulphur compounds by a technique employing pyrohydrolysis or organics followed by adsorption and ion chromatographic detection of the resultant sulphate. Pyrohydrolysis apparatus, combustion conditions and adsorption of the organic anions were optimised for the complete conversation of model compounds such as a thiobenzamide and toluene sulphonic acid and maximal anion recoveries.

6.6 Chlorobenzo sulphonic acid

Pyrolysis-gas chromatography–mass spectrometry with single-ion monitoring

Van Loon et al. [10] have used this technique to determine chlorolignosulphonic acids of molecular weight greater than 1000 in natural waters in amounts down to 0.1 $\mu g\,L^{-1}$. Between 18 and 310 $\mu g\,L^{-1}$ of this substance were present in River Rhine Waters.

6.7 Miscellaneous

Grate et al. [11] have discussed the smart sensor system for detecting down to 0.5 $mg\,M^{-1}$ organosulphur compounds in environmental waters. This consisted of a temperature controlled array of surface acoustic wave sensors, with automated sample preparation and pattern recognition.

Organosulphur compounds, for example tetrahydrothiophen have been preconcentrated on XAD-2 and XAD-4 macroreticular resins prior to solvent desorption and analysis by headspace gas chromatography [12].

Chen and Naidu [13] detected sulphur species in water by coelectroosmotic capillary electrophoresis with direct and indirect UV detection.

Henatsch and Hunter [14] determined ethane diol, dimethyl sulphide and carbon disulphide in environmental waters.

References

[1] C. Leek, L.E. Bagander, Anal. Chem. 60 (1988) 1680.
[2] L. Duane, J.T. Stock, Anal. Chem. 50 (1978) 1891.
[3] R. Simo, J.O. Grimalt, J. Albaiges, Anal. Chem. 68 (1996) 1493.
[4] M.O. Andreae, Anal. Chem. 52 (1980) 150.
[5] R.S. Braman, J.M. Ammons, J.L. Bricker, Anal. Chem. 50 (1978) 992.
[6] C. Salgado-Petinal, R. Alzaga, C. Garcia-Janes, M. Llompart, J.M. Bayona, Int. J. Environ. Anal. Chem. 85 (2005) 543.
[7] Vad De Poll, G.G. Vershuis-Haas, O.J. Wilde, J. Oceanogr. 643 (1993) 163.
[8] G. Brandt, A. Kettrup, Int. J. Environ. Anal. Chem. 31 (1987) 129.
[9] G. Brandt, A. Kettrup, Fresenius Z. Anal. Chem. 326 (1987) 213.

[10] W.M.G.M. Van Loon, J.S. Bron, B. De Groot, Anal. Chem. 65 (1993) 1726.
[11] L. Grate, S.L. Pehrsson, D.L. Vene, M. Khasy, H. Wahitjen, Anal. Chem. 65 (1993) 1868.
[12] C. Garlucci, L. Aroldi, R. Farelli, J. Chromatogr. 287 (1984) 425.
[13] Z. Chen, R. Naidu, Int. J. Environ. Anal. Chem. 83 (2003) 749.
[14] L.L. Henatsch, E. Hunter, J. Chromatogr. 445 (1988) 97.

Chapter 7

Surface active agents in nonsaline waters

Chapter Outline

7.1 **Nonionic surface active agents** **127**
Gas chromatography 127
Column chromatography 128
Ion-exchange chromatography 129
Spectrophotometric methods 130
Atomic absorption spectrometry 132
Miscellaneous 133
7.2 **Anionic surface active agents** **133**
Gas chromatography 133
High-performance liquid chromatography 134
Miscellaneous 135
7.3 **Cationic surface active agents** **136**
High-performance liquid chromatography 136
Gas chromatography 136
Titration methods 136
Spectrophotometric methods 136
Miscellaneous 137
Titration methods 138
Spectrophotometric methods 138
References **139**

7.1 Nonionic surface active agents

Gas chromatography

Favretto et al. [1] applied gas–liquid chromatography to an evaluation of the polydispersity of polyoxyethylene nonionic surfactants. The water sample was extracted with 1,2-dichloroethane, and the combined organic layers were extracted with 0.10 M sulphuric acid and then with 1.10 M sodium hydroxide solution. The purified extract was concentrated under vacuum transferred into a glass minivial, evaporated to 10 mL by means of a stream of nitrogen. An aliquot of this solution was injected into the gas chromatographic column.

Stephenou et al. [2] identified nonionic surfactants in natural water by gas chromatography coupled with chemical ionisation mass spectrometry. Tertiary octylphenol and lauryl alcohol ethoxylates were qualitatively detected using their chemical ionisation mass spectra, and the results used to compare the chemical ionisation and electron impact mass spectrometry techniques. In the case of alkylphenol ethoxylates, electron-impact-induced mass spectra can provide reliable structural identification for the lowest homologues, and

Determination of Toxic Organic Chemicals In Natural Waters, Sediments and Soils.
DOI: https://doi.org/10.1016/B978-0-12-815856-2.00007-2

although gas chromatography–chemical ionisation mass spectrometry gives more reliable molecular weight information on the higher homologues, ease of operation still favours the gas chromatography–electron impact approach. In the case of linear alcohol ethoxylates, electron impact spectra are of less value, while the chemical ionisation technique allows the reliable determination of molecular weight and some structural characteristics.

Column chromatography

Huber et al. [3] have described a column chromatographic method for the rapid separation and determination of nonionic detergents of the polyoxyethylene glycol monoalkylphenyl type and applied it to the analysis of water samples. Such additions ranging in chain length from 1 to 20 ethylene oxide units were separated on columns of PE on Spherosil with three mobile phases. Use of the first mobile phase permitted separation of the oligomers produced by the reaction between 1 mol of phenol and from 3 to 7 mol of ethylene oxide, and of the second mobile phase, those produced similarly from 7 to 14 mol of ethylene oxide (but with no separation on the shorter oligomers). With the use of the third mobile phase, there was no separation of oligomers, but for each sample, a peak of Poisson distribution was obtained that represented the oligomers distribution of the sample. Detection was by ultraviolet absorption; the relative precision was 0.4%, and the detection was 0.2 μg.

Otsuki and Shiraishi [4] used reversed-phase adsorption liquid chromatography and field desorption mass spectrometry to determine polyoxyethylene alkylphenyl ether nonionic surfactants in water.

From the field desorption mass spectra of standard samples, a table for identification of poly(oxyethylene)alkylphenyl ethers and determination of the degree of polymerisation of ethylene oxide was constructed. When the field desorption mass spectrum having a peak pattern with the difference of 33 *m/z* was obtained such as the peaks at 484, 528, 572, 616 and 660 *m/z*. These peaks are due to poly(oxyethylene)-nonylphenyl ethers with the degree of polymerisation of $m = 6-10$.

Otsuki and Shiraishi [4] concluded that mixtures of poly(oxyethylene)-alkylphenyl ether could be separated by the affinity between the chain length of the alkyl group in the poly(oxyethylene) ethers and the octadecyl group of the bonded stationary phase, and the solubility in the solvent at the mobile phase composition during gradient elution. This reversed-phase adsorption chromatography involves the injection of the sample either by direct injection or by pumping a certain volume through the column.

Evans et al. [5] carried out a quantitative determination of linear primary alcohol ethoxylate surfactants in environmental waters by thermospray liquid chromatography–mass spectrometry.

Ion-exchange chromatography

Nickless and Jones [6] and Musty and Nickless et al. [7,8] evaluated Amberlite XAD-4 resin as an extractant for down to milligram per litre levels of polyethylated secondary alcohol ethoxylates, [$R(OCH_2CH_2)nOH$] surfactants and their degradation products from water samples. This resin was found to be an effective adsorbent for extraction of polyethylated compounds from water except for polyethylene glycols of molecular weight less than 300.

Nickless and Jones [6] used in this study fairly pure secondary ethoxylate standards [$R(OCH_2CH_2)nOH$], where *n* is 3, 5, 7, 9 and 12, and also alkylphenol ethoxylate standard [$RC_6H_4(OCH_2CH_2CH)$] where *n* is 5, 7, 8 and 9, and polyethylene glycols [$H(OCH_2CH_2)nOH$] is where *n* is 2, 3, 9 and 22.

Results obtained by this procedure indicated a fairly rapid drop in adsorption efficiency for the resin of polyethylene glycols between 9EO and 3EO. It is not clear if the drop in efficiency is roughly linear with a shortening in chain length, but it appears that the resin might be limited to the study of polyethylene glycols in water for chain length greater than 7EO.

Jones and Nickless [9] continued their study of Amberlite XAD-4 resins by examining polyethoxylated materials before and after passage through sewage works and from the adjacent river were subjected to a three-stage isolation procedure, and the final extracts were separated into a nonionic detergent and a polyethylene glycol. The nonionic detergent concentration was 100 times lower (8 ng mL^{-1}) in river than in sewage effluent. Thin-layer chromatography and ultraviolet, infrared and NMR spectroscopy were used to identify the nonionic detergent compound, alkylphenol and ethoxylates (the most persistent), secondary alcohol ethoxylate and primary alcohol ethoxylate.

Cassidy and Niro [10] have applied high-speed liquid chromatography combined with infrared spectroscopy for the analysis of polyoxyethylene surfactants and their decomposition products in industrial process waters. Molecular sieve chromatography combined with infrared spectrometry gives a selective method for the analysis of trace concentrations of these surfactants. These workers found that liquid–solid chromatography and reversed-phase chromatography are useful for the characterisation and analysis of free fatty acids.

Okada [11] used indirect conductiometric detection of poly(oxyethylene) surfactants following chromatography separation.

Zgola-Grzeskovich and Grzeskowiaki [12] have described a simple and robust method for analysis of octyl- and nonylphenol as well as their short-chained ethoxylates in river water. Quantification of these analytes was performed by high-performance liquid chromatography with fluorescence detection (HPLC–FLD) after isolation using solid-phase extraction with polytetrafluoroethylene sorbent. The method allowed one to obtain about 80%–100% recovery for octylphenol and its ethoxylates and 70%–80% for nonylphenol and its ethoxylates. Also, there was no need for additional sample cleaning before chromatographic analysis. The limit of detection was 0.01 $\mu g L^{-1}$ for octylphenol and its ethoxylates and 0.03 $\mu g L^{-1}$ for nonylphenol together with their short-chained ethoxylates. Nonylphenol, nonylphenol mono- and diethoxylates were detected at concentrations ranging from 0.12 to 0.53 $\mu g L^{-1}$. Octylphenol, octylphenol mono- and diethoxylates were detected in 4 out of 11 samples at concentrations ranging from 0.03 to 1.17 $\mu g L^{-1}$. High concentrations of nonylphenol and its ethoxylates were found in the samples, despite the fact that their use in European countries was forbidden several years ago.

Spectrophotometric methods

Brown and Hayes [13] and Grieff [14] have described spectrophotometric methods for nonionic surfactants based on the ammonium cobaltothiocyanate.

Various other workers [15–18] have investigated solvent extraction–spectrophotometric methods for the determination of polyoxyethylene nonionic surfactants in water. Dozanska [19] has described a spectrophotometric method involving reaction with cobalt thiocyanate or tungstophosphoric acid for the determination of nonionic detergents. Hey and Jenkins [20] have also reported on spectrophotometric methods utilising cobalt thiocyanate.

Crabb and Persinger [21] have determined the molecular extinction coefficients of the cobalt thiocyanate complexes of nonylphenol-ethylene oxide used in the spectrophotometric determination of the latter. These workers showed that compounds containing fewer than three ethylene oxide units do not form a colour, and also that for compounds of low molecular weight, the apparent molecular extinction coefficient does not vary rectilinearly with chain length.

Belen'skii et al. [22] have applied the ammonium cobaltothiocyanate spectrophotometric method for the determination of nonionic surfactants of the oxythylated alkylphenol type. Activated charcoal removes from the solution both the dyes and the surfactants; the surfactants can then be separately desorbed by treatment of the charcoal with benzene–chloroform–light-petroleum–ethanol (4:2:2:1) and determined by the ammonium cobaltothiocyanate method.

The barium chloride–phosphomolybdic [23] and spectrophotometric [24] method for nonionic surfactants involves the formation of a nonstoichiometric complex between the polyethoxylated compound, barium chloride and phosphomolybdic acid. The complex precipitates quantitatively form an aqueous acidic solution. It is then isolated by centrifugation and redissolved in 2-methoxyethanol, and its absorbance is measured at 310 nm.

Favretto et al. [1,25,26] in a series of papers have described spectrophotometric methods for nonionic surfactants based on the formation of a sodium surfactant adduct.

In a modification of their original method [27] devised to overcome interference by anionics and cationics and improve sensitivity, Favretto et al. [1,25,26] consider particularly the analysis of polyoxyethylene-*n*-dodecyl, the composition of which has been evaluated by gas–liquid chromatography.

It is preferable to express the concentration of nonionics as potassium picrate active substances. The concentration of potassium picrate active substances is usually referred to a standard monodisperse, and Favretto et al. [25] suggested the use of $C_{12}E_6$ as a standard for the polyoxyethylene *n*-alkyl ethers.

In the pH range 5.2–12.8 the efficiency of the extraction of the surfactant (at a concentration of 1.0 mg L^{-1}) from water by means of this procedure is 98%–99%.

Anionic surfactants interfere in the spectrophotometric determination of nonionic surfactants, because the colourless surfactant anion competes with the yellow picrate in establishing equilibrium in the two-phase extraction system [28] causing negative errors. No interference is observed up to 1.0 mg L^{-1} of sodium dodecyl sulphate, but the absorbance of the organic extract decreases with the concentration of anionic surfactant.

The interference of anionic surfactants is eliminated if the nonionic surfactant is previously extracted with 1,2-dichloroethane. At least in the pH range 5.2–12.8 and up to 10 mg L^{-1} of anionic surfactants in the aqueous phase, this solvent does not extract the latter; the test for methylene blue active substances gives a negative response on the extraction residue redissolved in water. However, when the concentration of the anionic surfactant is greater than 10 mg L^{-1}, the recovery of the nonionic surfactant is good.

Stancher and Tunis [29] have investigated the reaction mechanism in the determination of nonionic surfactants in natural waters as potassium picrate active substances.

Crisp et al. [30] discuss a method for determining nonionic surfactants in natural waters, which is based on the extraction of surfactant molecules into 1,2-dichlorobenzene as a neutral adduct with potassium tetrathiocyanatozincate followed by spectrophotometric determination.

Petts and Sliney [31] have described an autoanalyser spectrophotometric determination of nonionic surfactants in water, using ammonium cobaltothiocyanate as the chromogenic reagent.

Zhu et al. [32] have described a novel direct spectrophotometric method to determine traces of polyoxyethylene nonionic surfactants (such as Triton X-100) in environmental water. It is based on the formation of a ternary complex, mesotetra (3,5-dibromo-4-hydroxylphenyl) porphyrin [T(DBHP)P]–Pb(II)–Triton X-100 in sodium hydroxide medium. Under the optimum reaction conditions, mesotetra (3,5-dibromo-4-hydroxy phenyl)porphy–Pb II complexes react with Triton X-100 to form a yellow ternary complex immediately. The complex, which has a maximum adsorption peak at 479 nm, is stable for at least 24 hours. The main molar adsorption coefficient and the limit of detection for Triton X-100 are 1.1×10^4 L mol^{-1} cm^{-1} and 0.02 μg mL^{-1}, respectively. Beer's law is obeyed in the concentration range of 0–0.5 mg mL^{-1} Trion X-100. It is found that all the studied coexisting substances, especially cationic and anionic surfactants, which seriously interfere in some reported methods, can be tolerated in considerable amounts. The method is used for the direct determination of polyoxyethylene in environmental water and has good precision and accuracy.

Atomic absorption spectrometry

Greff et al. [33] have also described a widely used method for nonionic based on the benzene extraction of the ammonium tetrathiocyanatocobaltate(II) complex followed by atomic absorption spectrometry at 320 nm. The method is simple, rapid, well suited to the routine analysis of large numbers of samples, and applicable to surfactant concentration in the range 0.1–20 mg L^{-1}. With the extraction system, it is necessary to saturate the water samples with sodium chloride to improve extraction efficiency. A more serious disadvantage is interference by ionic surfactants.

Courtot Coupez and Le Bihan [34] determined the optimum pH (7.4) for extraction of nonionic surfactants with the tetrathiocyanatocobaltate(II) $(NH_4)_2[Co(SCN)_4])^-$ complex-benzene system. Cobalt in the extract is estimated by atomic absorption spectrometry after evaporation to dryness and dissolution of the residue in methyl isobutyl ketone. The method is applicable to surfactant concentrations in the range of 0.02–0.5 mg L^{-1} and is not seriously affected by the presence of anionic surfactants.

Arpino et al. [35] performed microdeterminations of nonionic ethoxylated surfactants in surface waters by precipitation with tungstophosphoric acid as described by Burttschell [36] and by the precipitations of tetrathiocyanatocobaltate(II) detergent ion-pair, followed by determination of cobalt in the precipitate by atomic absorption spectrometry as described by Courtot Coupez and Le Bihan [34]. Both the methods were found by Arpino et al. [35] to give erratic results.

Crisp et al. [37] have described a method for the determination of nonionic detergent concentrations between 0.05 and 2 mg L^{-1} in natural waters including sea water based on solvent extraction of the detergent-potassium tetrathiocynatozincate(II) complex followed by determination of extracted zinc by atomic absorption spectrometry. Surfactant molecules are extracted into 1,2-dichlorobenzene as a neutral adduct with potassium tetrathiocynatozincate(II) and the determination is completed by atomic absorption spectrometry.

This method is, however, subject to some interferences.

Thus the presence of up to 5 mg L^{-1} linear alkylbenzene sulphonate increases the apparent concentration of nonionic surfactants by a maximum of 0.07 mg L^{-1} (as Triton X-100). Cationic surfactants, however, form extractable ion-association compounds with the tetrathiocyanatozincate ion which interfere with the method.

Miscellaneous

Thin-layer chromatography of nonionic detergents is carried out by development with acetone or chloroform and spraying with Dragendorff's reagent to reveal pink spots on a pale yellow background [16].

Pojana et al. [38] studied the aerobic biodegradation of aliphatic alcohol polyethoxylate sulphate.

Aurlorite XA D-4 resin [6–8] has been used to preconcentrate nonionic surfactants.

Another method is to convert the nonionic detergent to the cobalt II complex which can then be extracted from aqueous solution with toluene [39].

Linhart [40] has described a method for separating detergents into cationic, anionic and nonionic types on a mixed bed of a cation-exchange resin, for example Lewatit S1020, H^+ form and an anion-exchange resin (Lewatit M5020, OH^- form). He then determined the detergents by a polarographic procedure. This method is based on the damping of the polarographic maximum of oxygen by the surfactants and on the fact that in 1 M potassium chloride, it is not the polarity but the percentage by weight of the surfactants that is proportional to this damping. The proportionality (which is rectilinear over the range 0–100 mg L^{-1}) applied equally to anionic, cationic and nonionic surfactants containing up to 90 ethoxy units and to the polyoxyethylene glycols containing >400 ethoxy units.

7.2 Anionic surface active agents

Gas chromatography

Combined gas chromatography–mass spectrometry has been used [41] to determine trace amounts of the individual components of alkylbenzene sulphonates as their methylsulphonate derivatives. Gas chromatography analysis was performed using a gas chromatograph equipped with a flame ionisation detector.

Waters and Garrigan [42] have described a microsulphonation–gas chromatographic technique for the determination of linear alkylbenzene sulphonates in UK rivers. Sample clean-up, particularly by extracting the linear alkylbenzene sulphonates as the 1-methyl-heptyl amine salt into hexane, resulted in gas chromatographic traces free from significant interferences in which individual linear alkyl-sulphonate isomers can be identified on the basis of relative retention times. The procedure has a limit of 10 μg linear alkyl-sulphonate L^{-1} and permits submicrogram levels of C–C15 homologues to be quantified.

High-performance liquid chromatography

Taylor and Nickless [43] have described a paired-ion high-performance liquid chromatographic technique for the separation of mixtures of linear alkylbenzene sulphonates and *p*-sulphophenylcarboxylate salts in river waters. Partially biodegradable linear alkylbenzene sulphonate was analysed by the same method. Structural information on measurable intermediates formed was provided by stopped flow ultraviolet spectra, comparison of retention behaviour with that of standards and analysis of collected fractions.

Samples (1.51) were concentrated for analysis by acidification with sulphuric acid to pH 2 followed by passage through a column containing 20 mL of XAD-4 resin at a flow rate of 7 mL min^{-1}. Compounds retained by the resin were eluted with 3 × 25 mL portions of methanol, and the combined eluates evaporated to dryness and then up to 2 μL.

High-performance liquid chromatography with fluorimetric detection has been used to determine alkylbenzene sulphonates in river waters at the 1 μg L^{-1} level [44].

Popenoc et al. [45] determined alkyl sulphonates and alkyl ethoxy sulphonates at the ppb level in river waters using ion-spray liquid chromatography–mass spectrometry.

Pojana et al. [38] have described a method of the routine-specific determination of the anionic surfactant alcohol polyethoxylate sulphate in environmental aqueous samples. An enrichment/fractionation of the target analytes in water samples was performed by solid-phase extraction on graphitised carbon black (recoveries: 90%–103%), followed by hydrolysis/derivatisation with fluorescent reagents and separation/detection by reversed-phase HPLC–FLD. The procedure was applied to the study of aerobic biodegradation of alcohol polyethoxylates sulphate under laboratory conditions and to the 10-month monitoring of the alcohol polyethoxylates sulphate as well as of linear alkylbenzene sulphonates, nonylphenol polyethoxylates and alcohol polyethoxylates surfactants in the Po river (northern Italy).

Gagnon [46] has described a rapid and sensitive atomic absorption spectrometric method, for the determination of anionic detergents at the microgram per litre level in natural waters. The method is based on determination by atomic absorption spectrometry using the bis(ethylenediamine) copper II ion. The method is suitable for detergent concentrations of 0.3 μg L^{-1} level.

Benoit and Lamathe [47] compared two methods for determining anionic surfactants in natural water by flameless atomic absorption spectroscopy.

One method depends on the formation of the detergent/orthophenanthroline/cupric complex, extractable with methyl isobutyl ketone. The other method depends on the chloroform-extractable detergent/diamine/cupric complex. The extraction conditions were examined, and the determination of the extracted complexes by flameless atomic absorption spectrometry studied. The detection limit for both methods was of the order of 10 $\mu g\ L^{-1}$.

Miscellaneous

Ultraviolet spectroscopy has been used for the determination of alkylbenzene sulphonates [48,49].

Higuchi et al. [50] determined detergents in river waters spectrophotometrically using 1-(4-nitrobenzyl)-4-4(4-diethylaminophenylazo)pyridinium bromide as a cationic dye. This technique is faster and simpler than the methylene blue procedure.

Schneider et al. [51] identified cationic and anionic detergents in the river Rhine by combined field desorption-collisionally activated decomposition mass spectrometry. The results demonstrated that three types of surfactants desorbed at distant emitter heating currents and a partial separation of nonionic, cationic–anionic surfactants could be achieved.

This inhibitory effect of anionic detergents on the adsorptive accumulation of the (dimethyl/glyoximate) nickel complex on a hanging mercury trap electrode is the basis of an indirect method for determining anionic detergents [52]. Adachi and Kobayashi [53] have compared various methods for determining the ionic and nonionic detergents in natural waters.

Preconcentration in XAD-7 resin has been used in the case of alkyl ethoxylated sulphates followed by derivatisation to alkyl bromides gas chromatography was used to determine these substances in surface and wastewaters [54,55]. Alkylbenzene sulphonate surfactants have been preconcentrated on XAD-8 resin and also desorbed with methyl alcohol prior to examination by ^{13}C nuclear magnetic resonance spectrometry.

Chen et al. [56] have developed a sensitive and rapid method which does not involve extracting procedures for the determination of anionic detergents expressed as sodium dodecylbenzene sulphonate in a method based on the resonance light scattering technique at a pH 4.3. Britton–Robinson buffer or Victoria Blue B reacted with sodium dodecylbenzene sulphonate to produce large particles which resulted in enhancement of the intensity of resonance light scattering. The enhanced resonance light scattering intensity of the assay system was proportioned to the concentration of sodium dodecylbenzene sulphonate in the range 0.08–3.0 $mg\ L^{-1}$, and the correlation coefficient was $n = 0.09996$. The detection limit was 0.013 $mg\ L^{-1}$. This method was applied to the determination of anionic surfactants in surface water samples. It was shown that the interaction between Victoria Blue B and sodium dodecylbenzene sulphonate was mainly governed by electrostatic effect and the TI–IT stacking effect.

7.3 Cationic surface active agents

High-performance liquid chromatography

De Ruiter et al. [57] described a high-performance liquid chromatography method for cationic surfactants based on the on-line ion-pair extraction of the cationic surfactant with the counter ion of methyl orange or 9,10-dimethoxyanthracene-2-sulphonate sodium salt into the organic phase. UV or fluorescence spectrometry was used to monitor the ion-pair of methyl orange and 9,10-dimethoxyanthracene-2-sulphonate sodium salt, respectively. A new sandwich-type phase separator, as part of the extraction detector, was used successfully. Detection limits for dialkyldimethylammonium chloride in river water were 2 $\mu g\ L^{-1}$ and 10 $ng\ L^{-1}$ for methyl orange and 9,10-dimethoxyanthracene-2-sulphonate sodium salt, respectively.

Gas chromatography

In the gas chromatographs of alkylbenzene sulphonate methyl esters in a river sample [58] the pattern of the gas chromatogram is analogous to that of the linear alkylbenzene sulphonate standard. The assignment of the peaks was performed on the basis of the retention times and mass spectrum, and the individual components of alkylbenzene sulphonate were determined by mass fragmentography. For overlapped peaks on the gas chromatogram, more than two mass spectra were recorded. The total amounts of alkylbenzene sulphonate determined by mass fragmentography were in good accord with those determined by gas chromatography. Desulphonation gas chromatography has been applied to the analysis of partially degraded linear alkylbenzene sulphonate mixtures.

Titration methods

Wang [59] has described a simple titration method applicable to the analysis of cationic surfactants. Methyl orange and azure A were used as primary dye and secondary dye, respectively. The method is free from interference by inorganic salts even at high concentrations. This method, however, cannot accurately measure the surfactants at a concentration below 1 $mg\ L^{-1}$.

Spectrophotometric methods

Wang and Langley [60] devised a more sensitive colorimetric method for analysing cationic surfactants of the quaternary ammonium-type in fresh water. Methyl orange will react with an ionic surfactant to form a chloroform-soluble coloured complex. The colour intensity of the chloroform layer at 415 nm is proportional to the concentration of the 'dye–ionic surfactant complex', and can then be measured by making spectrophotometric readings of the chloroform solution at the optimum wavelength.

Subsequently, Wang et al. [61] devised a more sensitive spectrophotometric method for analysing cationic surfactants of the quaternary ammonium-type in natural water. Methyl orange will react with an ionic surfactant to form a chloroform-soluble coloured complex in the presence of chloroform. The colour intensity of the chloroform layer is proportional to the concentration of the 'dye–ionic surfactant complex' and can then be measured by making spectrophotometric readings of the chloroform solution at the optimum wavelength.

Kawase and Yamasake [62] have described a continuous solvent extraction procedure for the spectrophotometric determination of cationic surfactants.

The British Standard method [63] for cationic detergents is based on the addition to the sample of an excess of a standard anionic detergent and determination of excess anionic by a spectrophotometric methylene blue method.

Segmental injection techniques [64] have been used in conjunction with spectrophotometry to determine trace cationic detergents in water.

Lavaronte et al. [65] described an automatic flow-analysis procedure for the spectrophotometric determination of cationic surfactants in surface water using a solenoid micropump for propelling solutions of reagents and sample. The method is based on a ternary formation complex between chromazurol S, the Fe(III) ion and the cationic surfactant. The flow network comprised four solenoid micropumps controlled by a microcomputer, which performed the sampling step by loading a reaction coil with sample and reagent solutions and displacing the sample zone through the analytical path. The system is simple, easy to operate and very flexible, with sufficient sensitivity to determine cationic surfactants in water without any preconcentration or separation step. After determining the best operational conditions, favourable features such as a linear response between 0.34 and 10.2 mg L^{-1} of surfactant ($R = 0.999$), a relative standard deviation of 0.6% ($n = 11$) for a sample containing 3.4 mg L^{-1} of surfactant, a detection limit of 0.035 mg L^{-1} of surfactant and a sampling throughout of 72 determinations per hour were achieved. The system was used to determine cationic surfactant in river-water samples, and recovery values between 91% and 106% were achieved.

Miscellaneous

Other methods that have been employed for the determination of cationic surfactants include ultraviolet spectroscopy [48], flow injection analysis [66] and ion pairing [67]. The flow injection method is based on spectrophotometry following reaction with the anionic azo compound 3-(2-hydroxy-3-carboxyanilide-1-azonaphthalene)-4-sulphonic acid at pH 9.8.

Simms et al. [68] determined trace levels of cationic surfactants in natural waters by fast atom bombardment mass spectrometry.

Titration methods

Wang et al. [69,70] discuss indirect two-phase titration methods for the determination of anionic detergents. Wang et al. [71–73] also describe titration procedure for anionics involving titration with 1,5-dimethyl-1,5-diazoundecamethylene polymethobromide. Li and Rosen [72] have described a two-phase mixed indicator titration method for anionic surfactants, including sodium decanesulphonate, sodium dodecanesulphonate and sodium tetradecanesulphonate. The method was found not to be quantitative for surfactants with alkyl chains containing less than 12 carbon atoms but could be applied to them using chloroform, 1-nitropropane (2:3) as the organic phase and a multiple extraction and titration techniques.

Tsubouchi and Mallory [73] described a different determination of anionic and cationic surfactants in natural waters by two-phase titration.

Spectrophotometric methods

Motomizu et al. [74] describe a spectrophotometric technique for determining anionic surfactants in water using ethyl violet as the reagent. The absorption maximum occurs at 615 nm. It was found that traces of anionic surfactants could be extracted into benzene and toluene and then determined by spectrophotometry. The technique is claimed to be simple, rapid and highly sensitive.

Del Valle et al. [75] describe a continuous solvent extraction flow injection analysis automated system for the routine determination of anionic surfactants in river and treated waters in the concentration range 0.94–3.5 $\mu g\ mL^{-1}$. The method was based on an ion-pair extraction reaction with methylene blue in chloroform. Results compared favourably with those obtained by a standard batch extraction method. Those anions that caused the greatest interference (perchlorate, thiocyanate and nitrate) were the same ones interfering with the standard batch method. Interference was generally lower at pH 2 than at pH 7.

Evstifeev et al. [76] determined the anionic surfactant sodium lauryl sulphate in natural waters by complexation with a pyrylium salt and subsequent extraction of the resultant complex with toluene. The optical density of the toluene layer was measured at 420 nm.

Motomizu et al. [77] determined the anionic surfactant sodium dodecyl sulphate present in river water using a spectrophotometric flow injection analysis system coupled with solvent extraction. Methylene blue (cationic dye) and 1,2-dichlorobenzene (solvent) were the most efficient. The ion associate of anionic surfactant and methylene blue was extracted into the organic phase and its absorbance measured at 658 nm. A detection limit of 5 $\mu g\ L^{-1}$ sodium dodecyl sulphate was obtained.

References

[1] L. Favretto, B. Stancher, R. Thurus, Analyst 103 (1978) 955.
[2] A. Stephenou, Chemosphere 13 (1984) 43.
[3] J.F. Huber, F.F.M. Kloder, J.M. Miller, Anal. Chem. 44 (1972) 105.
[4] A. Otsuki, H. Shiraishi, Anal. Chem. 51 (1979) 2329.
[5] K.A. Evans, S.T. Pubey, L. Kravetz, I. Dzidic, J. Gumulka, R. Muller, et al., Anal. Chem. 66 (1994) 699.
[6] G. Nickless, P.J. Jones, J. Chromatogr. 156 (1978) 87.
[7] P.R. Musty, G. Nickless, J. Chromatogr. 89 (1974) 185.
[8] P.R. Musty, G. Nickless, J. Chromatogr. 120 (1976) 369.
[9] P. Jones, G. Nickless, J. Chromatogr. 156 (1978) 99.
[10] R.M. Cassidy, C.M. Niro, J. Chromatogr. 126 (1976) 787.
[11] T. Okada, Anal. Chem. 62 (1990) 734.
[12] A. Zgola-Grzeskovich, T. Grzeskowiaki, Int. J. Environ. Anal. Chem. 91 (2011) 57.
[13] N. Brown, O. Hayes, Anal. Abstr. 3 (1956) 1432.
[14] P. Grieff, Anal. Abstr. 13 (1996) 3650.
[15] N.T. Crabb, H.F. Persinger, J. Am. Oil Chem. 41 (1964) 752.
[16] S.J. Patterson, E.C. Hunt, K.B.E. Tucker, J. Proc. Inst. Sewerage Purif. (1996) 190.
[17] S.J. Patterson, E.C. Hunt, K.B.J. Tucker, J. Am. Oil Chem. Sci. 44 (1967) 407.
[18] L. Favretto, Petroldi, G. Marlett, L. Fabrielli, Ann. Chim. 62 (1972) 478.
[19] W. Dozanska, Rocz. Panstw. Zakl. Hig. 20 (1969) 137.
[20] A.E. Hey, S.H. Jenkins, Water Res. 3 (1969) 887.
[21] N.T. Crabb, H.F. Persinger, J. Am. Oil Chem. Soc. 45 (1968) 611.
[22] S.M. Belen'skii, N.Y. Bel'fon, L.P. Glazyrina, N.A. Keitina, Zavod Lab. 34 (1968) 1441.
[23] N. Heatly, E.J. Page, Water Sanit. 3 (1952) 46.
[24] M.J. Rosen, H.A. Goldsmith, Systematic Analysis of Surface Active Agents, second ed., Wiley, New York, 1972.
[25] L. Favretto, B. Stancer, R. Tunis, Analyst 104 (1979) 241.
[26] L. Favretto, B. Stancher, R. Tunis, Analyst 105 (1980) 833.
[27] L. Favretto, F. Tunis, Analyst 101 (1976) 198.
[28] B. Stancher, L. Favretto, J. Chromatogr. 150 (1978) 447.
[29] B. Stancher, R. Tunis, Analyst, London 104 (1979) 242.
[30] P.T. Crisp, J.M. Eckbert, N.A. Gibson, I.J. Webster, Anal. Chim. Acta 123 (1981) 355.
[31] K.W. Petts, I. Sliney, Water Res. 15 (1981) 129.
[32] Z. Zhu, A. Li, Y. Liu, Int. J. Environ. Anal. Chem. 84 (2004) 267.
[33] R.A. Greff, E.A. Setzkorn, W.D.J. Leslie, J. Am. Oil Chem. Soc. 42 (1965) 180.
[34] J. Courtot Coupez, A. Le Bihan, Anal. Lett., London 2 (1969) 567.
[35] A. Arpino, C. Calatroni, G. Jacini, Ital. Sostanze Grasse 51 (1974) 140.
[36] O. Burttschell, Anal. Abstr. 14 (1967) 6490.
[37] P.T. Crisp, J.M. Eckert, W.A. Gibson, Anal. Chim. Acta 104 (1979) 93.
[38] G.L. Pojana, G. Cassani, A. Marcomini, Int. J. Environ. Anal. Chem. 84 (2004) 729.
[39] K. Inaba, Int. J. Environ. Anal. Chem. 31 (1987) 43.
[40] L.K. Linhart, Tenside 9 (1972) 241.
[41] H. Hon-Nami, T.J. Hanya, J. Chromatogr. 161 (1978) 205.
[42] J. Waters, J.T. Garrigan, Water Res. 17 (1983) 1549.
[43] P.W. Taylor, G. Nickless, J. Chromatogr. 178 (1979) 259.
[44] A. Nahai, K. Tsuji, M. Yamanka, Anal. Chem. 52 (1980) 2275.

[45] P.D. Popenoc, S.T. Morris, P.S. Harris, K.T. Norwood, Anal. Chem. 66 (1994) 1620.

[46] M.J. Gagnon, Water Res. 13 (1979) 53.

[47] O. Benoit, J. Lamathe, Bull. Liaison Lab. Ponts Chaussees 115 (1981) 25.

[48] M. Uchiyama, Water Res. 11 (1977) 205.

[49] M. Uchiyama, Water Res. 13 (1979) 847.

[50] K. Higuchi, Y. Shimoishi, H. Miyata, K. Toei, Analyst 105 (1980) 768.

[51] E. Schneider, K. Leveson, A.J.H. Boerboom, P. Kistemaker, S.A. McLuckey, M. Przybylski, Anal. Chem. 56 (1984) 1987.

[52] B. Pihlar, B. Gopenc, D. Petric, Anal. Chim. Acta 189 (1986) 229.

[53] A. Adachi, T. Kobayashi, Eisei Kagaku 34 (1988) 1.

[54] T.A. Neubecker, Environ. Sci. Technol. 19 (1985) 1232.

[55] E.M. Thurman, T. Willoghby, L.B. Barbier, K.A. Thorn, Anal. Chem. 9 (1987) 1798.

[56] Z.G. Chen, R. Peng, F. Xie, W.Y. Zou, H.D. Qui, J.H. Chen, Int. J. Environ. Anal. Chem. 90 (2010) 573.

[57] C. De Ruiter, J.C.H.T. HJefkns, U.A.R. Brinkman, R.W. Frei, M. Evers, F. Matthews, Int. J. Environ. Anal. Chem. 31 (1987) 325.

[58] H. Hon Nami, T. Hanya, J. Chromatogr. 113 (1998) 747.

[59] L.K. Wang, Evaluation of Improved Two-Phase Titration Methods and a Field Test Kit for Analysing Ionic Surfactants in Water and Wastewater, Technical Report No. ND-5296-M-3, Calspan Corporation Buffalo, New York, p. 66.

[60] L.K. Wang, D.F. Langley, J. Environ. Contam. Toxicol. (1977) 5447.

[61] L.K. Wang, M.H. Wang, J.T. Kao, Water Air Soil Pollut. 9 (1978) 337.

[62] L. Kawase, M. Yamasaki, Analyst 104 (1979) 750.

[63] British Standard Institution, BS 2690 Part II, 1971.

[64] S. Feng, X. Chen, J. Fan, Z. Ha, Int. J. Environ. Anal. Chem. 85 (2005) 63.

[65] A.F. Lavaronte, C.K. Pires, A. Morales-Rubio, M. Dela Guardia, B.F. Reis, Int. J. Environ. Chem. 86 (2006) 723.

[66] K. Toei, T. Zaitsu, C. Igarashi, Anal. Chim. Acta 174 (1985) 369.

[67] Z.M. Shakhser, W.R. Seitz, Anal. Chem. 62 (1990) 1758.

[68] J.R. Simms, T. Keough, S.R. Ward, B.L. Moore, M.M. Baudurraga, Anal. Chem. 60 (1988) 2613.

[69] I.K. Wang, P.K. Panzardi, J. Pedro, W.W. Shuster, D.B. Autenbach, J. Environ. Health 38 (1957) 159.

[70] L.K. Wang, S.I. Kao, H. Wang, J.K. Kao, A.L. Lashkin, Ind. Eng. Prod. Res. Dev. 17 (1978) 186.

[71] L.K. Wang, J.Y. Yang, R.G. Ross, M.H. Wang, Water Resolut. Bull. b11 (1975) 267.

[72] Z. Li, M.L. Rosen, Anal. Chem. 53 (1981) 1516.

[73] M. Tsubouchi, J.H. Mallory, Analyst 108 (1983) 636.

[74] S. Motomizu, S. Fujiwara, A. Fujiwara, K. Toei, Anal. Chem. 54 (1982) 392.

[75] M. Del Valle, J. Alonso, J. Bartoli, Analyst 113 (1988) 1677.

[76] M.M. Evstifeev, A.D. Semenov, K.N. Bagdasarov, L.V. Kzaraenko, Y.M. Gavrilko, E.P. Oleknovich, et al., J. Water Chem. Technol. 8 (1986) 81.

[77] S. Motomizu, M. Oshima, T. Kuroda, Analyst 113 (1988) 747.

Chapter 8

Volatile organic compounds in nonsaline waters

Chapter Outline

References 143

The determination of mixtures of volatile organic compounds (VOCs) in environmental waters presents its own particular problems, which are now discussed.

Solid-phase microextraction followed by capillary gas chromatography (GC)−mass spectrometry (MS) is a particularly useful technique and has been applied to the simultaneous analysis of 23 priority pollutants, as defined by the FDA, in environmental waters [1].

Most water contaminants with VOCs are traceable to leaking underground fuel reservoirs, solvent storage vessels, agricultural practices, industrial residues and deficient wastewater treatment and disposal. In order to perform an effective monitoring of such organic micropollutants in a straightforward manner, Guimaraes et al. [1] developed a multiresidue method for the determination of 23 VOCs (trihalomethanes and chlorinated solvents) in water using GC−MS. This group also includes methyl-*tert*-butyl, ether, epichlorhydrin and vinyl chloride which present additional analytical difficulties. Different fibres were assayed, which are 7-μm polydimethylsiloxane (PDMS) and 100- and 75-μm carboxen−PDMS, and the extraction conditions were optimised. The best results for the majority of the analytes and mainly for those with the lowest signals were obtained using the carboxen−PDMS fibre after 15 minutes of extraction in the headspace mode at a room temperature of 20°C ± 2°C. The analytical sensitivity, linearity, precision, accuracy and uncertainties were studied for method validation in agreement with the international standard ISO/IEC 17025:2005. The limits of detection achieved with this method (0.06−0.17 $\mu g\ L^{-1}$) are adequate to determine the VOCs at the restrictive levels established by the European legislation. This was a decisive achievement which enabled the analysis of all VOCs listed under the drinking water directive in a single assay. The method exhibits performance capabilities suitable for the

Determination of Toxic Organic Chemicals In Natural Waters, Sediments and Soils.
DOI: https://doi.org/10.1016/B978-0-12-815856-2.00008-4

routine analysis of VOCs in drinking water by quality-control laboratories as enforced by EU directives.

Thermal desorption–GC–MS (DTD–GC–MS) is a technique that is finding application in the characterisation of the semivolatile organic carbon fraction of ambient and combustion source particularly matter collected on filters. In a study conducted by Graham et al. [2], particularly three DTD–GC–MS methods were developed. He compared them with results obtained by a conventional solvent extraction method for the analysis of a mixture of target analytes in solution and of diesel particulate matter collected on quartz filters. The target analytes included *n*-alkanes, hopanes, steranes and polycyclic aromatic hydrocarbons. This study shows that while three DTD–GC–MS methods were generally comparable to the solvent extraction method, the choice of calibration strategy and calibration materials has a significant impact on the measured accuracy of a method. Very large variations were seen in all the methods for the more volatile compounds such as C_{10}–C_{13} *n*-alkanes and naphthalene. Accuracy, defined as the difference from the known concentration of a liquid sample, ranged from 5% to 32%. Precision, defined as the relative standard deviation (RSD), ranged from 4% to 16%. The average difference of DTD–GC–MS results from the solvent extraction results for the diesel particulate matter filters ranged from 20% to 40%. This difference was driven by the large number of target analytes present at relatively low concentrations ($<25\ pg\ mm^{-2}$) and their corresponding higher variability. Differences in performance among the compound classes were noted. Minimum detection limits for the DTD–GC–MS method were in the order of $0.1-1\ pg\ mm^{-2}$ and were as good as or better than those obtained for the solvent extraction method.

Pans et al. [3] monitored the evolution of VOCs in groundwaters of the Najerilla River Basin, Spain. The contaminants originated from past uncontrolled spills from painting and varnishing industries. The analytical method is based on headspace-solid-phase microextraction using a 75-μm carboxen–PDMS fibre. Quantification was carried out by GC with flame ionisation detection. This method has permitted the determination of 13 VOCs identified in the polluted underground samples and provided good sensitivity (LOD between 0.1 and $6.0\ ng\ mL^{-1}$) and reproducibility (r.s.d. less than 10%).

Gal and Nutin [4] investigated an extraction devise for the separation and preconcentration of a series of VOCs ($CHCl_3$, $CHCl_2$, Br $CHClBr_2$ and $CHBr_3$) in aqueous matrices. The device consisted of a microporous membrane system utilising a hollow fibre tube filled with organic solvent directly immersed into the sample solution. The hollow fibre containing 160 μL organic solvent was immersed in a glass vial with 10 mL capacity, and the extraction took place through diffusive transport between the aqueous sample and the small amount of solvent. For validation of the method, some operational conditions, such as extraction solvent, temperature, stirring rate and separation time, were optimised. Limit of detection was at low ppb levels, with GC–MS analysis under selected ion monitoring, whereas enrichment factors between 22 and 35 were obtained. Good reproducibility with RSDs between 7.2% and 9.8% and large linear dynamic ranges with R^2 between 0.996 and 0.998 were also achieved. In addition, the performance of the membrane-assisted solvent extraction system was compared with two existing configurations: a nonporous membrane separation device, as well as with a comparable microporous configuration. The comparison considered the extraction mechanism and the underlying transport process. The application to real samples showed a good agreement with classical analytical methods.

References

[1] A.D. Guimaraes, J.J. Carvalho, C. Gon Calve, M. de Fatima, Alpendnraat Int. J. Environ. Anal. Chem. 88 (2008) 151.

[2] L.A. Graham, A. Tong, G. Poole, L. Ding, Ke Fu, D. Wang, et al., Int. J. Environ. Anal. Chem. 90 (2010) 511.

[3] B. Pans, M.S. Fernandez-Torrola, G. Prtiz, M.T. Tena, Int. J. Environ. Anal. Chem. 83 (2003) 495.

[4] C. Gal, R. Nutin, Int. J. Environ. Anal. Chem. 88 (2008) 447.

Chapter 9

Multiorganic compounds in nonsaline waters

Chapter Outline

9.1 Preliminary extraction of organic compounds **146**
9.2 Determination of organic compounds **149**
Gas chromatography 149
High-performance liquid chromatography 149
Infrared spectroscopy 150
Miscellaneous 154
References **155**
Further reading **158**

Samples of water are often screened to determine the types of organic compounds present using gas-chromatographic and high-performance liquid chromatographic (HPLC) techniques. Other techniques that have found limited application include Fourier transform spectroscopy, proton magnetic resonance spectroscopy and nuclear magnetic resonance spectroscopy. Raman spectroscopy combines ultraviolet (UV)–visible (Vis) fluorescence spectroscopy, cyclic voltammetry, photoelectric spectrometry and gas purging kinetics.

Borsdorf and Roland [1] developed an integrated submersible sensor probe capable of simultaneously measuring UV/Vis and fluorescence spectra. The probe transfers data electrically to the surface and operates at almost any depth due to its compact construction with light sources, flow cell, detection system and data acquisition/processing unit in a waterproof case. The variability in the measuring techniques allows a wide variety of chemical compounds to be analysed within a broad concentration range. Borsdorf and Roland tested the new submersible sensor probe ex situ on a laboratory scale and found the performance resembled that of stationery measuring instruments. In an in situ application, the sensor successfully monitored the migration of chemical substances during a tracer experiment in groundwater. The concentration of the tracer compound uranin (sodium fluorescein) was measured in the range of $5-500\,\mu g\,L^{-1}$ using fluorescence spectroscopy, while the content of benzene and toluene were continuously detected in the same groundwater wells using the UV/Vis sensor channel. The researchers used UV spectroscopy and headspace gas chromatography (GC) to determine the presence of benzene and toluene in groundwater.

Determination of Toxic Organic Chemicals In Natural Waters, Sediments and Soils.
DOI: https://doi.org/10.1016/B978-0-12-815856-2.00009-6

A detailed in situ characterisation was carried out of the concentration changes during a combined tracer experiment in a contaminated aquifer. This provided a unique insight into the hydrogeological conditions at the site. In particular, uranin and toluene were applied as hydrophilic and hydrophobic tracers, reflecting a different extent of interaction with the matrix. The two compounds could be simultaneously detected using both UV/Vis and fluorescence spectroscopy. During the short duration (approximately 5 minutes) of the analysis, the probe provided a large amount of analytical data within short time intervals, while allowing the detection of relatively fast groundwater flows. The in situ measurement provided a good basis for the modelling of the groundwater flows and for the prognosis of the natural attenuation process at the contaminated site.

9.1 Preliminary extraction of organic compounds

Various researchers have discussed preliminary extraction of organo compounds from water samples (see Table 9.1).

Ehnhalt et al. [22] investigated the use of supercritical fluid carbon dioxide for isolating and concentrating traces of organic compounds from aqueous solution to obtain a representative selection of such compounds. The use of carbon dioxide avoids the introduction of organic solvents or other contaminants, which could interfere with subsequent tests on the isolated compounds.

Von Bavel et al. [23] developed a solid-phase carbon trap that they used in conjunction with high-performance supercritical carbon liquid chromatography for the simultaneous determination of polychlorobiphenyls, polychlorodibenzofurans, polychlorodibenzodioxins and pesticides in environmental water. They found materials that could be used as a solid-phase trap in commercial supercritical fluid extraction instruments. They used PX-21 active carbon in a solid-phase trap to separate the planar fraction containing polychlorodibenzofurans and polychlorodibenzodioxins from the nonplanar fraction containing polychlorobiphenyls and organochlorine pesticides.

The researchers injected the concentrated fractions on to a gas-chromatographic column equipped with a mass spectrometric detector without the need for sample clean-up.

Kopfler et al. [24] described various methods capable of isolating gram quantities of organic matter from water samples. They determined the ability of each method to recover a set of model compounds possessing a wide variation in polarity functional groups, water solubility and molecular weight. No single method appeared to be superior overall on the basis of model solutes recovery. However, some methods could be eliminated from field application because the absorbents required were not commercially available. The researchers applied two methods in the field and tested the samples in several bioassays.

TABLE 9.1 Gas chromatography of multicomponent organic mixtures in natural water.

Substances	GLC column	GLC detector	Sensitivity	Extraction	Reference
Methanol, acetaldehyde, hexafluoroacetone, acetic acid	Direct gas chromatography Chromosorb104	–	–	None	[2]
Misc.	SE-30, OV-M on dodecyl phthalate	–	$\mu g\ L^{-1}$	Misc.	[3]
Misc.	Capillary	EC	–	Et_2O	[4]
		NP		CS_2	
		FID		CH_2C1_2	
		FID			
Benzene to 1,4-dimethylnaphthalene	Steam chromatography	FID	$10\ \mu g\ L^{-1}$	None	[5]
Misc.	Steam chromatography	–	–	None	
Misc.		–	–	Misc.	[6]
12 Compounds in landfill water	Capillary	FID	–	Misc.	[7]
Boiling point range 77°C–238°C	–	–	–	Steam distillation	[8]
27 Compounds including methyl dichloroethanoate, 2,3-dichloro-2-methy-l butane, dodecane, ethenylbenzene, 2-ethoxypropane, 3-ethyl-4-methyl-furan-2,5-dione, methyl and ethylnaphthalenes, hexane, hex-1-ene, 1-methyl-1-H-indene, 2-methylpentane, 1,1,1-trichloro-propane-2-one, 1,1,3-trichloro-propan-2-one and undecane	Gas chromatography–mass spectrometry	Mass	–	Sorption on resin column and separation into basic, neutral and acidic fractions	[9]

(*Continued*)

TABLE 9.1 (Continued)

Substances	GLC column	GLC detector	Sensitivity	Extraction	Reference
Misc.	–	MS	–	Misc.	[10]
Misc.	–	MS	–	Misc.	[11]
Misc.	Steam solid GLC	MS	–	Misc.	[12]
21 Volatile compounds	Charcoal	MS	–	Organic concentrator	[13]
100 Compounds	–	MS	$0.1\ \mu g\ L^{-1}$	Misc.	[14]
Misc.	–	MS	–	Gas purging	[15]
78 Compounds	–	MS	–	Misc.	[16]
Volatile organic compounds and aromatic, haloforms, chloroaromatic	–	MS	–	–	[17]
Benzothiazole–naphtholnols, 3-chloro-ethylphosphate	–	MS	–	–	[18]
Misc.	Spray and trap	MS	–	–	[19]
Misc.	–	MS	–	–	[20]
23 Priority pollutants	Microextraction gas chromatography–mass spectrometry	MS	–	–	[21]

9.2 Determination of organic compounds

Gas chromatography

Matamoros et al. [25] described a procedure based on comprehensive two-dimensional GC (GC × GC) coupled with time-of-flight mass spectrometry (TOF-MS) for the simultaneous determination of 97 organic contaminants at trace concentrations in river water. The target analytes included 13 pharmaceuticals, 18 plasticizers, 8 personal care products, 9 acid herbicides, 8 triazines, 10 organophosphorus compounds, 5 phenylureas, 12 organochlorine biocides, 9 polycyclic aromatic hydrocarbons (PAHs) and 5 benzothiazoles. The researchers obtained the best resolution of the target analytes in the contour plots when using a nonpolar stationary phase in the first dimension and a polar one in the second. However, in the opposite configuration, the retention time in the second dimension exhibited a strong correlation with the log K_{ow} ($P < .01$); Matamoros et al. proposed this as an additional identification criteria. The developed methodology is based on a polymeric solid-phase extraction followed by a GC-postmethylation and GC × GC/TOF-MS determination. Moreover, limits of detection and quantification ranged from 0.5 to 100 ng L^{-1} and from 2 to 185 ng L^{-1}, respectively. Repeatability was always lower than 20%.

Other applications [26–79] of GC and GC coupled with MS are reviewed in Table 9.1.

High-performance liquid chromatography

In a recent work Haot et al. [80] carried out an optimisation study of a multiresidual method based on liquid chromatography/tandem MS and isotope dilution MS for the determination of waterborne organic pollutants in environmental waters.

The method was validated for the analysis of 38 pharmaceutically active, 10 endocrine disrupting and three perfluoroalkylated compounds. Haot et al. discussed method performance parameters, including sample preservatives, pH values used in the solid-phase extraction, sample storage, sample extract storage time and matrix effects, for different aquatic matrices, including wastewater and surface water. To improve data quality, the researchers used isotope-labelled compounds as either injection internal standards or isotope dilution quantitation standards. These standards also investigated the behaviour of matrix effects during solid-phase extraction, sample preparation and liquid chromatography/MS–MS analysis and validated isotope dilution mass spectrometric determination of selected compounds. Method detection limits were in the low nanogram per litre range for the compounds evaluated. Haot et al. applied this method to the analysis of effluents and of samples downstream of a wastewater treatment plant and quantified more than 35 target organic pollutants. Matrix effects were dependent on modes of electrospray and ionisation and could not be removed via solid-phase extraction during clean-up procedures. ^{13}C- and ^{2}H-labelled isotope dilution quantitation standards could be added to samples before sample extraction to correct matrix effects in LC/MS–MS organic pollutant analysis.

Graham [81] has reviewed HPLC instrumentation, techniques and methodologies for the determination of trace organic compounds in water. The review includes approaches to sample clean-up or analyte isolation for those compounds likely to be candidates for analysis by HPLC. The researcher discussed column technology as it contributes to the use of HPLC for trace organic analyses and of various techniques for quantitative and qualitative detection of analytes.

Ouyang et al. [82] applied a liquid chromatography membrane ionisation MS with on-line immune affinity extraction to determine a group of 23 organic compounds. This method uses an immunoaffinity column for on-line clean-up and enrichment, a 5-μm C_{18} trapping column for analyte focusing, a 3-μm C_{18} analytical column for separation and a membrane introduction mass spectrometer for quantification. The researchers evaluated the immunoaffinity column in terms of binding capacity, recovery and enrichment factor. This method showed no matrix interference and an excellent detection limit. This method determined the presence of gasoline in contaminated water samples.

Applications of HPLC are reviewed in Table 9.2.

Infrared spectroscopy

Gomez-Taylor et al. [83] applied Fourier transform infrared spectroscopy to the on-line identification of trace organics separated by GC and HPLC in water at the μg level. To illustrate the feasibility of GC–infrared spectroscopy for this application, 25 L of distilled water were spiked with 50 μg of several organics (anisole, butyl ether, chlorobenzene, diethyl malonate, diethyl oxalate and salicylaldehyde). The water sample was then passed through a column of Amberlite XAD-2 macroreticular neutral polystyrene resin (Mallinckrodt Chemical Co., St Louis, MO) onto which the organic solutes were sorbed. They were subsequently eluted using 100 mL of diethyl ether and the resultant solution was evaporated to 1 mL A 10 μL aliquot was injected into the chromatograph and the spectra of each of the above was obtained. Each component is present at a maximum level of 500 ng (assuming 100% sample recovery).

Using a single beam gas chromatograph Fourier transform interface, Gurka et al. [84] determined values for a minimal identifiable quantity of 52 organic pollutants in test water and sediment samples. These were determined and compared with those previously reported for a commercial capillary GC–FT-IR interface. The 52 minimum identifiable quantities were 20–120 ng on average, 13 times more sensitive than previously reported

TABLE 9.2 High-performance liquid chromatography of organic mixtures in natural waters.

Type of organic	Type of water	Column packing	Elution solvent	Detection derivatisation	Limit	Reference
Misc. organics	Natural	–	–	UV	–	[26]
Misc.	Natural	–	–	Spectroscopic-electrochemical flame ionisation, mass spectrometric	–	[27]
Misc. organics	Natural	–	–	–	–	[28]
Misc. organics	Natural	–	–	–	–	[29]
Formaldehyde, acetaldehyde, propionaldehyde and benzaldehyde	Natural	Low capacity ion-exchange resin	Acetonitrile	4-Dinitrophenyl hydrazones	0.6 $\mu g\ L^{-1}$	[30]
Aliphatic aldehydes			*n*-Pentane	As 2,4-dinitorphenyl hydrazones	Spectro 1 $\mu g\ L^{-1}$ at 365 nm	[31]
Carbonyl compound, halogenated anilines, benzidines, nitrocompounds, phenols, polychlorinated compounds, alkylbenzene sulphonates, alkyl sulphonates and alkylbenzene sulphonates carbamates	Natural	–	–	–	–	[32]
Benzothiazoles aromatic amines, chlorinated phenols	Natural	–	–	–	–	[33]

(*Continued*)

TABLE 9.2 (Continued)

Type of organic	Type of water	Column packing	Elution solvent	Detection derivatisation	Limit	Reference
PAH	Lake	–	–	Tune resolved fluorimetry	1.8 fg	[34]
PAH	Natural	–	–	UV at 254 nm	–	[35]
PAH	Natural	Silica gel	–	Fluorescence	–	[36]
Chlorination production of PAHs	Natural	–	–	–	$\mu g\ L^{-1}$	[37]
PAH	River	Chemically bonded phthal-imidoropyl trichlorosilane stationary phase	–	–	–	[38]
PAH	Natural	–	–	Spectrofluorimetry 0.1 $ng\ L^{-1}$	0.1 $\mu g\ L^{-1}$	[39]
Alkylbenzene	River	LiChrosorb	Sodium perchlorate in methanol–water	Fluorimetric at 225 m	0.1 $mg\ L^{-1}$	[40]
Sodium alkylbenzene sulphonates	Natural	–	–	–	–	[41]
Cationic detergents	Wastewater	–	–	–	–	[42]
Alkylphenols, chlorophenols, dihydroxy-benzenes biphenyls	Natural	–	–	–	–	[43]
Phenols	Natural	–	2-Fluoren sulphonyl chloride	UV	0.05 ng	[44]

Phenols	Natural	Reverse phase	–	Electron chemical	–	[45]
Phenol, hydroquinone, pyrocatechol, hydroxyhydroquinone, resorcinol and phloroglucinol chlorinated	Natural	–	–	–	–	[46]
	Natural	(1) Silica, (2) amino propyl silica and (3) octadecyl or silica	–	–	–	[47]
Nitrophenols	Natural	Reverse phase ion pair	–	UV 280 nm	1 mg L^{-1}	[48]
Aromatic amines	Natural	–	–	Amperometric	–	[49]
Amino acids	Natural	–	–	Fluorimetric	–	[50]
Amino acids	Surface	–	–	Fluorimetric	–	[51]
Nitrobenzene	Surface	–	–	–	–	[52]
Dyes, azo, diazo and anthraquinone	Natural	–	–	Thermo spray detector	0.01 μg L^{-1}	[53]
Ethyl parathion methyl parathion	Surface	–	–	–	–	[54]
Organophosphorus type, trimethylphosphate and diethylmethyl phosphonothioate	Natural	–	–	Molecular emission cavity detector	–	[55]

PAH, Polycyclic aromatic hydrocarbon; *UV*, ultraviolet.

values. The researchers provided FT-IR group frequencies of typical environmental contaminants to test the new interface. By using selected frequency regions, the rapid screening of extracts for the presence of various compound classes was possible.

Miscellaneous

Bustamonte et al. [85] used various methods to measure levels of PAHs, polychlorinated biphenyls, methylmercury and butyltin in a natural UNESCO reserve in the Bay of Biscay.

The method was evaluated for the on-line determination of 4-fluorobenzoic acid, 3,5-difluorobenzoic acid, 2-chlorophenol, *p-tert*-butylphenol and dimethyl sulphoxide in water. The selectivity of the in-membrane preconcentration technique for semivolatile organics in the presence of volatile organic compounds was demonstrated. Cryotrapping and a rapid gas-chromatographic separation were added between the membrane and the mass spectrometer ion source for the determination of semivolatile organics in complex mixtures.

Amaral et al. [86] developed a method for the vacuum extraction of volatile organic compounds from water samples for ultratrace determination of carbon isotopic signatures. This method permits compound-specific stable carbon isotope analysis at concentrations as low as $0.03-1.3\ \mu g\ L^{-1}$. The researchers developed and used vacuum extraction to extract and preconcentrate volatile organics for subsequent carbon-stable carbon isotope analysis. This method used standard technique purge-and-trap analysis coupled to an isotope-ratio mass spectrometer. Even without complete extraction, the $\sigma^{13}C$ signatures of the volatile organic compounds were in good agreement (deviation %) with signatures determined by purge-and-trap-isotope-ratio mass spectrometric. This indicated that vacuum extraction does not cause isotopic discrimination. Limits of quantification for $\sigma^{13}C$ analysis were $0.03-0.06\ \mu g\ L^{-1}$ for benzene, toluene, *o*-xylene, *m-p*-xylene and ethylbenzene.

Lin and Pawliszyn [87] described a method of sorbent extraction using membrane extraction with a sorbent interface coupled to a portable gas chromatograph system. The main components of this system include a membrane module, a microtrap and a control unit for the heater and cooler. The membrane module, as an on-line sample-introduction device for this system, can be manipulated in different configurations, allowing for the selective permeation of analytes across the membrane into the carrier/stripping gas. The analytes were trapped and concentrated onto a microtrap, which serves as an injector for GC separation. A concentration pulse of the trapped analytes was generated through direct electrical heating of the microtrap. The characteristics of this system were explored, and its applicability and effectiveness have been demonstrated in field monitoring applications including the analysis of toluene in water. This system is very promising, as it is a simple, fast and portable tool for on-site process environmental monitoring.

Bruckner et al. [88] used a flame ionisation detector to detect volatile organic compounds that have been separated by water-only reversed phase liquid chromatography (WRPLC). The mobile phase is 100% water at room temperature without use of organic solvent modifiers. An interface between the liquid chromatography and detector is presented, whereby a helium stream samples the vapour of volatile components front individual drops of the liquid chromatographic eluent, and the vapour-enriched gas stream is sent to the flame ionisation detector. The design of the drop headspace cell is simple because the water-only nature of the liquid chromatographic separation obviates the need to do any organic solvent removal prior to gas phase detection. Despite the absence of an organic modifier, hydrophobic compounds can be separated in a reasonable time due to the low phase volume ratio of the WRPLC columns. The drop headspace interface manages liquid chromatographic column flows of 1 mL min^{-1}.

Tan [89] developed a method for the determination of volatile organic compounds in water based on solid-phase microextraction G.C.

Creaser and Weston [90] described a method for the on-line determination of volatile and semivolatile compounds using membrane inlet MS with in-membrane preconcentration. Semivolatile organic compounds in aqueous samples are preconcentrated in a flow-through silicone hollow-fibre membrane inlet held in a GC oven. The sample stream is replaced with air, and the semivolatile organics are thermally desorbed into the mass spectrometer by rapid heating of the membrane.

Lin et al. [91] developed a single extraction procedure for the determination of semivolatile organic compounds in water.

Gal and Nutin [92] used a compound membrane assisted liquid microextraction device with a series of organic compounds in aqueous environmental samples.

Graham et al. [93] compared direct thermal desorption methods with solvent extraction for the gas chromatograph–mass spectrometric analysis of semivolatile organic compounds.

Li et al. [94] used headspace single-drop microextraction coupled to GC to determine volatile halocarbons in water samples.

Pons et al. [95] monitored the evolution of pollution by volatile organic compounds in Spanish groundwaters.

References

[1] H. Borsdorf, U. Roland, Int. J. Environ. Anal. Chem. 88 (2008) 279.

[2] R.L. Grob, O.J. Kaiser, Environ. Sci. Health A11 (1976) 623.

[3] I.H. Suffet, E.R. Glazer, J. Chromatogr. Sci. 16 (1978) 12.

[4] T.A. Rooney, R.R. Freeman, Hewlett Packard Ltd, Avondale, USA, technical paper no 69, 1977, analysis of organic contaminants in surface water using high resolution gas chromatography and selective detectors, in: 174th American Chemical Society National Meeting,

Chicago, IL, August 28–September 2, 1977. Also in High Resolution Gas Chromatography, S.P. Cram, 1978.
[5] J. Teply, M. Dressler, J. Chromatogr. 191 (1980) 221.
[6] J.C. Peterson, G. Guiochom, C. Demorate, M. Dugang, Int. J. Environ. Anal. Chem. 14 (1983) 23.
[7] S. Liu, D.J. Chichester-Constable, J. Huball, S.R. Smith, J.D. Stuart, Anal. Lett. 20 (1987) 2073.
[8] K.D. Dix, J.S. Fritz, J. Chromatogr. 408 (1987) 201.
[9] M. Koga, R. Shinohara, A. Kido, S. Etoh, T. Hori, T. Akoyama, Jpn. J. Water Pollut. 1 (1978) 23.
[10] S.R. Heller, J.M. McGuire, W.L. Budde, Environ. Sci. Technol. 9 (1975) 210.
[11] A. Yasuhara, H. Shirashi, M. Tsuji, T. Okuno, Environ. Sci. Technol. 15 (1981) 570.
[12] C.L. Guillemin, F. Marinez, S. Thialt, J. Chromatogr. Sci. 17 (1979) 677.
[13] W.H. Pereira, B.A. Hughes, J. Am. Water Works Assoc. 72 (1980) 220.
[14] S.P. Scott, N. Sutherland, R.J. Vincent, Anal. Proc., London 21 (1984) 179.
[15] R.P. Kozloski, B.L. Sawnhey, Bull. Environ. Contam. Toxicol. 29 (1982) 1.
[16] D.R. Scott, Anal. Chim. Acta 211 (1988) 11.
[17] M.A. Le Pack, J.C. Ton, C.G. Enke, Anal. Chem. 62 (1990) 1265.
[18] L.D. Burkhard, E.J. Durham, M.T. Lucasewycz, Anal. Chem. 63 (1991) 277.
[19] G. Matz, P. Kesners, Anal. Chem. 65 (1993) 2366.
[20] G. Bayut Avoit, Anal. Chem. 64 (1992) 677.
[21] A.D. Guimaraes, J.J. Carvello, C. Concalves, M. De Fatima Alpendurada, Int. J. Environ. Anal. Chem. 88 (2008) 151.
[22] D.H. Ehnhalt, K. Trun, C. Epping, F. Ringhard, Int. J. Environ. Anal. Chem. 13 (1983) 219.
[23] B. Von Bavel, M. Jaremo, L. Karisson, G. Lindstreom, Anal. Chem. 68 (1966) 1279.
[24] F.C. Kopfler, H.P. Ringhard, R.G. Miller, Adv. Chem. 214 (1986) 425.
[25] V. Matamoros, E. Jover, J. Bayona, Anal. Chem. 82 (2010) 699.
[26] J.B. Reust, V.R. Meyer, Analyst 107 (1982) 673.
[27] B.K. Afghan, G.E. Batley, Eau du Quebec 14 (1981) 204.
[28] J.E. Haky, A.M. Young, J. Liq. Chromatogr. 7 (1984) 675.
[29] E.C.D. Gillyon, Lab. Pract. 28 (1979) 1194.
[30] K. Takani, K. Kuwata, A. Sugimae, M. Nakamoto, Anal. Chem. 57 (1985) 243.
[31] F. Van Hoof, A. Wittock, E. Van Buggenhart, J. Jansseus, Anal. Chim. Acta 169 (1985) 419.
[32] W.F. Smith, J.A. Healy, Sci. Total Environ. 37 (1984) 71.
[33] D.E. Games, Anal. Proc. 21 (1984) 174.
[34] N. Furuta, A. Otsuki, Anal. Chem. 55 (1983) 2407.
[35] J.J. Black, T.F. Hart, P.J. Black, Environ. Sci. Technol. 16 (1982) 247.
[36] S.S. Ho, H.T. Butler, C.J.J. Poole, J. Chromatogr. 281 (1983) 330.
[37] A.R. Oyler, D.L. Bodenner, K.J. Welsh, R.J. Liukkomen, R.M. Carlson, Anal. Chem. 50 (1978) 837.
[38] N.T. Crosby, D.C. Hunt, Anal. Proc. 17 (1980) 381.
[39] G. Cartoni, F. Coccioli, M. Ronchetti, L. Simonetti, L. Zoccolillo, J. Chromatogr. 370 (1986) 157.
[40] A. Nakae, K. Tsuji, M. Yamanaka, Anal. Chem. 52 (1980) 2275.
[41] T. Saito, K. Higashai, K. Hagiwara, Fresenius Z. Anal. Chem. 21 (1982) 313.
[42] V.T. Wee, Water Reserve 18 (1984) 223.
[43] E. Tesarova, V. Pakokova, Chromatography 17 (1983) 269.

[44] R.M. Carlson, T.A. Swanson, A.R. Dyler, M.T. Lukasewycs, R.J. Liukkonen, K.S. Voelkner, J. Chromatogr. Sci. 22 (1984) 272.
[45] D.N. Armentrout, J.D. McLean, M.W. Lon, Anal. Chem. 51 (1979) 1039.
[46] S. Hashimoto, T. Miyata, M.M. Washino, W. Kawakamis, Environ. Sci. Technol. 13 (1979) 71.
[47] K. Ugland, E. Lundanes, E. Griebrokle, A. Bjorseth, J. Chromatogr. 213 (1981) 83.
[48] H. Roseboom, C.J. Berkoff, R.C.C. Wegman, J. Chromatogr. 208 (1981) 331.
[49] J.R. Rice, P.T. Kissiner, Environ. Sci. Technol. 16 (1982) 263.
[50] S. Khim-Heang, Int. J. Anal. Chem. 15 (1983) 309.
[51] C. LeCloirec, P. LeCloirec, J. Morvan, G. Martin, Rev. Frank. Sci. Leau 2 (1983) 25.
[52] W. Golkiewicz, C.E. Werkhoven-Goewie, U.A. Brinkman, R.W. Frei, H. Colin, G.J. Guichon, J. Chromatogr. Sci. 21 (1983) 27.
[53] R.D. Voyksner, Anal. Chem. 57 (1985) 2600.
[54] G.J. Clark, R.R. Goodwin, J.W. Smiley, Anal. Chem. 57 (1985) 2233.
[55] M.J. Cope, Anal. Proc. 17 (1980) 273.
[56] A.A. Podowski, M. Feroz, P. Mentens, M.A.Q. Khan, Bull. Environ. Contam. Toxicol. 32 (1984) 301.
[57] T.A. Bellar, T.A. Buddevi, Anal. Chem. 60 (1988) 2076.
[58] L.H. Wright, M.D. Jackson, R.G. Lewis, Bull. Environ. Contam. Toxicol. 28 (1982) 740.
[59] W.P. Cochrane, M. Languette, S. Trudeau, J. Chromatogr. 243 (1982) 307.
[60] C.J. Miles, J.J. Delfino, J. Chromatogr. 299 (1984) 275.
[61] K.K. Hill, R.H. Holowell, L.A. Dal, Anal. Chem. 56 (1984) 2465.
[62] K. Levsen, L.H. Schafer, F. Freudenthal, J. Chromatogr. Chromatogr. Rev. 271 (1983) 51.
[63] I. Staber, H.R. Schulten, Sci. Total Environ. 16 (1980) 249.
[64] C. Geowie, P. Kwakman, R.W. Frei, W.A.T. Brinkman, W. Mansfield, T. Sechadri, et al., J. Chromatogr. 289 (1984) 73.
[65] M.W.F. Nielen, G. Koomen, R.W. Frei, U.A.T. Brinkman, J. Liq. Chromatogr. 8 (1985) 315.
[66] S.H. Hoke, E.E. Bryeggemann, L.J. Baxter, T. Trybus, J. Chromatogr. 357 (1986) 429.
[67] R. Hamann, A. Kettrup, Chemosphere 16 (1987) 527.
[68] A. Akerblom, J. Chromatogr. 319 (1985) 427.
[69] C.N. Vaughan, Anal. Chim. Acta 131 (1981) 307.
[70] R. Bushway, J. Chromatogr. 303 (1984) 263.
[71] D.R. Lauren, M.P. Agnew, J. Chromatogr. 303 (1984) 206.
[72] M. Chiba, R.P. Singh, J. Agric. Food Chem. 34 (1986) 108.
[73] P.G. Falkowski, J. Sucher, J. Chromatogr. 213 (1981) 349.
[74] G. Becker, G.E. Carlberg, E.T. Gjessing, J.K. Hongslo, S. Monarta, Environ. Sci. Technol. 19 (1985) 422.
[75] C.J. Miles, P.L. Brezonik, J. Chromatogr. 259 (1983) 499.
[76] D.E. Richardson, B. O'Grady, J.B. Bremner, J. Chromatogr. 268 (1983) 341.
[77] J. Hojzlar, Water Res. 21 (1987) 1311.
[78] I. Clark, Chromatogr. Int. 22 (1987) 12.
[79] D.P. Silgouer, M. Grasserkanel, D. Barelo, Anal. Chem. 69 (1987) 2756.
[80] C. Haot, X. Zhao, S. Tabac, P. Yang, Environ. Sci. Technol. 42 (2008) 4068.
[81] J.A. Graham, Adv. Chem. 214 (1987) 97.
[82] S. Ouyang, Y. Xu, Y.H. Chan, Anal. Chem. 70 (1998) 931.
[83] M.M. Gomez-Taylor, D. Kyehl, P.R. Griffiths, J. Environ. Anal. Chem. 5 (1978) 103.
[84] D.F. Gurka, R. Titus, P.R. Griffiths, D. Henry, A. Giogretti, Anal. Chem. 59 (1987) 2362.

[85] J.B. Bustamonte, A. Abisu, A. Bortolome Preito, A. Atubxa, S. Arrasote, E. Anakabe, et al., Int. J. Environ. Anal. Chem. 90 (2010) 722.
[86] H.J.F. Amaral, M. Berg, M.S. Brennwald, M. Hafer, R.K. Eawag, Environ. Sci. Technol. 44 (2010) 1020.
[87] X. Lin, J. Pawliszyn, Int. J. Environ. Anal. Chem. 85 (2005) 1189.
[88] C.A. Bruckner, S.T. Eckers, R.E. Synovec, Anal. Chem. 69 (1997) 3465.
[89] S. Tan, Proceedings Water Quality Technol Conference, 1998.
[90] C.S. Creaser, D.J. Weston, Anal. Chem. 72 (2010) 2730.
[91] M.S. Lin, V. Him, Y.I. Mazur, Proceedings Water Quality Technol Conference, 1998.
[92] C. Gal, R. Nutin, Int. J. Environ. Anal. Chem. 88 (2008) 447.
[93] L.A. Graham, A. Tong, G. Poole, L. Ping, F. Ke, D. Wang, et al., Int. J. Environ. Anal. Chem. 90 (2010) 511.
[94] X. Li, X. Xu, X. Wang, L. Ma, Int. J. Environ. Anal. Chem. 84 (2004) 633.
[95] B. Pons, M.A. Fernandez-Torrobe, G. Ortiz, M.T. Tena, Int. J. Environ. Anal. Chem. 83 (2003) 495.

Further reading

A. Agostiona, M. Caselli, M.A. Provesnzano, Water Air Soil Pollut. 19 (1983) 309.
M. Ahnoff, B. Josefsson, Anal. Chem. 48 (1976) 1268.
R.C. Chang, J.S. Fritz, Talanta 25 (1978) 659.
A. Bacaloni, G. Goretti, A. Lagona, B.M. Petronio, M. Rotatori, Anal. Chem. 52 (1980) 2033.
K. Beyermann, W.Z. Exkrich, Fresenius Z. Anal. Chem. 265 (1974) 1.
R.D. Blanchard, J.K. Hardy, Anal. Chem. 57 (1985) 2349.
H.T. Bodings, C. De Jong, R.P.M. Dooper, J. High Resol. Chromatogr. 8 (1985) 755.
H. Boren, A. Grimvall, J. Palmborgi, R. Sauerhed, B. Wiglius, J. Chromatogr. 348 (1985) 67.
P. Burchill, A.A. Herod, K.M. March, C.A. Pirt, E. Pritchard, Water Res. 17 (1983) 1891.
A.K. Burham, G.V. Calder, J.S. Fritz, G.A. Junk, H.J. Svec, R. Willis, Anal. Chem. 44 (1972) 139.
A. Cailleux, P. Turcant, P. Aelain, D. Toussaint, J. Gaske, A. Roux, J. Chromatogr. 391 (1987) 280.
J. Chudaba, E. Chelbkova, F. Tusek, Vodno Hospodarstvi 27 (1977) 236.
W.E. Coleman, J.W. Munch, R.W. Slater, R.G. Melton, F.C. Kopfler, Environ. Sci. Technol. 17 (1983) 571.
R.W. Countant, G.W. Keighley, Anal. Chem. 60 (1988) 2836.
S.A. Daignault, D.K. Noot, D.T. Williams, P.M. Huck, Water Res. 22 (1988) 803.
H.J. Danz, E. Jackwerth, Fresenius Z. Anal. Chem. 318 (1984) 22.
R. De Groat, H_2O 12 (1979) 333.
A.D. Dietrich, D.S. Millington, Y.H. Seo, J. Chromatogr. 436 (1988) 229.
B. Dowty, L. Green, J.T. Laster, J. Chromatogr. Sci. 14 (1976) 187.
J.W. Eichelberger, K.H. Kerns, P. Olynk, J.L. Budde, Anal. Chem. 55 (1983) 1471.
M. Fielding, T.M. Gibson, H.A. James, K. McLoughlin, C.P. Steel, Water Research Centre, Medmenham. Technical Report, TR 159, Organic Micropollution in Drinking Water, February 1981.
J.S. Fritz, Ind. Eng. Prod. Dev. Res. 14 (1975) 95.
J.S. Fritz, ACC Chem. Res. 10 (1977) 67.
M.C. Goldberg, L. LeLong, L. Kahn, Environ. Sci. Technol. 5 (1971) 161.
C. Gomella, J.P. Belle, J. Auvray, Technol. Sci. Municipal 71 (1976) 439.
J.W. Graydon, K. Grob, F. Zuercher, W. Giger, J. Chromatogr. 285 (1984) 307.

D.E. Gryder-Boutet, M. Kenniwsh, J. Am. Water Works Assoc. 80 (1988) 52.
K. Haberer, F.Z. Schredelseker, Wasser Abwasser Forsch. 17 (1984) 206.
W.D. Hammers, H.F.P.M. Bosman, J. Chromatogr. Sci. 14 (1976) 187.
R.L. Harris, R.J. Huggett, H.D. Stone, Anal. Chem. 52 (1980) 779.
R. Ishiwatari, H. Hamara, T. Machihara, Water Res. 14 (1980) 1257.
C.M. Josefson, J.B. Johnston, R. Trubey, Anal. Chem. 56 (1984) 764.
G.A. Junk, J.J. Richard, Anal. Chem. 60 (1988) 451.
J.R. Kaczvinsky, K. Saitoh, J.S. Fritz, Anal. Chem. 55 (1983) 1210.
J.A. Leenheer, E.W.D. Huffman, J. Res. US Geol. Surv. 4 (1976) 737.
D. Levesque, V.N. Mallet, Int. J. Environ. Anal. 16 (1983) 139.
V. Lopez Avila, R. Wood, M. Flannagan, R. Scott, J. Chromatogr. Sci. 25 (1987) 286.
F. Mangani, G. Crescentini, P. Palma, F. Bruner, J. Chromatogr. 452 (1988) 527.
M.F. McNally, R.L. Grob, J. Chromatogr. 260 (1983) 232.
R.G. Melcher, V.J. Caldecourt, Anal. Chem. 52 (1980) 875.
T.A. Misharina, I.L. Shuravleva, R.V. Golonnya, J. Anal. Chem. USSR 42 (1987) 459.
R.A. More, F.W. Karasek, Int. J. Environ. Anal. Chem. 17 (1984) 187.
C. Munz, P.U. Roberts, J. Am. Water Works Assoc. 79 (1987) 62.
A.S. Narange, G. Eadon, Int. J. Environ. Anal. Chem. 11 (1982) 167.
A. Noorsdij, H_2O 12 (1979) 167.
R. Otsen, D.T. Williams, Anal. Chem. 54 (1982) 942.
R. Otson, C. Chau, Int. J. Environ. Anal. Chem. 30 (1988) 275.
J.F. Pankow, M.E. Rosen, L.M. Izabelle, K.M. Hart, Anal. Chem. 60 (1988) 40.
T.L. Peters, Anal. Chem. 54 (1982) 1913.
J. Rayer, C. Hennequin, Tech. Sci. Municipal 77 (1982) 25.
G.A.V. Rees, L. Au, Bull. Environ. Contam. Toxicol. 22 (1979) 561.
M.R. Rice, H.S. Gold, Anal. Chem. 56 (1984) 1436.
J.P. Riley, D. Taylor, Anal. Chim. Acta 46 (1969) 307.
J.L. Robinson, J. John, A.J. Safa, K.A. Kirkes, P.E. Griffiths, J. Chromatogr. 402 (1987) 201.
J. Schultz, Vom Wasser 69 (1987) 49.
R. Smith, J. High Resol. Chromatogr. Commun. 10 (1987) 60.
S.I. Stepan, J.F. Smith, Water Res. 11 (1977) 339.
S.F. Stepan, J.F. Smith, U. Flego, J. Renkers, J. Water Res. 12 (1978) 447.
S.J. Theron, D.W. Hassett, Water South Afr. 12 (1986) 31.
D.R. Thielen, P.S. Foreman, A. Davis, R. Yyerth, Environ. Sci. Technol. 21 (1987) 145.
E.M. Thurman, R.L. Malcolm, G.R. Aiken, Anal. Chem. 50 (1978) 775.
V. Totor, M. Popl, Fresenius Z. Anal. Chem. 322 (1985) 419.
S.A. Vandergrift, J. Chromatogr. 26 (1988) 513.
L. Weil, Gas-u. Wass. Fach. 113 (1972) 64.
L. Weil, Anal. Abstr. 24 (1973) 1259.
B. Wiglius, H. Boren, G.E. Carlberg, A. Grimvall, B.V. Loudgren, R. Savenhed, J. Chromatogr. 391 (1987) 169.
P.L. Wyllie, J. Am. Water Works Assoc. 80 (1988) 65.

Chapter 10

Pesticides and herbicides in nonsaline waters

Chapter Outline

10.1 Organochlorine insecticides **162**
Gas chromatography 162
Mixtures of chlorinated insecticides and polychlorinated biphenyls 166
Gas chromatography–mass spectrometry 167
High-performance liquid chromatography 168
Thin-layer chromatography 169
Other techniques 169
Preconcentration 170
10.2 Organophosphorus insecticide **170**
Extraction procedures 172
Gas chromatography 173
High-performance liquid chromatography 176
Thin-layer chromatography 177
Spectrometric methods 177
Electrochemical methods 180
Miscellaneous 181
10.3 Urea herbicides **182**
Gas chromatography 183
High-performance liquid chromatography 184
10.4 Sulphonylurea herbicides **186**
10.5 Phenoxyacetic acid–type herbicides **186**
Thin-layer chromatography 189
Paper electrophoresis 189
Miscellaneous 190
10.6 Triazine type **191**
Gas chromatography 191
High-performance liquid chromatography 193
Thin-layer chromatography 194
Miscellaneous 194
10.7 Carbamate type **195**
Gas chromatography 195
High-performance liquid chromatography 196
Thin-layer chromatography 197
Enzymic assay 198
Miscellaneous 199
10.8 Pyrethroids **199**
10.9 Other insecticides **201**
10.10 Miscellaneous herbicides **202**
Microextraction of herbicides 202
10.11 Pesticide survey **206**
References **206**
Further reading **217**

Determination of Toxic Organic Chemicals In Natural Waters, Sediments and Soils.
DOI: https://doi.org/10.1016/B978-0-12-815856-2.00010-2

10.1 Organochlorine insecticides

Gas chromatography

Researchers determine the presence of chlorinated insecticides with gas chromatography (GC), using different types of detection systems; less commonly used methods include liquid chromatography and thin-layer chromatography. By its nature, GC controls the analysis of complex mixtures of chlorinated insecticides. Much of the published work discussed herein focuses on the analyses of mixtures of different types of chlorinated insecticides found in environmental samples. Working on the determination of individual insecticides is reported at the end of this section.

Table 10.1 illustrates various procedures to determine the organochlorine type in insecticides found in environmental waters.

Brodtmann [37] carried out a long-term study on the qualitative recovery efficiency of the carbon adsorption method versus that of continuous liquid–liquid extraction method for determining several chlorinated insecticides. Comparative results obtained by electron-capture GC indicate that the latter method may be more efficient.

In this method, river water samples were passed through carbon for a predetermined period. An extract was then obtained by chloroform extraction of the carbon in a modified Soxhlet apparatus for 36 hours using glass distilled, pesticide grade petroleum ether. The neutral fraction of the chloroform extract was then prepared for GC by the methods of Breidenbach [38]. Brodtmann [37] used a continuous liquid–liquid extraction apparatus, as described by Khan and Wayman [39], for the extraction of nonpolar solutes from the river water. Pesticide grade petroleum ether was recycled internally (initial solvent charge of 359 mL), thereby exposing fresh solvent to the river water throughout the procedure.

The researchers employed a florisil clean-up step using sequential elutions with 6% and 15% ethyl ether–petroleum ether solutions.

Mangani et al. [40] also employed columns of graphitised carbon black to extract organochlorine insecticides from water.

Kongovi and Grochowski [41] discussed the problems that arise during the analysis of pesticide and herbicide by GC with electron-capture detection. During a routine run of pesticide standards (lindane, endrin and methoxychlor), five peaks were obtained. A study followed to determine the molecular structure of the endrin contaminants and the decomposition of the Endrin molecule in relation to temperature (220°C–235°C) and nitrogen flow rate. The researchers found that the retention period alone does not serve as an absolute identifying criterion, particularly for esters of low-molecular-weight acids. Other confirmation, such as mass spectroscopy, is required.

Researchers [42] used electron-capture GC coupled with glass capillary columns to separate 15 chlorinated insecticides. They used purified hexane

TABLE 10.1 Extraction solvents for concentration of chlorinated insecticides prior to gas chromatographic analysis.

Solvent	Insecticides mentioned	Gas chromatography	Reference
n-Hexane	α-, β-Endosulphan and heptachlor	SE-30 on Chromosorb W	[1–5]
n-Hexane	α-BHC, β-BHC, α-BHC, aldrin, *o,p′*, α-endosulphan, *p,p′*-DDE, dieldrin, *o,p′*-DDD, endrin, β-Endosulphan, *p,p′*-DDD, *o,p′*-DDT and *p, p′*-DDT	Capillary column	[6]
n-Hexane	*p,p′*-DDD, *p,p′*-DDD, *p, p′*-DDE, *p,p′*-DDT, aldrin, dieldrin, endrin, heptachlor, heptachlor epoxide, lindane, isodrin, methoxychlor, chlordane, chlordene, hexachlorobicycloheptadiene and hexachlorocyclopentadiene		[7–11]
Hexane, benzene and hexane–toluene	Dichlorvos		[12]
Hexane	α-BHC, γ-BCH, *o,p′*-DDD, *pp,p′*-DDD, dieldrin, endrin heptachlor and heptachlor epoxide	Electron-capture detection	[13]
n-Hexane or acetonitrile	α-BHC, β-BCH, γ-BCH, δ-BCH, heptachlor, heptachlor epoxide, aldrin, endrin, *p,p′*-DDE, *p,p′*-TDE, *p,p′*-DDT, *a,p′*-DDE, *p,p′*-TDE, *o,p′*-DDT, mirex, methoxychlor and photodieldrin	Electron capture and capillary column	[14]
n-Hexane	15 Organochlorine pesticides		[15]
Petroleum ether	DDT	DC11 on Chromosorb WAWSMCS	[15]
Petroleum ether	DDT, γ-BHC, heptachlor, epoxide, dieldrin and methoxychlor		[16]

(Continued)

TABLE 10.1 (Continued)

Solvent	Insecticides mentioned	Gas chromatography	Reference
Petroleum ether	Dieldrin	DC-200 on Chromosorb WHMDS flame ionisation and electron-capture detectors	[17]
Petroleum ether	α-BHC, β-BHC, γ-BHC, *p,p′*-DDE, *o,p′*-DDT, *p,p′*-DDD and *p,p′*-DDT	1.5% Silicone OV-17 plus fluoroalkyl-siloxane on Chromosorb W, electron-capture detection	[18–22]
Petroleum	8 Organochlorine pesticides γ-BHC, methychlor	Electron-capture detection	[23,24]
Benzene	8 Organochlorine pesticides		[25]
Benzene	15 Organochlorine pesticides		[26]
Benzene dichloromethane and 15% dichloromethane in *n*-hexane	9 Organochlorine pesticides		[27]
15% Methylene chloride in *n*-hexane	15 Organochlorine pesticides		[28]
Methylene chloride	Aldrin, α-BHC, β-BHC, γ-BHC, chlorobenside, chlordane (Kepone), *p,p′*-DDD, *p,p′*-DDE, *o,p′*-DDT, *p,p′*-DDT, dichlone, dieldrin, endrin, endosulphan, heptachlor, hexachlorobenzene, 1-hydroxchlordene, methoxychlor and mirex capture detector	5% OV-210 or 1%–5% O-17/ 1.95% OV-210 on Chromosorb, electron	[29]
Diethyl ether	Miscellaneous organochlorine insecticides	3% QF-1 or 2% OV-1 electron-capture detection	[30]
15% Diethyl ether in *n*-hexane	19 Aminochlorine insecticides		[31]

(*Continued*)

TABLE 10.1 (Continued)

Solvent	Insecticides mentioned	Gas chromatography	Reference
Liquid–liquid extraction methods	General discussion, lindane, hexachlorobenzene		[32,33]
Liquid–liquid	Lindane, hexachlorocyclohexane and α, β, γ hexachlorobenzene		[9,34]
Goulden extractor dispersive	Miscellaneous		[35]
Liquid–liquid	Microextraction	Electron-capture detection	[36]

as the extractant, emphasising that both the capillary column and the electron-capture detector require extremely pure carrier gas.

Erney et al. [43] employed ultraviolet (UV) irradiation followed by the gas–liquid chromatographic detection of the photodecomposition product patterns of 33 chlorinated insecticides, and they successfully confirmed the identity of selected organochlorine insecticides. In this procedure the degree of photochemical reaction depends on wavelength intensity, time of irradiation and the physical state of the chlorinated insecticides. This degree of photochemical reaction must be controlled to give consistent reactions suitable for identification. This technique offers a means of independent identification of insecticides.

Godefroot et al. [44] described a simple technique for the quantitative determination of sub-microgram per litre levels of organochlorine pesticides and polychlorinated biphenyls (PCBs) in water. The method uses a microsteam distillation/extraction apparatus. Capillary GC analyses the resulting extract.

An automatic sampler has been used [45] to observe an endosulphan wave in the river Rhine. The automatic sampler at the gauging station was programmed to provide 1 hour mixed samples composed of 125 mL portions taken every 10 minutes. The resulting samples together with the 24 hour mixed samples for the proceeding and succeeding periods were analysed for endosulphan by gas–liquid chromatography. The results show a distinct peak in the endosulphan concentration with a maximum recorded value of 276 mg L^{-1} or about 100 times the normal background level. An estimation of the total mass flow aided by flow measurements indicated that 40–50 kg of endosulphan passed Koblenz, Germany, during August, compared with a baseline value of 5 kg.

Fuka et al. [46] have studied the decomposition and determination of trichlorfon and dichlorvos in pond water. Gas chromatographic evidence indicates that trichlorfon undergoes a partial conversion to dichlorvos in a few days.

Harless et al. [47] used GC–mass spectrometry (MS) involving chemical ionisation MS to detect and confirm the presence of Kepone (chlorodecone), Kepone photoproducts and a conversion product of Kepone in environmental samples.

Various researchers described gas chromatographic procedures for the multiclass, multiresidue analysis of organochlorine insecticides in water [23,25,26,29,31,48,49].

Thompson et al. [29] described a gas chromatographic procedure for the multiclass, multiresidue analyses of organochlorine insecticides in water. It involves extraction with methylene chloride, separation into groups on a partially deactivated silica gel column and sequential elution with different solvents. GC with electron-capture detection determines the final halogenated compounds and derivatised carbamates. A flame photometric detector is used for organophosphorus compounds. This study included 42 organochlorine insecticides, 33 organophosphorus insecticides and 7 carbamate herbicides.

Gas chromatographic methods specific to a particular insecticide have been described: aldrin [49], dieldrin [49], endrin and diendrin [50–52], dichlorvos [12,53], DDT [15,51,54], heptachlor, chlordane, lindane, *p*, *p*′-DDT [52,55–59], methoxychlor, [53,60–62] trichlorfon [53,61] and toxaphene [63].

Mixtures of chlorinated insecticides and polychlorinated biphenyls

In actual practice, environmental samples that are contaminated with PCBs are also highly likely to be contaminated with chlorinated insecticides. Many reports have appeared discussing the cointerference effects of chlorinated insecticides in the determination of PCBs and vice versa. Much of the more recently published work analyses both types of compounds.

Bacaloni et al. [64] used capillary column GC and graphitised capillary columns for the analysis of the mixtures of chlorinated insecticides, PCBs and other pollutants. Depending on the amount of stationary phase coated on the walls, these columns operate in gas–solid, gas–liquid and gas–liquid–solid chromatographic modes. Bacaloni et al. [64] used columns coated with PEG 20M to show that selective columns can give different performances suitable for particular applications.

Bacaloni et al. [64] described GC of a mixture of 15 chlorinated pesticides on columns of different lengths and loadings of stationary phase under the same operation conditions at 160°C. They concluded that a glass capillary column loaded with a large amount of PEG 20M is suitable for the analysis of volatile chlorinated compounds, phenols and amines. A column with a low loading of stationary phase is suitable for the analysis of chlorinated compounds including PCBs. The researchers separated mixtures of chlorinated insecticides and PCBs (Aroclors 1242, 1256 and 1260) on a 70 m column having 210,000 theoretical plates for hexadecane at 100°C and 160,000 theoretical plates for γ-BHC at 160°C.

Zitko and Choi [65] and other workers [63,66–82] reviewed methods for the codetermination of organochlorine insecticides and polychlorobiphenyls.

Leoni et al. [83,84] described a fast and automated sample pretreatment technique to determine the presence of lipophilic organic compounds in water samples. They determined the PCB levels in environmental river water that might also contain organochlorine insecticides and other substances. The technique uses a miniaturised microporous membrane liquid–liquid extraction coupled online to GC with electron-capture detection. The device carries a miniaturised membrane extraction card attached to an electrically and mechanically designed instalment, mounted directly over a gas chromatogram injector, for fully automated injection of the extract. The researchers extracted 10 PCB congeners from 1-mL water samples (after addition of 40% acetonitrile) with an extraction time of 10 minutes.

The main features of this optimised methodology include good linearity (in the dynamic concentration range of 5 ng L^{-1} to 1 mg L^{-1}), enrichment factors of 33–40 times, repeatable extractions [relative standard deviations (RSDs) 2%–5%, $n = 4$] and low detection limits (2–3 ng L^{-1}). Acetonitrile was added to the samples to minimise the overall memory effect and to overcome the influence of PCB adsorption on the repeatability of extraction and enrichment. Overall memory effect and carry-over depended on the concentration of the organic solvent added to the sample and the solvent used in the washing procedure. When an optimised washing procedure was applied, the overall memory effect was ~0.2% at high concentrations (1 μg L^{-1}). When each extraction occurred in a new extraction card, no overall memory effect was detected. In addition, no significant adsorption onto a glass surface or a matrix effect on extraction was noticed. The total consumption of organic (nonchlorinated) solvents was less than 5 mL per sample.

Yamoto et al. [85] described a combination of GC with field desorption mass spectrometric analysis. They applied this technique to the determination of 5(4-benzthiocarb), (*S*-chlorobenzyl-*N*,*N*-diethylthiocarbamate), oxidiazon [2-*t*-butyl-4-(2,4-dichloro-5-isopropoxyphenyl) 2-1,3,4-oxadiazolin-5-one] and CNP (2,4,4-dichlorophenyl-4′-nitrophenylether) herbicides in water and environmental samples. They demonstrated its usefulness for screening unknown compounds in the natural environment.

Gas chromatography–mass spectrometry

A combination of GC and MS has been used to identify and determine Kepone (chlorodecone) [47], picloram (4-amino-3,5,6-trichloropicolinic acid) [86] and 4-(chloromethyl)sulphenyl bromobenzene herbicide [87].

Sabik et al. [88] discussed a method using simultaneous filtration and solid-phase extraction combined with large sample volume injection GC–MS to the ultratrace analysis of polar pesticides in water.

Sichilongo et al. [89] discussed signal losses in pesticide analysis using broadband isolation wave form in a commercial quadrupole ion trap to accomplish precursor ion isolation in tandem MS [87].

High-performance liquid chromatography

Suzuki and Novagoshi [90,91] separated chlorinated insecticides into two groups by column chromatography and then into three groups by thin-layer chromatography. Individual insecticides were then isolated and determined by GC. The members of the two groups separated by column chromatography were group 1 first-division insecticides – aldrin, DMC ethylene [1,1-bis-(4-chlorophenyl)ethylene], chlorfensulphide, DDS [bis-(4-chlorophenyl) disulphide], *o,p′*- and *p,p′*-DDT, heptachlor, quintozene, isobenzan and tetrasul and group 2 second-division insecticides – α-, β- and γ-BHC, triflurin, nitrofen and dieldrin.

Millar et al. [92] studied a method for the recovery of 18 organochlorine pesticides and 7 PCBs from water. The researchers obtained the best results when using liquid–solid column chromatography with florisil or alumina columns and when extracting with dichloromethane and 15% dichloromethane in hexane at pH 2, 7 and 10. They investigated the effects of pH, temperature and residual chlorine on the preservation of spiked samples and recommended the most suitable storage conditions.

Pocaro and Shabiaw [93] determined the presence of the insecticide hexachlorophene. The researchers converted hexachlorophene into its *p*-anisate ester, subjected it to chromatography on a column packed with Sil-X silica (36–40 μm) using 1-chlorobutane as the solvent and detected it by UV absorption. Calibrations are made over the range 40–280 ng of the derivative. The coefficient of variation at the 10 ng level was 1.0%.

Carpenter et al. [94] described a method for the assay of two metabolites within the herbicides dimethyl tetrachloroterephthalate, monomethyl tetrachloroterephthalate and tetrachloroterephthalic acid via high-performance liquid chromatography (HPLC) with ion pairing. The researchers analyse samples via direct injection without preparation and detect analytes with an UV photodiode-array detector. The metabolites are extracted from positive samples with a petroleum trifluoroacetamide and confirmed by GC–MS. The high-performance liquid chromatographic analysis of spiked water samples yielded a recovery range of 92%–106% with a mean recovery of 101% for tetrachloroterephthalate acid and a recovery range of 92%–101% with a mean recovery of 95% for monomethyl tetrachloroterephthalate. The minimum detection limits for these two metabolites were 2.4 and 2.7 $\mu g\ L^{-1}$, respectively.

Willmott and Dolpim [95] applied their HPLC–electron-capture detector approach to the determination of chlorinated insecticides in surface waters. The concentration of pesticides in most surface waters is less than 1 $\mu g\ L^{-1}$, so some form of extraction and concentration from large volumes of water was necessary prior to analysis. These workers applied a conventionally coated chromatographic support as a reverse liquid–liquid partitioning filter for the extraction of pesticides. They used uncoated polyether polyurethane foam; 0.5 g of the flexible foam was inserted into a 10 mm i.d. quartz tube and cleaned by washing with consecutive 100 mL aliquots of acetone, *n*-hexane, ethanol and distilled water.

Ye et al. [96] determined chlorotoluron and thiophanate-methyl in water samples using single-drop microextraction and HPLC–UV detection. Several parameters, such as solvent type, sale concentration, stirring rate, extraction time, pH and organic drop volume, were investigated. The optimum experimental conditions were 20 μL 1-octanol, 10% (w/v) NaCl, 600 rpm stirring rate, 40 minutes extraction time, neutral sample pH and 5 mL water sample. Under the optimum conditions the enrichment factors were 45.3- and 107.0-fold for thiophanate-methyl and chlorotoluron, respectively. The method exhibited a wide linear range (1–100 $\mu g\ L^{-1}$), reasonable detection limits (0.35 $\mu g\ L^{-1}$) and suitable repeatability (RSD 0.6%) for both analytes. The proposed method was validated with three real water samples fortified at two levels, and reasonable spiked recoveries were achieved in the range 84%–110.3%. The experimental results indicated that single-drop microextraction was a simple, reliable and convenient technique and practical for the enrichment of other pollutants.

Thin-layer chromatography

Several researchers used thin-layer chromatography to confirm the identity of chlorinated pesticides [97–103]. Arias et al. [101] described a method for the determination of organochlorine insecticide residues in river water. The method involves extraction, florisil column clean-up and analysis of the extract. Analysis by thin-layer chromatography requires silica gel G or alumina with hexane or hexane–acetone (49:1) as the solvent. Analysis by GC requires a polar column of 10% DX-200 on Chromosorb W HMDS or a semipolar column of 5% of DC-200 plus 7.5% of QF-1 on Chromosorb W, followed by electron-capture detection.

Other techniques

Less commonly used techniques to determine the organochlorine pesticides in water include nuclear magnetic resonance spectroscopy [104], flow injection spectrophotometry for chloramben [105,106]and selective electrodes for chlormequat [106].

Preconcentration

Techniques to preconcentrate organochlorine insecticides in environmental waters include microsteam distillation [107], gel permeation chromatography [44], adsorption of XAD resins [108,109], Carbopack B [110], C_{18} silica bonded resins [111–113], online liquid chromatography [114,115], adsorption on reverse-phase adsorption [116], adsorption bonded silica [117], adsorption on activated carbon [112,113,118–129], adsorption on polyethylene film [130], adsorption on Tenax [84,131] and adsorption on Amberlite [132–134].

10.2 Organophosphorus insecticide

The most commonly used organophosphorus insecticides are listed later in alphabetical order. There is a wide variety of phosphorus-based compounds in use. GC has played a major part in the development of suitable methods for the determination of submicro amounts of such compounds in environmental samples, such as crops, animal tissues and water.

Abate	*O,O,O′,O′*-Tetramethyl-*O,O′*-(thiodo-*p*-phenylene diphosphorothioate)
Acciothion (sumithion, fenitrothion and folithion)	*O,O*-Diemthyl-*O*-4-nitro-*m*-ethyl phosphorothioate
Amidithion	*S*-(*N*-2-Methoxyethylcarbamoylmethyl)dimethyl phosphorothiolothionate, *S*-(*N*-2-methocyethylcarbamoylmethyl)-*O,O*-dimethyl phosphorodithioate
Amiton	*S*-(2-(Diethylamino)ethyl)-*O,O*-diethyl phosphorothioate
Azinphos-ethyl	*S*-(3,4-Dihydro-4-oxobenzo(*d*)-(1–3)triazine-3-ymethyl) diethyl phosphorothiolothionate *S*-(3,4-dihydro-4-oxobenzo(*d*)-(1–3)triazine-3-ymethyl)-*O,O*-diethyl phosphorodithioate
Azinphos-methyl	*S*-(3,3-Dihydro-4-oxobenzo(*d*)-(1–3)triazine-3-ymethyl)-dimethyl phosphorothiolothionate *S*-(3,4-dihydro-4-oxobenzo(*d*)-(1–3)triazine-3-ymethyl-*O,O*-dimethyl phosphorodithioate
Azothoate	4-(4-Chlorophenylazo)phenyldimethyl phosphorothionate *O*-4(4-chlorophenylazo)phenyl-*O,O*-dimethyl phosphorothioate
Bensulide	*N*-2-(*O,O*-Di-isopropylphosphorothiolothioyl)ethyl benzenesulphonamide di-isopropyl *S*-(2-phenylsulphonylaminoethyl)-phosphorothiothionate *O*, *O*-di-isopropyl *S*-(2-phenylsulphonylamino-ethyl)-phosphorodithioate
Bromophos	4-Bromo-2,5-dichlorophenyldimethyl phosphorothionate *O*-4(4-bromo-2,5-dichlorophenyl)-*O,O*-dimethyl phosphorothioate
Bromophos-ethyl	(4-Bromo-2,5-dichlorophenyldiethyl) phosphorothionate *O*-(4-bromo-2,5-dichlorophenyl)-*O,O*-diethyl phosphorothioate

Butonate carbophenothion	Dimethyl-1-butyryloxy-2,2,2-dichloroethylphosphate *S*-(4, chlorophenylthiomethyl)diethyl phosphorothiolothionate *S*-(4-chlorophenylthiomethyl)-*O,O*-dimethyl phosphorodithioate
Chlorfenvinphos	2-Chloro-1(2,4-dichlorophenyl)vinyldiethyl phosphorodithioate
Chlorphonium	Tributyl-2,4-dichlorobenzylphosphonium (2-chloro-α-cyanobenzylideneamino)diethyl phosphorothionate *O*-2-chlorocyanobenzylideneamino-*O,O*-diethyl phosphorothioate 2-chloro-α-(diethoxyphosphinothioyloxyimino)-phenylacetonitrile
Coumaphos	3-Chloro-4-methyl-7-coumarinyldiethyl) phosphorothionate *O*-(3-chloro-4-methyl-7-coumarinyl)-*O,O*-diethyl phosphorothionate
Coumithoate	3-Chloro-4-methyl-7-coumarinyldiethyl phosphorothioate *O*-(3-chloro-4-methyl-7-coumarinyl)-*O,O*-diethyl phosphorothioate
Coumithoate	Diethyl-7,8,9,10-tetrahydro-6-oxobenzo(*c*)chroman-3-yl phosphorothionate *O,O*-diethyl-*O*-(7,8,9,10-tetrahydro-6-oxobenzo(*c*)-chroman-3-yl) phosphorothionate *O,O*-diethyl-*O*-(7,8,9,10-tetrahydro-6-oxo-6-didibenzo(bd) pyran-3-yl) phosphorothionate
Crotoxyphos	Dimethyl *cis*-1-methyl-2-(1-phenylethoxycarbonyl) vinyl phosphate 1-methylbenzyl-3-(dimethyoxyphosphinyloxy) isocrotonate
Cyanophos	4-Cyanophenyl phosphorothionate *O*-4-cyanophenyl-*O,O*-dimethyl phosphorothionate
Dasanite	*O,O*-Diethyl-*O*-(4-methylsulphinylphenyl) phosphorothioate
Demephion	A mixture of Demephion-*O* and Demephion-*S* (see later)
Demephion-*O*	Diemthyl-2-(methylthio)ethyl phosphorothionate *O,O*-dimethyl-*O*-(2-(methylthio)ethyl) phosphorothioate
Demephion-*S*	Dimethyl-*S*-(2-methylthio)ethyl phosphorothioate *O,O*-dimethyl-*S*-(2-methylthio)ethyl phosphorothioate
Demeton	A mixture of Demeton-*O* and Demeton-*S* (see later)
Demeton-methyl	A mixture of Demeton-*O*-methyl and Demeton-*S*-methyl (see later)
Demeton-*O*	Diethyl-2-(ethylthio)ethyl phosphorothionate
Demeton-*O*-methyl	2-(Ethylthio)ethyldimethyl phosphorothionate *O*-(2-(ethylthio)ethyl-*O,O*-dimethyl phosphorothioate
Demeton-*S*	Diethyl-*S*-(2-(ethylthio)(ethyl) phosphorothiolate *O,O*-diethyl-*S*-(2-(ethylthio)ethyl) phosphorothioate
Demeton-*S*-methyl	*S*-(2-(Ethylthio)ethyl)dimethyl phosphorothiolate *S*-(2-ethylthio)ethyl)-*O,O*-dimethyl phosphorothioate
Diazinon	Diethyl-2-isopropyl-6-methyl-4-pyrimidinyl phosphorothionate *O,O*-diethyl-*O*-(isopropyl-6-methyl-4-pyrimidyl) phosphorothioate
Dichlorvos	2,2-Dichlorvinylmethyl phosphate
Disulphoton	*O,O*-Dimethyl-*S*-(2-ethylthio)ethyl) phosphorodithioate
Dursban (chlorpyrifos)	*O,O*-Dimethyl-*O*-(3,5,6-(trtiethylchloro-2-pyridyl) phosphorothioate

Edifenphos	Ethyl-*S,S*-diphenyl phosphorodithiolate thionate *O*-ethyl-*S,S*-diphenyl phosphorothioate
Ethion	*S,S*-Methylene-*O,O,O′,O′*-tetraethyl phosphorodithioate
Fenchlorphos (ronnel)	*O,O*-Diemthyl-*O*-2,4,5-trichlorophenyl phosphorothioate
Fenthion	*O,O*-Dimethyl-*O*-(4-methylthio)-*m*-tolyl) phosphorothioate
Fonofos	Ethyl-*S*-phenylethyl phosphorothiolothionate Ethyl-*S*-phenylethyl phosphorodithioate
GS-13005	*S*-(5-Methoxy-2-oxo-1,3,4-thiadazaolin-3-yl)-*O,O*-dimethyl phosphorodithioate
Methocrotophos	Dimethyl-*cis*-2-(*N*-methoxy-*N*-methyl-carbamoyl)-1-methylvinyl phosphate 3-(dimethyloxyphosphinyloxy)-*N*-methoxy-*N*-methylisocrotonamide
Monitor	*O,S*-Dimethyl phosphoroamidothioate
Nemacur	Ethyl-4-(methylthio)-*m*-tolyl isopropyl phosphoramidate
Paraoxon	The oxygen analogue of parathion
Parathion	*O*-(4-Nitrophenyl phosphorothioate) parathion-ethyl
Phorate	*O,O*-Dimethyl-*S*-(ethylthio)methyl phosphorodithioate
Primiphos-ethyl	2-Diethylamino-6-methypyrimidin-4-yl diethyl phosphorothionate *O*-(2-diethylamino-6-mthylpyrimidin-4-yl)-*O,O*-diethyl phosphorothioate
Primiphos-methyl	2-Diethylamino-6-methylpyrimidin-4-yl-dimethyl phosphorothionate *O*-(2-diethylamino-6-methylpyrimidin-2-yl)-*O,O*-dimethyl phosphorothionate
Prothoate	*O,O*-Dimethyl-*S*-*N*-isopropyl carbamoyl methyl phosphorothiolothionate
Salithion	2-Sulphide of 2-methoxy-4-*H*-benzo-1,2,3-dioxaphosphrin
Tetrachlorvinphos	*trans*-2-Chloro-1(2,4,5-trichlorophenyl)vinyldimethyl phosphate
Velsicol VC-506	*O*-(4-Bromo-2,5-diethylchorophenyl)-*O*-methylphenyl phosphorothioate

Extraction procedures

In recent studies, researchers used microextraction procedures to extract organophosphorus insecticides from water samples prior to analysis [135–139]. This includes studies on fenitrophion and methyl parathion [124], methomyl [138] and 1,3-diisoactylimidazolium hexafluorophosphates [137]. Various workers studied the application of macroreticular resins, such as XAD-2, to recover organophosphorus insecticides from water samples [140–144].

Organophosphorus insecticides vary greatly in their polarity and in their extraction ability from water. The solvent used to extract the insecticide determines its extraction ability. Insecticides that are readily retained by nuclear carbon include azinphos-ethyl, azinphos-methyl, coumaphos, dichlorvos, maloxon, menazon, phosalone and vamidothion. These insecticides are best removed with alumina [145] or magnesium oxide [146].

Kotrikis et al. [147] studied the effect of mercury II on the extraction efficiency of organophosphorus herbicides in water.

Berhanu et al. [136] described a new design of equilibrium hollow fibre liquid-phase microextraction to determine the presence of three freely dissolved organophosphorus pesticides: diazinon (*O*,*O*-diethyl-*O*-2-isopropyl-4-methyl-6-pyrimidyl thiophosphate), chlorpyrifos (*O*,*O*-diethyl-*O*-[3,5,6-trichloro-2-pyridyl] phosphorothioate) and fenthion (*O*,*O*-dimethyl-*O*-4-methlthio-*m*-tolyl phosphorothioate). In this design the researchers used a 1.2–1.4 cm hollow fibre attached to the end of a 20 cm copper wire, impregnated with organic solvent, to extract the freely dissolved concentration of organophosphorus pesticides in various water samples. The limit of detection in reagent water using GC–MS in the selected ion mode was in the range of 15–80 ng L^{-1}. The RSD of the analysis (inter- and intraday) was 8.7%–30%.

Berhanu et al.'s [136] method was applied to the extraction of spiked lake and ground water samples. The ground water sample was spiked at 0.1 and 0.2 μg L^{-1} concentrations for the study analytes. The average extraction efficiency for the two concentrations was below 1%, showing the nondepletive nature of the extraction; the freely dissolved concentrations were measured as opposed to measuring the total concentrations. Good linearity was obtained for all the analytes in both reagent water and lake water samples with correlation coefficients, R^2, ranging from 0.991 to 0.996 for the concentration ranges of 25–400 ng L^{-1}. This method was simple and capable of running hundreds of samples in parallel with very minimal expense.

Gas chromatography

Recent work in GC involved developing suitable detectors that were ultrasensitive and specific for phosphorus in the presence of other elements, such as carbon, hydrogen, oxygen, halogens, nitrogen and sulphur. Electron-capture, flame ionisation, flame photometric, microcoulmometric, thermionic and electrolytic conductivity detectors were studied to detect organophosphorus insecticides that contained halogens, nitrogen or sulphur. Alkali flame and flame photometric detectors made the quantitative analysis of organophosphorus pesticide residues much easier.

The Karmen–Guiffrida detector (the thermionic detector) has a flame ionisation detector with an alkali salt ring placed on the flame tip, which enhances the sensitivity and selectivity. The device detects to the order of 10^4–10^5 times greater for the phosphorus-containing compounds compared to the equivalent carbon compounds. The selectivity is between 10^2 and 10^3 times greater for halides and nitrogen, 10^2 for sulphur and about 10 for arsenic. The response for phosphorus increases with increasing hydrogen flow, with some increase in the background current as well. A variety of alkali salts are used: potassium chloride, caesium bromide and rubidium sulphate. The type and shape of the tip and anode and the flow rates depend on the researcher's choice. De Loach and Hemphill [148,149] discussed the design of the rubidium sulphate detector. Under optimum conditions, they claim a sensitivity of 1 pg. A charcoal column clean-up was used before injection to prevent contamination within the column.

Brazhnikov et al. [150] developed a highly stable thermionic detector (the thermaerosol detector), which combined a conventional flame ionisation detector with an aerosol generator of alkali metal salts to analyse organophosphorus insecticides. The detector avoided the limitations of previous thermionic detector, such as rapid exhaustion of the alkali metal source, difficulty in replacing one salt by another and dependence on the flow rate of hydrogen, air and carrier gas for the detector's sensitivity and reproducibility. Brazhnikov et al.'s [150] detector had a simple design, was capable of rapidly replacing the salt without dismantling the detector, and operated for long periods (several thousand hours) without replacing the salt reservoirs.

Researchers used flame ionisation detectors to detect 0.05–0.5 mg kg^{-1} sulphur and halogen [151–154]. Helium plasma microwave emission spectrometry was used to determine phosphorus and sulphur-containing insecticides [151]. Microcoulometric detectors were used to determine insecticides containing phosphorus, sulphur and halogen [155–158]. Alkali metal salt flame photometric detectors were used to determine insecticides containing phosphorus, chlorine, nitrogen, sulphur and arsenic [159–169]. Electron-capture detectors were used to determine amounts down to 3×10^{-12} g of insecticides containing phosphorus [153–155,170]. Flame photometric detectors were used to determine amounts down to 0.005 ppm of insecticides containing phosphorus and sulphur [156].

Askew et al. [171] developed a general method to determine organophosphorus insecticide residues and their metabolites in river waters and sewage effluent utilising GC. The researchers used polar solvents, such as chloroform.

Mallet et al. [172] used Amberlite XAD-2 resins to preserve water samples containing traces of organophosphorus insecticides. This technique was instrumental in contaminated water sample preservation.

Puijker et al. [173] separated organic compounds containing phosphorus and sulphur from water on to XAD resin and then reduced the compounds with hydrogen sulphide. Phosphorous and sulphur were separated on a chromatographic column and detected at 526 and 384 nm, respectively, in a flame photometer. Their detection limits were 0.1 and 1 ng, respectively.

Cassita and Mallet [174] simultaneously determined fenitrothion, aminocarb and their corresponding phenolic hydrolysis products in water. Phenols were derivatised to the corresponding esters with acetic and anhydride directly in water. The researchers extracted the esters with methylene chloride and simultaneously analysed the esters with gas–liquid chromatography using a nitrogen–phosphorus selective detector. The limit of detection was 0.1 $\mu g\ L^{-1}$.

Suffet and Faust [175] applied the *P*-value approach to the liquid–liquid extraction of diazinon, parathion, malathion, fenthion and their oxygen analogues. They hydrolysed products from water samples prior to their analyses by GC on Reoplex-400 with electron-capture and flame ionisation detection.

Bargnoux et al. [176] carried out a comparative chromatographic study, utilising thin-layer and GC with phosphorus and sulphur dual detectors to detect parathion and malathion residues in water. They discuss the application of two low-temperature methods, lyophilisation and cryoconcentration, for the recovery of these insecticides from water. The use of lyophilisation and cryoconcentration improves the analysis of some particularly labile phosphorus and sulphur pesticides over previous methods by significantly reducing the risks of degradation in aqueous solution.

Thompson et al. [29] developed a multiclass, multiresidue gas chromatographic method for the determination of insecticides (organophosphorus, organochlorine and carbamate types) and herbicides in water samples. They extracted these compounds from water with methylene chloride and concentrated the extract with an evaporative technique utilising reduced pressure and low temperature. Compounds were segregated into groups using a column of partially deactivated silica and sequential elution with four different solvent systems. Carbamate residue, converted to its 2,4-diniorphenyl ether derivatives, was gas chromatographed via electron-capture detection into organohalogen compounds. Organophosphorus compounds were determined by GC using a flame photometric detector. Recovery studies were conducted on 42 halogenated compounds, 38 organophosphorus compounds and 7 carbamates. A total of 58 of 87 compounds tested produced recoveries in excess of 80%, 13 compounds yielded recoveries exceeding 60% and the remaining 16 compounds yielded recoveries below 60%.

GC was used [177] to determine the following organophosphorus insecticides at the microgram per litre level in water and waste water samples: azinphos-methyl, demeton-*O*, demeton-*S*, diazinon, disulphoton, malathion, parathion-methyl and parathion-ethyl. This method offered several analytical alternatives, dependent on the analyst's assessment of the nature and extent of interferences and on the complexity of the pesticide mixture found. Specifically, the procedure used a mixture of 15% v/v methylene chloride in hexane to extract organophosphorus insecticides from the aqueous sample. The method used column chromatography and liquid–liquid partition to eliminate nonpesticide interference and to preseparate pesticide mixtures. Identification was made by selective gas chromatographic separation and could be corroborated with two or more unlike columns. Detection and measurement were best accomplished by flame photometric GC using a phosphorus-specific filter. The electron-capture detector, though nonspecific, could also be used for target compounds. Confirmation of the identity of the compounds was made by GC–MS, only when a new or undefined sample type was being analysed and the concentration was adequate for such determinations.

Organochlorine insecticides, PCBs and phthalate esters interfere with the analysis of organophosphorus insecticides by electron-capture GC. Phosphorus-specific flame photometric detectors overcome this interference. Elemental sulphur interferes with the determination of organophosphorus insecticides by flame photometric and electron-capture GC. The elimination of elemental sulphur is discussed in detail.

Verweij et al. [178] described a procedure for the determination of phosphine containing insecticides in surface water. The researchers hydrolysed the insecticide to methylphosphinic acid, concentrated the acid by anion exchange and converted the acid to dimethyl ester. After clean-up on the microsilica gel column, the ester is analysed by GC using a thermionic phosphorous-specific detector. The detection limit is 1 nmol L^{-1}.

Insecticides determined by GC include paraquat and diquat [179,180], fenitroxon [181–183], dursban [181] and amino fenitrophion [184].

Kawahara et al. [185] described a procedure for the determination of phosphorothioate insecticides including parathion [*O*(-4-nitrophenyl phosphorothioate)] and parathion-methyl (dimethyl-*p*-nitrophenyl phosphorothioate). The researchers extracted solvent and used thin-layer chromatography in silica gel G (0.25 mm layer) for clean-up, and they identified the compounds with GC on an aluminium column (1.2 m × 0.6 cm o.d.) packed with equal portions of acid-washed Chromosorb P (supporting 5% of DX-200 silicone oil) and unwashed Chromosorb W (supporting 5% of DOW-11 silicone). The column operated at 180°C, with argon–methane (9:1) as a carrier gas (120 mL min^{-1}). They used electron-capture detection and confirmed the identity by infrared spectrometry.

High-performance liquid chromatography

Locorte and Barcelo [186,187] described a procedure for the determination of nanogram per litre levels of organophosphorus pesticides in ground waters using automated online liquid–solid extraction followed by liquid chromatography. The liquid chromatography detector was an atmospheric pressure chemical ionisation mass spectrometer using negative- and positive-ion modes of operation. They used the technique to determine several organophosphorus pesticides in ground water, such as mevinphos, dichlorvos, azinphos-methyl, azinphos-ethyl, parathion-methyl, parathion-ethyl, malathion, fenitrothion, fenthion, chlorfenvinphos and diazinon. This method required 100 mL of water combined with a prior automated online liquid–solid extraction step on OSP-1 autosampler containing C_{18} cartridges. The limit of quantification varied between 5 and 37 ng L^{-1} in positive-ion mode. Under negative-ion mode, only the parathion group (parathion and fenitrothion) had a better sensitivity, with a limit of quantitation of 5–515 ng L^{-1}. The remaining pesticides had two to four times higher limits of quantitation under negative-ion mode compared to those in positive-ion mode. Selected ion monitoring of the organophosphorous pesticide group–specific fragments $[(CH_3O)_2PO_2]^-$ and $[M+H]^+$ ions occurred under both modes. Optimal fragmentation was obtained for most of the studied analytes when sample voltage was between 10 and 30 V. The system was used for the certification of a ground water sample spiked at the nanogram per litre level with organophosphorus pesticides.

Barcelo [187–189] characterised 10 organophosphorus insecticides by HPLC–MS using acetonitrile:water:chloroacetonitrile (69:30:1) as the mobile phase. The negative-ion chemical ionisation spectra were dominated by the $[M-R]^-$ ion, with R being a methyl or ethyl group. At lower source temperatures $(M+C1)^-$ ions were formed. The negative-ion chemical ionisation sensitivity was approximately an order of magnitude higher than in the positive-ion mode.

Barcelo et al. [187] characterised selected pesticides by negative-ion chemical ionisation thermospray HPLC–MS. Ions observed in the negative-ion chemical ionisation spectra corresponded to the mechanisms of anion attachment [M + acetate], electron capture ($[M]^-$) and dissociative electron capture ($[M-R]^-$). Sensitivity was lower in the negative-ion chemical ionisation mode than in the positive-ion mode.

Recently, Wu et al. [190] applied dispersive liquid–liquid extraction combined with HPLC DAD detector for the determination of sulphenyl urea in water samples.

Rotich et al. [191] described HPLC with solid-phase extraction for the analysis of organophosphorus pesticide residues in water. Octadecylsilica disks (47-mm diameter) were used for solid-phase extraction. The researchers investigated parameters that affect both separation and extraction of methyl parathion, parathion and phoxim including mobile-phase composition, ionic strength temperature, pH and breakthrough volume. The application of optimised HPLC–solid-phase extraction to environmental samples gave reproducible results with low detection limits of $5\ \mu g\ L^{-1}$ for methyl parathion and parathion per litre. A precision of less than 8% was obtained for water.

HPLC has been used to determine various thermally labile organophosphorus insecticides in water [192–195].

Thin-layer chromatography

Various thin-layer chromatographic systems were described for the identification and determination of organophosphorus insecticides in natural waters Table 10.2.

Spectrometric methods

Venkataraman and Sathy Murthey [205] developed a simple and direct spectrophotometric method for the determination of parathion in water.

TABLE 10.2 Thin-layer chromatography of organophosphorus insecticides.

Compounds separated	Extraction solvent from water sample	Thin-layer plate	Migration solvent	Detection reagent	Detection limit	Reference
Malathion, dimethoate and ethion	Chloroform or dichloroethane	Silica gel G	Hexane-diethyl ether (9:2) (Malathion) or benzene-diethyl ether (9:2) (Malathion) or benzene-ethyl acetate (9:2) (Malathion), hexane-diethyl ether (9:2) (Dimethoate) or benzene-diethyl ether (5:1) (Dimethoate), benzene (Ethion) or hexane (Ethion)	Bromine vapour then 3,5-dibromo-*p*-benzoquinone-chloroamine in dimethyl formamide or bromophenol blue in acetone	0.01–0.2 μg, 0.4–0.45 dimethoate 0.3–0.35 ethion	[196]
Malathion, trichlorfon, dichlorvos	Chloroform	Silica gel G containing silver nitrate or, in the case of malathion fluorescein	Chloroform–acetate (9:1)	Spray with butter yellow then spot in UV radiation or, in the case of malathion exposure to bromine vapour	1–2 μg, Trichlorfon 0.1–0.12, dichlorvos 0.55–0.6 and malathion 0.5–0.6	[193]
Malathion	Dichloromethane	Silica gel F	Hexane	Palladium (II) chloride		[197,198]
Parathion, parathion-methyl and bromophos			Dichloromethane (1:1)	Aqueous		

23 Organophosphorus insecticides	Miscellaneous	Alumina G (Merck type E)	Acetone–hexane (7:93)	(1) Enzymic choline–erase inhibition method or (2) tetrabromomethane in acetone or citric acid in acetone		[198–200]
Parathion, parathion-methyl, phosphamidon and azinphos-methyl	Miscellaneous	Silica gel G containing 10% copper sulphate and 2% aqueous ammonia (5:1)	Hexane-diethyl ether 7:3	Heating for 10 m at 100°C	4–25 μg	[201,202]
Abate		Reverse-phase plate		Densitometry		[193,203]
	Chloroform extraction of acidified sample activated at 90°C	Silica gel previously	Hexane–acetone	1.N.N. Dimethyl PC phenyl azo/ aniline	9 μg	[203,204]

They hydrolysed a benzene extract of parathion and reduced it with zinc dust in acid solution. The resulting amino derivative was diazotised with sodium nitrite and hydrochloric acid. The diazo compound was coupled with naphthylethylenediamine hydrochloride, producing a magenta dye that was evaluated spectrophotometrically. This method is likely useful for other compounds.

Frobe et al. [206] compared hydrolytic degradation and oxygen combustion methods of decomposition for the organophosphorus pesticides, such as malathion, parathion and phosalone. The pesticides were decomposed to orthophosphoric acid then were spectrophotometrically determined as phosphomolybdenium blue. Pesticide adsorption on Amberlite XAD-4 resin prior to oxygen combustion increased the sensitivity of this method and eliminated inorganic phosphates and some nonpesticide organophosphorus compounds. A tandem approach involved increasing phosphorus content and using a positive cholinesterase inhibition test to determine the total organophosphorus pesticides and to differentiate among phosphorous sources in surface waters.

Electrophoresis using polypyrrole electrodes determined the presence of *N*-cetyltrimethyl and glyphosphate. A glyphosate concentration of 70 $\mu g\ L^{-1}$ gives a distinct test for glyphosate nitrosamine and aids in the determination of this compound in water.

Zen et al. [207] used square-wave polarography at a perfluorinated ionomer–clay-modified electrode (MCME) to determine the presence of paraquat down to 0.5 ppb in natural waters. The researchers found nontronite (Sw-1, ferruginous smectite) performed best when used in the electrode. The electrochemical behaviour of paraquat gave a cathodic peak at −70 V versus Ag/AgC1, which permits adequate quantification of the analyte. Linear calibration curves were obtained over the 0–80 ppb range with a detection limit of 0.5 ppb in pH 8 phosphate buffer solution for a 4 minute preconcentration time.

Calderbank and Youens [208] and Pope and Bennett [209] described a spectrophotometric method for the determination of paraquat in water in amounts down to 0.1 ppm.

Electrochemical methods

Polarographic methods were used to determine glyphosate [210–212], morfamquat and diquat [210] in natural waters.

A method based on electrophoresis analysed glyphosate [212–214]. In this method an electrospray condensation nucleation light scalloping detector was coupled with capillary electrophoresis. An *N*-acetyl trimethylammonium bromide prerinsing capillary electrode method was employed to reduce the separation time and the adsorption of glyphosphate on the capillary electrophoresis electrode. The protocol consisted of 15 minutes prerinsing of the capillaries and 5 minutes with ammonium nitrate buffer at pH 2.8 before analysis with *N*-acetyl trimethyl-ammonium bromide solution.

The resultant capillary linear wall coating lasted up to 10 hours without bleeding to interfere with the electrospray condensation nucleation light scattering detector signal. Calibration dates were linear over two orders of magnitudes, with an instrumental detection limit of 60 $\mu g\ L^{-1}$ and a method detection limit of 200 $\mu g\ L^{-1}$.

Miscellaneous

Other techniques to detect organophosphorus insecticides in environmental waters include flow injection analysis (dimethoate, malathion and parathion) [215], Raman spectrometry (parathion-methyl) [216], ion-mobility spectrometry (dimethoate and malathion) [217–219], enzymic assay (parathion-methyl, fenitrothion and fenitroxon) [220–222], isopropyl methyl phosphorofluoridate [sarin,*O*-ethyl-*S*(2-diproplyaminot)ethyl methyl phosphonofluoridate and parathion] [223–227], spectrofluorimetry (coumaphos and bayrusil) [227,228] and spectroscopy (glyphosate and toxaphene) [229].

Vogler [230] described a method for the separate determination of orthophosphoric acid esters (COP-phosphate) and condensed phosphates (POP-phosphate) in natural water samples. The sample (100 mL) of filtered natural water was treated with 30% aqueous hydrogen peroxide (0.2 mL) and exposed to UV radiation below 23°C under conditions of maximum irradiation intensity and minimum time. Orthophosphates were determined by standard procedures to obtain a measure of COP-phosphate. The other half of the sample was treated in a polythene flask with 10 N sulphuric acid (0.4 mL) on a steam bath for 4 hours. Excess hydrogen peroxide was then removed with potassium permanganate solution, and the orthophosphate was determined after 20 minutes. The value obtained was for total dissolved phosphate, POP-phosphate was obtained by subtraction.

Bagnoux et al. [231] used preconcentration at low temperatures to extract organophosphorus insecticides from water, prior to GC.

Volpe and Mallet [232] developed a method to determine fenitrothion (down to a concentration of 0.5 ng) and five fenitrothion derivatives in water by adsorption on XAD-4 and XAD-7 resins. They then used solvent elution and gas–liquid chromatography of the extract.

Aranjo and Compas-McCanela [219] discussed the degradation of methyl parathion in natural and sterilised waters. Experiments were prepared using natural waters collected at two aquatic systems in Rio de Janeiro, Brazil. The exposure to sunlight occurred using experimental bottles without headspace immersed in a swimming pool for temperature control. The results from the lake water experiments indicated the degradation of methyl parathion into the first half-lives under direct sunlight and shade was 4.41 and 6.89 days, respectively. The kinetic curve for methyl parathion degradation in river waters showed no difference when samples were sterilised and placed in direct sunlight or shade: the half-lives ranged from 2.75 to 5.37 days for sunlight and shade conditions, respectively. Photolysis plays an important role in the decomposition of methyl parathion in aquatic environments.

UV irradiation was used to decompose organophosphorus compounds prior to infrared and mass spectrometric analyses [233,234].

Anigbogl et al. [235] described mixed-mode electrokinetic capillary to determine neutral and charged cyclodextrins and resolve enantioseparations. The conditions mimic a mixture of micellar electro capillary chromatographic and dual cyclodextrin electrokinetic capillary chromatographic conditions. The researchers mixed sodium dodecylsulphate, carboxy methyl-β cyclodextrin, hydroxypropyl-β cyclodextrin and organic modifiers at various concentrations to achieve enantiomers separations of the three organophosphonil pesticides, such as cruformate (ruelene), malathion and fensulthion. Ruelene enantiomers were resolved using a mixture of 70 mm sodium dodecyl sulphate/15, carboxymethyl β cyclodextrin/45 mm hydroxypropyl-β-cyclodextrin/10% (v/v) acetonitrile in 20 mm borate buffer at pH 8.61 with applied voltage qf. 25 kV at 25°C. Malathion enantiomers were resolved using either 10 mm sodium dodecyl sulphate/50, carboxymethyl cyclodextrin/40 mm hydroxypropyl-β-cydodextrin or 50 mm carboxymethyl-β-cyclodextrin/50 mm hydroxypropyl-β-cyclodextrin/20% v/v methanol in 20 mm borate buffer. Fensulphothion enantiomers were resolved using a mixture of 75 m sodium/dodecylsulphate/12.5 mm carboxymethyl-β-cyclodextrin/45 mm hydroxypropyl-β-cyclodextrin in 20 mm borate buffer. The results demonstrate the mixed-mode electrokinetic capillary chromatography technique was capable of separating complex compounds, such as the organophosphoamidate enantiomers. The technique offers options for selectivity control by combining three or more pseudostationary phases in the background electrolyte. Optimisation in mixed-mode electrokinetic capillary chromatography is more simple than using solid stationary phases(s) in high-performance chromatography. In theory the separation selectivity of a mixed-mode electrokinetic capillary chromatographic system can be modified to exclude micellar electrokinetic capillary electrophoresis and the dual cyclodextrin electrokinetic capillary chromatography of the chiral selectors.

10.3 Urea herbicides

Some substituted urea herbicides are listed below:

Buturon	3-(4-Chlorophenyl)1-methyl-1(1-methylprop-2-ynyl)-urea
Chlorotoluron	3-(4-Bromo-3-chlorophenyl)-1-methocyl-1-methylurea
Chlorotoluron	3-(3-Chlorotoluyl)-,1-dimethyl urea
Chlorooxuron	3-[4(4-Chlorophenoxyl)]-1,1-dimethylure
Diuron	3-[4(4-Chlorophenoxy)phenyl]-1,1-dimethylurea
Diuron	3-(3,4-Dichlorophenyl)-1,1-dimethylurea
Linuron	3-[3,4-Dichlorophenyl]-1-methoxymethylurea
Monolinuron	3(4-Chlorophenyl)-1-methoxymethylurea

Monouron	4(4-chlorophenyl)-1-dimethylurea
Siduron	1-(2-Methylcyclohexyl-3)-phenylurea
Fenuron	1,1-Dimethyl-3-phenylura
Neburon	1-Butyl-3-(3,4-dichlorophenyl)-1-methylurea
Metabenzthiazuron	
Metaxuron	
Metobromuron	

Gas chromatography

Phenylurea herbicides decompose easily with heat, making GC of these compounds difficult. These compounds can be gas chromatographed intact under careful conditions [210,211,213]. Alternatively, phenylurea herbicides can be hydrolysed to the corresponding substituted anilines and determined by GC directly [236] or as derivatives [237]. GC using a nitrogen-specific detector has also been used [213,214,236,238–240].

In an electron-capture method for estimating diuron [3-(3,4-dichlorophenyl)-1,1-dimethylurea] in surface waters, McKone and Hance [241,242] extracted the water sample (100 mL) with dichloromethane, and the lower layers were combined and washed with water. After filtering through cotton wool the organic solvent was evaporated under reduced pressure at 35°C. Saturated aqueous sodium chloride was added to the residue, and the mixture was shaken, then 2,2,4-otrimethyl-pentane was added and the mixture was shaken again. An aliquot of the organic layer was then subjected to GC in a stainless steel column packed with 5% of E301 (methyl silicone) on gas chrome Q (60–80 mesh). Recoveries from control and from pond, canal and river waters containing 0.001–1 ppm of diuron were about 94%; the coefficient of variation at the higher levels was about 70%.

In a method described by Rosales [243], 1-methoxyl-1-methyl-3-phenylurea herbicides were hydrolysed by phosphoric acid to give aniline and *N*-methodimethamine. These products and impurities in the commercial product were titrated with sodium nitrite solution. Aniline and certain by-products were determined separately by GC on a glass column of 10% silicone oil OV-17 on Chromosorb Q (80–100 mesh) from 100°C to 200°C increasing at 4.5°C min^{-1}. They used thermal conductivity detection and helium carrier gas.

Delue et al. [244] used two-dimensional thin-layer chromatography and GC to separate and identify 10 urea herbicides, metouron, fenuron, monouron, isoproturon, chlorotoluron, diuron, metabenzuron and buturon in concentrations down to 4 $\mu g\ L^{-1}$ in river waters. Four different adsorptions were compared. The eluting solutions were diethyl ether-toluene (1:3), diethyl ether-toluene (2:1), equal volumes of 1 and 2 and chloroform-nitromethane (3:1).

De Kok et al. [245] described two complete schemes for the analysis of 15 phenylurea herbicides and their degradation products. They allowed the selective determination of each separate class of compounds and anilines in each other's presence. They give descriptions of procedures involved in extractions, clean-up, catalytic hydrolysis of herbicides to anilines derivatisation with heptafluorobutyric anhydride and liquid or gas chromatographic fraction and/or separation.

Cohen and Wheals [246] used GC equipped with an electron-capture detector to determine 10 substituted urea and carbamate herbicides in river water, in concentrations down to 0.001–0.05 ppm. The methods applied to urea and carbamate herbicides that can be hydrolysed to yield an aromatic amine. A solution of the herbicide was detected on the silica gel G plate at herbicide standards (5–10 μg) and developed with chloroform or hexane–acetone (5:1). The plate containing the separated herbicide or the free amines was sprayed with 1-fluoro-1,4-dinitrobenzene (4% in acetone) and heated at 190°C for 40 minutes to produce the 2,4-dinitrophenyl derivative of the herbicide amine moiety. Acetone extracts of the areas of interest were subjected to GC on a column of 1% of XE-60 and 0.1% of Epikote 1001 and Chromosorb C (AE-DCMS) (60–80 mesh) at 215°C.

High-performance liquid chromatography

Sulphonylurea forms a class of herbicides introduced in the 1980s that contain labile and weakly acidic compounds. Compared to older herbicides, sulphonylurea herbicides have much lower use rates and are more rapidly degraded in soil. When present, very low concentrations (low parts per trillion) of these herbicides are expected in environmental waters. Monitoring of these herbicides is a particularly challenging problem. Although various separation techniques, such as chromatography–MS of sulphonylurea derivatives [247], supercritical fluid chromatography [247] and capillary electrophoresis [248], have been proposed to analyse sulphonylurea in various matrices, liquid chromatography is ideal [249–266]. Liquid chromatographic methods have advanced with the introduction of robust and sensitive devices, such as thermospray and electrospray, which interface liquid chromatography to MS.

Volmer et al. [255] evaluated the performances of both thermospray and electrospray ion sources to determine trace levels of sulphonylurea herbicides in water. They concluded that electrospray ion sources combined with a tandem mass spectrometer were superior to thermospray sources.

Smith and Lord [256] used liquid chromatography for the determination of chlorotoluron [3-(3-chlorotouyl)-1,1-dimethylurea] residues. Diuron and monouron interfered in the chromatographic system used, so their technique was not successful.

Farrington et al. [257] and others [258–265] described HPLC procedures for the determination of residues of the following phenylurea herbicides in water:

- Chlorobromuron [3-(4-bromo-3-chlorophenyl)-1-methoxy-1-methylurea]
- Chlorooxuron [3-(4,4-chlorophenoxy)phenyl-1,1-dimethylurea]
- Chlorotoluron [3(3-chlorotoluyl)-1,1-dimethylurea]

- Diuron [3-(3,4-dichlorophenyl)-1,1-dimethylrea]
- Linuron [3, (3,4-dichlorophenyl)-1-methoxy-1-methylurea]
- Metobromuron
- Monolinuron [3(4-chlorophenyl)-methoxy-1-methylurea]
- Monuron [3(4-chlorophenyl)-1,1-dimethylurea]

The lower limits of detection were estimated at 0.01 ppm for river water, coextractives and signal noise interfered with levels below this concentration. Temperature was controlled to remain below 55°C to prevent degradation at higher temperatures. A clean-up procedure was required to remove coextractives.

Ruberu et al. [267] described the application of HPLC methods to determine the presence of phenyl urea herbicides in water. The target compounds included chlorotoluron, diuron, fluometuron, isoproturon, linuron, metobromuron, metoxuron, nonuron, heburon and siduron. The researchers subjected water to solid-phase extraction using either automated solid-phase extraction with 47 mm C_{18} Empore disks or online precolumn concentration. Herbicides were separated on a C_{18} reversed-phase column with an acetonitrile–water gradient and were detected with either a diode-array detector or a postcolumn photolysis and derivatisation detector system. Photolysis converted the phenyl ureas to monoalkylamines that were derivatised to fluorescent isoindoles by reaction with *o*-phthalaldehyde and 2-mercaptoethanol. The diode-array detector monitoring at 245 nm was linear over three decades with instrument detection limits of ~0.01 mg L^{-1}. Solid-phase efficiency was between 48% and 70% in laboratory reagent water, but use of the internal standard quantitation method improved accuracy. High total dissolved solids and total organic carbon values in surface water improved recoveries relative to laboratory reagent water for all of the phenyl ureas. In Colorado River water, spiked at 1 or 50 μg L^{-1}, mean recoveries ranged from 74% to 104%. Method detection limits ranged from 4 to 40 ng L^{-1} (parts per trillion) with the diode-array detector instrument. Postcolumn photolysis-derivatisation detection was highly specific, but resulted in a slight loss in chromatographic efficiency; average detection limits were ~5 times higher using a single set of detection conditions. Methods based on solid-phase extraction followed by HPLC with diode-array detection have practical utility for trace analysis of phenyl ureas in drinking water or surface waters.

Wu et al. [190] discussed the use of dispersive liquid–liquid microextraction combined with HPLC with a diode-array detector to determine sulphonylurea herbicides in water samples. The herbicides detected were metsulphuron-methyl, chlorsulphuron, bensulphuron-methyl and chlorimuron-ethyl. Parameters that affect the extraction efficiency included type and volume of extraction and disperser solvent, extraction time and salt addition. The researchers investigated and optimised these parameters. Under optimum conditions the enrichment factors were in the range between 102 and 216, and the method detection limits were 0.2–0.3 μg L^{-1}.

Saha and Kulshresthe [268] studied the hydrolysis kinetics of the sulphonylurea herbicide sulphosulphuron. They investigated the kinetics of hydrolytic degradation of sulphosulphuron to predict the fate of the herbicide in an aqueous environment. The study revealed that the hydrolytic degradation of sulphosulphuron followed first-order kinetics and was dependent on pH and temperature. Hydrolysis rate was faster in acidic conditions ($t_{1/2}=9.24$ days at pH 4.0) than alkaline conditions ($t_{1/2}=14.14$ days at pH 9.2). A several fold increase in the degradation rate was found when temperature was increased from 10°C ± 1°C ($t_{1/2}=518$ hours) to 50°C ± 1°C ($t_{1/2}=10$ hours). Activation energy (E_a) for hydrolytic degradation of the molecule was calculated at 63.87 kJ mol^{-1}. Media pH and temperature effects were coupled to derive a complex equation estimating the overall effect of these two abiotic factors. The major degradation mechanism was cleavage of the sulphonylurea bridge to yield sulphonamide and aminopyrimidine. The researchers discuss the possible significance of these results to the persistence of the herbicide in the field.

Both Partisil aluminosilicate [269] and cellulose [270] have been used to preconcentrate urea herbicides in water samples.

10.4 Sulphonylurea herbicides

Si et al. [271] applied coupled liquid chromatography–MS to hydrolyse the herbicide ethametsulphuron-methyl [methyl-2-[[[[(4-ethoxy-6-methylamino-1,3,5-triazin-2-yl)amino]carbonyl]amino]sulphonyl]benzoate] in aqueous buffers at different pH values. The reaction was first-order and pH dependent. Ethametsulphuron was more persistent in neutral or weakly basic solutions than in acidic solutions. Eleven degradation products were detected and tentatively identified by liquid chromatography and mass spectrometric analysis. At all pH values studied the primary pathway of degradation was the cleavage of the sulphonylurea bridge. However, minor degradation pathways were observed, such as O-di-ethylation, N-demethylation and opening of the triazine ring.

10.5 Phenoxyacetic acid–type herbicides

Some phenoxyacetic acid–type herbicides are listed below:

2,4-D ester and sodium salts (2,4-dichlorophenoxyacetic acid)
2,4-DP (dichloroprop); (2,4-dichlorophenoxypropionic acid)
2,4-DB (2-dichlorophenoxyl-butyric acid)
MCPA (4-chloro-2-methylphenoxyacetic acid)
MCPB (4-(chloro-2-methyl phenoxyl) butyric acid)
Silvex (1(2,4,5-trichlorophenoxyl)-propionic acid)

MCPP (mecoprop; mixture of mecoprop and 2-(2 chloro-4-methylphenoxyl)propionic acid)
2,5,5-T (2,4,5-trichlorophenoxyacetic acid)
Dicamba
Trifluralin (2-methoxyl-3,6-dichlorobenzoic acid)
Methoxychlor
Fenoprop

Devine et al. [272] adjusted a water sample (1 L) to pH 2 with hydrochloric acid and extracted phenoxyacetic acid using benzene. The extract was dried over sodium sulphate and methylated with diazomethane in ethyl ether. Acetone was added, an aliquot was analysed by GC with nitrogen as the carrier gas and detection was by electron capture. The minimum detectable amount of pesticide in water was two parts per 10^9 for MCPA (4-chloro-2-methyl-phenoxyacetic acid) and 0.01–0.05 parts per 10^9 for 2,4-D(2,4-dichlorophenoxyacetic acid) and its esters [2,4,5-T (2,4,5-trichlorophenoxyacetic acid)], dicamba, trifluralin (2-methoxy-3,6-dichlorobenzoic acid) and fenoprop. Recoveries were 50%–60% for MCPA and dicamba and 80%–95% for the other compounds.

Croll et al. [273] reviewed back-flushing with electron-capture GC to determine phenoxyacetic acid–type herbicides in water. The researchers successfully used a variety of stationary phases, temperatures ranging from 25°C to 225°C and nitrogen flow rates ranging from 25 to 220 mL min^{-1}.

Colas et al. [274] described methods to separate and identify phenoxyalkanoic acid herbicides down to approximately 1 ppm in concentration. Those present as salts or esters were hydrolysed by heating the water samples (1 L) under reflux with sodium hydroxide for about 1 hour, and the resulting free acids were extracted at pH 2 with chloroform or dichloromethane. After complete evaporation of the solution the residual free acids were dissolved in acetone and treated with diazomethane. The methyl esters were then analysed on a temperature-programmed column (1.5 m × 6 mm) containing 5% silicone DOW-710 on Chromosorb WAW (45–60 mesh). Helium was used as the carrier gas. Results were presented for MCPA, MCPP and 2,4-D. Possible interferences from phenols, chlorinated biphenyls and surfactants were discussed.

Larose and Chau [275] claimed that alkyl esters were subject to incorrect identification if several herbicides were present due to similar retention times for several common phenoxyacetic acid–type herbicides. They stated that electron-capture detection of alkyl esters by some herbicides, such as MCPA and MCPB, had poor sensitivity, so this method was inadequate for identifying these compounds in water. In addition, the methyl ester of MCPA had a very short retention time close to the solvent front and was prone to interference from sample coextractives, which usually appeared in the solvent front. The authors claimed the MCPA methyl ester is difficult to detect at higher levels because it overlaps with other coextraction peaks during GC using the same parameters as those to detect organochlorine pesticides. Other derivatives have been considered.

Carnac [276] studied a modified technique of dynamic distribution in liquid–liquid systems to concentrate traces of organic substances. The technique was used to determine the presence of phenoxyalkanecarboxylic acids by gas–liquid chromatography with a flame ionisation detector and chloroform as the solvent, the limit of detection was 50–10 μg L^{-1}.

Chau and Terry [277,278] discussed the disadvantages of the GC of methyl esters produced with diazomethane. They developed reaction conditions for forming 2-chloroethyl and pentafluorobenzyl esters of phenoxyacetic acids.

Agemian and Chau [279] reported a method to determine low levels of 4-chloro-2-methylphenoxyacetic acid and 4-(4-chloro-2-methylphenoxyl)-butyric acid in water by derivatisation with pentafluorobenzyl bromide and other reagents [280] followed by GC. Agemian and Chau [279,281] founded that the few organochlorine pesticides eluted in the same fraction as the pentafluorobenzyl derivatives of the phenoxyacetic acid herbicides did not interfere because they had distinct retention times. They also founded that eluting with organophosphorus pesticides did not interfere in the derivatives.

Lopez-Avila et al. [282] used isotope dilution GC–MS to determine dicamba and 2,4-D in natural waters at the low microgram per litre level. Labelled stable isotopes were added to the sample before extraction, and the ratio of unlabelled isotopes to labelled isotopes was used to quantify the unlabelled compounds. Average recoveries exceeded 84%, and the RSD was better than 19%.

An acid-catalyst esterification reaction with boron trifluoride–trifluoroethanol identified triclophon (3,5,6-trichloro-2-pyridyloxyacetic acid) in river water in amounts down to 5 ng L^{-1} [283]. Triclophon was first extracted from acidified river water with diethyl ether, and the resulting concentrate was esterified with nitrogen at 80°C for 1 hour. The trifluoroethyl ester product was cleaned with silica gel column chromatography and determined by GC with electron-capture detection. Recoveries from actual river waters were 90%–93% with coefficients of variation of less than 4%. With slight modifications at the clean-up stage the method could simultaneously determine MCPA, 2,4-D and 2,4,5-T.

Lee et al. [284] developed a multiresidue method with a low detection limit for 10 commonly used acid herbicides in nonsaline waters. The herbicides were dicamba, MCPA, 2,4-DP, 2,3,6-TBA, 2,4-D, silvex, 2,4,5-T, MCPB, 2,4,5-DB and picloram. The method involved solvent extraction and the formation of pentafluorobenzyl esters. The derivatives were quantified by capillary column GC with electron-capture detection. The detection limit was 0.05 μg L^{-1}. Recoveries of herbicides from spiked Ontario Lake water (0.5–1.0 μg L^{-1}) measured 73%–108%, except for picloram recovery, which was 59% at the 0.1 μg L^{-1} level.

Thin-layer chromatography

Bogacka and Taylor [285] determined the presence of 2,4-D and MPCA herbicides in water using thin-layer chromatography. In this method a 1 L sample of filtered water was treated with 50 g of sodium chloride and 5 mL of hydrochloric acid. The herbicides were then extracted into ethyl ether. The extract was dried with anhydrous sodium sulphate, concentrated to a few millilitres and passed through a column (180 mm × 15 mm) of silica acid with 90% methanol–acetic acid (9:1) as the stationary phase. The herbicides were eluted with 150 mL of light petroleum saturated with the methanol–acetic mixture. The eluate was evaporated to dryness, and the residue was dissolved in ether and concentrated to about 0.1 mL. The researchers performed thin-layer chromatography on silica gel G–Kieselgel G (2:3) (activated for 30 minutes at 120°C) with light petroleum–acetic acid–liquid paraffin (10:1:2) as the solvent. The developed plates were air-dried, sprayed with 5% silver nitrate solution and dried. This product was sprayed with 2 M potassium hydroxide–formaldehyde (1:1), dried at 130°C–135°C for 30 minutes, sprayed with nitric acid and observed with UV illumination. The spots were compared with standards to determine the compounds present. Evaluation by the method of standard addition for 2,4-D and MCPA gave recoveries of 95.1% and 88.8%, respectively, with standard deviations of 14.2% and 14.3%, respectively.

Bogacka and Taylor [280] also examined thin-layer chromatography of 2,4-DP (dichlorprop) and MCPP [mixture of mecoprop and 2-(2-chloro-4-methylphenoxy) propionic acid]. In this method the ethyl ether extract of the sample was purified on a column of silicic acid. The herbicides were separated by thin-layer chromatography on silica gel–Kieselgel (2:3) with light petroleum–acetic acid–kerosene (10:1:2) as the solvent. The sensitivity for both compounds was measured at 3 $\mu g\,L^{-1}$. The average recoveries of dichlorprop and MCPP were 85.7% and 87.44%, respectively, and the corresponding standard deviations were 13.9% and 15.5%.

Meinard [286] described a new chromogenic reagent to detect phenoxyacetic acid herbicides on thin-layer plates. The separated phenoxyacetic acids were detected as violet spots on a white background after spraying the plate with a solution of chromotrophic acid (4 g) in water (490 g) and sulphuric acid (56 g) then heating at 160°C. The researchers sprayed the plates with silver nitrate reagent then exposed them to UV radiation and determined that the limits of detection for 2,4-D, 2,4,5-T and MCPA ranged from 0.05 to 0.2 μg.

Paper electrophoresis

Purkayastha [287] examined paper electrophoresis to identify nine ionisable chlorinated phenoxyacetic acid–type herbicides including 2,4-D, 2,4,5-T, MCPA, fenoprop, dicamba, trifluralin (2-methoxy-3,6-dichlorobenzic acid) and Picloram (4-amino-3,5,6-trichloropicolinic acid). The researchers applied solutions to paper moistened with pyridine–acetic buffer solution at pH 3.7, 4.4 or 6.5 and applied a voltage of 2–4 kV. After 30 minutes the paper was air-dried, sprayed with ammoniacal silver nitrate solution and exposed to UV radiation. Increased spot mobility was observed when the applied voltage varied from 2 to 4 kV (potential gradients of 50–1000 V cm^{-1}) and when a foreign electrolyte (e.g. potassium nitrate) was added to the buffer. Addition of methanol to the buffer resulted in decreased mobility and greater variance in the relative mobilities of the compounds.

Miscellaneous

Arjmand et al. [288] determined the presence of dicamba using solid-phase extraction and ion-pair HPLC. The detection limit was 1.6 μg L^{-1}.

Marshall [289] used two methods for the infrared analysis of dicamba–MCPA and dicamba–2,4-D formulations. The 'indirect' method involved precipitation of the herbicides with hydrolytic acid and extraction with chloroform. The chloroform extract was evaporated to dryness, the residue was dissolved in acetone and the herbicides were determined by measuring infrared extinctions at the relevant wavelengths. The 'direct' method involved dissolving the sample in acetone and measuring infrared extinctions. Although both methods gave good precision, the 'indirect' method was more accurate.

Bogacka [290] used 4-aminiphenazone as a reagent for the spectrophotometric determination of phenoxyacetic acid herbicides (2,4-D, dichlorprop and MCPA) in water. The herbicides were extracted from an acidified 1 L sample of water with ethyl ether. The extract was evaporated, and the residue was eluted for 1 hour with 10 g of pyridine hydrochloride at 207°C–210°C for 2,4-D or at 225°C–230°C for dichlorprop and MCPA. The resulting phenol derivative was steam distilled into aqueous ammonia (IM), acidified, extracted with light petroleum and reextracted into 0.05 m aqueous ammonia for coupling with 4-aminophenazone and potassium ferricyanide. The extinction of this solution was measured at 515, 505 or 151 nm, for 2,4-D, dichlorprop or MCPA, respectively. The respective sensitivities were 20, 20 and 80 μg L^{-1} of water; the corresponding standard deviations of the recovery were 4.0%, 8.5% and 3.5%. The method cannot determine the presence of mixed herbicides.

Suffet [291] evaluated liquid–liquid extraction techniques for separating phenoxyacetic acid herbicides from river water. He used the *P*-value concept (defined as the fraction of the total solute that distributed itself in the nonpolar phase of an equivolume solvent pair) to develop equations based on liquid–liquid extraction theory. He related the number of extractions to the water-to-solvent ratios for maximum recovery of the herbicide. Calculations showed that a pesticide with a *P*-value of 0.90 or greater in an aqueous system can be 95% extracted from the aqueous phase by up to five extractions with a total volume of solvent up to 50 mL^{-1}. Suffet [291] also showed that the best solvents for 2,4-D and its butyl and isopropyl esters were ethyl ether or ethyl acetate; the best solvent for 2,4,5-T and its butyl and isopropyl esters was benzene. A 90% recovery of 2,4-D from 1 L of an aqueous solution was obtained by a two-stage serial extraction with 200 and 50 mL of ethyl acetate under *P*-value conditions.

The most commonly used solvents for extracting phenoxyalkanoic acids have been ethyl ether [292–295] and chloroform, benzene has also been used [272,291–295].

Chemical derivatisation of phenoxyalkanoic acid herbicides has been used to decrease polarity and increase volatility of compounds for gas–liquid chromatographic analysis. Alkyl esters [296–304] have been used extensively for this purpose. Devine et al. [272] and Colas et al. [274] discussed the examples of earlier work utilising diazomethane to form methyl esters from phenoxyalkanoic acid herbicides.

Kasna and Sklodal [305] described a four-channel electrochemical immunosensor to detect phenylalkanoic acid herbicides.

10.6 Triazine type

Some triazine herbicides are listed below:

- Atrazine (2-chloro-4-ethylamino-6-isopropylamino-1,3,5-triazine)
- Propazine
- Simazine (2-chloro-4,6-bis-ethylamino-1,3,5-triazine)
- Prometon
- Prometryne
- Atraton (2-ethylamino-4-isopropylamino-6-methoxy-1,3,5-triazine)
- Ametryne (2-ethylamino-4-isopropylamino-6-methylthio-1,3,5-triazine)
- Terbutryne
- Terbutylazine (4-*tert*-butylamino-2-chloro-5-ethylamino-1,3,5-triazine)
- GS26571 (2-amino-4-*tert*-butylamino-5-methoxyl-1,3,5-triazine)
- GS30033 (2-amino-4-chloro-5-ethylamino-1,3,5-triazine)
- Terbumeton
- Secbumeton

Gas chromatography

GC is the preferred method for detecting triazine herbicides. McKone et al. [306] compared GC methods to detect atrazine (2-chloro-4-ethylamino-6-isopropyl-amino-1,3,5-triazine), ametryne (2-ethylaminio-4-isopropylamino-6-methylthio-1,3,5-triazane) and terbutryne in water. The herbicides were extracted from water with dichloromethane, and the dried extracts were evaporated to dryness at a temperature below 35°C. The three herbicides were separated by GC on a glass column (1 m × 4 mm) of 2% neopentyl glycol succinate on Chromosorb W (80–100 mesh) operated at 195°C with a RbBr-tipped flame ionisation detector. The researchers detected 0.001 ppm of each herbicide. This method was superior to spectrophotometric methods.

Purkayastha and Cochrane [307] compared electron-capture and electrolytic conductivity detectors in the gas chromatographic determination of prometon, atraton, (2-thylamino-4-isorpopylamino-6-emthoxy-1,3,5-triazine), propazine, atrazine, (2-chloro-4-ethylamino-6-isopropylamino-1,3,5-triazine), prometryne, simazine (2-chloro-4, 6-bis-ethylamino,1,3,5-trianxine) and ametryne (2-ethylamino-4-isorporpylamino-6-methylthio-,1,3,5-triazine) in inland water samples. They found that the electrolytic conductivity detector had a wider application than a ^{63}Ni electron-capture detector. Use of the ^{63}Ni electron-capture detector necessitated a clean-up stage for all the study samples. The conductivity detector could be used in water analysis without sample clean-up. The researchers observed efficient recoveries of atrazine added to water by extraction with dichloromethane.

Ramsteiner et al. [308] compared alkali flame ionisation microcoulmometric, flame photometric and electrolytic conductivity detectors to determine the presence of triazine herbicides in water. The researchers cleaned methanol extracts on an alumina column and detected 12 herbicides by GC with conventional columns containing 3% Carbowax 20M on 80–100 mesh Chromosorb G.

Hormann et al. [309] monitored various European rivers for levels of atrazine, terbumeton, dealkylated metabolites GS26571 (2-amino-4-tertbutylamino-5-methoxy-1,3,5-triazine) and GS30033 (2-amino-4-chloro-5-ethylamino-1,3,5-triazine). The compounds were extracted into dichloromethane and quantitated by GC with nitrogen-specific detection. Selected results were verified by GC with mass fragmentographic detection. The limit of detection was usually 0.4 $\mu g\ L^{-1}$.

The US Environmental Protection Agency [305] issued a gas chromatographic method to determine the following herbicides at the microgram per litre level in water and waste water: ametryne, atraton, atrazine, prometon, prometryne, propazine, secbumeton, simazine and terbutylazine. The method described an efficient sample extraction procedure and provided a method to eliminate nonpesticide interferences and prepared pesticide mixtures using column chromatography. Identification was made by nitrogen-specific gas chromatographic separation, and measurement was accomplished with an electrolytic conductivity detector or a nitrogen-specific detector. Steinheimer and Brooks [310] developed a multiresidue method for the simultaneous determination of seven triazine herbicides in surface and ground water with a detection limit at the micrograms per litre level. The technique used solvent extraction, gas chromatographic separation and nitrogen-elective detection devices. The researchers examined solid-phase extraction techniques using chromatographic grade silicas with chemically modified surfaces for three natural water samples. Solid-phase extraction was found to produce a rapid and efficient concentration with quantifiable recovery. Solid-phase extraction was presented as an alternative to liquid–liquid partition.

Jahda and Marha [311] investigated the isolation of *s*-triazines from water using continuous steam-distillation extraction followed by gas–liquid chromatography. They report the recovery of seven triazine herbicides from water at pH values of 5.7 and 9: propazine, tertbutylazine, atrazine, prometryne, terbutryne, desmetryne and simazine. Recovery rates were independent of pH but generally improved when steam-distillation extraction time was increased from 1 to 3 hours. Low recovery rates were obtained for simazine. Atrazine only produced decent recovery rates after 3 hours of steam-distillation extraction.

Lee and Stokker [312] developed a multiresidue procedure for the quantitative determination of 10 triazines in natural waters by GC using a nitrogen–phosphorus detector. The researchers analysed ametryne, atraton, atrazine, cyanazine, prometon, prometryne, propazine, simazine, simetone and simetryne. All compounds were successfully quantified on the Ultrabond 20 M and 3% OV-1 columns. Extraction was by methylene chloride and clean-up was by florisil. Recoveries of triazines at 10, 1.0 and 01 $\mu g\ L^{-1}$ were between 87% and 108%, and simetone and simetryne at 0.1 $\mu g\ L^{-1}$ were only 80%.

Isotope dilution GC–MS was used [313] to detect 0.1–1 $\mu g\ L^{-1}$ of atrazine, lindane, diazinon and pentachlorophenol in natural water. An accuracy of 86% and a precision of 8% were demonstrated.

Zangwei et al. [314] determined atrazine in water at sub-ppt levels using solid-phase extraction and GC–high-resolution MS. They used a C_{18}-bonded cartridge followed by column chromatography on florisil to remove interfering substances.

Deleu and Copin [315] used GC–MS to determine the presence of atrazine in water down to 0.1 $\mu g\ L^{-1}$.

High-performance liquid chromatography

This method has been used to determine microgram per litre levels of triazine herbicides (simazine, atrazine [316], prometryn and terbutryne) and their degradation products and *N*-methylcarbamate insecticides (propoxur, carbamyl and methiocarb [317]) in surface water.

Topuz [317] used HPLC. The researchers preconcentrated a 0.5 L water sample by passing it through a 1 g C_{18} solid-phase extraction cartridge. The retained compounds were eluted with 5 mL of methanol from the cartridge. The pesticides were separated and quantified by reversed-phase HPLC with UV diode-array detection. Analytical separation was performed using concave gradient elution with acetonitrile and water on a C_{18} column. Prometryn and terbutryn were determined at 240 nm, propoxur and methiocarb were determined at 204 nm and the remainder were determined at 220 nm. Recoveries varied from 85% to 102% for concentrations between 0.125 and 0.2 $\mu g\ L^{-1}$. The limits of detection for the compounds investigated are in the range of 0.005–0.012 $\mu g\ L^{-1}$.

Thin-layer chromatography

Researchers have applied thin-layer chromatography to detect triazine herbicides in water samples [309–323].

Zawadzka et al. [319] and Abbott et al. [320] used thin-layer chromatography to determine the presence of simazine, atrazine and prometryne herbicides in water with dichloromethane or ethyl ether at pH 9. The organic extract was condensed and applied to a column of basic aluminium oxide (activity III), and the herbicides were eluted with ether containing 0.5% water. The eluate was condensed and applied to a layer of silica gel G impregnated with fluorescein. The chromatograms were developed with chloroform–acetone (9:1). The plates were dried, and the spots were detected with a spray of 0.5% Brilliant Green (CI Basic Green 1) in acetone and with exposure to bromine vapour. The plates were evaluated planimetrically. For example, samples containing 5–100 μg of herbicide per litre exhibited recoveries between 83% and 97%.

Thin-layer chromatography was used to determine ring-labelled [^{14}C] Ametryne in water [284]. The herbicide was applied to silica gel plates as a methanol solution. Following development, the researchers located the spots under UV radiation, the spots were removed and treated with liquid scintillation solution for counting. The limit of detection was 5 ng.

Smith and Fitzpatrick [295] described a thin-layer method for the detection of 0.01 $mg\ L^{-1}$ of herbicide residues in water. The residues included barban, diuron, linuron, monouron, simazine, trifluralin, bromoxynil, dalapon and dicamba.

Miscellaneous

Enzymic immune assay [323,324] and high-resolution fast atom bombardment MS [325,326] have been used to detect triazine type herbicides.

Baker's yeast cells (*Saccharomyces cerevisiae*) were used to isolate and enrich desisopropylatrazine, desethylatrazine, hydroxyatrazine, simazine, cyanazine, atrazine, carabaryl, propanil, linuron and fenamiphos. The pesticides were preconcentrated using online solid-phase extraction on silica gel petroleum with immobilised yeast cells followed by liquid chromatography with diode-array detection. The degree of selectivity was evaluated by comparing the chromatograms obtained after online sample preconcentration on the yeast precolumn with those obtained by online solid-phase extraction on a C_{18}-filled precolumn. The RSD to detect pesticides at the 0.3 μg L^{-1} concentration level ranged from 1% to 9%, which depended on the pesticide and the type of water. Detection limits, within the range 0.01–0.5 μg L^{-1}, were obtained by percolating 25 mL of water sample without an additional clean-up step.

10.7 Carbamate type

Some carbamate types of herbicides are listed below:

Aminocarb (Metacil)
Metalkinate
Carbaryl (Sevin)
Carbofuran (Fududan)
Methiocarb (Mesurol)
Propoxur (Baygon)
Mexacarbate (Zetran)
Aldicarb
M–*S*-Butyl phenyl methyl(phenylthio)carbonate
Methanoyl
Propham
Carbine (barban)
Benthiocarb

Gas chromatography

Crosby and Bowarg [327] described a method to detect carbaryl (Sevin) (1-maphthyl-*N*-methylcarbamate). The researchers heated the 0.5 g sample under reflux for 1 hour with acetone, 0.1 M $N_aB_4O_7$ and either 2-chloro-a, a,a-5-nitroltoluene (1 mmol) or 4-chloro-a,a,a-trichloro-3,3-dinitroluene (0.7 mmol) to convert its amine moiety into an *N*-substituted nitro(trifluoromethyl)aniline derivative. The derivative was subjected to GC on a stainless steel column with 3% SE-30 or 3% FFAP on HMDS treated Chromosorb G at a temperature from 150°C to 250°C. Nitrogen (30 mL min^{-1}) was used as the carrier gas. The detection was made by electron-capture or flame ionisation detection. Electron-capture detection can identify volatile derivatives down to 50 pg.

Cohen et al. [328] used an electron-capture GC to determine carbamate insecticides in their 2,4-dinitrphenyl derivatives in water and plant material. After isolation from the sample material by the appropriate extraction method, the carbamates were reacted with 1-fluoro-2,4-dinitrobenzene. The resulting 2,4-dinitrphenyl ethers were subjected to hexane and identified with GC. The recovery of carbamates added to water was in the range of 82%–100%. Carbaryl and butacarb added to water exhibited recoveries of 63% and 40%–65%, respectively. The determination limits for carbamates in water was 0.005 ppm.

Coburn et al. [329] reported a procedure for the extraction and analysis of *N*-methylcarbamate in environmental water. The researchers began their extraction using solvent partitioning with methylene chloride at pH 3–4.

They then salted out with sodium sulphate and hydrolysed the extracts using methanolic potassium hydroxide to form the corresponding phenols. Reextraction of the phenols at a pH of 2 or extraction with methylene chloride was then performed. Chemical derivatisation with pentafluorobenzyl bromide was employed to form the ether derivatives, and clean-up was on a silica gel microcolumn. Final analysis by GC included electron-capture detection. The recoveries were 87%–98% for propoxur, carbofuran, 3-ketocarbofuran, metmercapturon, carbaryl and mobam. This procedure failed to detect mexacarbate and aminocarb as the phenolic products obtained during hydrolysis were not extracted from the acidification media.

Thompson et al. [29] devised a multiresidue scheme using silica gel column chromatography followed by GC to analyse the mixtures of organochlorine, organophosphorus and carbamate types of insecticides. A mixture of the three compound types is fractioned into groups on a partially deactivated silica column with three sequential elutions. A mixture of 0.5 mL 1-fluoro-2,4-dinitrobenzene (1% in acetone) and 5 mL sodium borate ($Na_2B_4O_7 \cdot 10H_2$) buffer solution (1 M solution at pH 0.40) was added to the tubes containing 0.1 mL concentrates of the carbamate fractions. The reagents were added to an empty tube to serve as a reagent blank. The tubes were tightly capped, heated at 70°C for 1 hour in a water bath and cooled to room temperature. A volume of 5 mL of hexane was added to each tube, and each tube was shaken vigorously for 3 minutes. The layers separated and 4 mL of hexane (upper layer) were transferred to a tube and stoppered tightly. Propoxur, carbofuran, aminocarb, mexacarbate, metacarbate carbaryl and methiocarb were successfully separated by this procedure. Final determinations were made by GC using a carbon-capture detector for the halogenated compounds and derivatised carbamates and a flame photometric detector for the organophosphorus compounds.

GC has identified other carbamate insecticides, including aminocarb [330], carbaryl [330–339], mexacarb [332], propoxur [332] and aldicarb [333].

High-performance liquid chromatography

Liquid chromatography with an electrochemical detector was used [340,341] to estimate 2–7 ppm of various carbamate insecticides in water.

He et al. [342] applied dispersive liquid–liquid microextraction followed by HPLC to determine carbamate pesticides in water samples.

Al-Degs and Al-Ghouti [343] preconcentrated and detected highly leachable pesticide residues in water using solid-phase extraction coupled with HPLC. Primel et al. [344] applied solid-phase extraction coupled with HPLC to determine herbicides in surface waters.

He et al. [342] used a simple, rapid and efficient dispersive liquid–liquid microextraction method in conjunction with HPLC to determine the carbamate pesticides methomyl, carbofuran and carbaryl in river and rain water samples. In the extraction process a mixture of 35 μL chlorobenzene (extraction solvent) and 1.0 mL acetonitrile (disperser solvent) was rapidly injected into 5.0 mL aqueous sample containing the analytes. After centrifuging for 5 minutes at 4000 rpm, a fine droplet of chlorobenzene was sedimented at the bottom of the tube. The sedimented phase (20 μL) was analysed with HPLC. The researchers investigated and optimised types and volumes of extraction solvent and disperser solvent, extraction time and salt addition. Under optimum extraction conditions the enrichment factors and extraction recoveries ranged from 148% to 189% and 74.2% to 94.4%, respectively. The methods yielded a linear range in concentration from 1 to 1000 $\mu g\ L^{-1}$ for carbofuran and carbaryl and from 5 to 1000 $\mu g\ L^{-1}$ for methomyl. The limits of detection for carbofuran, carbaryl and methomyl were 0.5, 0.9 and 0.1 $\mu g\ L^{-1}$, respectively. The RSDs for the extraction of 500 $\mu g\ L^{-1}$ carbamate pesticides were in the range of 1.8%–4.6% ($n = 6$).

Thin-layer chromatography

N-Methylcarbamate and *N,N*′-dimethylcarbamates were determined in water samples by hydrolyses with sodium bicarbonate [345]. The resulting amines reacted with 4-chloro-7-nitrobenzo-2,1,3-oxadiazole in isobutyl methyl ketone solution to produce fluorescent derivatives. These derivatives were separated with thin-layer chromatography on silica gel G or alumina with tetrahydrofuran–chloroform (1:49) as the solvent. The fluorescence was then measured in situ with an excitation at 436 nm and an emission at 528 and 537 nm for the methylamine and dimethylamine derivatives, respectively. The method was applied to natural water samples containing parts per 10^9 levels of carbamate. This method is incapable of differentiating among carbamates of any one class.

Thin-layer chromatographic methods were used to determine *N*-arylcarbamate and urea herbicides (barban, chloroprophan, diuron, fenuron, fenuron TCA, linuron, monuron, monuron TCA, neburan, propham, siduron and swep) in natural water and to determine *O*-arylcarbamate pesticides (aminocarb, carbaryl, methiocarb, mexacarbate and propoxur) in water [346]. The researchers detected these herbicides by extracting and concentrating water samples with methylene chloride, then cleaning with a florisil column. Appropriate fractions from the column were concentrated and separated by thin-layer chromatography. The *N*-arylcarbamates were hydrolysed to primary amines on the layer, and the diazonium salts hydrolysis products were sprayed with 1-napththol to yield coloured spots. The *O*-arylcarbamates were hydrolysed on the layer, and the hydrolysis products were reacted with 2,6-dibromoquinone chlorimide to yield specific coloured products; identification was confirmed by changing the pH of the layer and observing colour changes in the reaction products. The researchers visually compared the sample extract responses to the standard responses on the same thin-layer plate to quantify their measurements.

Direct interferences may be encountered from aromatic amines in the *N*-arylcarbamates solution. Direct interference by phenols in the *O*-arylcarbamates solution could cause reactions with chromogenic reagents to yield reaction products similar to those of the carbamates.

Carbofuran and its degradation products, carbofuranphenol, 3-ketocarbofuran, 3-hydroxycarbofuran, were detected in water samples by thin-layer chromatography of ether extracts [347].

Frei et al. [348] employ a thin-layer chromatographic method to determine carbamate and urea herbicides in water at the parts per 10^9 level. The researchers extracted a 500 mL sample with dichloromethane and evaporated it to dryness at 40°C. The residue was dissolved in acetone and 0.5 mL of sodium hydroxide, then heated to 80°C, cooled and shaken with 0.2 mL of hexane. The layer was applied to a 0.25 mm layer of silica gel G-$CaSO_4$. The researchers applied 0.2% dansyl chloride in acetone to the sample spot and developed the chromatogram by the ascending technique with benzene–triethylamine–acetone (75:24:1). The plate was sprayed with 20% triethanolamine in isopropyl alcohol or 20% liquid paraffin in toluene and then dried. The fluorescence of the spots of the dansyl derivatives of the aniline moieties was measured in situ. Results were reported for carbamate pesticides (propham, chloropropham and barban) and urea pesticides (linuron, diuron, chlorobromuron and fluometuron). Detection limits are about 1 ng. Two-dimensional chromatography was used to eliminate interference.

Enzymic assay

Yamani et al. [349] described an automated method to detect pesticides using immobilised butyrylcholinase derived from horse serum. Detection limits ranged from 1 to 3 $\mu g\ L^{-1}$.

Rule et al. [350] described a method to determine the presence of carbofuran using online immunoaffinity chromatography with coupled-column liquid chromatography–mass spectrometric detection.

Siew et al. [351] showed that monoclonal antibodies were useful in immunoassays of aldicarb.

Ni et al. [352] described a simultaneous enzyme kinetic determination of carbamate pesticides in water. This method used enzymatic reaction kinetics and spectrophotometric measurements, and the results were interpreted with chemometrics. The analytical method relied on the inhibitory effect of the pesticides on acetylcholinesterase. The method uses 5,5′-dithiobis (2-nitrobenzoic) acid as a chromogenic reagent for the thiocholine iodide released from the acetylthiocholine iodide substrate. The complex rate equation for the formation of the chromogenic product, P, was solved under certain experimental conditions, creating a direct association between the absorbance ($A_{p'}$ at $\lambda_{max} = 412$ nm) and concentration of the mixtures for the three pesticide inhibitors. The detection limits of the enzymatic kinetic spectrophotometric procedures for the determination of oxamyl, aldicarb and aminocarb were 0.81, 2.23 and 1.25 ng mL^{-1}, respectively. Calibration models were constructed for principal component regression, partial least squares and radial basis function–artificial neural network and verified with synthetic samples of the three pesticides. The prediction performance of these models showed generally satisfactory results.

Miscellaneous

Other techniques with limited applications to determine carbamate insecticides in water include spectrophotometry (barban, 4-chlorobut-2-ynyl-3-chloro-carbamate [353], carboxyl [354–365], beniocarb [354], carbofuran [366], aldicarb [367], propham [368]), spectrofluorimetry (carbaryl [356]), flow injection analysis (carbaryl [357,358,369]) and amperometry (dithiocarbamates, aminocarb [358], benthiocarb, carbofuran [370]).

Waseem et al. [359] discussed the determination of carbaryl in natural waters by a flow injection luminol chemiluminescence inhibition detection method. The method has an inhibitory effect on luminol-cobalt (II) chemiluminescence in alkaline medium in the presence of hydrogen peroxide. The calibration data over the range $5.0 \times 10^7 - 20 \times 10^{-6}$ M gave a correlation coefficient (r^2) of 0.9972 with RSD ($n = 4$) in the range of 1.0%–2.1% and a limit of detection (3 × blank noise) of 2.37×10^{-7} M for carbaryl. The researchers studied the effects of some carbamates, anions and cations on the luminol chemiluminescence inhibition system for detecting carbaryl.

Hon et al. [371] determined carbofuran in water samples by measuring the quenching effect on resonance light scattering. Costa and Mateus [360] discussed improvements in a method for to determine the presence of dithiocarbamates. Bufo et al. [372] used laser microprobe MS to determine herbicides in water samples.

XAD-7 cation-exchange resin was used to preconcentrate Aminocarb prior to solvent extraction and estimate by GC using a nitrogen–phosphorus detector [361].

10.8 Pyrethroids

GC detected pyrethrin down to 0.2 μg in water [373]. A hexane extract of the sample was washed with aqueous sodium chloride solution. Detection was by chromatography on a helical glass column with 5% SE-30 on AW-DMCS Chromosorb W (60–80 mesh), operating at 190°C with nitrogen (40 mL min^{-1}) as the carrier gas and flame ionisation detection. This permitted the simultaneous determination of pyrethrin (liner response range 0.2–2.2 μg), the synergistic piperonyl butoxide (range 0.6–5.6 μg) and *N*-(2-ehtylhexyl)-norborn-5-ene-2,3-dicarboximide (range 0.6–1.8 μg). Recoveries averaged 93%–94% [374–376].

You and Lydy [377] simultaneously determined pyrethroid organophosphorus and organochlorine pesticides in water. The method was based on headspace solid-phase microextraction followed by GC with electron-capture detection. The parameters affecting headspace solid-phase microextraction of pesticides from water were optimised, including extraction temperatures, samples, headspace volume and the amount of added sodium chloride. The effects of desorption temperature, desorption time and position of the fibre in the gas chromatograph inlet were also investigated. Extraction temperature was the most important factor affecting the recoveries of analytes, and the optimised temperature was 90°C. The addition of salt did not increase extraction efficiency of the pesticides from water. The optimum desorption conditions in the gas chromatograph were as follows: a desorption time of 10 minutes, desorption temperature of 260°C and a portion of the fibre located in the inlet. The method detection limits were in the low nanogram per litre level with linearity ranges of 50–100 ng L^{-1} for the organochlorine pesticides, 50–100ng L^{-1} for organophosphorus insecticides and 50–20,000 ng L^{-1} for pyrethroid. Headspace solid-phase microextraction sensitively determines pyrethroid, organochlorine and organophosphorus pesticides in water.

Sneehar et al. [378] studied the polar behaviour of cyfluthrin and a-cynoester pyrethroid. They used a dropping mercury electrode and hanging mercury drop electrode in methanolic Britton–Robinson (B–R) buffer of pH 2.0–12.0 with different ionic media. The nature of the electrode process was examined, the number of electrons was evaluated and the reduction mechanism was proposed. Quantitative determination was achieved in the concentration range of 6.0×10^{-1}–1.15×10^{-1} mol dm^{-3} using a different pulse polarographic method with a lower detection limit of 2.4×10^{-8} mol dm^{-1}. The proposed method was successfully applied to determine cyfluthrin in spiked water samples.

Dowd et al. [379] showed that recombinant glutathione-*S*-transferase (9ST) can be used as an analytical tool to develop simple insecticide quantification assays. This assay explored pyrethroid's ability to promote inhibition of the glutathione-*S*-transference catalysing the chloro-2,4 dinitrobenzene/glutathione GSH conjugation reaction. The glutathione-*S*-transferase reaction sensed the pH charge in a weak buffer system and measured it using a spectrophotometer and the dye indicator bromothymol blue (616 nm). Practical use depended on the recognition affinity of the glutathione-*S*-transferase for insecticides, the inhibition kinetics in enzyme stability and the compatibility with the study's detection assay.

Dowd et al. [379] compared the recombinant AgGSTD1-6 and AgGSTD1-1 from the mosquito vectors *Anopheles gambiae* and *Anopheles dirus*, respectively. They found the recombinants' high affinity for pyrethroids was suitable for detecting pyrethroids in vector disease control programmes. The results showed that AgGSTD1-6 was the most suitable enzyme with the best structural stability at higher temperatures ($T_m = 57$°C) and pH optima in the alkaline range (pH 7.7). Using AgGSTD1-6, Dowd et al. [379] developed a pH-change colorimetric assay for detecting pyrethroids. Linear calibration curves were obtained for deltamethrin ($R^2 = 0.99$) with useful concentration ranges of 0–50 μg mL^{-1}. The effect of temperature in the range of 25°C–40°C on the pyrethroid quantification assay was negligible. The assay was validated with extracts from insecticides sprayed on surfaces and found to be reproducible and reliable when compared to the standard reversed-phase HPLC method. The researchers discussed the assay's potential to monitor insecticide residues for insecticide-based malaria control interventions.

Bjamholt et al. [380] described a high-performance liquid chromatographic method with online solid-phase microextraction preconcentration to determine the presence of permethric acid.

10.9 Other insecticides

Li et al. [381] described a sample preparation technique termed dispersive liquid-phase microextraction for the preconcentration and determination of 2,2,2-trichloro-1,1-bis(4-chlorophenyl)ethanol (dicofol) and its degradation products (2-(2-chlorophenyl)-2-(4-chlorophenyl)-1,1-dichlorothene(2,4′-DDE), 1,1-dichloro-2,2-bis(4-chlorophenyl)ethane(4,4′-DDE) and 1,1,1-trichloro-2(2-chlorophenyl)-2-(4-chlorphenyl)ethane(2,4′-DDT)) in water samples. They coupled this method with GC–MS, with a new ionic liquid 1,3-diisoactylimidazolium hexafluorophosphate [D(i-C_8)IM] (PF_6) used as an extraction solvent. For each extraction, 1.00 mL of the methanol solution containing 40 μL of the ionic liquid was sprayed into 25.00 mL of water sample. The ionic liquid was finely dispersed into the aqueous phase, and analytes were rapidly migrated into the ionic liquid. After the solution was centrifuged for 2 minutes at 5000 rpm, the droplets of the ionic liquid (30.0 ± 0.1 μL) sunk to the bottom of the conical test tube. The researchers investigated and optimised the volume of ionic liquid disperser solvent, extraction time, sample pH and ionic strength to increase extraction efficiencies. Under optimal conditions the enrichment factors for the extraction were between 550 and 725 with an extraction efficiency ranging from 66% to 87% for each analyte. Finally, 1.0 μL of the ionic liquid collected from the abovementioned extraction was injected into the injector block of the GC–MS instrument for analysis. The detection limit ($S/N = 3$), the RSDs for 2.0 μm L^{-1} of the standard analyte ($n = 5$) and the linearity in a calibration range were found to be 3–8 ng L^{-1}, 1.0%–2.7% and 10–3000 ng L^{-1}, respectively. Good spiked recoveries over the range of 92.0%–13.5% were obtained. This method is simple and quick and produces good extraction efficiencies and enrichment factors.

Primel et al. [344] studied the pollution in surface and drinking waters caused by five herbicides and nine metals in two sampling locations in the Municipal Water Supply System (CORSAN) in Rio Grande city, RS, Brazil.

The analytical determination was performed by solid-phase extraction, HPLC-photodiode-array detection and liquid chromatography coupled with electrospray ionisation tandem MS for herbicides and graphite furnace atomic absorption spectrometry for metals. The concentrations of herbicides in the surface water were very low; however, the herbicide clomazone was detected in concentrations exceeding 1.0 $\mu g\,L^{-1}$ in more than 50% of the samples. The concentration of metals was below the maximum contaminant level set by the Brazilian regulation.

Al-Degs and Al-Ghouti [343] developed a multiresidue procedure to analyse four common pesticides in water using solid-phase extraction and HPLC in combination with UV detection at 230 nm. The investigated pesticides (atrazine, dicloram, metazachlor and simazine) were highly leachable and easily migrated within the soil. Multiwalled carbon nanotube adsorbent outperformed C_{18}-bonded silica and graphitised carbon black for preconcentration of the pesticides from solution. The optimum experimental conditions for pesticide extraction were studied and optimised. The optimum preconcentration conditions included sample volume of 700 mL, adsorbent mass of 300 mg, solution pH 5.0, flow rate of 3.0 $mL\,min^{-1}$ and elution medium of methanol. A quantitative recovery for pesticides was reported, and a high preconcentration factor (1400) was attained. The detection limits were in the range of 5–15 $ng\,L^{-1}$. The dynamic range for simazine and atrazine determination was extended from 15 to 1000 $ng\,L^{-1}$, while the range for metazachlor and dicloran was within 500–1000 $ng\,L^{-1}$. Five replicate determinations of 70 ng of pesticide mixture in 700 mL solution produced good results with RSD values between 2.5% and 4.6%. The method gave recoveries from 92.7% to 95.3% for determination of 100 $ng\,L^{-1}$ of pesticides in tap water and recoveries from 85.3% to 87.0% for pesticides in well water with satisfactory RSD values ($\leq$6%).

10.10 Miscellaneous herbicides

Methods to determine various types of herbicides not previously discussed are reviewed in Table 10.3.

Microextraction of herbicides

Wei et al. [138] compared the efficiencies between single-drop microextraction and continuous flow microextraction for the determination of methomyl in natural water.

Albanis et al. [425] compared GC and MS methodologies using solid-phase microextraction to determine multiple pesticide residues in surface waters. Solid-phase microextraction techniques were used to determine the presence of parathion [381] and dicofol and its degradation products [415].

TABLE 10.3 Analysis of miscellaneous types of herbicides.

Herbicide	Type of water	Method	Sensitivity	Reference
Picloram (4-amino-3,5,6-trichloro-picolinic acid)	Natural	EC GLC	0.1 μg^{-1}	[43]
Acarol (isopropyl-4,4′-D dibromo benziliate)	–	GL radio chromatography	–	[180,382]
Dicamba (2-methocy-3,6-dichlorobenzoic acid)	Natural	GLC	–	[383,384]
Paraquat (1,1-dimethyl-4,4-bipyridylium bromide)	Natural	GLC	–	[384]
	Natural	GLC	–	[385]
Paraquat (1,1-dimethyl-4,4-bipyridylium bromide)	Natural	Ringoven	–	[386]
Paraquat and diquat	Natural	Pyrolysis-GLC	–	[180]
Fenitrothion dichlorobenil	Natural	Preconcentration	1 $\mu g\ L^{-1}$	[387]
(2,6-Dichlorobenzonitrile)	Ponds	Persistence studies	–	[388]
Kepone hydrate	Natural	Column chromatography	–	[389]
Tricyclazole	River	GLC	–	[389,390]
3-Henyl-4-hydroxyl 6-chloro-pyridazine	Natural	GLC	–	[391]
Fluridone	Natural	GLC	–	[392]
Chlorosulphon	Natural	GLC	–	[393]

(*Continued*)

TABLE 10.3 (Continued)

Herbicide	Type of water	Method	Sensitivity	Reference
Mirex	Natural	GLC–MS	–	[394]
Squoxin (1,1-methylene di-2-naphthol)	Natural	GLC	–	[395]
Miscellaneous	Surface	GLC–MS	–	[395]
Substitute of urea and carbamate	River	GLC	–	[396]
Lindane, endrin methylchlor	Natural	GLC	–	[397]
Dicamba MPCP, 2,4-D, silvex 2,4,5-T MCPB, 24,5-DB Picloram	Natural	GLC	–	[284]
Chlormethoxynil, bifenox and butachlor	River	GLC–MS	–	[381]
Benthiocarb [*S*(4-chloro-benzyl)-*N*,*N*′ diethylthiol carbamate], oxidiazon [2-*t*-butyl-4-(2,4-dichloro-5-isopropoxy phenyl)]	River	GLC–MS	–	[398]
2,1,3,4-Oxidiazolin-5-(2,4-dichlor-1), CND (2,4,6-trichlorophenyl/phenyl-4′-nitrophenyl-4′-nitrophenyl ether)	Natural	GLC–MS	–	[85,399]
12 Acidic and 19 nitrogenous herbicides (carbamates substituted ureas, triazines)	Natural	TLC	–	[400]
Atrazine, barban, diuron, linuron, nonouron, simazine, trifluralin, bromoxynil, dalapon, dicamba, MCPR, mecoprop dichloran, 2,4-D, 2,4, DB, dichloroprop, 2,4,4T,2,3,6-trichlorobenzoic acid	Natural	TLC	–	[271,295,386, 401–407]
Acidic, basic and neutral herbicides, incl. atrazine, barban, diuron, linuron, monouron, simazine, trifluralin, dalapon, dicamba, MCPA, MCPB, mecoprop, dichloram, 2,4,D,2,4, DB, dichlorocrop, 2,4,5-T,2,3,6-trichlorobenzoic acid				[354]

Carbamate and urea herbicides incl. propham, chloropropham and fluometrium	Natural	TLC	–	[354]
Carbamates (e.g. carmbamoylorimes, carbamothionic acids), substituted ureas (dithiocarbamate, phenylamides, phenyl carbamates), substituted aromatic pesticides, aldicarb sulphonide, aldicarb, propoxur and thiram neburon	Natural	HPLC	–	[408]
Propoxur, carbofuran chloropham, barban and butyrate	Natural	HPLC	–	[409]
35 Pesticides incl. oxamyl, methomyl, phoxan, 2,4,6-trichloro phenoxyacetic acid, 2,4-DB, MCPA	Natural	HPLC	–	[410]
27 Pesticides incl.	Natural	HPLC	0.01 μg L^{-1}	[411–413]
Dichlorvos, methoate, oxamyl, methomyl, 52, acidic, basic and neutral pesticides	Natural	HPLC–MLS		[414]
20 Acidic pesticides incl. 2,4-dinitro-phenoxy acid mecoprop and bromoxynil	Natural	HPLC–MS	2.5–200 ng	[191,197,415]
Various pesticides and their metabolites	Natural	μg	–	[198,372,396, 416–419]
Tributyrin, aldicarb, sulphone, propoxur and carbofuran	Natural	MLS	–	[138,139,191, 197–202,235, 370,371,381, 415,420–424]

Sakellorides et al. [139] studied the photocatalytic degradation of the organophosphorus insecticides fenitrothion and methyl parathion. The researchers used solid-phase microextraction in aqueous titanium dioxide suspensions under simulated solar irradiation. The degradation kinetics followed a pseudo-first-order reaction and was monitored by headspace solid-phase microextraction and gas chromatographic techniques. Degradation of both insecticides was a rapid process with half-lives varying between 3.7 and 12.9 minutes depending on the titanium dioxide concentration and the compound's surface. The transformation by-products formed during the process were extracted by solid-phase microextraction in two approaches: by direct immersion in the filtered samples and by exposing the fibre in the sample headspace. The intermediates were formed predominately via oxidation process and were identified by GC and MS techniques.

10.11 Pesticide survey

Surveys have reported water sample composition from various locations around the world, including San Francisco Bay [418], Yazoo River Basin [419], Spanish Agricultural Region [426] and the Bioxo Region of Portugal [427]. Matoriotas and Albanis [428] discussed the effect of land store on the run-off of herbicides from cropped and uncropped plots of land. Coupe [421] used a watershed model to characterise the transport and fate of fluometuron herbicide from land to surface water. Korasoli et al. [422] studied the application of quality control to detect the pesticides fenthion and trifluralin.

References

[1] A.D. Murray, J. Fish. Res. Board Can. 32 (1975) 457.

[2] S. Gorbech, R. Harrington, W. Knauf, H.J. Werner, Bull. Environ. Contam. Toxicol. 6 (1971) 40.

[3] A.S.Y. Chau, K. Terry, J. Assoc. Off. Anal. Chem. 55 (1972) 1288.

[4] A.S. Chau, K. Terry, Anal. Abstr. 23 (1973) 2513.

[5] A.S.Y. Chau, J. Assoc. Off. Anal. Chem. 5 (1972) 1232.

[6] Chromopack News, June 20, 1979.

[7] American Public Health Association, Standard Methods for the Examination of Water and Waste Water, Method 509, A,P 293, 1990, 15th ed.

[8] Method S73, Supplement to the 15th ed. of Standard Methods for the Examination of Water and Waste Water, Selected Analytical Methods Approved and Cited by US Environmental Protection Agency, American Public Health Association, American Water Works Association, Water Pollution Control Federation.

[9] W.L. Lamar, D.F. Goerlitz, L.M. La, Identification and Measurement of Chlorinated Organic Pesticides in Water by Electron Capture Gas Chromatography, US Government Survey Water Supply, Paper 1817-B, 1965, 12 pp.

[10] W.I. Lamar, D.F. Goerlitz, L.M. Law, Determination of Organic Pesticides in the Environment, American Chemical Society, Advances in Chemistry, Series 60, 1966, p. 87.

[11] US Federal Water Pollution Control Administration FWPCA, Method for Chlorinated Hydrocarbon Pesticides in Water and Wastewater, Cincinnati Federal Water Pollution Control Administration, 1969. 29 pp.
[12] L.A. Shevchuk, Y.G. Dubchenco, Y.S. Nordenova, Sov. J. Water Chem. Technol. 7 (1985) 73.
[13] Y. Kahanovitch, N. Lahav, Environ. Sci. Technol. 8 (1974) 762.
[14] M. Susuki, Y. Yamato, T. Wanatabe, Environ. Sci. Technol. 11 (1977) 1109.
[15] R. Taylor, T. Bagacka, K. Krasnicki, Chem. Anal. 19 (1974) 73.
[16] L. Weil, K.E. Ernst, Gas-u. Wass. Fach. 112 (1971) 184.
[17] J. Simal, J. Crous Vidal, A. Maria-Charro Aria, et al., Anal. Bromat. (Spain) 23 (1971) 1.
[18] M. Sackmauereva, O. Pal'usova, A. Szokolay, Water Res. 11 (1977) 551.
[19] M. Sackmauereva, O. Pal'usova, E. Hluchan, Vod. Hospod. 10 (1972) 267.
[20] A. Szokolay, J. Uhnak, A.A. Madaric, Chem. Zvesti 25 (1971) 453.
[21] J. Janak, M. Sackmauereva, A. Szokolay, A. Madaric, Chem. Szvesti 27 (1973) 128.
[22] J. Janak, M. Sackmauereva, A. Szokolay, O. Pal'usova, J. Chromatogr. 91 (1974) 545.
[23] J.G. Konrad, H.B. Pionke, G. Chesters, Analyst 94 (1969) 490.
[24] H.B. Pionke, Anal. Abstr. 17 (1969) 2442.
[25] H.B. Pionke, J.G. Konrad, J.G. Chesters, D.E. Armstrong, Analyst, London 93 (1968) 363.
[26] A.M. Kadoum, Bull. Environ. Contam. Toxicol. 3 (1968) 65.
[27] J.D. Millar, R.E. Thomas, H.J. Schattenberg, Anal. Chem. 53 (1981) 214.
[28] N.V. Bodtmann, Bull. Environ. Contam. Toxicol. 15 (1976) 33.
[29] J.F. Thompson, S.J. Reid, E.J. Kantor, Arch. Environ. Contam. Toxicol. 6 (1977) 143.
[30] G. Lauren, Bull. Environ. Contam. Toxicol. 5 (1970) 542.
[31] D.G. Ballinger, 58 pp Methods for Organic Pesticides in Water and Waste Water, US Govt. Printing Office, Washington, DC, 1971.
[32] US Government Printing Office, Methods for Organic Pesticides in Water and Waste Water, 1872-759-31/2113, Region 5-11, US Government Printing Office, Washington, DC, 1971.
[33] B. Stachel, K. Bactjer, M. Cetinskaya, et al., Anal. Chem. 53 (1981) 1469.
[34] M. Malaiyandi, M. Jenkins, E. Lee, M.J. Bowron, Environ. Sci. Health A20 (1985) 219.
[35] G.D. Gister, P.M. Gates, W.I. Foreman, S.T. Kenzie, F.A. Rinellei, Environ. Sci. Technol. 27 (1993) 1911.
[36] H. Fajari, M. Helalizadeh, Int. J. Environ. Anal. Chem. 90 (2010) 869.
[37] N.V. Brodtmann, J. Am. Water Works Assn. 67 (1975) 558.
[38] A.W. Breidenbach, The Identification and Measurement of Chlorinated/Hydrocarbon Pesticides in Surface Waters, US Government Printing Office, Washington, DC, 1968. -0-315-842.
[39] L. Khan, C.H. Wayman, Anal. Chem. 36 (1964) 1340.
[40] F. Mangani, G. Crescentini, F. Brumer, Anal. Chem. 981 (1627) 53.
[41] R.R. Kongovi, R.J. Grochowski, Am. Labor. (1981) 30.
[42] Hewlett Packard, Chromopak News No. 20 (1979).
[43] D.R. Erney, Anal. Lett. 12 (1979) 501.
[44] M. Godefroot, M. Stechale, P. Sandra, M. Verzele, J. High Resolut. Chromatogr. Chromatogr. Commun. 5 (1982) 75.
[45] H. Bergman, H. Hellman, Dtsch. Gewasserkundeliche Mitt. 24 (1981) 31.
[46] T. Fuka, V. Janda, J. Triska, Vodni Hospod. Ser. B 33 (1983) 245.
[47] R.L. Harless, D.E. Harris, W. Sovocool, R.D. Zehr, N.K. Wilson, E.O. Oswald, Biochem. Mass Spectrom. 5 (1978) 232.

[48] L.C. Johnson, Bull. Environ. Contam. Toxicol. 5 (1970) 542.
[49] J.F. Lester, J.W. Smiley, Bull. Environ. Contam. Toxicol. 7 (1972) 43.
[50] D.W. Woodham, C.D. Loftis, C.J. Collie, Agric. Food. Chem. 20 (1972) 163.
[51] K.H. Denhart, Bull. Environ. Contam. Toxicol. 5 (1970) 379.
[52] K.I. Astila, J.M. Carron, A.S.Y. Chau, J. Assoc. Anal. Chem. 60 (1977) 1097.
[53] T. Fuka, V. Jonda, Vodni Hospod. Ser. B 33 (1983) 245.
[54] P.G. Balayanis, J. Chromatogr. 90 (1974) 198.
[55] J.R.W. Miles, C.M. Tu, O.R. Harris, J. Econ. Entomol. 62 (1969) 1334.
[56] A.S.Y. Chau, J.D. Rosen, W.P. Cochrane, Bull. Environ. Contam. Toxicol. 6 (1971) 225.
[57] A.S.Y. Chau, J. Am. Oil Colour Chem. Assoc. 57 (1974) 585.
[58] J.F. Thompson (Ed.), Analysis of Pesticide Residue in Human and Environmental Samples, US Environmental Protection Agency, Research Triangle Park, NC, sec, 10A 12/2/74, p. 9.
[59] J.W. Elchelberger, J.L. Lichtenberg, Environ. Sci. Technol. 5 (1971) 541.
[60] F.J.H. Fredeen, J.G. Saha, M.H. Balha, Pestic. Monit. J. 8 (1975) 241.
[61] M.J. Devine, Agric. Food. Chem. 21 (1973) 4095.
[62] Analytical Methods Manual, Inland Waters Directorate, Water Quality Branch, Ottawa, ON, 1974.
[63] A.S.Y. Chau, J. Assoc. Off. Anal. Chem. 55 (1972) 519.
[64] A. Bacaloni, G. Goretti, A. Lagano, B.M. Petronio, J. Chromatogr. 175 (1979) 169.
[65] V. Zitko, P. Choi, PCB and Other Industrial Halogenated Hydrocarbons in the Environment, Fish. Res. Board Can. Technical Report No 272, Biological Station, St Andrews, NB, 1971.
[66] W.L. Oller, M.F. Cramer, J. Chromatogr. Sci. B13 (1975) 296.
[67] R. Edwards, Chem. Ind., London 1340 (1970).
[68] L. Fishbein, J. Chromatogr. 58 (1972) 345.
[69] E. Schulte, H.P. Their, L. Acker, Dtsch. Lebensm.-Rundsch. 72 (1976) 229.
[70] G. Goke, Deut. Lebensm.-Rundsch. 71 (1975) 309.
[71] B. Ahling, S. Jensen, Anal. Chem. 42 (1970) 1483.
[72] J.W. Dolan, R.C. Hall, Anal. Chem. 45 (1973) 2198.
[73] P.R. Musty, G. Nickless, J. Chromatogr. 89 (1974) 185.
[74] G. Elder, J. Chromatogr. 121 (1976) 269.
[75] A. Södergren, J. Chromatogr. 160 (1978) 271.
[76] A. Södergren, J. Chromatogr. 71 (1972) 532.
[77] L. Blomberg, Chromatographia 71 (1972) 532.
[78] O. Thomas, Anal. Chem. 62 (1990) 1667.
[79] V. Leoni, J. Chromatogr. 62 (1971) 63.
[80] B. Bavel, M. Jaaremo, L. Karisson, G. Lindstream, Anal. Chem. 68 (1996) 1279.
[81] R. Goetz, P. Enge, P. Friesel, K. Roch, L.V. Kjeller, S.E. Kuly, et al., Chemosphere 28 (1994) 63.
[82] R.A. Hughes, G.F. Lee, Environ. Sci. Technol. 7 (1973) 934.
[83] V. Leioni, G. Pucetti, R.J. Columbo, P. Ovidio, J. Chromatogr. 125 (1976) 399.
[84] V. Leoni, G. Pucetti, A. Grella, J. Chromatogr. 106 (1975) 119.
[85] Y. Yamoto, M. Susuki, T. Wanatabe, Biomed. Mass Spectrom. 6 (1979) 205.
[86] S.D. Abbott, R.C. Hall, G.S. Giam, J. Chromatogr. 45 (1969) 317.
[87] H. Shiraiski, A. Otsuki, Water Res. 21 (1987) 843.
[88] H. Sabik, B. Rondean, K. Dohrendorf, P. Gagnon, Int. J. Environ. Anal. Chem. 83 (2003) 457.

[89] K. Sichilongo, Int. J. Environ. Anal. Chem. 86 (2006) 973.
[90] T. Suzuki, H. Nayayoshi, T. Kashiwa, Agric. Biol. Chem. 38 (1974) 279.
[91] T. Suzuki, Anal. Abstr. 26 (1974) 2374.
[92] J.D. Millar, R.T. Thomas, H.J. Schattenby, Anal. Chem. 53 (1981) 21.
[93] P.J. Pocaro, P. Shabiaw, Anal. Chem. 44 (1972) 1865.
[94] R.A. Carpenter, R.H. Hollowell, K.M. Hill, Anal. Chem. 69 (1977) 3314.
[95] F.N. Willmott, R.J. Dolpim, J. Chromatogr. Sci. 12 (1974) 695.
[96] C. Ye, Q. Zhou, Z. Wang, Int. J. Environ. Anal. Chem. 88 (2008) 461.
[97] N.V. Fehringer, J.E. Westfall, J. Chromatogr. 57 (1971) 397.
[98] T. Chen, Jo-Yun, R.W. Dority, J. Am. Oil Color Chem. Assoc. 55 (1972) 15.
[99] O. Hutzinger, W.D. Jamieson, S. Safe, J. Am. Oil Color Chem. Assoc. 54 (1971) 178.
[100] J.A. Kawastski, D.L. Frash, J. Assoc. Off. Anal. Chem. 52 (1960) 1108.
[101] C. Arias, A. Vidal, C. Vidal, J. Maria, J. Anal. Brom (Spain) 22 (1970) 273.
[102] R.G. Achari, S.S. Sandhu, W.J. Warren, Bull. Environ. Contam. Toxicol. 13 (1975) 94.
[103] A. Berenue, J.W. Keller, J. Chromatogr. 54 (1971) 71.
[104] ASTM, Tentative Method of Test for Organochlorine Pesticides in Water, Designation D3086-72T, 1972.
[105] H. Keith, A.I. Alford, J. Assoc. Off. Anal. Chem. 53 (1970) 1018.
[106] M.R. Jan, J. Shah, N. Bashir, Int. J. Environ. Anal. Chem. 88 (2008) 27.
[107] M. Goreti, F. Sales, N.F.M. Lino, C.B. Paiga, Int. J. Environ. Anal. Chem. 83 (2003) 295.
[108] K.R. Griffiths, J.C. Craun, J. Assoc. Off. Anal. Chem. 57 (1974) 168.
[109] A.K. Burnham, Anal. Chem. 44 (1972) 139.
[110] J.F. Richards, J.S. Fritz, Talanta 21 (1974) 91.
[111] F. Mangani, G. Crescentini, F. Bruner, Anal. Chem. 53 (1981) 1627.
[112] I. Sherma, J. Liq. Chromatogr. 11 (1988) 2121.
[113] N.V. Brodtmann, J. Am. Water Works Assn. 63 (1975) 25.
[114] G.R. Harvey, I Report, US Environmental Protection Agency, EPA-R2-73-177, 1973, 32 pp.
[115] L. Weil, Gas-u-Wassfach 113 (1972) 64.
[116] T. Noy, E. Weiss, T. Herps, H. Van Crutchen, J. Rijks, J. High Resolut. Chromatogr. Chromatogr. Commun. 11 (1988) 181.
[117] E. Noroozian, F.A. Moris, M.W.F. Nielen, R.W. Rei, G.J. De Jong, H.A.T. Brinkman, J. High Resolut. Chromatogr. Commun. 10 (1987) 17.
[118] W.A. Aue, S. Kapila, C.R. Hastings, J. Chromatogr. 73 (1972) 99.
[119] E.D. Goldberg (Ed.), A Guide to Marine Pollution, Gordon and Breach, New York, 1972 (Chapters 12 and 4).
[120] H. Braus, F.M. Middleton, G. Watton, Anal. Chem. 23 (1951) 1160.
[121] F.M. Middleton, W. Grant, A.A. Rosen, Ind. Eng. Chem. 48 (1956) 268.
[122] F.J. Ludzok, F.M. Middleton, E.B. Ettinger, Sewerage Ind. Wastes 30 (1958) 662.
[123] R.C. Delange, S. Megergian, J. Am. Water Works Assn. 50 (1958) 1214.
[124] F.M. Middleton, J. Lichtenberg, J. Ind. Eng. Chem. 52 (1960) 99A.
[125] Standard Methods for Examination of Water Waste and Waste Water. 13th ed., APHA, AWWA, WPCF, Washington, DC, 1971, p. 103.
[126] O.J. Sproul, D.W. Ryckman, Water Pollut. Control Fed. 33 (1961) 1188.
[127] A.A. Rosen, F.M. Middleton, Anal. Chem. 31 (1959) 1729.
[128] R.L. Booth, J.N. English, G.N. McDermot, J. Am. Water Works Assn. 57 (1965) 215.
[129] A. Eichlberger, J.J. Litchenberg, J. Am. Water Works Assn. 25 (1971) 63.
[130] J.W. Eichlberger, J. Litchenberg, J. Am. Water Works Assn. 63 (1971) 25.
[131] R. Bagnati, E. Bentenati, W. Davoli, R. Fanelli, Chromosphere 17 (1988) 59.

[132] U. Niederschulte, K.Z. Ballschmiter, Anal. Chem. 269 (1974) 350.
[133] G.A. Junk, J.J. Richard, M.D. Greser, J.L. Witiaki, M.D. Anguello, R. Vick, et al., J. Chromatogr. 99 (1974) 745.
[134] D.A. Kurtz, Bull. Environ. Contam. Toxicol. 17 (1977) 391.
[135] K. Bevermann, W. Eckrich, Z. Anal. 265 (1974) 1.
[136] T. Berhanu, N. Megersa, J.A. Jonsson, T. Solomon, Int. J. Environ. Anal. Chem. 88 (2008) 933.
[137] Z. Li, Y. Fang, P. Chen, A. Wang, L. Wong, G. Ren, et al., Int. J. Environ. Anal. Chem. 90 (2010) 856.
[138] G. Wei, Y.I.X. Wang, Int. J. Environ. Anal. Chem. 88 (2008) 397.
[139] T.M. Sakellorides, V.A. Sakkas, D.A. Lombropoulou, T.A. Albans, Int. J. Environ. Anal. Chem. 84 (2004) 161.
[140] V.N. Mallet, G.L. Brun, R.N. MacDonald, K. Berkane, J. Chromatogr. 160 (1978) 81.
[141] K. Berkane, G.E. Caissie, V.N. Mallet, J. Chromatogr. 139 (1977) 386.
[142] J.G. Zakrevski, V.N. Mallet, J. Chromatogr. 132 (1977) 315.
[143] J.A. Coburn, J.A. Valdamanis, A.S.Y. Chau, J. Assoc. Off. Anal. Chem. 60 (1977) 224.
[144] K. Berkane, G.E. Caissie, V.N. Mallett, Proc. Symp. On Fenitrothion, NRC, (Canada) Association Comm. Sci. Crit. Environmental Quality, Report 16073, NRCC, Ottawa, ON, 1977, p. 95.
[145] E.P. Laws, D.J. Webley, Analytic 86 (1961) 249.
[146] J.A.A.R. Bates, Analyst 90 (1965) 453.
[147] A. Kotrikis, N.S. Thomaidis, T.D. Lekkas, Int. J. Environ. Anal. Chem. 86 (2006) 553.
[148] H.K. De Loach, D.D. Hemphill, J. Assoc. Off. Anal. Chem. 52 (1969) 333.
[149] H.K. De Loach, D.D. Hemphill, J. Assoc. Off. Anal. Chem. 53 (1970) 1129.
[150] V.V. Brazhnikov, V.V. Poshememsky, I.K. Sadodynskii, J. Chennnjakin, J. Chromatogr. 175 (1979) 221.
[151] S.S. Brody, J.E. Chaney, J. Gas Chromatogr. 4 (1966) 42.
[152] C.A. Bache, D.J. Lisk, Anal. Chem. 37 (1965) 1477.
[153] C.H. Hartmann, Aerograph Res. Notes (1966) 1–6.
[154] J. Kanazawa, T. Kawahara, Nippon Nogei Kagaju Kaishi 40 (1966) 178.
[155] A.J. Kamen, Gas Chromatogr. 3 (1965) 336.
[156] H.V. Claborn, H.D. Mann, D.D. Vehler, J. Assoc. Off. Anal. Chem. 51 (1968) 1243.
[157] H.P. Burchfield, J.W. Rhoades, R.J. Wheeler, J. Agric. Food Chem. 13 (1965) 511.
[158] H.P. Burchfield, D.E. Johnson, J.W. Rhoades, R.J. Wheeler, J. Gas Chromatogr. 3 (1965) 28.
[159] N.F. Ives, L. Suiffrida, J. Assoc. Off. Anal. Chem. 50 (1967) 1.
[160] M. Dressler, J. Janak, Collect. Czech. Chem. Commun. 33 (1968) 3970.
[161] D.E. Wooley, Anal. Chem. 40 (1968) 210.
[162] A.V. Novak, H.V. Malmstadt, Anal. Chem. 40 (1968) 1108.
[163] M.C. Bowman, M. Berzoa, Anal. Chem. 40 (1968) 1448.
[164] M.C. Bowman, M. Beroza, K.R. Hill, J. Assoc. Off. Anal. Chem. 54 (1971) 346.
[165] A. Kamen, J. Chromatogr. Sci. 7 (1969) 541.
[166] M. Skolnick, J. Chromatogr. Sci. 8 (1970) 462.
[167] H.W. Grice, M.I. Yates, D.J. David, J. Chromatogr. Sci. 9 (1970) 90.
[168] V. Svojanousky, R. Nebola, Chemickchisty 67 (1973) 295.
[169] R. Greenhalgh, W.P. Cochrane, J. Chromatogr. 70 (1972) 37.
[170] L. Guiffrida, N.F. Ives, D.C. Bostwick, Assoc. Off. Anal. Chem. 49 (1966) 8.
[171] J. Askew, J.H. Ruzick, B.B. Wheals, Analyst 94 (1969) 275.

[172] V.N. Mallet, K. Berkane, G.E. Caissie, J. Chromatogr. 139 (1977) 386.
[173] L.M. Puijker, G. Veendaal, H.M.J. Janssen, B. Griepink, Z. Fresenius, Anal. Chim. 306 (1981) 1.
[174] A. Cassita, V.N. Mallet, Chromatographia 16 (1984) 305.
[175] I.H. Suffet, S.D. Faust, J. Agric. Food Chem. 20 (1972) 52.
[176] H. Bargnoux, P. Papin, J.L. Chahard, F. Vedrini, J. Petit, J.A. Berger, Analysis 5 (1977) 170.
[177] Supplement of the 15th edition of Standard Method for the Examination of Water, and Waste Waters Selected Analytical Methods, approved and edited by US Environmental Protection Agency American Public Health Association Method, 1978, p. 551.
[178] A. Verweij, C.F.A. Regenhardt, H.I. Boter, Chemosphere 8 (1979) 115.
[179] C.J. Soderquist, D.G. Crosby, Bull. Environ. Contam. Toxicol. 8 (1972) 363.
[180] A.J. Cannard, W.J. Criddle, Analyst 100 (1975) 848.
[181] J.R. Rice, H.L. Dishberger, J. Agric. Food. Chem. 16 (1968) 867.
[182] B.D. Ripley, J.A. Hall, S.S.Y. Chau, Environ. Lett. 7 (1974) 97.
[183] N. Grift, W.L. Lockhart, Assoc. Off. Anal. Chem. 57 (1974) 1282.
[184] NRC, Associate Committee on Scientific Criteria for Environmental Quality, Fenitrothion: The Effects of Its Use in Environmental Quality and its Chemistry, NRCC No. 14104, 1975, p. 106.
[185] F.K. Kawahara, J. Water Pollut. Control Fed. 39 (1967) 572.
[186] S. Locorte, D. Barcelo, Anal. Chem. 68 (1966) 2464.
[187] D. Barcelo, D. Maris, R.B. Geerdink, et al., J. Chromatogr. 394 (1987) 65.
[188] D. Barcelo, Liq. Chromatogr. Gas Chromatogr. 6 (1988) 324.
[189] D. Barcelo, J. Chromatogr. 643 (1993) 117.
[190] Q.H. Wu, Y.P. Hi, C. Li, C.X. Wa, Z.M. Liu, Y.Y. Hou, et al., Int. J. Environ. Anal. Chem. 90 (2010) 891.
[191] H.K. Rotich, Z. Zhang, J. Li, Int. J. Environ. Anal. Chem. 83 (2003) 851.
[192] D. Barcelo, F.A. Maris, R.H. Geerdink, R.W. Frei, U.A.J. Brinkman, J. Chromatogr. 394 (1987) 65.
[193] D. Zycinski, Rocz. Panstw. Zakl. Hig. 22 (1971) 189.
[194] H.A. Moye, J. Chromatogr. Sci. 13 (1975) 268.
[195] D.C. Paschal, R. Bieknell, D. Dresbach, Anal. Chem. 49 (1977) 1551.
[196] T. Rodica, Rev. Chem. 20 (1969) 259.
[197] J. Geng, X. Niu, Y. Wong, M. Edwards, D. Glindemann, Int. J. Environ. Anal. Chem. 90 (2010) 737.
[198] D.A. Iavanovic, Z. Proic, Acta Pharm. 22 (1971) 189.
[199] V. Leoni, G. Pucciti, Farmaco, Ed. Prat. 26 (1971) 383.
[200] D. Schultsman, R. Barthel, Anal. Abstr. 18 (1966) 3743.
[201] T.F. Bidleman, B. Nowlan, R.W. Frei, Anal. Chim. Acta 60 (1972) 13.
[202] K.C. Gygen, A. Aktugla, Frzacilik Bull. 14 (1972) 44.
[203] S. Sherman, J.L. Boymel, J. Chromatogr. 247 (1982) 201.
[204] S.H. Hove, J. Agric. Food Chem. 17 (1969) 401.
[205] S. Venkataraman, V. Sathy Murthey, J. Indian Water Works Assoc. 11 (1979) 353.
[206] Z. Frobe, V. Drevenkar, B. Stengl, Z. Stefanac, Anal. Chem. 206 (1988) 299.
[207] J.M. Zen, S.H. Jeng, H.J. Chay, Anal. Chem. 68 (1996) 498.
[208] A. Calderbank, O. Youens, Analyst 90 (1965) 99.
[209] J.D. Pope, J.F. Bennett, J. Assoc. Off. Anal. Chem. 57 (1974) 202.
[210] J. Volke, V. Velokovc, Collect. Czech. Chem. Commun. 34 (1969) 2037.
[211] J.O. Brostad, O. Hakon, J. Freistad, Analyst 101 (1976) 820.

[212] J. Yon, M. Kalyurand, J.A. Koropchak, Int. J. Environ. Anal. Chem. 83 (2003) 797.
[213] C.E. McKone, R.J. Hance, J. Chromatogr. 36 (1968) 234.
[214] S.U. Khan, R. Greenhalgh, W.P. Cochrane, Bull. Environ. Contam. Toxicol. 13 (1975) 602.
[215] J.L. Burguera, M. Burguera, Anal. Chim. Acta 179 (1986) 497.
[216] R.J. Thibeau, L. Van Haverbeke, C.W. Brown, Appl. Spectrosc. 32 (1978) 98.
[217] Y. Ting Lon, Y. Guo, M. Qinhu, Anal. Chem. 70 (1998) 347.
[218] N.I. Wolfe, R.G. Zepo, G.A. Gordon, G.L. Baughman, D.M. Clive, Environ. Sci. Technol. 11 (1977) 88.
[219] T.M. Aranjo, M.N.N. Compas-McCanela, Int. J. Environ. Anal. Chem. 87 (2007) 937.
[220] W. Russell-Everet, G.A. Rechnitz, Anal. Chem. 70 (1988) 807.
[221] C. Molina, M. Honing, D. Boreio, Anal. Chem. 66 (1994) 444.
[222] A. DiCorcia, L. Crescenzi, R. Samperi, Am. Chem. Soc. J. Environ. Chem., Preprints of Papers Presented at National Meeting Anaheim, CA, April 2–7, 1995.
[223] R.M. Gauson, D.W. Robinson, A. Goodman, Environ. Sci. Technol. 7 (1973) 1137.
[224] B.A. Karlhuber, D. Eberle, Anal. Chem. 47 (1975) 1094.
[225] L.H. Goodson, W.B. Jacobs, A. Davis, Anal. Biochem. 51 (1973) 362.
[226] O. Bauman, Anal. Abstr. 15 (1968) 5550.
[227] V.N. Mallet, G.L. Brun, Bull. Environ. Contam. Toxicol. 12 (1974) 739.
[228] R.I. Glais, Anal. Chem. 53 (1983) 921.
[229] D. Hempel, R. Liebman, A. Hellwig, Fortschr. Wasserchem. Ihrer Grenzgeb. 13 (1971) 181.
[230] E. Vogler, Liminologica 7 (1970) 309.
[231] A. Bagnoux, D. Pepin, J.L. Chahard, F. Vedrine, J. Petit, J.A. Berger, Analysis 5 (1977) 170.
[232] G.G. Volpe, V.N. Mallet, J. Int. Environ. Anal. Chem. 80 (1981) 291.
[233] L.T. Goosen, J.G. Kloostenoer, Anal. Chem. 50 (1978) 707.
[234] V. Zetko, T.D. Cunningham, Bull. Environ. Contam. Toxicol. 14 (1975) 19.
[235] V.C. Anigbogl, H. Wolddeab, A.Y. Garrison, J.K. Avants, Int. J. Environ. Anal. Chem. 90 (2010) 89.
[236] J.J. Kirkland, Anal. Chem. 34 (1962) 428.
[237] H. Lolke, Pestic. Sci. 5 (1974) 749.
[238] H.J. Jarczyk, Pflanzenschutz-Nachr. Bayer 25 (1972) 21.
[239] H.J. Jarczyk, Pflanzenschutz-Nachr. Bayer 28 (1975) 334.
[240] W.K. Lowen, W.E. Bleinder, J.J.W.K. Lowen, W.E. Bleinder, J.J. Kirkland, H.L. Pease, et al. (Eds.), Analytical Methods for Pesticides, Plant Growth Regulators and Food Additives, Volume IV, Herbicides, Academic Press, New York, 1964.
[241] C.E. McKone, R.J. Hance, Bull. Environ. Contam. Toxicol. 4 (1960) 31.
[242] C.E. McKone, R.J. Hance, Anal. Abstr. 17 (1969) 3849.
[243] J.Z. Rosales, Anal. Chem. 256 (1971) 194.
[244] R. Delue, J.P. Barthelemy, A. Copin, J. Chromatogr. 134 (1977) 483.
[245] A. De Kok, Y.J. Vos, C. Van Garderen, T. Se Jong, M. Van Opstal, R.W. Frei, et al., J. Chromatogr. 288 (1984) 71.
[246] J. Cohen, R.B. Wheals, J. Chromatogr. 43 (1969) 233.
[247] P. Klaffenbach, P.T. Holland, J. Agric. Food Chem. 41 (1993) 358.
[248] A. Berger, Chromatographia 41 (1995) 133.
[249] F. Garcia, J. Henion, J. Chromatogr. 606 (1992) 237–247.
[250] R.W. Raiser, A.C. Barefoot, R.F. Dietrich, A.J. Fogiel, W.R. Johnson, M.T. Scott, J. Chromatogr. 553 (1991) 91.
[251] A.L. Howard, L.T. Taylor, J. Chromatogr. Sci. 30 (1992) 374.

[252] G.E. Schneider, M.K. Koeppe, M.V. Naidu, P. Horne, A.M. Brown, C.F.J. Mucha, J. Agric. Food Chem. 41 (1993) 2404.
[253] G.C. Galletti, A. Bonetti, G. Dinelli, J. Chromatogr. 692 (1995) 27.
[254] J.P. Cambon, J. Bastide, J. Agric. Food Chem. 41 (1996) 333.
[255] D. Volmer, J.G. Wilke, K. Leveson, Rapid Commun. Mass Spectrom. 9 (1995) 767.
[256] W.E. Smith, K.A. Lord, J. Chromatogr. 107 (1975) 407.
[257] D.S. Farrington, R.G. Hopkins, J.A.H. Ruziche, Analyst 102 (1977) 377.
[258] R. Gotz, Facum Stadt Hygiene 29 (1978) 10.
[259] N.K. Shtivel, L.I. Gipolova, Z.S. Smirnova, Soc. J. Water Chem. Technol. 7 (1985) 66.
[260] J. Mallevialle, M. Lefebvre, C. Roussau, C. Sagbier, Rev. Fronc. Sci. L'eau 1 (1982) 25.
[261] O. Hunter, Environ. Sci. Technol. 9 (1975) 241.
[262] E. Stottmeister, P. Hendel, H. Hermenau, Chemosphere 17 (1988) 801.
[263] J.T. Gomez-Belinchou, J. Albaiges, Int. J. Environ. Anal. Chem. 30 (1987) 183.
[264] F.P. Schwarz, S.P. Wasik, Anal. Chem. 48 (1976) 525.
[265] S. Monarca, B.S. Causey, F.G. Kirkbright, Water Res. 13 (1979) 503.
[266] G. Nilve, M. Knutsson, J.A. Jonsson, J. Chromatogr. 688 (1994) 75.
[267] J.R. Ruberu, W.H. Draper, L.K. Perera, J. Agric. Food Chem. 48 (2000) 4109.
[268] S. Saha, G. Kulshresthe, Int. J. Environ. Anal. Chem. 88 (2008) 891.
[269] N.N. Senin, V.S. Filippov, N.F. Toliskins, G.A. Simolyanninov, S.A. Volkov, V.S. Kurkushkin, J. Chromatogr. 364 (1986) 315.
[270] H. Geissbuhler, D. Gross, J. Chromatogr. 27 (1967) 296.
[271] Y. Si, L. Zhang, K. Takagi, Int. J. Environ. Anal. Chem. 85 (2005) 73.
[272] J.N. Devine, G. Zweig, B. Zip, J. Assoc. Off. Anal. Chem. 52 (1969) 187.
[273] B.J. Croll, Analyst 96 (1971) 810.
[274] A. Colas, A. Lerneard, J. Rover, J. Chim. Anal. 54 (1972) 7.
[275] R.H. Larose, A.S.Y. Chau, J. Assoc. Off. Anal. Chem. 56 (1973) 1183.
[276] V.D. Carnac, Zh. Anal. Khim. 30 (1975) 2444.
[277] A.S.Y. Chau, K. Terry, J. Assoc. Off. Anal. Chem. 58 (1975) 1294.
[278] A.S.Y. Chau, K. Terry, J. Assoc. Off. Anal. Chem. 59 (1976) 633.
[279] H. Agemian, A.S.Y. Chau, J. Assoc. Off. Anal. Chem. 59 (1976) 732.
[280] T. Bogacka, R. Taylor, Chem. Anal. 16 (1971) 215.
[281] H. Agemian, A.S.Y. Chau, J. Assoc. Off. Anal. Chem. 60 (1977) 1070.
[282] V. Lopez-Avila, V. Hirta, S. Kraska, J.H. Taylor, J. Agric. Food Chem. 34 (1986) 530.
[283] T. Tsukoika, R. Takeshita, T. Murakami, Analyst 111 (1986) 145.
[284] H.B. Lee, D. Stokker, A.S.Y. Chau, J. Assoc. Off. Anal. Chem. 69 (1986) 557.
[285] T. Bogacka, R. Taylor, Chem. Anal. 15 (1970) 143.
[286] C. Meinard, J. Chromatogr. 61 (1971) 173.
[287] R. Purkayastha, Bull. Environ. Contam. Toxicol. 4 (1969) 246.
[288] M. Arjmand, N.N. Spittler, R.O. Mumma, J. Agric. Food Chem. 36 (1988) 492.
[289] M. Marshall, J. Assoc. Off. Anal. Chem. 54 (1971) 706.
[290] T. Bogacka, Chem. Anal. 10 (1971) 59.
[291] I.H. Suffet, J. Agric. Food Chem. 21 (1973) 288.
[292] D.W. Woodham, W.G. Mitchell, C.D. Loftis, W.C.W. Collier, J. Agric. Food Chem. 19 (1971) 186.
[293] C.A. Bache, D.J. Lisk, M.A. Loos, J. Am. Oil Color Chem. Assoc. 47 (1964) 438.
[294] P.E. Mattson, W.J. Kirsten, J. Agric. Food Chem. 16 (1968) 908.
[295] A.E. Smith, A. Fitzpatrick, J. Chromatogr. 57 (1971) 303.
[296] W.R. Meagher, J. Agric. Food Chem. 14 (1966) 374.

[297] C.E. McCone, R.J. Hance, J. Chromatogr. 69 (1972) 204.
[298] G. Yip, J. Am. Oil Color Chem. Assoc. 45 (1962) 367.
[299] G. Yip, Am. Oil Color Chem. Assoc. 47 (1964) 1116.
[300] S.H. Suzuki, M. Malina, J. Am. Oil Color Chem. Assoc. 48 (1965) 1164.
[301] D.F. Goerlitz, W.L. Lamar, Determination of Phenoxy Acid Herbicides in Water by Electron-Capture and Microcoulometric Gas Chromatography, Geological SurveyWater-Supply Paper 1817-C, US Government Printing Office, Washington, DC, 1967.
[302] G. Yip, J. Am. Oil Color Chem. Assoc. 54 (1971) 966.
[303] S.F. Howard, G. Yip, J. Am. Oil Color Chem. Assoc. 54 (1971) 970.
[304] L.G. Johnson, J. Assoc. Off. Anal. Chem. 56 (1973) 1503.
[305] A. Kasna, S. Sklodal, Int. J. Environ. Anal. Chem. 83 (2003) 101.
[306] C.E. McKone, T.M. Byast, R.J. Hance, Analyst 97 (1972) 653.
[307] R. Purkayastha, W.P. Cochrane, J. Agric. Food Chem. 21 (1973) 93.
[308] K. Ramsteiner, W.D. Hoermann, D.O. Eberle, J. Assoc. Off. Anal. Chem. 57 (1974) 192.
[309] W.D. Hormann, J.C. Tournayre, H. Egli, Pestic. Monit. 13 (1979) 128.
[310] T.R. Steinheimer, M.G. Brooks, Int. J. Environ. Anal. Chem. 17 (1984) 97.
[311] V. Jahda, K. Marha, J. Chromatogr. 329 (1985) 186.
[312] H.B. Lee, Y.D. Stokker, J. Assoc. Off. Anal. Chem. 69 (1986) 568.
[313] V. Lopez-Avila, P. Hirata, S. Kraska, H. Flanage, J.H. Taylor, Anal. Chem. 57 (1985) 2797.
[314] Lai Zangwei, O. Sodagopa, N.M. Ramanujam, D. Giblin, M.L. Gross, Anal. Chem. 65 (1993) 21.
[315] R. Deleu, A. Copin, Bull. Res. Agron. Gembloux 27 (1987) 121.
[316] J. Sherma, J. Liq. Chromatogr. 9 (1986) 3433.
[317] S. Topuz, B. Alperunge, Int. J. Environ. Anal. Chem. 83 (2003) 787.
[318] Method S60, Method for Triazine Pesticide in Water and Waste Water, Method for Benzidine, Chlorinated Organic Compounds, Pentachlorophenol and Pesticides in Water and Waste Water (Interim Pending Issuance of Methods for Organic Analysis of Water and Wastes), Environmental Protection Agency, Environmental Monitoring and Support Laboratory (EMSL), 1978.
[319] H. Zawadzka, M. Adamezewska, H. Elbanowska, Chem. Anal. 18 (1993) 327.
[320] O. Abbott, Anal. Abstr. 13 (1966) 5917.
[321] L. Fishbein, Chromatogr. Rev. 12 (1970) 167.
[322] J. Sherma, J. Chromatogr. 9 (1986) 2422.
[323] R.J. Buckway, S.T. Lekansi, B. Perkins, S.A. Savage, B.S. Ferguson, Bull. Environ. Contam. Toxicol. 40 (1988) 647.
[324] E.M. Thurman, M. Meyer, M. Powes, C.A. Perry, A.B. Schwab, Anal. Chem. 62 (1990) 2043.
[325] K.A. Cadwell, V.M. Ramannjam, Z. Cai, M.L. Gross, Anal. Chem. 65 (1993) 2372.
[326] A. Martin Estehon, P. Fernandez, C. Cameea, Anal. Chem. 69 (1997) 3267.
[327] D.G. Crosby, O. Bowarg, J. Agric. Food Chem. 16 (1968) 839.
[328] I.C. Cohen, J. Norcrop, J.H.A. Kizicke, B.B. Wheals, J. Chromatogr. 48 (1970) 215.
[329] J.A. Coburn, B.D. Ripley, A.S.Y. Chau, J. Assoc. Off. Anal. Chem. 59 (1976) 188.
[330] W.Z. Zhang, A.T. Lemley, J. Spalik, J. Chromatogr. 299 (1984) 269.
[331] Private Communication.
[332] K.M.S. Sundaram, S.Y. Szeto, R. Hindle, J. Chromatogr. 177 (1979) 29.
[333] M.L. Trehy, R.A. Yosf, J. McCreay, Anal. Chem. 56 (1984) 1281.
[334] W.E. Westlake, I. Monika, F.A. Gunther, Bull. Environ. Contam. Toxicol. 8 (1972) 109.

[335] R.G. Reeves, D.W. Woodham, J. Agric. Food Chem. 22 (1974) 76.

[336] J.W. Rolls, A. Cortes, J. Gas Chromatogr. 2 (1964) 132.

[337] E.R. Holden, W.M. Jones, M. Beroze, J. Agric. Food Chem. 17 (1969) 56.

[338] W.H. Gutenmann, D.J. Lisk, J. Agric. Food Chem. 3 (1965) 48.

[339] D.L. Lewis, D.F. Paris, J. Agric. Food Chem. 22 (1974) 148.

[340] J.L. Anderson, K.A. Whiten, J.D. Breseter, T.Y. Ou, W.K. Nonidez, Anal. Chem. 57 (1985) 1366.

[341] J.L. Anderson, Chesney, Anal. Chem. 52 (1980) 2156.

[342] L. He, C. Wang, Y. Sun, X. Luo, J. Zhong, K. Lu, Int. J. Environ. Anal. Chem. 89 (2009) 439.

[343] Y.S. Al-Degs, M.A. Al-Ghouti, Int. J. Environ. Anal. Chem. 88 (2008) 487.

[344] E.G. Primel, M.R. Milani, A. Demuliner, L.F.H. Meincheski, A.L.V. Escarrone, Int. J. Environ. Anal. Chem. 90 (2010) 1048.

[345] J.F. Lawrence, R.W. Frei, Int. Chem. 44 (1972) 2046.

[346] Supplement of the 15th Edition of Standard Methods for the Examination of Water and Waste Water, Selected Analytical Methods Approved and Cited by the US Environmental Protection Agency, American Public Health Association, American Water works Association, Water Pollution Control Federation, 1978. Methods S60 and S63. Methods for Benzidine, Chlorinated Organic Compounds, Pentachlorophenol and Pesticides in Water and Waste Water (Interim Pending Issuance of Methods for Organic Analysis of Water and Wastes, 1978), Environmental Protection Agency, Environmental Monitoring and Support Laboratory (EMSL).

[347] C.C. Ya, G.H. Booth, D.J. Hanson, J.P. Larsen, J. Agric. Food Chem. 22 (1974) 431.

[348] R.W. Frei, J.F. Lawrence, D.S. Le Gau, Analyst 98 (1973) 9.

[349] H. Yamani, C. Iran-mihn, C. Charanne, Sens. Actuators 15 (1988) 193.

[350] G.S. Rule, A.V. Mordechei, J. Henion, Anal. Chem. 66 (1994) 230.

[351] L.K. Siew, L.A. Winger, J.A. Spoors, J.L. Renai, L. Jennens, C.H. Self, Int. J. Environ. Anal. Chem. 83 (2003) 417.

[352] Y. Ni, N. Deng, S. Kolecof, Int. J. Environ. Anal. Chem. 89 (2009) 939.

[353] E.N. Bosyakova, A.S. Bukharbaeva, N.R. Utebekova, I.M. Shabanov, Trudy Institute Braev, Patal, -VOZ, Drovookir Kaz SSR 16 (1969), 28. Ref. Zh-Khim 19GD a6 Abstract No. 169-210.

[354] S.K. Handa, A.K. Dikshit, Analyst 104 (1979) 1185.

[355] E.E. Venesch, M.H.C.K. Riveros, J. Assoc. Off. Anal. Chem. 54 (1971) 128.

[356] R.W. Frei, J.F. Lawrence, D.E. Bellivean, Anal. Chem. 254 (1971) 271.

[357] B. Fernandez, M.D. Linares, M.D. Luque de Castros, M. Valearcel, Anal. Chem. 63 (1991) 1672.

[358] M. Suzuki, Y. Yamoto, T. Akiyama, Water Res. 11 (1977) 275.

[359] A. Waseem, M. Yaqoob, A. Nabi, A. Siddique, Int. J. Environ. Anal. Chem. 87 (2007) 825.

[360] J.L. Costa, M.C.D.A. Mateus, J. Environ. Anal. Chem. 89 (2009) 83.

[361] D. Revesque, V.N. Mallet, Int. J. Environ. Anal. Chem. 16 (1983) 139.

[362] A. Salinas, J.F. Fernandez, A. Segura, A.F. Gutierrez, Int. J. Environ. Anal. Chem. 85 (2005) 443.

[363] J. Cheng, Y. Zhou, M. Zuo, L. Dai, X. Guo, Int. J. Environ. Anal. Chem. 90 (2010) 84.

[364] M.E. Getzenander, J. Assoc. Off. Anal. Chem. 52 (1969) 824.

[365] P.A. Frank, R.J. Demint, Environ. Sci. Technol. 3 (1969) 69.

[366] R. Kumar, C.B. Sharma, Anal. Lett. 20 (1987) 777.

[367] D.P. Condo, G.E. Janauer, Analyst 112 (1987) 102.

[368] F. Garcia, C. Blanco, Anal. Chem. 58 (1986) 73.

[369] J.M. Van der Poll, R.H. de Vos, J. Chromatogr. 187 (1980) 244.

[370] M. Goreti, F. Sales, M. Carino, V.F. Vaz, C. Deleren-Matos, S.A.A. Alemida, et al., Int. J. Environ. Anal. Chem. 88 (2008) 37.

[371] S. Hon, Y. Cui, F. Dur, W. Kong, S. Song, M. Li, et al., Int. J. Environ. Anal. Chem. 89 (2009) 59.

[372] G.A. Bufo, A. D'Anria, L. Sorano, R. Toghil, Int. J. Environ. Anal. Chem. 84 (2004) 39.

[373] R. Miskus, H.T. Gordon, D.A. George, J. Agric. Food Chem. 7 (1959) 613.

[374] Z. Stransky, J. Chromatogr. 320 (1985) 219.

[375] J.E. Dretz, A.T. Scribner, M.T. Meyer, W. Koplin, Int. J. Environ. Anal. Chem. 85 (2005) 1141.

[376] V. Homen, A. Alves, L. Santos, Int. J. Environ. Anal. Chem. 90 (2010) 1063.

[377] J. You, M.J. Lydy, Int. J. Environ. Anal. Chem. 86 (2006) 381.

[378] M. Sneehar, M.T. Reddy, K. Balaji, J.S. Raddy, Int. J. Environ. Anal. Chem. 86 (2006) 757.

[379] A. Dowd, E. Maron, A. Stevens, H.M. Ismail, M. Habron, J. Hemmingway, et al., Int. J. Environ. Anal. Chem. 90 (2010) 922.

[380] N. Bjamholt, B. Svensmirchang, H.C.B. Hancen, Int. J. Environ. Anal. Chem. 84 (2004) 303.

[381] Z. Li, Y. Fang, P. Cheu, Z. Wang, G. Ren, Y. Huang, Int. J. Environ. Anal. Chem. 90 (2010) 856.

[382] R.D. Cannizzara, T.E. Cullen, R.T. Murphey, J. Agric. Food Chem. 18 (1970) 728.

[383] L.A. Norris, M.L. Montgomery, Bull. Environ. Contam. Toxicol. 13 (1975) 1.

[384] W.R. Payne, J.D. Pope, J.E. Benner, J. Agric. Food Chem. 22 (1974) 79.

[385] C.J. Soderquist, B. Crosby, Bull. Environ. Contam. Toxicol. 8 (1972) 363.

[386] O. Coha, Anal. Lett. 2 (1969) 623.

[387] G.G. Volpe, N.N. Mallett, Int. J. Environ. Chem. 8 (1980) 291.

[388] C.P. Rice, H.C. Sikka, R.S. Lynch, J. Agric. Food Chem. 22 (1974) 533.

[389] T. Cairs, E.G. Sigmund, G.M. Doose, Anal. Chem. 54 (1982) 953.

[390] Tsuikoika, Analyst 113 (1988) 193.

[391] W. Auer, H. Malissa, Chromatographia 25 (1988) 817.

[392] V.S. Bondarev, V.G. Spirdonov, V.G. Shestakov, B.Y. Chvertkin, J. Anal. Chem. USSR 42 (1987) 1040.

[393] S.D. West, G. Turner, J. Assoc. Off. Anal. Chem. 71 (1988) 1049.

[394] F. Onuksa, M.E. Comba, J.A. Coburn, Anal. Chem. 52 (1980) 2272.

[395] U. Kiiemagi, J. Burnard, L.C. Terrier, J. Agric. Food Chem. 23 (1975) 717.

[396] H.R. Schulten, D.H. Beckey, J. Agric. Food Chem. 21 (1973) 272.

[397] R.E. Kongovi, R. Grochowski, Am. Lab. 30 (1981) 30.

[398] M. Ishibashi, M. Suuzuki, J. Chromatogr. 456 (1988) 382.

[399] Y. Yamato, M. Suzuki, T. Wanatabe, J. Assoc. Anal. Chem. 61 (1978) 1135.

[400] D.C. Abbott, P.J. Wagstaff, J. Chromatogr. 43 (1969) 361.

[401] I. Moreno, G. Repetto, E. Carballal, A. Gago, A.M. Camean, Int. J. Environ. Anal. Chem. 85 (2005) 461.

[402] N. Drescher, Methodensamminiung zur Ruckstandardsanalytik von Pflansenschutsmittelm, Verlag Chimie, Weinheim.

[403] N. Drescher, Bestimmung der Ruckstande von Pyramin in Pflanze und Boden, 8–9 January, BASF, Ludwigshafen and Rhein, 1964, p. 78.

[404] W. Zaborowska, I. Witkowska, H. Kozak, Rocz. Panstw. Zakl. Hig. 24 (1973) 735.

[405] O. Palusova, M. Sackmauerova, A. Madarix, J. Chromatogr. 106 (1975) 405.

[406] B. Crathorne, C.D. Watts, J. Chromatogr. 169 (1979) 436.
[407] Y. Kawono, A. Bevenuz, J. Chromatogr. 72 (1972) 51.
[408] C.J. Miles, H.A. Moye, Anal. Chem. 60 (1988) 220.
[409] C.H. Marvina, I.D. Brindle, C.D. Hall, M. Chiba, Anal. Chem. 62 (1990) 1495.
[410] E.O. Jones, Anal. Chem. 63 (1991) 580.
[411] A. Di Corcia, R. Samperi, A. Mariomini, S. Stilluto, Anal. Chem. 65 (1993) 907.
[412] T.A. Bellar, W.L. Budde, Anal. Chem. 60 (1988) 2076.
[413] C. Crescenzi, A. Do Corcia, S. Marches, R. Sampori, Anal. Chem. 68 (1996) 1968.
[414] A. Cappiello, G. Famiglini, F. Burner, Anal. Chem. 66 (1994) 1416.
[415] F.G. Primel, M.R. Milani, A. Demoliner, L.P.H. Neincheski, A.L.V. Escarrone, Int. J. Environ. Anal. Chem. 90 (2010) 1048.
[416] H.R. Schulten, H. Prince, H.D. Beckey, W. Tomberg, F. Korte, Chemosphere 2 (1979) 22.
[417] H.R. Shulten, J. Agric. Food Chem. 24 (1976) 743.
[418] K.M. Kuivila, R.E. Jennings, Int. J. Environ. Anal. Chem. 87 (2007) 897.
[419] R.H. Coupe, H.I. Welch, A.B. Pell, M. Thurman, Int. J. Environ. Anal. Chem. 85 (2005) 1127.
[420] Y. Lin Hung, R.D. Vyksner, Anal. Chem. 65 (1993) 451.
[421] R.H. Coupe, Int. J. Environ. Anal. Chem. 87 (2007) 883.
[422] H. Korasoli, H. Angostfoppoulos, H. Ekonomopoulos, A. Hourdakis, Int. J. Environ. Anal. Chem. 84 (2004) 55.
[423] O. Kovacs, Anal. Abstr. 196 (2016) 13.
[424] V. Leopni, G. Pucciti, Farmaco, Ed. Prat. 26 (1971) 383.
[425] T.A. Albanis, D.G. Hela, D.A. Lonbropoulon, A.S. Vasilos, Int. J. Environ. Anal. Chem. 84 (2004) 1079.
[426] F. Hernandex, J.M. Marin, O.J. Pazo, J.V. Sancho, F.J. Lopez, I. Morell, Int. J. Environ. Anal. Chem. 88 (2008) 409.
[427] E. Silva, S. Batista, P. Viana, P. Antunes, L. Serodio, A.T. Cardoso, et al., Int. J. Environ. Anal. Chem. 86 (2006) 955.
[428] G.A. Matoriotas, T.A. Albanis, Int. J. Environ. Anal. Chem. 84 (2004) 103.

Further reading

A. Di Corcie, C. Crezenzi, E. Samperi, L. Scappaticeio, Anal. Chem. 69 (1997) 2819.
S.K. Handa, J. Assoc. Off. Anal. Chem. 7 (1988) 51.
C. Hu, G. Gan, Z. He, L. Song, Int. J. Environ. Anal. Chem. 88 (2008) 267.
H. Kuang, Y.N. Wu, X. Hou, M. Miao, G. Zhong, J.Z. Sher, et al., Int. J. Environ. Anal. Chem. 89 (2009) 423.
P. Manisankar, G. Selvanathan, C. Kechi, Int. J. Environ. Anal. Chem. 85 (2005) 409.
J.H. Onley, G. Yip, J. Assoc. Off. Anal. Chem. 52 (1969) 526.
J.R. Procopio, M.T.S. Excribino, L.H. Hernandez, Z. Anal. Chem. 331 (1988) 27.
G.E. Schneider, M.K. Koeppe, M.V. Naidu, et al., J. Agric. Food Chem. 41 (1993) 2404.

Chapter 11

Miscellaneous organic compounds in nonsaline waters

Chapter Outline

11.1 Plant pigments 219
High-performance liquid chromatography 219
Thin-layer chromatography 220
Spectrophotometric and spectrofluorimetric methods 220
11.2 Humic and fulvic acid 221
Polarography 223
Gel permeation chromatography 223
Ultraviolet spectroscopy 223
Fluorescence spectroscopy 224
11.3 Geosmin 225
11.4 Mestranol and ethynyloestradiol 226
11.5 Cobalamin (vitamin B12) 226
11.6 Algal toxins and blooms 226
11.7 Microcystins 226
11.8 Anthropogenic, ostragenic and oestrogenic hormones 227
11.9 Antibiotics 228
11.10 Pharmaceuticals 229
11.11 Miscellaneous pollutants 230
References 230

11.1 Plant pigments

High-performance liquid chromatography

The first application of high-performance liquid chromatography (LC) (HPLC) to plant pigments was by Evans et al. [1] and other [2–12] who separated phaeophytins a and b on Corasil II with a mobile phase consisting of 1:5 (v/v) mixture of ethyl acetate and light petroleum. Eskins et al. [13] have employed columns of C_{18}-Porasil B for preparative separation of plant pigments by means of a programmed stepwise elution with methanol–water–ether. However, the method is of little value for routine application because the time required and also because of the chlorophyll degradation products, other than phaeophytin, are not separated.

Shoaf [14] has used HPLC to separate the chlorophylls a and b of the pigment extract from which the carotenoids had been previously removed. Good resolution of the two pigments and several of their unspecified degradation products was achieved on a 23 cm column of Partisil PXS 1025 by elution with aqueous 95% methanol; however, chlorophylls were not determined quantitatively.

Determination of Toxic Organic Chemicals In Natural Waters, Sediments and Soils.
DOI: https://doi.org/10.1016/B978-0-12-815856-2.00011-4

Mantoura and Llewelyn [15] developed a reversed-phase high-performance liquid chromatographic system for a rapid (about 20 m) separation and quantification of 14 chlorophylls and their breakdown products and 17 carotenoids from acetone extracts of natural waters. An ion-pairing reagent is included to achieve good resolution with the acidic chloropigments (chlorophylls) and phaeophorbides. Fluorescence and absorption detectors are used to quantifying chloropigments and carotenoids, respectively, with a detection limit of 0.01–0.2 ng for these pigments and 200–600 ng for carotenoids.

Garside and Riley [16] applied HPLC to the determination of a range of algal pigments.

Tests carried out with a high-performance liquid chromatographic technique endorsed the claim that it causes negligible degradation of both the chlorophylls and the xanthophylls. A more efficient separation of plant pigments could be achieved on a silica stationary phase than on a C_{18} reversed-phase medium. Excellent separation of algal pigments such as β carotene, fucoxanthin, chlorophyll b could be achieved by means of a mixture containing light petroleum (b.p. 60°C–80°C, acetone, methanol and dimethyl sulphoxide in the ratio 30:40:27:27 by volume, respectively).

Thin-layer chromatography

Thin-layer materials employed have included the following: layers of kieselguhr G impregnated with triolein, castor oil or paraffin oil [17], layers prepared from glucose shaken with ether [18], plates coated with kieselguhr G impregnated with peanut oil dissolved in iso-octane [19] and layers of powdered confectioner's icing sugar containing 5% cornflour suspended in light petroleum.

Shoaf and Lium [20] used thin-layer chromatography to separate algal chlorophylls from their degradation products. Chlorophyll was extracted from the algae with dimethyl sulphide and chromatographed on commercially available thin-layer cellulose sheets using 2% methanol and 98% petroleum ether as solvents, before determination by either spectrophotometry or fluorimetry.

Chlorophylls and degradation products such as phaeophytin, chlorophyll a, phaeophytin, chlorophyll b, phaeophytin c and chlorophyll c each had a distinct Rf value.

Spectrophotometric and spectrofluorimetric methods

Probably the most widely used analytical technique depends upon spectrophotometric measurements of extinctions of 90% acetone extracts at three different wavelengths [21]. Solution of appropriate simultaneous equations (trichromatic) provides estimates of chlorophylls a, b and c. Phaeophytin a can also be estimated by this method after acidification of the extract solution. Various improvements to the trichromatic equations have been published by a number of authors [22,23]. The simplest and most widely adopted technique for chlorophylls is that developed by Richards and Thompson [24] and subsequently modified by others. In this, spectrophotometric measurements are made at selected wavelengths on 90% acetone extracts of the separated phytoplankton. With simultaneous equations [25–27], it is possible to calculate the amounts of chlorophylls, a, b and c and total carotenoids present in the sample.

The major criticism is that in natural plankton extracts, spectrally similar chlorophyll breakdown products are frequently present [28–31]. Thus chlorophyll cannot be accurately determined by this method. One way to avoid the problem is to chromatographically separate the breakdown products from the chlorophylls before measurement. Separation of the spectrally similar chlorophylls a and b (as well as degradation products) will result in a more accurate determination of the chlorophylls [2–5].

Fluorescence spectroscopy [21,32,33] has also been applied to the determination of carotenoids.

Recoveries of pure chlorophylls a and b were 98% and 96%, respectively.

Boto and Bunt [34] also used thin-layer chromatography for the preliminary separation of chlorophylls and phaeophytins and combined this with selective excitation fluorimetry for the determination of the separated chlorophylls a, b and c and their corresponding phytoplankton components. An advantage of the latter technique is that appropriate selection of excitation and emission wavelengths reduces the overlap between the emission spectra of each pigment to a greater extent than is possible with broadband excitation and the use of relatively broadband filters for emission.

Garside and Riley [16] have used thin-layer chromatography to achieve a preliminary separation of chlorophylls in solvent extracts of water and algae prior to a final determination by spectrophotometry or fluorimetry. Garside and Riley [16] filtered sea water samples (0.5–5 L) through Whatman GF/C glass fibre coated with a layer, 1–2 nm thick, light magnesium carbonate.

11.2 Humic and fulvic acid

Canadian rivers and lakes [35] revealed that the majority of dissolved organic carbon was of natural origin with fulvic acid, tannins and lignin being the major components.

Humic and fulvic acids are among the most widely distributed products of plant decomposition on the Earth's surface, occurring in soil, water, sediments and a variety of other deposits [6,7]. They are amorphous, yellow-brown or black, hydrophilic, acidic, polydisperse substances of wide-ranging molecular weight, namely <10,000 for the fulvic acid and 10,000–300,000 for humic acid are the usual ranges, although some values outside of these ranges have been reported by Rashid and King [36]. Humic and fulvic acids are responsible for the yellow-brown colouration of many lakes and rivers. By definition, humic acid is the fraction of organic material soluble at pH 9 but insoluble at pH 2, while fulvic acid is soluble at both pH 9 and pH 2 [37].

Tannins and lignins are high-molecular-weight polycyclic aromatic compounds widely distributed through the plant kingdom. Pearl [38] defined lignin as 'the incrusting material of plants which is built up of methoxy- and hydroxyphenylpropane units'. It is not hydrolysed with acids but is readily oxidised and soluble in hot alkali and bisulphite. The exact chemical composition of lignin-like compounds commonly found dissolved in natural waters is not known. Tannins, according to Geissman and Crout [39] are either polymers of gallic acid linked to carbohydrate residues (hydrolysable tannins) or polymeric flavonoid compounds (condensed tannins). Both tannins and lignins are highly resistant to chemical and biological degradation. How closely these compounds resemble their laboratory counterparts, tannic acid and lignosulphonic acid, is not known.

In natural waters, dissolved humic compounds are an important source of fluorescence. The presence of such soluble fluorescence could cause serious errors in the in vivo measurement of chlorophyll using Lorenzen's fluorimetric method.

Anoxic waters often contain both green sulphur bacteria and organisms with chlorophyll a, including algae which have sunk from oxic surface waters and cyanobacteria which grow in anoxic waters. Calculating pigment concentrations in waters containing both the bacterial chlorophylls Pheo-GSB and Chl-GSB and algal chlorophyll a was difficult, but equations are given that allow the calculation of algae and green surface bacteria, hence both chlorophyll a and Chl-GSB in anoxic waters.

Stevenson et al. [40–42] used potentiometric titration to study the nature of divalent copper, lead, cadmium and zinc complexes of humic acids. They showed at least two major sites were involved in complex formation, and differences in the results using different humic acids were negligible. Stability constants decreased in the order copper, lead, cadmium, zinc.

These methods used a sequential titration procedure in which sequential additions of the metal ion were made to solutions of the humic acid at constant pH. The pH was returned to the initial starting point after each addition, using carbon dioxide–free potassium hydroxide.

Results obtained by sequential additions of Cu^{2+} to the soil humic acid at pH values of 4, 5 and 6 show that each addition of metal ion depressed the pH to a lesser extent than the previous addition, and less and less base was required to neutralise the liberated protons. Not all acidic groups had reacted at pH 4, apparently due to competition of protons for the ligand. For Cu^{2+}, more protons were liberated at pH 5 and 6 than could be accounted for by acidic functional groups. This can be accounted for by release of protons from hydration water of the metal ion held in 1:1 complexes. In the case of the reactions carried out at pH 6, some of the extra protons undoubtedly resulted from the formation of insoluble oxide hydrates.

Polarography

Cosovic et al. [43] used an electrochemical method employing mercury electrodes (measurements of the suppression of the polarographic maxima and AC polarography) to study surface active substances in samples of river water and ground water. The shape and intensity of the electrochemical responses were compared with those obtained using model surfactants to give a rough identification of the principal surface active compounds present. Humic substances predominated. Fulvic acid reduces mercury (II), iron (III), iodine and triiodide ions under conditions characteristic of natural water. A reduction potential of approximately 0.5 V (versus normal hydrogen electrode) for fulvic acid has been estimated by Skogerboe and Wilson [44].

Rubenstein and Sil'chenko [45] also studied the application of polarography to the determination of fulvic acids in natural waters.

Gel permeation chromatography

Chian and De Walle [46] used gel permeation gas chromatography (GC) on Sephadex G-75 and G-200 ultrafiltration and GC to characterise soluble organic matter in heavily polluted ground water samples and leachates from landfills. The largest fraction consisted of three volatile fatty acids, and the next largest was a fulvic-like material with relatively high carboxyl and aromatic hydroxyl group density. There was also a small percentage of a high-molecular-weight humic carbohydrate-like complex.

Gel filtration chromatography using Sephadex G100 as column packing and ultraviolet (UV) detectors have been used in studies carried out on the elution of humic acid [47] and in characterisation studies on secondary sewage effluents [48] and in organic substances in river waters [49].

Ultraviolet spectroscopy

Wagner and Hoyer [50] and others [51–53] used UV spectroscopy for the simultaneous determination of humic acid and lignosulphonic acids in natural waters. These workers also compared absorption characteristics in the wavelength range 250–300 nm. To eliminate interfering compounds, especially highly polar organic acids, the extraction was performed using a mixture of trioctylamine and chloroform. Equations were derived which enabled the concentration of both humic and lignosulphonic acids to be determined from measurements of the extinction coefficient at three wavelengths.

Sontheimer and Wagner [54] determined humic and lignosulphonic acids in water by measuring the differences in the UV spectra of the sample between 250 and 300 nm. The workers confirmed that this method could be used for the approximate determination of these constituent in water, following a preliminary extraction and concentration procedure. Tests on several natural waters showed that the method could be used for the preliminary assessment of water quality.

Lawrence [55] described a simple UV spectroscopic method for semi-quantitative determining of fulvic acid, tannic acid and lignin in natural waters. This method is based on the fact that the concentrations of each component can be determined by measuring the absorbance at three different wavelengths.

Eberle [56] extracted humic acids from water with trioctylamine and showed that the salts thus formed are readily soluble in aromatic or chlorinated hydrocarbons and can this be extracted from aqueous solution. By the use of 10% (v/v) trioctylamine solution in chloroform, humic acids are extracted from water at pH less than 5 with a partition coefficient of less than 100 and are reextracted quantitatively at a pH of greater than 9. He describes a procedure for the extraction-spectrophotometric determination of humic acids in water. Lignosulphonic acid behaves like humic acids with respect to extraction by trioctylamine, and all these species can be determined simultaneously by calculation from the extinction measured at two wavelengths.

Fluorescence spectroscopy

This technique has been applied to the determination of humic matter in natural waters [8,53,57–69].

Ewald et al. [60] examined the technical requirements involved in the measurement of the fluorescence spectra of waters and the relation between fluorescence and humic acid content. These studies showed the need for applying corrections when determining the fluorescent spectra.

Pennanen and Mannio [61] have studied dilution effects in fluorescence measurements of humic lake waters. Optical measurements were carried out using excitation and emission wavelengths of 350 and 455 nm of both toxic substances. Heat from industrial outfalls cast further doubt on the accuracy of biological indicator systems [9–12,70–81].

The use of 5β-chlorestane-β-ol is the generally accepted criteria of a good indicator of faecal pollution. It is believed that the only source of this compound is the faeces of higher animals including man. It is biodegradable and can be removed from domestic sewage by adequate treatment. Furthermore, it has been unambiguously proved that the concentration of coprostanol is highest in the overtly faecal polluted water, and there is a progressive decrease in the concentration of this compound in the lesser polluted waters. Since its isolation and identification is unaffected by chlorination or by heat and toxic substances discharged from industrial outfalls, the advantage of using a molecular rather than a biological indicator of faecal pollution is further demonstrated. Such a characteristic of coprostanol is especially significant in the current trend promoting disinfection of raw and treated water. Because of this unique property, coprostanol might also be a useful indicator in monitoring the source, course and extent of faecal pollution in the ocean or brackish waters where bacteriological evidence is often doubtful.

Dawson and Best [79] investigated the use of coprostanol as an internal tracer for faecal contamination in water by investigating the method of analysis and the effects of sewage treatment on the concentration of coprostanol, and also by marine surveys at sewage outfalls. Analysis time is greater for faecal bacteria, but immediate analysis is not necessary, and the results are less subject to error.

Singley et al. [80] have described a gas chromatographic method for the determination of coprostanol. This technique was used in extensive field studies, degradation studies and studies on treatment plant efficiency, and was also used as a standard for evaluating a colorimetric method that was developed and shown to be capable of determining coprostanol in polluted water levels of 1 $\mu g\ L^{-1}$. It was shown that there was good correlation between coprostanol and biochemical oxygen demand, chemical oxygen demand and total organic carbon.

Wun et al. [81] used XAD-2 resin for the analysis of coprostanol in river and lake water and secondary sewage treatment plants. They showed that extraction of coprostanol from water by absorption on a column of Amberlite XAD-2 resin is as efficient as conventional liquid–liquid extraction. Maximal recovery depends on the pH value of the sample, flow rate, resin mesh size and concentration of coprostanol.

In further work, Wun et al. [82] improved the efficiency of the extraction of coprostanol using an XAD-2 resin column by decreasing the extraction time using a 'closed' column technique and by determining the effects of sample pH on adsorption process.

Coprostanol was strongly adsorbed on polystyrene XAD-2 adsorbents at pH 2, with 100% retention, and the adsorbed sterol was easily eluted with acetone adjusted to pH 8.5–9.0 with ammonium hydroxide. It was also shown that with a closed column method, large volumes of water sample can be extracted in a relatively short time and with higher sensitivity than that of the liquid–liquid partitioning procedure.

11.3 Geosmin

Persson [83] has discussed the determination of geosmin and 2-methylisoborneol in water.

Yasuhara and Fuwa [84] have described a quantitative method for determining geosmin in river water using computer-controlled mass fragmentography using a JEOL Model JMS-D 100 mass spectrometer connected with a JEOL Model JCG-20K gas chromatograph and a JEOL Model JMA-2000 mass data analysis system. The detection limit of geosmin was 10 $mg\ mL^{-1}$.

Parinet et al. [85] developed an automated method, based on a solid-phase microextraction coupled with GC/mass spectrometry (MS), for the determination of geosmin, 2-methylisoborneol, 2-isopropyl-3-methyloxypyrazine, 2-isobutyl-3-methoxypyrazine and 2,4,6-trichloronisole in environmental water samples.

11.4 Mestranol and ethynyloestradiol

Okuno and Higgins [86] have described a procedure for determining residual levels of mestranol [17a-ethynyl-3-methoxyestra-1,3,5,(10)-trien-17β-ol] animal damage control chemosterilant and its 3-hydroxy homologue, ethynyloestradiol, in water samples. The lower limits of detection were 0.01 mg^{-1} for water samples. After extraction in acidic medium, samples are cleaned by Florisil column chromatography. Water samples can then be analysed by gas–liquid chromatography. This layer chromatography was used to confirm the results obtained by GC.

11.5 Cobalamin (vitamin B12)

Beck [87] has described a competitive intrinsic factor–binding method for the determination of benzyl alcohol extractable cobalamins in amounts down to 1 pg L^{-1} in natural waters. He compares results of assays with those determined by high-pressure LC.

11.6 Algal toxins and blooms

Gago-Martinez et al. [88] applied capillary electrophoresis to the determination of algal toxins in the aquatic environment.

In this method, capillary electrophoresis with UV detection was applied to the analysis of different natural toxins produced in the aquatic environment. This is an alternative to other chemical techniques such as HPLC, and the optimisation of analytical methodologies was carried out for diverse marine toxins including paralytic and amnesic and some polyether toxins such as yessotoxins, as well as for certain microcystin toxins produced by cyanobacteria present in freshwaters. Sample preparation steps were optimised and adequate electrophoretic conditions developed for achieving a complete separation of compounds with similar structures involved in such contamination; the influence of the biological matrices was studied, and the use of capillary electrophoresis as a tool for monitoring these aquatic toxins was all studied.

11.7 Microcystins

Various workers have discussed methods for the determination of microcystins in water [89–91], Hu et al. [89] described a chemiluminescent immunoassay method based on gold nanoparticles for the detection of microcystins. The immunoassay method included three main steps: indirect competitive immunoreaction, oxidative dissolution of gold nanoparticles and indirect determination of microcystins with Au^{3+}-catalysed luminol chemiluminescent system. The method has a wide working range (0.05–0.99 μg L^{-1}). The limit of detection was 0.024 μg L^{-1}, which is much lower than the World Health Organisation (method 1 μg L^{-1}) for drinking water. The proposed method was applied to microcystins analysis in natural water samples; results were in agreement with the conventional indirect competitive enzyme-linked immunosorbent assay method. This chemiluminescent immunoassay was sensitive, reliable and suitable for microcystins analysis in natural water.

Mareno et al. [91] determined microcystins and cyanobacteria in Spanish river water, and Ouahid et al. [90] discuss the determination of a single microtoxin in a toxic cystic bloom taken from the river Rio de la Plata, Argentina.

11.8 Anthropogenic, ostragenic and oestrogenic hormones

Dell Bubba et al. [92] measured the horizontal and vertical distribution of biogenic and anthropogenic organic compounds in the Ross area, Antarctica.

Schwarzbauer et al. [93] determined 13c/12c ratios of anthropogenic contaminants in river waters by a method based on GC/MS.

This study describes the application of a common analytical procedure adapted for compound-specific carbon isotope analyses of riverine contaminants. To evaluate the sensitivity of the analytical method and the precision of the isotope date obtained, a set of numerous substances at different concentration levels were measured. For most of the anthropogenic contaminants investigated (including chlorinated aliphatics and aromatics, musk fragrances, phthalate-based plasticisers and tetrabutyl tin), acceptable carbon isotope analyses could be obtained down to amounts of approximately 5 ng absolutely applied to the gas chromatograph. These amounts correspond to concentrations in water samples at a natural abundance level of approximately 50–200 ng L^{-1} (low-to-medium contaminated river systems). However, it has to be considered that the precision and the sensitivity of the analytical method depend partially on the chemical properties of the substances measured. Recovery experiments were conducted to assess changes in carbon isotope ratios during sample preparation and measurement. The compounds selected for these experiments are known riverine contaminants. Isotope shifts or higher variations of the isotope ratios as a result of the analytical procedures applied were observed only for a couple of contaminants. Furthermore, compound-specific carbon isotope analyses were performed on eight water extracts of the Rhine River. By comparing the variation of the data of several individual compounds with the deviations obtained from the recovery experiments, it was possible to differentiate contaminants with unaffected isotope ratios and substances with significant alterations of the $\delta^{13}C$-values.

Pojana et al. [94] developed an analytical method for the simultaneous determination of oestrogenic compounds of natural (oestradiol, oestriol, oestrone) and synthetic origin, both steroids (ethinyloestradiol, mestranol) and nonsteroidal (benzophenone, bisphenol-A, diethylstilbestrol, octylphenol, nonylphenol, nonylphenol monoethoxylate carboxylate), in environmental aqueous samples by HPLC coupled with ion trap–MS via electrospray interface. Quantitative MS detection was performed in the negative mode for all compounds except mestranol and benzophenone, which were detected under positive ion compounds. Very low method detection limits, between 0.1 and 2.6 ng L^{-1}, were achieved in coastal lagoon water samples, while the developed solid-phase-extraction procedure permitted simultaneous recovery of all analytes from spiked water samples with yields >70% [7–11 relative standard deviation (RSD)%], except oestriol and benzophenone, which were recovered with 60% (9 RSD%) and 50% (11 RSD%) yields, respectively. This method was applied to the analysis of Venice (Italy) lagoon waters, where average concentrations of selected compounds in the 2.8–33 ng L^{-1} concentration range were found.

11.9 Antibiotics

The recent discovery of antibiotics in streams across the United States has raised awareness and demonstrated the need to monitor and determine sources of antibiotics and their persistence in the environment. It has been estimated that approximately 50% of the annual production of antibiotics in the United States is for human health and 50% is for agriculture and aquaculture practices. The US Food and Drug Administration has approved antibiotics for use in aquaculture to treat systemic bacterial infections in fish. The approved antibiotics include a combination drug containing ormetoprim and sulphadimethoxine, and oxytetracycline hydrochloride, which were approved for use on fish and salmonid. These drugs generally are administered directly to the water in medicated feed at fish hatcheries.

Dietz et al. [95] have conducted a 2-year study of extensive and intensive fish hatcheries to assess the general temporal occurrence of antibiotics in aquaculture. Antibiotics were detected in 15% of the water samples collected during the 2001–02 collection periods and in 31% of the samples during the 2003 collection period.

Antibiotics were detected more frequently in samples from the intensive hatcheries (17% and 39%) than in samples from the extensive hatcheries (14% and 4%) during the 2001–02 and 2003 collection periods, respectively. The maximum ormetoprim, oxytetracycline and sulphadimethoxine concentrations were higher in samples from the intensive hatcheries

(12, 10 and 36 μg L^{-1}), respectively, than in samples from the extensive hatcheries (<0.05, 0.31 and 1.2 μg L^{-1}), respectively. Sulphadimethoxine persisted for a longer period of time (up to 48 days) than ormetoprim (up to 28 days) and oxytetracycline (less than 20 days).

These ranged from 0.39 to 9.0 μg L^{-1}; the oxytetracycline concentration peaking between days 5 and 8. Samples were analysed for β-lactams, microlides, sulphonamides, quinolines and tetracyclines.

The antibiotics for each method were eluted and separated using an LC gradient with mobile phases A and B. In mobile phases A and B, phase A was 5 mM ammonium acetate for the β-lactones and microlides class, and 0.3% formic acid was used for the quinolines and oxytetracycline classes of antibiotics. Acetonitrile was used as mobile phase B for the β-lactam, microlides and quinolines classes, and methanol was used for the oxytetracyclines class of antibiotics. The initial flow rates of mobile phases A and B were decreased and contained a higher proportion of mobile phase B to elute the Prospekt SPE cartridge.

11.10 Pharmaceuticals

Wu et al. [96] applied a multiresidue method for the simultaneous determination of 36 pharmaceuticals (histamine receptor antagonists, psychoactive stimulant antiepileptics, antihypertensive, nonsteroidal antiinflammatory, analgesic and antipyretic, lipid regulator, antibiotics, antibacterial, skin care ingredient and metabolites of nicotine and lipid regulators) in surface water using solid-phase extraction (Strata-X at pH 5) and LC–tandem MS. Recoveries were greater than 70% with less than 20% SD for the majority of analytes. The instrumental quantification limit was between 2 and 181 pg, and method quantification limit varied from 0.5 to 98 ng L^{-1} in spiked water. The pH and sorbent dependence of matrix effects is discussed. The optimised method was used to determine the occurrence of target analytes in surface water from the coastal Lake Erie in Oregon, northwest Ohio. Seventeen analytes were detected with concentrations up to hundreds of nanograms per litre in stream and lake water samples.

This method eliminates the need for use of multiple extraction steps and laborious derivatisation associated with GC, or several LC gradients. Utilising the standard added method allowed for compensation for matrix effects present. The presented data allow researchers to use the method as a whole or subset list of analytes based on sorbent choice, pH, matrix and recovery data.

The proposed method was subsequently used to measure pharmaceuticals from several environmental matrices. Seventeen compounds were detected with concentrations up to the hundreds nanogram per litre in stream and lake water samples. The solid phase is a useful tool for the analysis of pharmaceuticals in the aquatic environment and can be used to study the transportation and fate of the pharmaceuticals.

Accinelli et al. [97] detected the antiviral drug oseltamivir in surface waters. The antiviral prodrug oseltamivir phosphate has received recent attention with regard to its possible use against the highly pathogenic H_5N_1 virus. This preliminary laboratory study investigated the persistence of the active antiviral drug, oseltamivir carboxylate, in water samples taken from an irrigation canal. After an initial rapid decrease, oseltamivir carboxylate concentrations slowly decreased during the remaining incubation period. Approximately 65% of the initial oseltamivir carboxylate amount remained in water at the end of the 36-day incubation period. A small amount of oseltamivir carboxylate was lost both from sterilised water and from sterilised water/sediment samples, suggesting a significant role for microbial degradation.

Simulating microbial processes by the addition of sediments resulted in reduced oseltamivir carboxylate persistence. Presence of oseltamivir carboxylate (1.5 μg mL^{-1}) did not significantly affect the metabolic potential of the water microbial population, estimated by glyphosate metolachlor mineralisation. In contrast, oseltamivir carboxylate caused an initial transient decrease in the size of the indigenous microbial population of water samples.

11.11 Miscellaneous pollutants

Other miscellaneous materials that have been determined in water samples include fluorescent optical whiteners [97–101], carboxymethyl succinate detergent [102], squoxin pesticide [103] and organic products resulting from ionisation of water in water treatment plants [104,105].

Sarafraz-Yazdi et al. [106–108] carried out a comparative study of direct immersion and headspace single-strip microextraction techniques combined with GC–flame ionisation detection for the determination of organic pollution in water.

Monisanbar et al. [107] reviewed the application of square-wave stripping voltammetry to the determination of organic pollutants in water.

Flavin et al. [108] have compared various organic materials as a means of preconcentrating organics in water using AT/FTIR sensing.

References

[1] N. Evans, D.E. Games, A.H. Jackson, S.A. Matlin, J. Chromatogr. 115 (1975) 325.
[2] S.O. Ryding, Vatten 31 (1975) 327.
[3] H. Rai, Arch. Hydrobiol. Suppl. 14 (1980).
[4] R.E. Youngman, Report No. TR82. Water Research Centre, Medmenham, Marlow, Bucks, 1978.
[5] E.A. Nusch, Wasser Abwasser Frosch 17 (1984) 89.
[6] F. Raghi-Atri, Gesundheits Ingenieur 99 (1978) 380.
[7] K.S. Baker, R.C. Smith, J.R. Nelson, Limnol. Oceanogr. 28 (1983) 1037.
[8] M.G. Snow, Proc. Anal. Div. Chem. Soc., Lond. 12 (1975) 253.

[9] M. Cetinkaya, J. Von Duszelin, B. Gabel, M. Gurlak, R. Kozicki, U. Lahl, et al., Z. Wasser Abwasser Forsch. 13 (1980) 213.
[10] J.H. Weber, S.A. Wilson, Water Res. 9 (1975) 1079.
[11] N. Tambo, T. Kamai, T. Nishihura, K. Fukushi, Jpn. Water Works Assoc. 532 (1979) 37.
[12] E.M. Hiraide, K. Tillekeratue, K. Otsuka, A. Mizike, Anal. Chim. Acta 172 (1985) 215.
[13] K. Eskins, C.R. Scholfield, H.H. Dutton, J. Chromatogr. 135 (1977) 217.
[14] W.T. Shoaf, J. Chromatogr. 152 (1978) 247.
[15] R.F.C. Mantoura, C.A. Llewelyn, Anal. Chim. Acta 151 (1983) 297.
[16] C. Garside, J.P. Riley, Anal. Chim. Acta 46 (1969) 179.
[17] J.P. Riley, T.R. Wilson, J. Mar. Biol. Assoc. 45 (1965) 583.
[18] J.C. Madgwick, Deep Sea Res. 12 (1965) 233.
[19] I.D. Jones, L.S. Butler, E. Gibbs, R.C. White, J. Chromatogr. 70 (1972) 87.
[20] W.T. Shoaf, B.I.W. Lium, J. Res. US Geol. Surv. 5 (1977) 263.
[21] O. Holm-Hansen, C.J. Lorenzer, R.W. Holmes, J.D.H. Strickland, J. Cons. Int. Explor. Mer. 30 (1965) 3.
[22] S.W. Jeffrey, G.J. Humphrey, Biochem. Physiol. Pflanz. 167 (1975) 191.
[23] J. Davies, J. Decaster, Hydribiologica 71 (1980) 19.
[24] F.A. Richards, T.F. Thompson, J. Mar. Res. 2 (1952) 150.
[25] G.I. Creitaz, F.A. Richards, J. Mar. Res. 14 (1955) 211.
[26] J.D.H. Strickland, T.F. Parsons, A Practical Handbook of Seawater Analysis, Fisheries Research Board of Canada, Ottawa, 1968.
[27] UNESCO Monographs on Oceanographic Methodology, No. 1, 1966.
[28] T.R. Parsons, J. Fish. Reserve Board Can. 18 (1961) 1017.
[29] N.W. Jeffrey, Biochem. Biophys. Acta 162 (1968) 271.
[30] R.J. Daley, C.B.J. Gray, S.R. Brown, J. Chromatogr. 76 (1973) 175.
[31] R.J. Daley, Arch. Hydrolysis 72 (1973) 400.
[32] M.E. Loftus, J.H. Carpenter, J. Mar. Res. 29 (1971) 319.
[33] N. Caraco, A.H. Puccoon, Limnol. Oceanogr. 31 (1986) 889.
[34] K.G. Boto, J.S. Bunt, Anal. Chem. 50 (1978) 392.
[35] J. Lawrence, National Inventory of Natural Organic Compounds – An Interim Report, Canada Centre for Inland Waters Unpublished Report, Burlington, ON.
[36] M.A. Rashid, L.H. King, Chem. Geol. 7 (1971) 37.
[37] M. Schnitzer, I.A. Khaus, Humic Substances in the Environment, Marcel Dekker, New York, 1972.
[38] I.A. Pearl, The Chemistry of Lignin, Marcel Dekker, New York, 1967.
[39] T.A. Geissman, D.H.G. Crout, Organic Chemistry of Secondary Plant Metabolism, Freeman Cooper, San Francisco, CA, 1969.
[40] F. Stevenson, J. Soil Sci. 123 (1967) 10.
[41] F.J. Stevenson, in: J.O. Nriagu (Ed.), Environmental Geochemistry, Ann Arbor Science Publishers, Ann Arbor, MI, 1976, pp. 96–106.
[42] F.J. Stevenson, K.M. Goh, Soil Sci. 117 (1974) 34.
[43] B. Cosovic, V. Vojvodic, T. Piese, Water Res. 19 (1985) 175.
[44] R.K. Skogerboe, S.A. Wilson, Anal. Chem. 53 (1981) 228.
[45] R. Rubenstein, O. Sil'chenko, Zhural Analitischeskoi Khim 12 (1975) 2448.
[46] E.S.K. Chian, F.B. De Walle, Exp. Sci. Technol. 11 (1977) 158.
[47] J. Aho, A. Lehto, Archiv. Hydrobiol. 101 (1984) 21.
[48] M. Manka, A. Mandlebaum, A. Boritinger, Environ. Sci. Technol. 8 (1974) 1017.
[49] J. Faure, P. Viallet, P. Picat, La Tribune de Cebendeau 28 (1975) 385.

[50] I. Wagner, O. Hoyer, Vom Wasser 45 (1975) 207.
[51] M.J. Miroslav, Water Pollut. Cent. Fed. 41 (1969) 1923.
[52] J.R. Moed, Limnol. Oceanogr. 16 (1971) 140.
[53] S.B. Eberle, H.K. Schweer, Vom Wasser 41 (1973) 27.
[54] H. Sontheimer, I. Wagner, Z. Wasser Abwasser Fursching 10 (1977) 77.
[55] J. Lawrence, Water Res. 14 (1980) 373.
[56] S.H. Eberle, Ber Kernsforschungzentrum Karlsrube, KFK-1731 UF, 1973, p. 18.
[57] W.R. Seitz, Trends Anal. Chem. 1 (1981) 79.
[58] J.G. Montalvo, C.G. Lee, National Bureau of Standards, Proceedings of the 8th International Symposium, Gaithersburg, MD, USA. Determination of Trace Organic Pollutants in Water by Spectrofluorescence after Treatment with Activated Carbon, p.97, November 1977.
[59] P.H. Boenin, P. Berckmann, V.L. Snoeyink, J. Am. Water Works Assoc. 72 (1980) 54.
[60] M. Ewald, P. Berger, C. Belin, Environ. Sci. Lett. 5 (1984) 31.
[61] V. Pennanen, J. Mannio, Aqua Ferrica 16 (1987) 143.
[62] W.A. McCrum, Anal. Proc., Lond. 23 (1986) 307.
[63] M.G. Snow, Water Treat. Exam. 24 (1975) 297.
[64] R.L. Weshaw, K.A. Thorn, D.J. Pincky, Environ. Technol. Lett. 9 (1988) 53.
[65] S.I. Cabannij, S. Shuman, Anal. Chem. 60 (1988) 2418.
[66] N.V. Blough, Environ. Sci. Technol. 22 (1988) 77.
[67] T. Ishbashi, Aqua Sci. Technol. Rev. 3 (1) (1980).
[68] S.A. Telang, Water Quality and Forest Management, Department of the Environmental Sciences Centre, University of Calgary, Kananaskis, 1991.
[69] C. Le Cloirec, P. Le Cloirec, M. Elmyhari, J. Morvan, G. Martin, Int. J. Environ. Anal. Chem. 13 (1983) 127.
[70] P.T. Hine, D.B. Bursill, Water Res. 18 (1984) 146.
[71] E.M. Hiraido, F.X. Ren, T. Tamura, A. Mizuike, Mikro Chim. Acta 416 (1987) 137.
[72] E.E. Geldreich, J. Am. Water Works Assoc. 63 (1971) 225.
[73] C.J. Kinchmer, 5β Cholestan-3β-ol: An Indication of Fecal Pollution (PhD thesis), University of Florida, 1971.
[74] B.J. Dutka, A.S.Y. Chau, J. Coburn, Water Res. 8 (1974) 1047.
[75] J.J. Murtaugh, R.L. Bunch, J. Water Pollut. Control Fed. 39 (1967) 404.
[76] L.L. Smith, R.E. Gouron, Water Res. 3 (1969) 141.
[77] H.H. Tabak, R.N. Bloomhuff, R.L. Bunch, Development in Industrial Biology, 13, Publication of the Society for Industrial Microbiology, American Institute of Biological Science, Washington, DC, 1972, pp. 296–307.
[78] R.G. Gould, R.P. Cook, in: R.P. Cook (Ed.), Cholesterol Chemistry, Biochemistry and Pathology, Academic Press, New York, 1958, pp. 289–292.
[79] J.P. Dawson, G.A. Best, Proc. Anal. Div. Chem. Soc. 12 (1975) 311.
[80] J.E. Singley, C.J. Kirhmer, R. Miura, US Environmental Protection Technology Series EPA-660/2-74-021, US Government Printing Office, Washington, DC, 1974, p. 126.
[81] C.K. Wun, L.W. Walker, W. Litsky, Water Res. 10 (1976) 955.
[82] C.K. Wun, R.W. Walker, W. Litsky, Health Lab. Sci. 15 (1978) 67.
[83] P.E. Persson, Water Res. 14 (1980) 1113.
[84] A. Yasuhara, F. Fuwa, J. Chromatogr. 172 (1979) 453.
[85] J. Parinet, M.J. Rodriguez, J. Serodes, F. Prouix, Int. J. Environ. Anal. Chem. 91 (2011) 505.
[86] I. Okuno, W. Higgins, Bull. Environ. Contam. Toxicol. 18 (1977) 428.
[87] R.A. Beck, Anal. Chem. 50 (1978) 208.

[88] A. Gago-Martinez, J.M. Leao, N. Pineiro, E. Carballai, E. Vatuero, M. Nogueiras, et al., Int. J. Environ. Anal. Chem. 83 (2003) 443.
[89] C. Hu, N. Gan, Z. He, L. Song, Int. J. Environ. Anal. Chem. 88 (2008) 267.
[90] Y. Ouahid, M.Z. Zaccaro, G. Zupla, M. Stomi, A.M. Stella, J.C. Bassop, et al., Int. J. Environ. Anal. Chem. 91 (2011) 528.
[91] I. Mareno, G. Repetto, E. Carballa, A. Gago, A.M. Camean, Int. J. Environ. Anal. Chem. 85 (2005) 461.
[92] M. Dell Bubba, A. Cincinella, L. Checchini, L. Lepri, P. Desideri, Int. J. Environ. Anal. Chem. 84 (2004) 1033.
[93] J. Schwarzbauer, L. Dsikivitzky, S. Hein, R. Littke, Int. J. Environ. Anal. Chem. 85 (2005) 349.
[94] G. Pojana, A. Bonfa, F. Busetti, A. Collarin, A. Marionini, Int. J. Environ. Anal. Chem. 84 (2004) 717.
[95] J. Dietz, E. Scribner, M.T. Meyer, D. Kolpin, Int. J. Environ. Anal. Chem. 87 (2007) 1141.
[96] C. Wu, A.L. Spongbury, J.D. Witter, Int. J. Environ. Anal. Chem. 88 (2008) 1033.
[97] E. Accinelli, A.Z. Coracciolo, P. Grenni, Int. J. Environ. Anal. Chem. 87 (2001) 579.
[98] A. Abe, H. Yoshima, Water Res. 13 (1979) 111.
[99] M. Uchiyama, Water Res. 13 (1979) 847.
[100] H. Hellmann, Z. Wasser Abwasser Forsch. 15 (1982) 229.
[101] M. Uchivama, Water Res. 13 (1979) 847.
[102] I.P. Viccara, E.L. Ambye, J. Am. Oil Chem. Soc. 50 (1973) 213.
[103] U. Kügemagi, T. Burnard, I.C. Terriero, J. Agric. Food Chem. 23 (1975) 717.
[104] R. Bruner, M.M. Bourdigot, B. Legube, M. Dore Agua, Sci. Tech. Rev. No. 4 (1980) 76.
[105] R.L. Jolly, R.B. Cummings, M.S. Denton, Oak Ridge National Laboratory Union Carbide Corporation Report ORNI/JM6555, 1985.
[106] A. Sarafraz Yazdi, S.H.K. Khaleghi-Miran, Z. Es'Haghi, Int. Environ. Anal. Chem. 90 (2010) 1036.
[107] P. Monisanbar, P.A. Sundari, R. Sasikumar, Int. J. Environ. Anal. Chem. 89 (2009) 245.
[108] K. Flavin, H. Hughes, V. Pobbyn, P. Kirwan, K. Murphy, H. Steiner, et al., Int. J. Environ. Anal. Chem. 86 (2006) 401.

Chapter 12

Organometallic compounds in nonsaline waters

Chapter Outline

12.1	**Organotin compounds**	**235**
	Gas chromatography–mass spectrometry	235
	Inductivity-coupled plasma mass spectrometry	238
	Miscellaneous	238
12.2	**Organomercury compounds**	**239**
	Gas chromatography	239
	Atomic absorption spectrometry	241
	Neutron activation analysis	243
	Miscellaneous	243
	Storage of mercury-containing samples	243
12.3	**Organolead compounds**	**244**
	Gas chromatography	244
	Polarography	245
	Atomic absorption spectrometry	245
	Preconcentration	246
12.4	**Organoarsenic compounds**	**246**
	Polarography	246
	Gas chromatography	247
	Atomic absorption spectrometry	247
	Ion-exchange chromatography	248
	Miscellaneous	248
12.5	**Organoantimony compounds**	**248**
12.6	**Organogermanium compounds**	**249**
12.7	**Organocopper compounds**	**249**
12.8	**Organoselenium compounds**	**249**
12.9	**Organosilicon compounds**	**249**
	References	**250**
	Further reading	**253**

12.1 Organotin compounds

Gas chromatography–mass spectrometry

Meinema et al. [1] have described a sensitive and interference-free method for the simultaneous determination of the tri-, di- and monobutyltin species in aqueous systems at tin concentrations of 0.01–5 $\mu g\ L^{-1}$. The species are concentrated from hydrobromic acid solutions into an organic solvent by extraction with 0.05% tropalone in benzene in the presence of a metal coordinating ligand. The butyltin species in the organic extract are transformed into butylmethyltin compounds by reaction with a Grignard reagent and analysed by a gas chromatography(GC)–mass spectrometry (MS) method. The inorganic tin (IV) species in the organic extract are butylated to tetrabutyltin which is detected by the same technique.

Determination of Toxic Organic Chemicals In Natural Waters, Sediments and Soils.
DOI: https://doi.org/10.1016/B978-0-12-815856-2.00012-6

The sensitivity of this procedure was demonstrated in the almost quantitative recovery of butyltin species $(Bu_3Sn)_2$ and $(Bu_3Sn)_2O$ from aqueous solutions containing dibutyltin dichloride and butyltin chloride at a concentration of 0.01 mg L^{-1}.

Mueller [2] determined down to 1 μg L^{-1} of tributyltin compounds in river and lake waters using capillary GC–MS. The tributyltin compounds were converted to tributyltin derivatives prior to GC.

Unger et al. [3] determined butyltins in natural water by GC with flame photometric detection and confirmation by MS. The sample was extracted with tropalone in *n*-hexane and *n*-hexyl magnesium bromide and organotin compounds derivatised with *n*-hexyl magnesium bromide to form tetraalkyltins. The *n*-hexyl derivatives of methyltin and butyltin species were easily separated and quantified relative to an internal standard (triphenyltin chloride), which was not found to be present in environmental samples and did not interfere.

Matthias et al. [4] have described a comprehensive method for the determination of aquatic butyltin and butylmethyltin species at ultratrace levels using simultaneous sodium borohydride hybridisation, extraction with GC–flame photometric detection and GC–mass spectrometric detection. The detection limits for a 1000 mL sample were 7 ng L^{-1} of tin for tetrabutyltin and tributyltin and 3 ng L^{-1} of tin for dibutyltin and 22 ng L^{-1} tin for monobutyltin.

Greaves and Ungar [5] used GC–MS with positive ion chemical ionisation to determine mono-, di- and tributyltin compounds in environmental waters in amounts down to 2 ng L^{-1}. In this method the organotin compounds were extracted with hexane-0.2% tropalone and derivitivised with hexyl magnesium bromide to form hexabutyltin. Neubert and Wirth [6] reported on quantitative determination of mono-, di-, tri- and tetraalkyltins compounds, present in a mixture by GC after alkylation to mixed tetraalkyltins. This technique was applied by Neubert and Andreas et al. [7] to the quantitative detection of tributyl and dibutyltin species present in dilute aqueous solution. Butyltin species were concentrated on a cation-exchange column, desorbed into diethyl ether-hydrogen chloride and determined by GC after methylation.

Soderquist and Crosby [8] have described a gas chromatographic procedure for the determination of down to 0.01 μg L^{-1} triphenyltin hydroxide and its possible degradation products (tetraphenyltin, diphenyltin benzene stannoic acid and inorganic tin) in water. The phenyltin compounds are detected by gas–liquid chromatography using an electron-capture detector after conversation to their hydride derivatives, while inorganic tin is determined by a novel procedure which responds to tin dioxide and aqueous solutions of tin salts. The basis for the method involves extraction of the phenyltin species from water followed by their quantitation as phenyltin hydrides by electron-capture GC and analysis of the remaining aqueous phase for inorganic tin Sn^{4+} plus SnO_2 by colorimetry GC and analysis of the possible degradation.

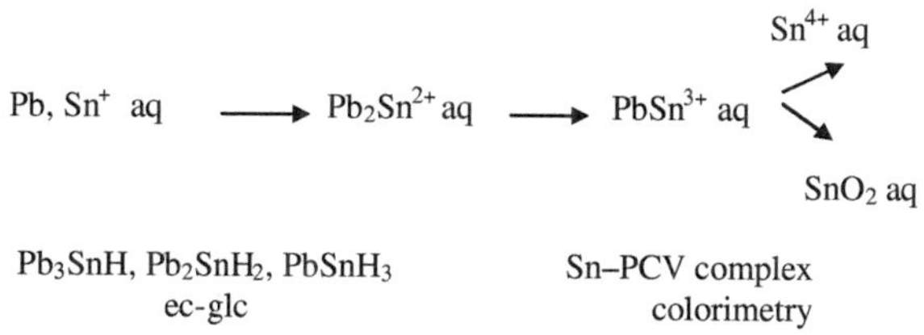

The sensitivity of this procedure ranges from $3\ \mu g\ L^{-1}$ (Pb^3Sn^{1+}, Pb_3Sn^{2+}, Pb_3Sn^{3+}) to $15\ \mu g\ L^{-1}$ (Pb_4Sn). Braman and Tompkins [9] have developed methods for the determination of trace amounts (ppb) of inorganic tin and methyltin compounds in river waters and, indeed, these workers were the first to confirm the presence of methyltin compounds in natural waters. Tin compounds are converted to the corresponding volatile hydride (SnH_4, $Ch_3)_2$, $(CH_3)_2SnH_2$ and $(CH_3)_3SnH$ by reaction with sodium borohydride at pH 6.5 followed by a separation of the hydrides and then atomic absorption spectrometry using a hydrogen-rich hydrogen-air flame emission type detector (Sn−H band). The technique described has a detection limit of 0.01 ng as tin; hence parts per trillion of organotin species can be determined in water samples.

Chau et al. [10] have described an improved extraction procedure for determining polar methyltin compounds, using benzene-containing tropolone, from water saturated with sodium chloride. Tetramethylbutyltin derivatives were prepared in the extracts and were separated by GC as well-defined peaks. The difference in sensitivity of the different species is attributed to difference of behaviour in the atomic adsorption furnace. A large number of samples can be analysed in large volumes of water. The overall recovery is satisfactory, coefficient of variation using six replicate samples is 5%−11% and a detection limit of $0.04\ \mu g\ L^{-1}$ was achieved.

Maguire and Huneault [11] determined bis(tri-*n*-butyltin) oxide and some of its dealkylation products in natural waters. Detection was by a modified flame photometric detector which has been shown to be sensitive to organotins. Mass spectra could be obtained with about 25 ng of the derivatives. The reproducibility of peak area with multiple injections was good.

Mueller [12] carried out trace-level determinations of organotin compounds using high-resolution capillary GC with flame photometric determination. In the procedure, butyltin compounds were ethylated prior to GC.

Hattori et al. [13] determined trialkyltin and triphenyltin compounds in river waters.

Inductivity-coupled plasma mass spectrometry

Garciea-Alonzo et al. [14–19] accomplished separation of all butyltin species in water using ion exchange chromatography and a low methanol content mobile phase. The low methanol content of the mobile phase allowed inductivity-coupled plasma MS and micellar spectrofluorimetry, to be used after lost column reaction with morin, for detection.

Miscellaneous

Various other techniques, which to a lesser extent have been used to determine organotin compounds, include spectrophotometry [20,21], spectrofluorimetry (triphenyltin) [21], thin-layer chromatography (di and monobutyltin) [22,23] and atomic adsorption spectrometry (inorganic tin, butyltin chlorides) [24], high-performance liquid chromatography (HPLC) [25], flame photometry [15–19,26] and atomic emission spectrometry.

Aguiler-Matrinez et al. [19] have carried out an assessment of the Chemeatcher passive sampler for the monitoring of organotin compounds in water. In this method, several configurations of receiving disks and diffusion membranes were tested for monitoring mercury and organotin compounds (monobutyltin dibutyltin, tributyltin and triphenyltin) in water with a passive sampler. This passive sampler is based on the diffusion of these compounds through a specific diffusion-limiting membrane and their subsequent accumulation on a specific receiving phase, all materials being commercially available. The proposed sampler for inorganic mercury comprises a 47-nm Empore chelating disk as receiving phase and polyethersulphone as diffusion membrane cellulose acetate. ICP-MS and GC-ICPMS/GC-FPD were used for inorganic mercury and the organotins analysis, respectively. The effects of environmental variables such as pH and salinity that could influence accumulation of test substances in receiving phase were studied. Linear uptake for all compounds was observed for at least 14 days of exposure at a constant aqueous analyte concentration in a flow-through system under controlled conditions of temperature, turbulence and analyte concentration. Compound-specific sampling rates at 11°C and simulated water turbulence of 40 cm s^{-1} varied between 0.018 and 0.137 L day^{-1}. Compounds collected by the sampler exhibited detection limits ranging between 0.7 and 5.9 ng L^{-1}. The feasibility of using these samplers in the field was tested in a polluted commercial harbour. The behaviour of the samplers to monitor target compounds was compared with those obtained from spot samples of water taken through the field development period. Data from laboratory studies and field trial support the feasibility of these samplers to measure the freely dissolved fraction of these important target analytes in water.

Methods for the preconcentration of organotin compounds have been described based on C_{18} discs and Menax [27].

12.2 Organomercury compounds

The two main techniques for the determination of mercury in water samples are atomic absorption spectrometry and GC. Mercury in water samples can exist in inorganic or organic form, or both. Normal methods of reducing inorganic mercury compounds to mercury with reagents such as stannous chloride do not work with organomercury compounds; hence organomercury compounds are not included in such determinations. Owing to the conversion of Hg^2 to CH_3Hg^+ in river waters, and to the presence of mercury in a large number of organic pollutants, it is often observed that a high percentage of the mercury is present in the form of organic compounds. Some organic mercurials such as methylmercury chloride and dimethylmercury may be reduced by a combination of cadmous chloride, but this method requires large quantities of reductants and the use of strong acid and strong alkali [28].

Organic mercury compounds can be decomposed by heating with strong oxidising agents such as potassium chromate or nitric acid–perchloric acid, followed by reduction of the formed divalent mercury to mercury vapour [29,30]. Both methods are rather time-consuming and not very suitable for automation. Potassium persulphate has also been used to aid the oxidation of organomercury to inorganic mercury compounds, and this forms the basis of an automated method [31].

Goulden and Afghan [32] have used ultraviolet irradiation as a means of decomposition following the original proposal of Armstrong [33]. After the photochemical oxidation the formed inorganic mercury is reduced to metallic mercury in the usual way by stannous chloride. This method reduced the consumption by oxidising agents and thus diminishes considerably the risk of contamination; this also leads to shorter analysis times. Determination with and without irradiation enables the separate determination of total and inorganic mercury respectively. Bennett et al. [34] later showed that acid permanganate alone did not recover three methyl mercuric compounds. The addition of a potassium persulphate oxidation step increased recoveries to 100%. El-Awady et al. [35] confirmed the low recoveries of methylmercury by acid permanganate. They showed that only about 3% methylmercury could be recovered by this method, while the use of potassium persulphate produced complete recovery.

Gas chromatography

Nishi and Horimoto [36,37] determined trace amounts of methyl-, and phenylmercury compounds in river waters. In this procedure the organomercury compound present at less than $0.4\ ng\ L^{-1}$ in sample (100–500 mL) is extracted with benzene (2×0.5 vol. relative to that of the aqueous solution). The benzene layer is then back-extracted with 0.1% L-cysteine solution (5 mL) and recovered from the complex by extracting with benzene in the presence of hydrochloric acid and submitted to GC using a stainless steel column packed with diethylene glycol succinate on Chromosorb W (60–80 mesh) nitrogen as carrier gas (60 mL min^{-1}) and an electron-capture detector.

The above method has been modified [37] for the determination of methylmercury (II) compounds in aqueous media containing sulphur compounds that affect the extractions of mercury. The modified method is capable of handling samples containing up to 100 mg of various organic and inorganic sulphur compounds per 100 mL.

Another application of GC to natural water analysis is that of Longbottom [38] who used a Coleman 50 mercury analyser system as a detector. A mixture of dimethyl-, dipropyl- and dibutylmercury (1 mg of each) was separated on a 6 ft. (1.8 m) column packed with 5% DX-200 and 3%, QF-1 on Gas Chrom Q and temperature programmed from 60°C to 180°C at 20°C min^{-1}. The mercury detector system was used after the column effluent and passed through a flame-ionisation detector; the heights of the resulting four peaks were related to the percentage of mercury in the compounds.

Dressman [39] also used the Coleman MAS 50 system in his determination of dialkylmercury compounds in river waters. These compounds were separated in a glass column (1.86 m × 2 mm) packed with 5% of DC-200 plus 3% of QF-1 on Gas Chrom Q and temperature programmed from 70°C to 180°C at 20°C min^{-1}, with nitrogen as carrier gas. The mercury compound eluted from the column was burnt in a flame-ionisation detector, and the resulting free mercury was detected by the analyser connected to the exit of the flame-ionisation instrument; down to 0.1 ng of mercury could be detected. River water (1 L) was extracted with pentane-ethyl ether (4:1) (2 × 60 mL). The extract was dried over sodium sulphate, evaporated to 5 mL and analysed as above.

Ealy et al. [40] have discussed the determination of methyl-, ethyl- and methoxymercury (II) halides in water. The mercury compounds were separated from the sample by leaching with 1 M sodium iodide for 24 hours and then alkylmercury iodides were extracted into benzene. These iodides were then determined by GC with 5% cyclohexanesuccinate on Anakron ABS (70–80 mesh) column with electron-capture detection. Good separation of chromatographic peaks was obtained for the mercury compounds as chlorides, bromides or iodides. The extraction recoveries were monitored by the use of alkylmercury compounds labelled with ^{208}Hg.

Zarneger and Mushak [41] have described a gas chromatographic procedure for the determination of organomercury compounds and inorganic mercury in natural water. The sample was treated with an alkylating or arylating reagent and the organomercury chloride is extracted into benzene. GC was carried out using electron-capture detection. The best alkylating or arylating reagents were pentacyano(meth)cobalt (III) and tetraphenylborate. Inorganic and organic mercury could be determined sequentially by extracting and analysing two aliquots of sample, of which only one had been treated with alkylating reagent. The limits of detection achieved in the method were 10–20 ng.

Cappon and Crispin-Smith [42] have described a method for the extraction, clean-up and gas chromatographic determination of organic (alkyl and aryl) and inorganic mercury in water. Methyl-, ethyl- and phenylmercury were first extracted as the chloride derivatives. Inorganic mercury was then isolated as methylmercury upon reaction with tetramethyltin. The initial extracts were subjected to thiosulphate clean-up, and the organomercury species were isolated as the bromide derivatives. Total mercury recovery ranges between 75% and 90% for both forms of mercury.

Bowles and Apte [43] and Umezaki and Iwamoto [44] have described a method for the determination of methylmercury compounds in natural waters using steam distillation followed by GC with an atomic fluorescence spectrometric detector. Mercury recoveries were 100% from river water supplies, precision was 0.2% at the 0.2 ng L^{-1} methylmercury chloride level.

Atomic absorption spectrometry

Umezaki and Iwamoto [44] differentiated between organic and inorganic mercury in river samples. They used the reduction-aeration technique described by Kimura and Miller [45]. By using stannous chloride in hydrochloric acid, only inorganic mercury is reduced, whereas stannous chloride in sodium hydroxide medium in the presence of cupric copper reduces both organic and inorganic mercury. The mercury vapour is measured at 254 nm. Ions that form soluble salts or stable complexes with Hg(II) interfere.

Doherty and Dorsett [46] analysed environmental water samples by separating the total organic and inorganic mercury by electrode position for 60–90 minutes on a copper coil in 0.1 M nitric acid medium and then determined it directly by flameless atomic absorption spectrophotometry [47,48]. The precision and accuracy are within ± 10% for the range 0.1–10 parts per 10^9. The sensitivity is 0.1 parts per 10^9 (50 mL sample).

A method has been described for the determination of mercury in water [49,50]. This method determines all forms of mercury provided that they are first converted to inorganic mercury. All forms of mercury in nonsaline waters are converted to inorganic mercury by prolonged oxidation with potassium permanganate. The inorganic mercury is determined by the flameless atomic absorption spectrometric technique using a method similar to that described by Osland [51]. Acid stannous chloride is added to the sample to produce elemental mercury.

$$Hg^{2+} + Sn^{2+} \rightarrow Hg^0 + Sn^{4+}$$

The mercury vapour is carried by a stream of air or oxygen into a gas cuvette placed in the path of the radiation from a mercury hollow-cathode lamp and the adsorption of this radiation at 253.7 nm.

Farey et al. [52] have discussed ultraviolet photochemical systems for the decomposition of organomercury compounds prior to analysis by cold-vapour atomic fluorescence spectroscopy. These workers compared the effectiveness of the bromination treatment for the liberation of mercury from organomercury compounds with a pretreatment procedure involving oxidation with a permanganate-sulphuric acid mixture recommended by workers at the Water Research Centre, United Kingdom [53]. The basis of the bromination technique, in which a bromated-bromide reagent in hydrochloric acid reagent is used to generate bromine in the sample, is that the bromine quantitatively cleaves both alkyl- and arylmercury compounds to inorganic mercury bromide.

Recoveries of inorganic mercury from distilled water spiked with phenylmercury (II) chloride, thiomersal, ethylmercury (II) chloride, methylmercury (II) chloride, phenylmercury (II) acetate and *p*-tolylmercury (II) chloride were greater than 95%. Standard deviation is $01-0.2\ \mu g\ L^{-1}$ at the $1-2\ \mu g\ L^{-1}$ level. Van Ettekoven [54] also applied ultraviolet radiation to the decomposition of organomercury compounds.

Kiemenezi and Kloosterboer [55] have described an improvement on the Goulden and Afghan [32] photochemical decomposition of organomercury compounds in the microgram per litre range in water prior to determination by cold-vapour atomic absorption spectrophotometry. Decomposition of the organomercurials is carried out by means of ultraviolet radiation of a suitable wavelength. The formed inorganic mercury was determined in the usual way by cold-vapour atomic absorption after reduction of divalent mercury to mercury vapour. Determinations with and without irradiation make possible separate determinations of total and inorganic mercury, respectively, in about 20 minutes.

Abo-Rady [56] has described a method for the determination of total inorganic plus organic mercury in nanogram quantities in water. This method is based on the decomposition of organic and inorganic mercury compounds with acid permanganate, removal of excess permanganate with hydroxylamine hydrochloride, reduction to metallic mercury with tin and hydrochloric acid, and transfer of the liberated mercury in a stream of air to the spectrometer. Mercury was determined by using a closed, recirculating air stream. Sensitivity and reproducibility of the 'closed system' was better, it was claimed, than those of the 'open system'. The coefficient of variation was 13.7%.

Other workers who have made earlier contributions to the determination of organically bound mercury compounds and inorganic mercury compounds by flameless atomic absorption spectrometry include Baltisberger and Knudsen [57].

Bisagni and Lawrence [58], Frimmel and Winckler [59], Chan and Saithoh [60], Stainton [61], Carr et al. [62], Fitzgerald et al. [63], Watling and Walting [64], Graf et al. [65], Simpson and Nickless [66], Lutz [67] and Goulden and Anthony [68].

Neutron activation analysis

Becknell and Marsh [69] converted the organomercury compound to mercuric acid using chlorine.

The mercury was then concentrated by removal as $HgCl^{2-}$ by passing the sample solution 500 mL, adjusted to be 0.1 M in hydrochloric acid through a paper filter disc loaded with SB-2 ion-exchange resin. The paper was then heat sealed in Mylar bags and irradiated for 2 hours in a thermal neutron flux of 1.3×10^{13} neutrons $cm^{-2}\, s^{-1}$, after which the concentration of mercury in the sample was determined by the comparison method using 77 keV γ-ray photopeak from the decay of ^{197}Hg.

Miscellaneous

Various other aspects of the determination of organomercury compounds in nonsaline natural waters include methods based on radio chromatography [70], a kinetic method [71], ultraviolet radiation decomposition [54], electrochemical methods [72,73], inductivity-coupled plasma, MS [74], photochorisin-induced photoacoustic spectrometry [75], X-ray fluorescence spectrometry, [76] cold-vapour atomic absorption spectrometry [77] and a comparison of distillation liquid–liquid extraction for the recovery of mercury [78].

Various methods have been used to preconcentrate samples containing organomercury compounds as a means of increasing the sensitivity of the subsequent analytical method, that is sulphur aniline resin [79], amalgamation. These include dithizone extraction [60,79,80], cold-trap preconcentration [63] adsorption into with noble metals [81–84] adsorption on polyurethane foam [85] or on vinyl dithiocarbamate resins [77,86].

Storage of mercury-containing samples

The problems of preserving mercury in solution are well known. Although controversy still exists over which preservative is best, agreement on several of the factors which affect the stability of mercury solutions seems to have been reached. For example it is agreed that low pH values, high ionic strengths and oxidising environments help in keeping mercury in solution. Acids such as sulphuric acid [87], nitric acid [88–94] and hydrochloric acid [95] have been widely used in different amounts. Oxidants such as permanganate [50,96–102] and dichromate [88–96] have been shown to prevent volatilisation of mercury. Sodium chloride and gold [101,102] have also been used as preservatives. Various workers have commented on the instability of mercury solutions when stored in polyethylene or polypropylene containers [92–94,103,104].

Carron and Agemian [105] have pointed out that whilst the majority of fresh water samples rarely contain mercury at levels over 0.5 $\mu g\ L^{-1}$ and in most cases 0.2 $\mu g\ L^{-1}$; most previous investigators of the stability of mercury solutions have carried out their tests at higher mercury levels. These workers studied the preservation methods which provide both low pH values as well as oxidising environments using both synthetic and natural samples in a variety of containers, in order to obtain a practical method which would be adaptable to routine analysis for mercury in natural waters at submicrogram per litre levels by the automated cold-vapour atomic absorption technique. The essential requirement was that the preservation method should maintain mercury in waters of low salt content (low conductivity such as distilled water) and of high salt content (high conductivity). The outcome of this work was that Carron and Agemian [105] and Harval and Bloom [78] recommended glass containers washed with concentrated nitric acid or chromic acid and a preservative consisting of a mixture of 1% sulphuric acid and 0.05% potassium dichromate. This preservative gives good accuracy, precision and low detection limits. It was also observed that the presence of methylmercury ion improves preservation efficiency.

12.3 Organolead compounds

Gas chromatography

All the ionic alkyllead compounds slowly degrade in the presence of light [106]. However, lake water samples enriched with dimethyllead chloride and trimethyllead acetate in 100 $\mu g\ L^{-1}$ level are stable over a period of at least 1 month in the laboratory when stored in the dark and refrigerated. There is no need to add any preservative to the sample. Storage in a cold dark room is recommended. Alternatively, the samples can be extracted, butylated and dried over anhydrous sodium sulphate.

Tetraalkyllead compounds (typically tetramethyllead, trimethyllead, dimethyldiethyllead, methyltriethyllead and tetraethyllead) have high vapour pressures and are seldom found in water unless they are adsorbed on sediments or particulate matter. If these compounds are present in the water sample, they can be extracted into benzene and included in the determination.

Chau et al. [107–110] have described a simple and rapid extraction procedure to extract the five tetraalkyllead compounds (Me_4Pb, Me_3Et Pb, Me_2Et_2Pb, $MeEt_3Pb$ and Et_4Pb) from natural water. The extracted compounds are analysed in their authentic forms by a gas chromatographic-atomic absorption spectrometry system. Other forms of inorganic lead do not interfere. The detection limit for water (200 mL) is 0.50 μg L^{-1}. The highly polar ions are quantitatively extracted into benzene from aqueous solution after chelation with dithiocarbamate. The lead species are butylated by Grignard reagent to the tetraalkyl form, namely R_n, Bu_{4-n} ($R = CH_3$) and Bu_4Pb, all of which can be quantified by a GC–atomic absorption spectrometry method. Molecular covalent tetraalkyllead species, if present in the sample, are also extracted and quantified simultaneously. Other metals coextracted by the chelating agent do not interfere.

Potter et al. [111] have applied GC and thin-layer chromatography to the detection and determination of alkyllead compounds and alkyllead salts in natural waters. The limit of detection of this method was 0.5–1.0 μg alkyllead salt.

Chakraborti et al. [112] analysed water samples for dialkyl and trialkyllead compounds by GC–atomic absorption spectrometry. Analysis of 500 mL of water enables the determination of 1.25 mg $L^{-1}$of lead as $PbMe_3$ and 2.5 ng L^{-1} lead for $PbEt_3$ and 2.5 ng L^{-1} for $_{+}PbEt_2$. Extraction recoveries are in excess of 90% for all compounds studied. Rapsomanikis et al. [113] speciated lead and methyllead ions in water by GC–atomic absorption spectrometry after elution with sodium tetraethylborate. This purge and trap technique has detection limits of 0.2 ng L^{-1} for trimethyllead and trimethyllead species in 50 mL water samples.

Quevanviller et al. [114] carried out ultratrace speciation analysis of organolead compounds in environmental waters by GC–atomic absorption after on-line preconcentration. Down to 0.1 ng (as Pb) organolead compounds were detected.

Polarography

Colombini et al. [115] have described a technique for the determination of organometallic species in natural waters based on selective organic phase extraction coupled with differential pulse polarography. The analytical procedure was applied to alkyllead compound speciation and found to be reliable for the individual detection and determination of organolead complexes at trace levels in natural waters including sea water.

Atomic absorption spectrometry

De Jonghe et al. [116] have described an extraction technique for determining traces of trialkyllead compounds in water samples, based on the salting out of the trialkyllead ions as neutral species into an inorganic solvent, followed by graphite-furnace atomic absorption spectrometry. A detection limit of 0.02 μg L^{-1} was achieved with samples.

Preconcentration

Mikac and Branica [106] preconcentrated dissolved dialkyllead and inorganic lead species by coprecipitation with barium sulphate. Alkyllead was then determined in the concentrate by differential pulse an inodic-stripping voltammetry.

12.4 Organoarsenic compounds

Large amounts of arsenic enter the environment each year because of the use of arsenic compounds in agriculture and industry as pesticides and wood preservatives. The main amount is used as inorganic arsenic (arsenite, arsenate) and about 30% as organoarsenicals such as monomethylarsinate and dimethylarsinite. The major organic arsenic compound in the environment is dimethylarsinite. Arsenic is known to be relatively easily transformed between organic and inorganic forms in different oxidation states biological and chemical action. Until recently, most of the analytical work has been concerned only with the total content of arsenic. But as the toxicity and biological activity of the different species vary considerably, information about the chemical form is of great importance in environmental analysis.

Polarography

Yamamoto [117] and Deitz and Perez [118] observed that dimethylarsinite has a strong affinity for acid-charged cation-exchange.

Elton and Geiger [119] used this fact to separate monomethylarsonate and dimethylarsinite prior to determination by different pulse polarography. The detection limits of 0.1 and 0.3 $\mu g\ L^{-1}$, respectively, were reported.

Henry and Thorpe [120,121] separated monomethylarsenic acid, dimethylarsinic acid. As(III) and As(IV) on an ion-exchange column from samples of pond water receiving ash from a coal-fired power station. They then determined these substances by differential pulse polarography. The above four arsenic species were present in natural water systems. Moreover, a dynamic relationship exists whereby oxidation-reduction and biological methylation–dimethylation reactions provide the pathways for the interconversations of the arsenicals.

Recoveries ranged from 4.4% (AsIII) to −2.4% (AsV) and between −2% and −3% for the organoarsenic compounds.

Henry et al. [121] reported a method for the determination of As(III), As(V) and total inorganic arsenic by different pulse polarography. As(III) was measured directly in 1 M perchloric acid or 1 M hydrochloric acid. Total inorganic arsenic was determined in either of these supporting electrolytes after the reduction of electroinactive As(V) with aqueous sulphur dioxide. As(V) was evaluated by difference.

Gas chromatography

The applications of GC are very limited. Andreae [122] and others [123–126] have detected inorganic and methylated arsenic species at the nanogram per litre level in natural waters by volatilisation and detection by atomic absorption electron-capture and flame-ionisation detectors.

Soderquist et al. [127] determined hydroxydimethylarsine oxide in water by converting it to iododimethylarsine using hydrogen iodide followed by determination at 105°C on a column [15 × 0.125 minutes (4.6 m × 3 mm)] packed with 10% of DC-200 on Gas Chrom Q (60–80 mesh) with nitrogen as carrier gas (20–30 mL min^{-1}) and electron-capture detection. The recovery of hydroxydimethylarsine oxide (O.15 mg L^{-1}) added to pure water was 92.3% with a standard deviation of ± 7.4%.

Atomic absorption spectrometry

Edmunds and Francesconi [128] estimated methylated organoarsenic compounds by vapour generation atomic absorption spectroscopy. Fishman and Spencer [129] and Agemian and Cheam [130] have described automated atomic absorption spectrophotometric methods for the determination of total arsenic in water. Fishman and Spencer [129] used an ultraviolet radiation or an acid persulphate digestion procedure to decompose the organoarsenic present prior to atomic absorption spectroscopy.

The automated method of Agemian and Cheam [130] uses hydrogen peroxide and sulphuric acid for the destruction of organic matter, combined with permanganate oxidation for the complete recovery of organoarsenic compounds. An automated system based on sodium borohydride reduction with atomisation in a quartz tube is used for the determination of the inorganic arsenic thus produced.

Agemian and Cheam [130] found that in the sodium borohydride reduction of inorganic arsenic to arsenic, concentrations from 0.5 to 1.5 M of hydrochloric acid gave the highest sensitivity; both As(III) and As(V) were equivalently detected. When the hydrochloric acid concentration was increased from 2 to 6 M, the sensitivity for both species decreased, particularly for As(V).

Stringer and Attrep [131] compared hydrogen peroxide–sulphuric acid digestion and ultraviolet photodecomposition methods for the decomposition of three organoarsenic compounds in water samples (triphenylarsine oxide, disodium methane arsenate and dimethylarsinic acid) to inorganic arsenic prior to reduction to arsine determination by absorption spectroscopy or by the silver diethyldithiocarbamate spectrophotometric method [132].

Recoveries of triphenylarsine oxide, disodium methane arsenate and dimethylarsenic acid were in the range of 84.4%–104.4%. Detection limits in the range of 100–300 μg L^{-1} were reported.

Ion-exchange chromatography

Grabinski [133] has determined the following species at the 10 μg L^{-1} level in lake water by ion-exchange chromatography with flameless atomic absorption spectrometric detection: arsenic (III), arsenic (IV), monomethylarsenic acid and dimethylarsinic acid.

Aqueous arsenic (III), arsenic (IV), monomethylarsenic acid and dimethylarsinic acid were separated by ion-exchange chromatography. The elution sequence was as follows:

0.006 M trichloroacetic acid (pH 2.5) yielding first arsenic (III) and then monomethylarsenic acid; 0.2 M trichloroacetic acid yielding arsenic (V); 1.5 M NH_4OH followed by 0.2 M trichloroacetic acid yielding dimethylarsinic acid. Separation was achieved on a single column containing both cation- and anion-exchange resins.

Miscellaneous

Other methods, which to a lesser extent, have been used in studies of organoarsenic compounds in water include MS [134], spectrophotometry [135], capillary zone electrophoresis [136], emission spectrometry [119,136–140], dimethylarsenic. Elton and Geiger [119] separated monomethylarsenic acid and dimethylarsinic acid prior to determinations of organoarsenicals by differential pulse polarography. This method is based on the fact that dimethylarsinic acid is strongly attracted to cation-exchange resins.

Dimethylarsinic acid is preconcentrated on a strong cation-exchange resin [141] by optimising the elution parameters and can be separated from the other arsenicals sample components, such as group I and II metals, which can interfere in the final determination. Graphite-furnace atomic absorption spectrometry is used as a sensitive and specific detector for arsenic. The described technique allows for the dimethylarsinic acid to be determined in sea water (20 mL) containing 10^5-fold excess of inorganic arsenic with a detection limit of 0.02 ng As mL^{-1}.

12.5 Organoantimony compounds

Andreae et al. [142] determined methylantimony species and antimony (III) and antimony (V) in natural water using atomic absorption spectrometry with hydride generation. The limit of detection was 0.3–0.6 μg L^{-1} for 100 mL water sample. Basada and Brahim [143] have described a polarographic method for the determination of antibilharzial organoantimony in river waters.

12.6 Organogermanium compounds

Braman and Tompkins [144] have described an atomic emission spectrometric method for the determination of inorganic germanium and methylgermanium (and inorganic antimony) in amounts down to 0.4 ng in environmental samples. These compounds are reduced to hydrides using sodium borohydride, then separated prior to atomic emission spectrography.

12.7 Organocopper compounds

Brown et al. [145] applied HPLC to the determination of organocopper speciation in soil-pore waters.

12.8 Organoselenium compounds

The determination of dimethylselenides, dimethyldiselenide and diethylselenide in scrubbed air samples has been discussed [23,24]. This method is based on a gas chromatographic-atomic absorption procedure and could, no doubt, be adapted for the analysis of waters.

Ng et al. [146] have reported a separation of organoleads and organoselenium compounds using β-cyclodextrin modified micellar electrokinetic chromatography with on-column ultraviolet detection (210 nm).

Laborde et al. [147] separated trimethylselenonium from inorganic selenium using HPLC with fraction collection and electrothermal atomic spectrometry. The reported detection limit was 112 μg of selenium.

12.9 Organosilicon compounds

Mahone et al. [148] have given details of a procedure to determine waterborne organosilicon substances such as silanols or silanols-functional materials by conversion to trimethylsilylated derivatives which can be determined by gas–liquid chromatography. The method gives good accuracy and precision in the milligram per litre range and with suitable precautions can be extended to the low microgram per litre range.

Van der Post [149] has described a method for the determination of silanols in water based on their ability to reduce nitrite or nitrate to ammonia at normal temperature. Individual silanols are identified by MS.

Wanatabe et al. [150] have described a simple and rapid method for the separation and determination of down to 10 $\mu g\ L^{-1}$ siloxanes in river water, using inductively coupled plasma emission spectrometry. The organosilicone compound is extracted with petroleum ether then evaporated by dryness. The damp residue is dissolved in methylisobutyl ketone and aspirated into the plasma.

References

[1] H.A. Meinema, T. Burger Wersina, G. Dehaan, E.S. Geners, Environ. Sci. Technol. 12 (1978) 288.
[2] M.D. Mueller, Fresenius Z. Anal. Chem. 317 (1984) 32.
[3] M.A. Unger, W.G. MacIntyre, J. Greaves, R.J. Huggert, Chemosphere 15 (1986) 461.
[4] C.I. Matthias, J.M. Bellama, G.J. Olsen, F.E. Brinkman, Environ. Sci. Technol. 20 (1986) 609.
[5] J. Greaves, M.A. Ungar, Bromed. Environ. Mass Spectrom. 15 (1988) 585.
[6] G. Neubert, H.O. Wirth, Z. Anal. Chem. 273 (1975) 19.
[7] G. Neubert, H. Andreas, Z. Anal. Chem. 280 (1976) 31.
[8] C.J. Soderquist, D.C. Crosby, Anal. Chem. 50 (1978) 1435.
[9] R.S. Braman, Ma Tompkins, Anal. Chem. 51 (1979) 12.
[10] Y.K. Chau, P.T.S. Wong, G.A. Bengert, Anal. Chem. 54 (1982) 246.
[11] R.J. Maguire, H. Huneault, J. Chromatogr. 209 (1981) 455.
[12] M.D. Mueller, Anal. Chem. 59 (1987) 617.
[13] Y. Hattori, A. Kobayaski, S. Takemoto, K. Takami, O. Kuges, A. Sugimae, et al., J. Chromatogr. 315 (1984) 341.
[14] J.L. Garciea-Alonzo, A. Seaz Medel, J. Ebdon, Anal. Chim. Acta 283 (1993) 261.
[15] H. Choi, E.Y. Kwon, D.S. Lee, Bull. Korean Chem. Soc. 14 (1993) 234.
[16] K. Takahashi, Nippon Kagaku Kaishi 6 (1993) 761.
[17] H. Ohnogi, T. Korenaga, M. Tanaka, S. Shinoda, Kankyo-Seigyo 14 (1992) 28.
[18] T.M. Dawling, P.C. Uden, J. Chromatogr. 644 (1993) 153.
[19] R. Aguiler-Matrinez, R. Greenwood, G.A. Mills, B. Vrana, M.P. Palacios-Corvillo, M.M. Gomez-Gomez, Int. J. Environ. Anal. Chem. 88 (2008) 75.
[20] B.M. Buskina, S.V. Svavtsillo, Obr. Prom Sanit. 186 (1969).
[21] S.J. Blunden, A.H. Chapman, Analyst 103 (1978) 1266.
[22] H. Waggon, D. Jehle, Die Nahrung 17 (1973) 739.
[23] H. Waggon, D. Jehle, Die Nahrung 19 (1975) 273.
[24] V.F. Hodge, S.L. Seidel, E.D. Goldberg, Anal. Chem. 51 (1979) 1256.
[25] C.W. Wang, I.L. Yang, Analyst 113 (1988) 1393.
[26] R. Campano, M. Grundador, C. Heal, M.D. Pratt, Anal. Chim. Acta 283 (1993) 272.
[27] M. Ceulemans, R. Lobinski, W.M.E. Dirkx, Fresenius J. Anal. Chem. 347 (1993) 256.
[28] L. Magos, Analyst 96 (1981) 847.
[29] Y. Kimura, V.I. Miller, Anal. Chim. Acta 27 (1962) 325.
[30] T.C. Rains, O. Menis, J. Assoc. Off. Anal. Chem. 55 (1972) 1339.
[31] Environmental Protection Agency, Methods for Chemical Analysis of Water and Waste Water, EPA Publication No. EPA-625/6-74-003, USEPA Office of Technology Transfer, Washington, DC, 1972, p. 118.
[32] P.D. Goulden, B.K. Afghan, Technical Bulletin, Water Department of Energy Mines and Resources, Ottawa, ON, 1974.
[33] E.A.J. Armstrong, P.M. Williams, J.D. Strickland, Nature 231 (1966) 481.
[34] T.B. Bennett, W.H. McDonald, R.N. Hemphill, Advances in Automated Analysis, Technical International Congress, vol. 8, Medical Inc, Tarrytown, NY, 1972.
[35] A.A. El-Awady, R.B. Miller, M.J. Carter, Anal. Chem. 48 (1976) 110.
[36] S. Nishi, H. Horimoto, Jpn. Anal. 17 (1968) 1247.
[37] S. Nishi, H. Horimoto, Jpn. Anal. 19 (1970) 1646.
[38] J.E. Longbottom, Anal. Chem. 44 (1972) 111.
[39] R.C. Dressman, J. Chromatogr. Sci. 10 (1972) 472.

[40] J. Ealy, W.D. Shultz, D.A. Dean, Anal. Chim. Acta 64 (1974) 235.
[41] L. Zarneger, P. Mushak, Anal. Chim. Acta 69 (1974) 389.
[42] C.J. Cappon, V. Crispin-Smith, Anal. Chem. 49 (1977) 365.
[43] K.C. Bowles, C.A. Apte, Anal. Chem. 70 (1998) 395.
[44] U. Umezaki, K. Iwamoto, Jpn. Anal. 20 (1971) 123.
[45] O. Kimura, O. Miller, Anal. Abstr. 10 (1963) 2943.
[46] P.E. Doherty, R.S. Dorsett, Anal. Chem. 43 (1971) 1887.
[47] O. Brandenberger, O. Bader, Anal. Abstr. 15 (1968) 5883.
[48] O. Brandenberger, O. Bader, Anal. Abstr. 17 (1969) 2617.
[49] Determination of Organomercury Compounds in Waters and Effluents, Department of the Environment and National Water Council, UK. HM Stationery Office, London (PF22 Ab ENV), 1978, 23 pp.
[50] S. Shimomura, A. Kise, Buneski Kagaku 18 (1969) 1412.
[51] R. Osland, Pye Unicam Spectrovision 24 (1970) 11.
[52] B.J. Farey, L.A. Nelson, M.J. Rolph, Analyst 103 (1978) 656.
[53] Report Water Pollution Research Laboratory Stevenage, Report No. 1272, 1972.
[54] K.G. van Ettekoven, H_2O 13 (1980) 326.
[55] A.N. Kiemenezi, J.G. Kloosterboer, Anal. Chem. 48 (1976) 575.
[56] Abo-Rady, Fresenius Z. Anal. Chem. 1299 (1979) 187.
[57] R.J. Baltisberger, C.L. Knudson, Anal. Chim. Acta 73 (1974) 265.
[58] J.J. Bisagni, A.W. Lawrence, Environ. Sci. Technol. 8 (1974) 850.
[59] F. Frimmel, H.A. Winckler, Z. Wasser Abwasser 8 (1975) 67.
[60] Y.K. Chan, H. Saithoh, Environ. Sci. Technol. 4 (1970) 839.
[61] M.P. Stainton, Anal. Chem. 43 (1971) 625.
[62] R.A. Carr, J.B. Hoover, P.W. Wilknis, Deep Sea Res. 19 (1972) 747.
[63] W.F. Fitzgerald, W.B. Lyons, C.D. Hunt, Anal. Chem. 46 (1974) 1882.
[64] R.J. Watling, H.R. Walting, Water S. Afr. 1 (1975) 113.
[65] E. Graf, L. Polos, L. Bezur, E. Pungor, Magy Kem. Foly. 79 (1973) 741.
[66] W.R. Simpson, G. Nickless, Analyst 102 (1977) 86.
[67] R.L. Lutz, Analyst 104 (1979) 979.
[68] P.D. Goulden, D.H.J. Anthony, Anal. Chim. Acta 120 (1986) 129.
[69] D.E. Becknell, B. Marsh, Anal. Chem. 43 (1971) 1230.
[70] M.C. Cullen, E.T. McGuiness, Anal. Biochem. 42 (1971) 455.
[71] P.J. Ke, R.T. Thibert, Microchim. Acta 3 (1973) 417.
[72] O. Evans, G.D. McKee, Analyst 113 (1988) 243.
[73] A. Rubel, Anal. Chim. Acta 115 (1980) 343.
[74] H.Y. Tong, D.E. Giblin, R.L. Lapp, S.J. Mouson, M.L. Grass, Anal. Chem. 63 (1991) 1772.
[75] N. Chen, R. Guo, E.P.C. Lac, Anal. Chem. 60 (1988) 2435.
[76] A. Kudo, H. Nagase, Y. Ose, Water Res. 16 (1982) 1011.
[77] T. Braun, M.N. Abbas, S. Torak, A. Zvakefalvi-Nagy, Anal. Chim. Acta 160 (1984) 277.
[78] M. Harval, N.S. Bloom, Anal. Chim. Acta 250 (1993) 282.
[79] H.J. Kramer, J. Neidhart, J. Radioanal. Chem. 37 (1977) 835.
[80] M. Schinta, T. Kauri, A. Coutu, K. Kudo, Ecotoxicol. Environ. Saf. 14 (1987) 208.
[81] M.J. Fishman, Anal. Chem. 42 (1972) 1462.
[82] V.I. Muscat, T.J. Vickers, Anal. Chem. 44 (1972) 218.
[83] K. Matstunaga, Mizushori 15 (1974) 431.
[84] J. Olafsson, Anal. Chem. 64 (1974) 68.
[85] T. Braun, M.M. Abbas, I. Bakos, A. Elek, Anal. Chim. Acta 131 (1981) 311.

[86] K. Minagama, Y. Takizawa, I. Kefune, Anal. Chim. Acta 115 (1980) 103.
[87] P.D. Goulden, B.K. Afghan, Technicon International Congress, vol. II, November 2–4, Futura, New York, 1970, p. 317.
[88] R.M. Rosain, C.M. Wai, Anal. Chim. Acta 65 (1973) 279.
[89] R.A. Carr, P.E. Wilkniss, Environ. Sci. Technol. 7 (1973) 62.
[90] G.N. Gaston, A.K. Lee, J. Am. Water Works Assoc. 66 (1974) 495.
[91] J.F. Lopp, M.C. Longbottom, L.B. Lobring, J. Am. Water Works Assoc. 20 (1972) 64.
[92] R.V. Coyne, J.A. Collins, Anal. Chem. 44 (1972) 1093.
[93] M.H. Bothner, J.A. Collins, Anal. Chem. 47 (1975) 592.
[94] H.V. Weiss, K. Chew, Anal. Chim. Acta 67 (1973) 444.
[95] M.S. Masri, A. Freidman, Environ. Sci. Technol. 7 (1970) 951.
[96] D.W. Newton, R. Ellis Jnr., J. Environ. Qual. 3 (1974) 20.
[97] I.R. Jonasson, J.J. Lynch, L. Trip, J. Geol. Surv. Can. (1973). Paper 73–21.
[98] T.Y. Shields, C.P.C. Shields, L. Koval, Talanta 17 (1970) 1025.
[99] P. Avotins, E.A. Jenne, J. Environ. Qual. 4 (1975) 515.
[100] C. Feldman, Anal. Chem. 46 (1974) 99.
[101] J.M. Lo, C.M. Wai, Anal. Chem. 4 (1975) 1869.
[102] D.R. Christman, J.D. Ingle, Anal. Chim. Acta 86 (1976) 53.
[103] R.W. Heiden, D.A. Aikens, Anal. Chem. 49 (1977) 668.
[104] R.C. McFarland, Radiochem. Radioanal. Lett. 16 (1973) 47.
[105] I. Carron, H. Agemian, Anal. Chim. Acta 92 (1977) 61.
[106] M. Mikac, M. Branica, Anal. Chim. Acta 212 (1988) 349.
[107] Y.K. Chau, P.T.S. Wong, G.A. Bengeut, O. Kramer, Anal. Chem. 186 (1979) 51.
[108] Y.K. Chau, P.T.S. Wong, P.D. Goulden, Anal. Chim. Acta 86 (1976) 85.
[109] Y.K. Chau, P.T.S. Wong, O. Kramer, Anal. Chim. Acta 46 (1983) 211.
[110] Y.K. Chau, P.T.S. Wong, P.D. Gould, Anal. Chim. Acta 85 (1976) 421.
[111] H.R. Potter, A.W.P. Jarview, R.N. Markell, Water Res. 76 (1977) 123.
[112] D. Chakraborti, W.R.A. De Jonghe, W.E. Mol, R.J.A. Von Cleuvenbergen, F.C. Adams, Anal. Chem. 56 (1984) 2692.
[113] S. Rapsomanikis, O.F.Y. Donard, J.H. Weber, Anal. Chem. 58 (1986) 35.
[114] P. Quevanviller, F. Martin, C. Belin, O.F.X. Donard, Appl. Organometal Chem. 7 (1993) 149.
[115] A.P. Colombini, R. Fuecco, M. Papoff, Sci. Total Environ. 37 (1984) 61.
[116] W.R.A. De Jonghe, W.E. Von Mol, E.C. Adams, Anal. Chem. 55 (1983) 1050.
[117] M. Yamamoto, Soil Sci. Soc. Am. Proc. 39 (1975) 859.
[118] E.A. Deitz, M.E. Perez, Anal. Chem. 48 (1976) 1088.
[119] R.K. Elton, W.E. Geiger, Anal. Chem. 50 (1978) 712.
[120] F.T. Henry, T.M. Thorpe, Anal. Chem. 52 (1980) 80.
[121] F.T. Henry, T.D. Kirch, T.M. Thorpe, Anal. Chem. 51 (1979) 215.
[122] M.O. Andreae, Anal. Chem. 49 (1977) 820.
[123] M.M. Kolthoff, R. Belcher, Volumetric Analysis, vol. 3, Interscience Publishers, New York, 1967, pp. 511–513.
[124] A.W. Fickett, E.H. Daughtrey, P. Mishak, Anal. Chim. Acta 79 (1975) 93.
[125] J.D. Lodmell, (Ph.D. thesis), University of Tennessee, Knoxville, TN, 1973.
[126] L.D. Johnson, K.O. Gerhart, W.A. Au, Sci. Total Environ. 1 (1972) 108.
[127] C.J. Soderquist, D.G. Crosby, J.B. Bowers, Anal. Chem. 46 (1974) 155.
[128] J.S. Edmunds, K.A. Francesconi, Anal. Chem. 48 (1976) 2019.
[129] M. Fishman, R. Spencer, Anal. Chem. 49 (1977) 1599.
[130] H. Agemian, V. Cheam, Anal. Chim. Acta 101 (1978) 193.

[131] C.E. Stringer, M. Attrep, Anal. Chem. 51 (1979) 731.
[132] D.C. Manning, At. Absorption Newsl. Perkin-Elmer 10 (1971) 6.
[133] R.A. Grabinski, Anal. Chem. 53 (1981) 966.
[134] Y.A.M. Bettencourt, M.A. Florence, F.F.N. Duart, I.R. Gomez, L.F.C.V. Bors, Appl. Organomet. Chem. 8 (1994) 43.
[135] S.S. Sandhu, P. Nelson, Anal. Chem. 50 (1978) 322.
[136] S.S. Sandu, Analyst 101 (1976) 856.
[137] J.F. Sandez-Lopez, M.B. Amdrau, M.D. Lakkis, F. Legarde, G. Rauret, M.J.F. Levry, Fresenius J. Anal. Chem. 348 (1994) 810.
[138] S.A. Peoples, J. Lakso, T. Lais, Proc. West. Pharmacol. Soc. 14 (1971) 178.
[139] M.G. Haywood, J.P. Riley, Anal. Chim. Acta 85 (1975) 219.
[140] T. Kamada, Talanta 23 (1976) 835.
[141] J.A. Persson, K. Irgum, Anal. Chim. Acta 138 (1982) 111.
[142] M.O. Andreae, J. Asmode, P. Foster, L. Vant'dok, Anal. Chem. 53 (1981) 1766.
[143] T.A. Basada, L.F. Brahim, Analyst 112 (1987) 549.
[144] R.S. Braman, M.S. Tompkins, Anal. Chem. 50 (1978) 1088.
[145] L. Brown, S.J. Haswell, M.M. Rhead, P. O'Neill, K.C. Bancroft, Analyst 108 (1983) 1511.
[146] C.L. Ng, H.K. Lee, S.F.Y. Li, J. Chromatogr. 652 (1993) 547.
[147] F. Laborde, D. Charaborti, J.M. Mir, J.R. Castello, J. Anal. At. Spectrom. 8 (1993) 643.
[148] L.G. Mahone, B.J. Garner, R.B. Buch, T.H. Lone, J.F. Tatera, R.C. Smith, et al., Environ. Toxicol. Chem. 2 (1983) 307.
[149] P.C. Van der Post, Water Pollut. Control 77 (1978) 520.
[150] N. Wanatabe, Y. Yasuda, K. Kalo, T. Nakamura, R. Funasaka, N. Shimokova, et al., Sci. Total Environ. 34 (1984) 169.

Further reading

R. Alzaga, J.M. Bayona, J. Chromatogr. 655 (1993) 51.

Chapter 13

Organic compounds in aqueous precipitation

Chapter Outline

13.1 Polycyclic aromatic hydrocarbons 255
13.2 Phenols 255
13.3 Carboxylic acids 256
13.4 Pesticides 256
13.5 Organomercury compounds 256
13.6 Organotin compounds 256
13.7 Organolead compounds 257
References 257

Pankow et al. [1] give the details of a sampler developed for collection of rainwater. The sampler is controlled electronically, provides in situ filtration of the sample and carries our preconcentration of nonpolar organic compounds by means of cartridges containing the sorbent Tenax GF. It is possible to incorporate cartridges of ion-exchange resin for preliminary concentration of organic acids. Samples of rainwater collected by this equipment from several rainfall events at two sites in Oregon were analysed by gas chromatography–mass spectrometry and 27 organic compounds were identified. The results were used in conjunction with available Henry's law constants to estimate the local atmosphere levels of these compounds.

13.1 Polycyclic aromatic hydrocarbons

Gurov and Navikov [2] have described a procedure for the determination of polycyclic aromatic hydrocarbons in snow.

Wanatabe et al. [3] identified trace amounts of phthalate esters, polyaromatic hydrocarbons, higher fatty acids and their esters in rainwater by means of a gas chromatography–mass spectrometry computer system.

13.2 Phenols

Cruezwa et al. [4] reported a method in which traces of phenols and cresols in rainwater were determined by continuous liquid–liquid extraction and normal phase high-performance liquid chromatography with ultraviolet fluorescence detection.

Determination of Toxic Organic Chemicals In Natural Waters, Sediments and Soils.
DOI: https://doi.org/10.1016/B978-0-12-815856-2.00013-8

13.3 Carboxylic acids

Kawamura and Kaplan [5] have described a sensitive method for measuring acids (C_1-C_7) in rain and fog samples using ρ-bromophenacyl esters and a high-resolution capillary gas chromatograph employing fused silica columns. Experiments showed that the measured concentrations of volatile acids in spiked rain samples increased linearly in proportion to the concentrations of volatile acids added. Relative standard deviations were less than or equal to 18% for C_1, C_2 and C_3 acids. The distributions of volatile acids in Los Angeles rain and fog samples are discussed.

Backman and Peden [6] used ion-exclusion chromatography to determine weak carboxylic acids in rainwater. Citrate, formate and acetate were identified.

A coupled enzymic high-performance liquid chromatographic method has been described for the determination of 4–10 μm of formate in water [7].

13.4 Pesticides

Engst and Knoll [8] used hexane extraction gas chromatography to study the occurrence in rainwater of down to 0.001 $\mu g\ L^{-1}$ of *p,p′*-DDE and *p,p′*-TDE. Samples were extracted by shaking with hexane. The combined hexane phases were dried with sodium sulphate and evaporated to between 1 and 5 mL in a rotary evaporator. The extract was injected, without further clean-up, on to a gas chromatographic assembly consisting of one column (1.8 m × 6 mm) packed with 5% QF-1 on Varaport 30, (100–120 mesh) and a second column (0.9 m × 6.6 m) packed with 4% of OV-17 on AW-DMCS Chromosorb W (60–80 mesh) both operated at 180°C with nitrogen as the carrier gas (30 $mL\ min^{-1}$) and a tritium electron capture detector.

13.5 Organomercury compounds

Ahmed et al. [9] carried out ultratrace analysis of mercury and methylmercury in rainwater using cold-vapour absorption spectrometry.

The cold-vapour atomic absorption spectrometric method involving preconcentration on dithiocarbamate resins described by Minagawa et al. [10] has been applied to the analysis of rainwater.

13.6 Organotin compounds

Braman and Tompkins [11] have developed methods for the determination of microgram per litre amounts of inorganic tin and methyltin compounds in rainwater and other waters. Tin compounds are converted to the corresponding volatile hydride [SnH_4, CH_3SnH_3 $(CH_3)_2SnH_2$ and $(CH_3)_3SnH$] by reaction with sodium borohydride at pH 6.5 followed by gas chromatographic separation of the hydrides and then atomic absorption spectroscopy using a hydrogen-rich hydrogen-air flame emission type detector (Sn−H band). The technique described has a detection limit of 0.01 ng as tin; hence parts per trillion or organotin species can be determined.

Braman and Tompkins [11] found that stannane (SnH_4) and methylstannanes [CH_3SnH_3, $(CH_3)_2SnH_2$ and $(CH_3)_3SnH$] could be separated very well on a column comprising silicone oil OV-3 (20% w/w) supported Chromosorb W.

An average total tin content of rainwater was found to be 25 ng L^{-1} and the methyltin form constituted 24% of that total.

13.7 Organolead compounds

Lobinski et al. [12] carried out speciation analysis of organolead compounds in Greenland snow at the femtogram per gram level by capillary gas chromatography using an atomic absorption detector. In this procedure the snow sample was mixed with EDTA buffered to pH 8.5 and extracted with hexane to preconcentrate organolead compounds. The extract was propylated using propyl magnesium chloride and the product analysed by capillary gas chromatography.

Et_3Pb^+ and Et_2Pb^{2+} could be detected in amounts down to 0.02 and 0.02−0.5 μg kg^{-1}, respectively. Neither Me_3Pb^+ nor Me_2Pb^{2+} was found in the snow samples analysed.

To preconcentrate trialkyllead species in rainwater, Blaszkewicz et al. [13] complexed interfering metal ions in rainwater with EDTA before adjustment of the pH to 10. Samples were pumped through an extraction column of silica gel to adsorb lead compounds which were then desorbed with acetate buffer containing methanol at pH 3.7. The eluate was diluted and adjusted to pH 8 with borate buffer before further concentration on a Nucleosil 10 C_{18} precolumn. Adsorbed trialkyllead compounds were eluted by back-flushing on to a RP−C_{18} column and separated with methanolic acetate buffer. On-line detection used a postcolumn chemical reaction detector. Detection limits for sample volumes of 500 mL were 15 μg L^{-1} for trimethyl- and triethyllead, respectively. Standard deviation was less than 4% for a sample containing 90 pg triethyllead per mL.

References

[1] J.F. Pankow, I. Isabelle, W.E. Asher, Environ. Sci. Technol. 18 (1984) 310.
[2] K.F. Gurov, Y. Novikova, Hyg. Sanit. 36 (1971) 409.
[3] K. Wanatabe, T. Nakanishi, J. Shizhido, Eisel Kagaku 34 (1988) 25.
[4] J. Cruezwa, C. Leuenberger, J. Tromp, W. Geiger, M.J. Abel, J. Chromatogr. 403 (1987) 233.
[5] K. Kawamura, I.R. Kaplan, Anal. Chem. 48 (1984) 1616.
[6] S.R. Backman, M.R. Peden, Water Soil Pollut. 33 (1987) 191.
[7] N. Minura, T. Kinoshita, Y. Yashida, A. Hetake, C. Nakai, Anal. Chem. 60 (1988) 2067.

[8] R. Engst, R. Knoll, Nahrung 17 (1973) 837.
[9] R. Ahmed, K. May, M. Stoeppler, Z. Anal. Chem. 326 (1987) 510.
[10] K. Minagawa, Y. Takizawa, I. Kefune, Anal. Chim. Acta 115 (1980) 103.
[11] B.S. Braman, M.A. Tompkins, Anal. Chem. 51 (1979) 12.
[12] R. Lobinski, C.F. Bontrom, J.P. Candelone, Anal. Chem. 65 (1993) 2510.
[13] M. Blaszkiewicz, G. Baumhoer, G. Neidbert, Int. J. Environ. Anal. Chem. 28 (1987) 207.

Chapter 14

Organic compounds in soil, solvent extraction

Chapter Outline

14.1 Conventional solvent extraction 259
14.2 Conventional solvent extraction from soil packed cartridges 259
14.3 Pressurised liquid extraction 259
14.4 Microwave-assisted extraction 259
14.5 Subcritical water extraction 263
14.6 Solid-phase microextraction 263
14.7 Subcritical fluid extraction 263
References 265
Further reading 268

Various procedures for the extraction of organic compounds from soils prior to their determination are now discussed below.

14.1 Conventional solvent extraction

Direct interaction of soil with a low-boiling solvent such as diethyl ether or methylene dichloride.

14.2 Conventional solvent extraction from soil packed cartridges

As Section 14.1 but with soil sample packed into cartridges.

14.3 Pressurised liquid extraction

As Sections 14.1 and 14.2 but extraction with water under pressure.

14.4 Microwave-assisted extraction

As Sections 14.1 and 14.2 but under microwaves. Advantages are a shorter extraction time and need smaller sample size.

Determination of Toxic Organic Chemicals In Natural Waters, Sediments and Soils.
DOI: https://doi.org/10.1016/B978-0-12-815856-2.00014-X

TABLE 14.1 Conventional solvent extraction procedures for organic compounds in soils.

Compound – pesticides and herbicides	Extractant	Clean-up	Analytical finish	References
Organochlorine insecticides	Polar solvents	Florisil		[1]
Alachlor	Methanol enrichment on C_{18} cartridge	–	–	[2,3]
Bromoxynil, ioxynil residues	Solvent portioning Methanol enrichment	Solvent partition	Gas chromatography with ion trap detector	[4]
Triazine-type herbicides	On C_{18} cartridges	–	Gradient C_{18} high-performance liquid chromatography	[5]
Imazapyr	Methanol	None	High-performance liquid chromatography detection at 250 nm	[6]
Terbuthylazine and degradation products	Hot acetone then cation-exchange solid-phase cartridge	–	High-Performance liquid chromatography with photodiode array detection	[7,8]
Norflurazon herbicide	Methanol	Filtration	C_{18} high-performance liquid chromatography	[9]
Isoproturon, dichlorprop-P bifenox	Solvent	Size exclusion chromatography	High-performance liquid chromatography	[10]
Fenpropimorph and metabolite Fenpropimorphic acid fungicide	Acetone–water partitioning with methylene chloride	Gel permeation chromatography	Methylation-gas liquid chromatography with NP detection and GC–MS	[11]
Fenoxaprop, Fenoxaprop-ethyl	Solvent	Florisil	High-performance Liquid chromatography	[12]

Polycyclic aromatic hydrocarbons	Soxhlet extraction and saponification	Silica gel	1 ppm detected in soil	[13]
Polycyclic aromatic hydrocarbons	BF_3 in methanol	–	–	[14]
Polycyclic aromatic hydrocarbons	Miniature liquid–liquid Extractor (100 μL solvents)	–	–	[15]
Polycyclic aromatic hydrocarbons	Automated Soxhlet extraction with ethyl acetate	–	–	[15]
Polycyclic aromatic hydrocarbons	Organic solvent, methanolic hydrolysis	–	–	[16]
Volatile organic compounds	Hot (70°C) methanol	–	–	[17]
Volatile organic compounds	Comparison of solvent extraction, headspace analysis and vapour partitioning, methanol extraction	–	–	[18]
Volatile organic compounds	Comparison of methanol extraction and purge and trap method	–	Hot method extraction is most effective	[19]
Polychlorodibenzo-*p*-dioxins and benzofurans	Comparison of Soxhlet extraction using 1. toluene and 2. methylene dichloride-acetone	–	–	[20,21]
Medium polar and polar analytes	Hot phosphate buffered water extraction	–	–	[22]

(*Continued*)

TABLE 14.1 (Continued)

Compound – pesticides and herbicides	Extractant	Clean-up	Analytical finish	References
Aromatic hydrocarbons	Solvent extraction with hydrolyses	–	–	[16]
Chlorophenols	Soxhlet extraction	–	Boiling point solvent mixtures at elevated temperatures of up to 200°C and pressure of up to 20 MPa [23] to maintain the solvent in a liquid state	[23–25]
Pesticides, polycyclic aromatic hydrocarbons volatile organic compounds, herbicides, polychlorodibenzo-*p*-dioxins	Soxhlet extraction	–	–	[26]
Aliphatic hydrocarbons	Pesticide	Carbon tetrachloride extraction	–	[27]

14.5 Subcritical water extraction

As Sections 14.1 and 14.2 but avoids contamination of extract with impurities present in organic solvents.

14.6 Solid-phase microextraction

Consists of extracting the soil sample with a solvent, usually subcritical water.

14.7 Subcritical fluid extraction

Extraction of soil sample from soils with carbon dioxide leads to higher diffusivity, faster extraction compared to Soxhlet extraction or sonication.

Conventional solvent extraction is the most commonly employed technique.

The extraction of various types of compounds from soil by this technique is reviewed in Table 14.1.

The application of various other extraction techniques is reviewed in Tables 14.1 and 14.2.

TABLE 14.2 Extraction of organic compounds from soils.

Type extraction	Compounds separated	References
Accelerated solvent extraction	Phenoxyacetic acid herbicides	[27–32]
	Polycyclic hydrocarbons	[33]
	Chlorophenols	[34]
	Miscellaneous	[35]
	Polychlorobiphenyl polycyclic hydrocarbons chlorobenzenes	[36]
	Organochlorine insecticides	[37]
Pressurised liquid extraction	Polycyclic aromatic hydrocarbons	[38–40]
Microwave-assisted extraction	Pesticides	[41–43]
	Polycyclic aromatic hydrocarbons	[43,44]
	Polychlorobiphenyls	[42–48]
	Hydrocarbons	[44,49–52]
	Volatile organic compounds	[53]
	Miscellaneous	[39,49,54–56]

(Continued)

TABLE 14.2 (Continued)

Type extraction	Compounds separated	References
Subcritical water extraction	Miscellaneous	[26,57–63]
	Phenoxyacetic acid and other herbicides	[32,64,65]
	Tetrabutyl azine and metabolites	[32]
	Polycyclic aromatic hydrocarbons	[66,67]
	Polychlorobiphenyls	[58,68]
Solid-phase microextraction	Miscellaneous	[32,64,65,68]
	Chlorophenols	
	Terabutyl	[32]
	Polycyclic aromatic hydrocarbons polychlorobiphenyls	[65,67]
	1-Chloronaphthalene, nitrobenzene and 2-chlorotoluene	[67]
	A solanine	[69]
Supercritical fluid extraction	Miscellaneous	[70,71]
	Aromatic compounds, polychlorobiphenyls	[72–78]
	Triacrylchloro dioxins	[78–87]
	Trialkyl phosphates	[88]
	Amines	[89]
	Pyridine hydrocarbons	[88,90]
	Volatile organic compounds	[91–94]
	Phenols	[95]
	Organic acids	[96]
	Enteroviruses	[97]
	Organochlorine pesticides	[98]
	Herbicides	[97,99–110]
	Triazines	[110–112]
	Sulphenyl ureas	[105–107]
	Organochlorine insecticides	[106–111]
	Flumetron herbicides dacthal	[113]
	Miscellaneous	[72,98,104,113–115]
	Polycyclic aromatic hydrocarbons	[69,116,117]

References

[1] D.E. Kimbraugh, R. Chin, J. Wakahuwa, Analyst 119 (1994) 1283.

[2] A.P. Schwab, P. Splichal, L.S. Sonon, ASTM STP 1162 (1993) 86.

[3] A.E.M. Chirnside, W.F. Ritter, ASTM STP 1162 (1993) 92.

[4] C. Sanchez-Brunete, A.I. Valcarcel, J.L. Tadeo, Chromatographia 38 (1994) 624.

[5] X. Qiao, R. During, H.E. Hummerl, Meded Fac Landbouwwet Rijksuniv, Gent 56 (1991) B949.

[6] W. Liu, A. Pusino, C. Gena, Sci. Total Environ. 39 (1992) 123.

[7] R. Schwes, F.X. Maide, G. Fishbeck, J. Lepschy, von Gleisenthall, S. Suess, Chromatography 641 (1993) 89.

[8] R. Schwes, S. Wuset, J. Pepschy, Von Gleisenthall, F.X. Maide, A. Suess, et al., Anal. Lett. 27 (1994) 487.

[9] W.T. Willianm, T.C. Mueller, AOAC Int. 77 (1994) 752.

[10] E. Licyois, Y. Dehon, B. De Brabant, P. Perry, D. Porteller, A. Copin, Sci. Total Environ. 17 (1992) 123.

[11] L. Celi, M. Gennari, Pest Sci. B38 (1993) 43.

[12] G. Coidina, M.T. Vaqucro, I. Comellas, F. Broto-Duig, Chromatography 673 (1994) 21.

[13] C. Lutermann, W. Dott, J. Hollender, Chromatogr. A 811 (1998) 151.

[14] L. Maos, J.J.U.A.I. Vreuis, Chromatogr. A 89 (2000) 273.

[15] H.S. Oliver, O.H.J. Szolar, H. Rost, R. Braun, A.P. Loiber, Anal. Chem. 74 (2002) 2379.

[16] A. Eisenback, M. Kaestner, R. Beirl, G. Schaefer, B. Mahro, Chemosphere 28 (1994) 683.

[17] W.P.R.D. Wilson, M.J. Brown, D.M. Mackay, Ground Water Monit. Rev. 17 (1997) 104.

[18] A.D. Hewitt, Environ. Sci. Technol. 32 (1998) 143.

[19] M.D.E. Askari, M.P. Maskarinec, S.M. Smith, D.M. Bean, C.C. Trevis, Anal. Chem. B 68 (1996) 3431.

[20] L.O. Kjeler, S.E. Kulp, B. Jonsson, C. Rappe, Toxicol. Environ. Chem. 39 (1993) 1.

[21] B.E. Ritcher, J.L. Ezzell, D.E. Knowles, F. Hoefler, A.K.R. Mattular, M. Schentwinkle, et al., Chemosphere 34 (1997) 975.

[22] C. Crecenzi, A. Di Corcia, M. Nazzari, R. Saunperi, Anal. Chem. 72 (2000) 3050.

[23] J.R. Dean, Anal. Commun. 33 (1996) 191.

[24] M.H. Tavendale, A.L. Wilkins, A.G. Langdon, K.L. Mackie, T.R. Stuthridge, P.N. McFarlane, Environ. Sci. Technol. 29 (1995) 1407.

[25] J. Paasivirta, H. Hakala, J. Knutienen, T. Otollinen, J. Sarkkla, L. Welling, et al., Chemosphere B21 (1990) 355.

[26] A. Miellet, A. Faisil, Expert Chim. Toxicol. 72 (1986) 864.

[27] US EPA, Test Methods for Evaluating Solid Waste, Method, 3545, US EPA SW-846, third ed., update 111. Washington DC, USGPO, 1995.

[28] J.L. Ezzell, B.E. Witter, W.D. Felix, S.R. Black, J.E. Meikle, Liq. Chromatogr.-Gas Chromatogr. 13 (1995) 390.

[29] B.E. Richter, B.A. Jones, J.L. Ezzell, N.L. Porter, N. Audalovic, C. Phol, Anal. Chem. 68 (1996) 1033.

[30] S.P. Frist, J.R. Dean, K.P. Evans, K. Harridine, C. Cary, M.H.I. Comber, Analyst 122 (2010) 895.

[31] E. Conte, R. Milani, G. Morali, F. Abballe, Chromatography 765 (1997) 121.

[32] A. Di Corcia, Private Communication, 1999.

[33] N. Saim, J.D. Dean, M.P. Abdulla, Z. Zakaria, Anal. Chem. 70 (1998) 720.

[34] I. Wennrich, P. Popp, M. Modee, Anal. Chem. 72 (2001) 546.

[35] E. Hofler, J. Ezzell, B. Richter, Labor Proxiz 4 (1995) 58.

[36] A. Hubert, K.D. Wensel, M. Manz, L. Weissflog, W. Englewald, G. Schűűrmann, Anal. Chem. 72 (2000) 1294.
[37] S. Pyle, A.B. Markus, Ass. Spec. 32 (1997) 897.
[38] S. Lundstedt, B. Von Bavel, D. Hagland, M. Tysklind, L. Oberg, Chromatogr. A 883 (2000) 151.
[39] S. Morales, J.L. Luque-Garcia, M.D.L. De Castro, Anal. Chem. 74 (2000) 4213.
[40] I. Ramos, J.J. Vreulis, U.A.T. Brinkman, Chromatogr. A 891 (2000) 275.
[41] V. Lopez-Avila, R. Young, W.F. Beckett, Anal. Chem. 66 (1994) 1097.
[42] T.G. Danis, T.A. Albanis, Toxicol. Environ. Chem. 53 (1996) 9.
[43] C. Molins, E.A. Hogendorf, H.A.G. Heusinokveld, P. Van Zoner, R.A. Bauman, J. Environ. Anal. Chem. 68 (1997) 155.
[44] M.L. Bac, F. Pantani, K. Barnieri, D. Buarnin, O. Grittine, J. Environ. Chem. 64 (1996) 23.
[45] V. Lopez-Avila, J. Benedicto, C. Cheron, R. Young, V.F. Beckett, Environ. Sci. Technol. 29 (1995) 2709.
[46] M.C. Alonso, D. Puig, I. Silgonoer, M. Grasserbauer, D. Barceló, J. Chromatogr. A 823 (1998) 231.
[47] L. Silgoner, R. Krska, E. Lombas, O. Gans, E. Rosenberg, M. Grasserbauer, Fresenius Anal. Chem. B362 (1998) B120.
[48] S.J. Sout, B.W. Dacunha, M.M. Safarpau, AOAC Int. 81 (1998) 1054.
[49] F. Sun, D. Littlejohn, M. Daved Gibson, Anal. Chim. Acta 364 (1998) 1.
[50] A. Pastor, E. Vazuez, R. Ciscar, M. De la Guardia, Anal. Chim. Acta 344 (1997) 241.
[51] R.W. Current, D.C. Tilotta, Chromatogr. A 785 (1997) 269.
[52] J. Hawari, A. Halasz, I. Sas, H.V. Tra, Environ. Anal. Chem. 66 (1997) 299.
[53] P. Dinaitoiu, I. Gorecki, B.L. Parker, Int. J. Environ. Anal. Chem. 86 (2006) 113.
[54] S. Jones, I.R. Calabeese, P.T. Kostecki, Hydrocarbon Contaminated Soils and Ground Water Arc, Boca Raton, FL, 1993, 3, 11.
[55] A.M. Illias, C. Jaeger, in: E.J. Calibreese, P.T. Kostecki, Hydrocarbon Contaminated Soils and Ground Waters, Boca Raton, FL, 1993, 3, 147.
[56] K.E. Karp, Ground Water Monit. Rem. 13 (1993) 101.
[57] S.B. Hawthorne, Y. Yang, D.J. Miller, Anal. Chem. 66 (1994) 2912.
[58] S.B. Hawthorne, C.B. Grabanski, K.J. Hageman, D.J. Miller, Chromatogr. A 814 (1998) 151.
[59] Z. Zhang, J. Pawliszyn, Anal. Chem. 65 (1993) 1843.
[60] Z. Zhang, J. Pawliszyn, Anal. Chem. 67 (1995) 34.
[61] F.J. Santos, M.N. Sarrió, M.T. Galceran, Chromatogr. A 771 (1997) 181.
[62] N.J. Harrich, Internal Reflection Spectroscopy, John Wiley, New York, 1967.
[63] V. Lopez-Avila, R. Young, W.F. Beckett, AOAC Int. 76 (1993) 864.
[64] C. Crecenzi, G. D'Ascenzo, A. DiCorcia, H. Nazzari, S. Marchese, R. Samperi, Anal. Chem. 71 (1999) 2157.
[65] V. Fernandex-Perez, M.D. Castro, Chromatogr. A 902 (2000) 357.
[66] Y. Yang, S. Bawadt, S.B. Hawthorne, J.D. Miller, Anal. Chem. 67 (1995) 4571.
[67] J. Yang, J.W. Her, Anal. Chem. 71 (1999) 4690.
[68] L. Wennrick, P. Pappe, M. Modee, Anal. Chem. Private Commun. 72 (2000) 546.
[69] P.H. Jensen, R.J. Hardie, B.W. Strobel, B. Svensmark, H.C. Horsen, Int. J. Environ. Anal. Chem. 1000, 87, 813.
[70] S. Bowadt, S.B. Hawthorne, Chromatogr. A 703 (1995) 549.
[71] R. Deuster, N. Lubahn, C. Freidrich, W. Kleiböhmer, Chromatogr. A 785 (1997) 227.
[72] J. Dunkers, M. Groenenboom, L.H. Scholtes, C. Van der Heiden, J. Chromatogr. 6421 (1993) 357.

[73] S. Reinde, F. Hoefler, Anal. Chem. 66 (1994) 1808.
[74] H.B. Lee, T.E. Pearte, R.L. Hong-You, R.D. Gere, J. Chromatogr. B653 (1993) 83.
[75] M.D. Burford, S.B. Hawthorne, D.J. Miller, Anal. Chem. 65 (1993) 1497.
[76] A. Meyer, W. Kleiböhomer, J. Chromatogr. 657 (1993) 327.
[77] F. Guo, X.Q. Li, J.P. Alcantara-Licudine, Anal. Chem., 199, 71, 1309.
[78] N. Alexandrou, J. Pawliszyn, Anal. Chem. 61 (1989) 2770.
[79] S.B. Hawthorne, J.J. Langenfield, J.J. Miller, M.D. Burford, Anal. Chem. 64 (1992) 1614.
[80] R. Fuoco, P.R. Griffiths, Anal. Chem., Rome 82 (1992) 235.
[81] F.I. Onuska, K.A. Terry, J. High Resolut. Chromatogr. 14 (1991) 830.
[82] F.I. Onuska, K.A. Terry, J. High Resolut. Chromatogr. 12 (1989) 357.
[83] B.W. Wright, C.W. Wright, J.S. Fruchter, Energy Fuels 3 (1989) 474.
[84] F.I. Onuska, K.A. Terry, J. High Resolut. Chromatogr. 12 (1989) 527.
[85] S.B. Hawthorne, K.A. Miller, J.J. Langenfield, J. Chromatogr. Sci. 28 (1990) 2.
[86] M. Loheit, K. Beechmann, J. Chromatogr. 5505 (1990) 227.
[87] T.G. Oostdyk, R.L. Grob, J.L. Snyder, M.E. McNally, Anal. Chem. 65 (1993) 596.
[88] M.D. Burford, S.B. Hawthorne, J.D. Miller, Am. Environ. Lab. 8 (1996) 1.
[89] R.J.B. Peters, M. Renesse, J.A.D. Duivenbode, Fresenius J. Anal. Chem. 348 (1994) 249.
[90] H. Gues, B.N. Zegers, H. Lingeman, U.A.T. Brinkman, Int. J. Environ. Anal. Chem. 56 (1994) 119.
[91] S. Laing, D.C. Tillota, Anal. Chem. 70 (1998) 616.
[92] Y. Yang, S.B. Hawthorne, D.J. Miller, J. Chromatogr. A 699 (1995) 265.
[93] J. Janku, V. Kunes, A. Machokova, M. Kuras, Anal. Chem. Bull. 3 (1994) 345.
[94] J.C. Futter, P. Wall, J. Planar Chromatogr. 6 (1993) 372.
[95] A. Langbehn, H. Steinhart, J. High Resol. Chromatogr. 17 (1994) 293.
[96] T.M. Staub, I.L. Pepper, M. Abbazzadegon, C. Berba, Appl. Environ. Microbiol. 60 (1994) 1014.
[97] W.C. Brumley, E. Latorre, V. Lelliner, A. Marcu, D.E. Knowles, Liq. Chromatogr. R T 21 (1998) 119.
[98] M.E.P. McNally, J.R. Wheeler, Chromatography 447 (1988) 53.
[99] J.M. Wong, Q. Li, B.D. Hammock, J.N. Seiber, J. Agric. Food Chem. 39 (1991) 1802.
[100] M.M. Jiminez-Carmona, J.J. Mancins, A. Montoya, M.D. Luque de Castro, J. Chromatogr. 785 (1997) 329.
[101] K.N. Reddy, M.A. Locke, Weed Sci. 42 (1994) 249.
[102] S. Papillod, W. Herdi, S. Chiron, D. Barcelo, Environ. Sci. Technol. 30 (1996) 1822.
[103] T.R. Stenheimer, R.L. Pfeiffer, K.D. Scoggins, Anal. Chem. 6 (1994) 645.
[104] M.E.P. McNally, J.R. Wheeler, J. Chromatogr. 435 (1988) 63.
[105] P. Klaffenbach, P.S. Holland, J. Agric. Food Chem. 41 (1993) 396.
[106] P. Klaffenback, P.J. Holland, Boil Mass Spectr. 22 (1993) 565.
[107] V. Lopez-Avila, N.S. Dodhiwela, W.F. Berkert, J. Agric. Food Chem. 41 (1993) 2038.
[108] A.J. Field, K. Monohan, R. Reed, Anal. Chem. 70 (1998) 1956.
[109] S. Pappilond, W. Haerdi, Chromatographia 38 (1994) 514.
[110] J.L. Snyder, R.L. Grob, M.E. McNally, T.S. Oosterdyk, J. Chromatogr. Sci. 31 (1993) 183.
[111] J.L. Snyder, R.L. Grob, M.E. McNally, T.S. Oostdyk, Anal. Chem. 64 (1992) 1940.
[112] M. Schanz, S. Bawadt, B.A. Benner, S.A. Wize, S.B. Hawthorne, Chromatography 816 (1998) 213.
[113] M. Notar, H. Lescovic, Fresenius J. Anal. Chem. 360 (2000) 846.
[114] A. Meyer, W. Kleiboehwer, K. Cammann, High Resolut. Chromatogr. 16 (1993) 491. C. Mangin, J. Bubroca, M. Barriuso, J. Chromatogr. A 19 (1996) 700.

[115] Concawe percent a/72, Hydrocarbons in Soil Method v/72-V11-6, November 1972.
[116] T.M. Fahing, M.E. Paulatic, D.M. Johnson, M.E.P. McNally, Anal. Chem. 65 (1993) 1462.
[117] J. Hollander, B. Koch, C. Laterman, W. Dott, Int. J. Environ. Anal. Chem. 83 (2003) 21.

Further reading

S. Deperon, P.M. Dudwerwel, D. Conturier, Analysis 25 (1997) 286.
A.P. Emery, S.N. Chesler, W.A. MacCrehan, J. Chromatogr. 606 (1992) 221.
S.B. Hawthorne, D.J. Miller, Anal. Chem. 66 (1994) 4005.
S.B. Hawthorne, S. Trembly, C.L. Monniot, C.R. Grabinski, D.T. Mille, J. Chromatogr. A 886 (2000) 237.
A. Koinecke, R. Krenzig, M. Bahider, J. Chromatogr. 786 (1997) 155.
M.A. Locke, Agric. Food Chem. 41 (1993) 1081.
B.E. Rechter, B.A. Jones, J.L. Ezzel, N.L. Poter, N. Avdalovic, C. Phol, Anal. Chem. 68 (1996) 1033.

Chapter 15

Determination of noninsecticidal compounds in soil

Chapter Outline

15.1 **Hydrocarbons** **269**
15.2 **Oxygen-containing compounds in soil** **273**
Oxalates 273
Nonylphenols 275
Organic acids and ketones 276
Methoxy groups 277
15.3 **Halogen-containing compounds in soil** **277**
Chlorinated organic compounds 277
Chlorinated aliphatic hydrocarbons 281
Chlorobenzoic acid 281
Perfluorooctane sulphonyl fluoride 282
15.4 **Nitrogen-containing compounds in soil** **283**
Nitro compounds 283
Polycyclic aromatic nitrogen heterocyclic 283
Hydrazines 285
Growth regulators 285
Miscellaneous 285
15.5 **Sulphur and phosphorus–containing compounds in soil** **285**
Sulphur compounds dimethyl disulphide 285
15.6 **Miscellaneous organic soil** **286**
Humic and fulvic acid 286
15.7 **Volatile organic compounds in soil** **288**
15.8 **Mixtures of organic compounds in soil** **290**
References **293**
Further reading **301**

15.1 Hydrocarbons

The determination of these compounds in soils is reviewed in Table 15.1.

More recently Risdon et al. [96] have carried out a study of weathering of nC_8-nC_{40} aliphatic hydrocarbons in soil.

The method covers the determination of total petroleum hydrocarbons including diesel range compounds, kerosene range compounds and mineral oil range compounds in soils. Further modification to the carbon banding is made as requested for risk assessment. These include a series of ranges known as Texas banding (from the Texas Risk Reduction Program) as well

Determination of Toxic Organic Chemicals In Natural Waters, Sediments and Soils.
DOI: https://doi.org/10.1016/B978-0-12-815856-2.00015-1

TABLE 15.1 Review of early published literature for the determination of hydrocarbons in soil.

Compound	Extraction procedure	Method of analysis	LD	Reference
Aliphatic hydrocarbons	Sonication or Soxhlet – supercritical fluid extraction	Infrared spectroscopy, GC	–	[1–17]
	–	Spectrofluorimetry	–	[18]
	–	Flourier transform spectroscopy	–	[19]
	Conventional extraction	–		[20–29]
	Supercritical fluid	–		
	Headspace analysis and GC/MS	–	5 ng g^{-1}	[20,21]
	Conventional solvent extraction, supercritical fluid extraction	GC–MS thermogravimetric MS		[22–29]
				[28]
Aromatic hydrocarbons	Purge and trap analysis	Pyrolysis – MS	–	[30]
	Curie point flash evaporation	Pyrolysis – GC–MS	Low μg kg^{-1}	[31]
	Misc.	Misc.	–	[32–37]
Polycyclic aromatic hydrocarbons	Misc.	Misc.	–	[38–50]
	Soxhlet extraction	–		[51,52]
	Pressurised liquid extraction	–	–	[53,54]
	Accelerated solvent extraction	–	–	[55]

Polycyclic aromatic	Supercritical water extraction	–	–	[56–68]
Polycyclic aromatic hydrocarbons	Benzene–water extraction Soxhlet extraction	GC pyrolysis	–	[69–73]
	–	MS	–	[74]
	Direct analyses of soil	Laser desorption (as per photoionization time of flight)	–	[71]
	Direct analysis	Real-time aerosol MS	–	[72]
Polycyclic aromatic	–	HPLC fluorobenzene detection	–	[59,66,75–77]
	Soxhlet extraction	TLC	–	[78–80]
Polycyclic aromatic hydrocarbons	Soxhlet extraction	Pyrolysis – 90 – MS		[31]
Polycyclic aromatic	Direct analysis of soil	Aerosol MS	–	[24,71,72]
Polycyclic aromatic hydrocarbons	–	HPLC	–	[59]
	Soxhlet extraction	TLC	–	[78–80]
	–	ELISA field screening	–	[81]
		Micellar electrokinetics capillary chromatography		[82]
	–	Supersonic jet laser-induced fluorescence	–	[83,84]
Polycyclic aromatic	–	Fluorescence quenching	–	[85]

(Continued)

TABLE 15.1 (Continued)

Compound	Extraction procedure	Method of analysis	LD	Reference
Polycyclic aromatic hydrocarbons	–	Thermal desorption GC–MS	–	[85,86]
	–	Microwave-assisted extraction	–	[87]
	–	Thermal desorption	–	[88]
Polycyclic aromatic	–	Immune chemical methods	–	[89]
Polycyclic aromatic hydrocarbons	–	Electrophoresis	–	[90]
	Solvent extraction with metholic alpha II boron trifluoride	–	–	[74,91,92]
	Origin of polycyclic aromatic hydrocarbons	–	–	[92–94]
	Field-based supercritical fluid extraction	–	–	[95]

GC, Gas chromatography; *HPLC*, high-performance liquid chromatography; *MS*, mass spectrometry.

as separation of the aliphatic and aromatic fractions. The method can be routinely used for measuring hydrocarbons down to 10 mg kg^{-1} in soil. Lower limits can be achieved by employing a suitable solvent concentration step following extraction; however, this would result in increased cycle time. Detection limits may vary for individual carbon ranges calculated on the percentage of the full range they cover. With an extraction efficiency and recovery between less than 95% and 99%, this method can be easily positioned as a good alternative to Soxhlet extraction and shows a good potential for implementation as a standard method potentially providing further insight to the contaminated land sector.

Terleak et al. [97] studied the determination of polycyclic aromatic hydrocarbons in soil pore waters.

Freely dissolved pore water concentrations are difficult to assess in complex matrixes such as soils or sediments. Terleak et al. [97] applied a negligible-depletion partitioning-based sampling technique to measure freely dissolved pore water concentrations. A polydimethylsiloxane (PDMS-coated glass fibre) was exposed to a slurry of a soil spiked with several PAHs at concentrations ranging from 2 to 2000 mg kg^{-1}. PAH-concentrations in the PDMS coating increased linearly with the total soil concentration until a certain maximum was reached. Freely dissolved pore water concentrations were calculated using PDMS–water partition coefficients, and the calculated maximum pore water concentrations corresponded with the aqueous solubility of the tested compounds. Furthermore, the sampling technique is very sensitive because it can detect freely dissolved pore water concentrations in the nanogram per litre range for the tested PAHs. Freely dissolved pore water concentrations are an important parameter for the exposure of organisms in soil. Saturation of the pore water with increasing soil concentrations should therefore be considered in soil toxicity testing. Sorption coefficients that were calculated from freely dissolved concentration were slightly higher than estimates based on octanol–water partition coefficients. These differences are discussed in relation to the effects of dissolved matter in soil pore water on the determination of sorption coefficients.

Earlier methods for the determination of hydrocarbons in soil are reviewed in Table 15.1.

15.2 Oxygen-containing compounds in soil

The determination of these in soil is reviewed in Table 15.2.

Oxalates

More recently Strobel et al. [115] have studied the distribution of oxalate in soils under rhubarb.

TABLE 15.2 Review of earlier published literature for the determination of oxygen-containing noninsecticidal chemicals in soil.

Compound	Extraction procedure	Method of analysis	LD	Reference
Alcohols, aldehydes and ketones	Methanol extraction, purgate absorption tenax column supercritical fluid extraction	Methoxy onto GC or MS	–	[30]
		Chromatography	–	[98]
Phenols	Supercritical carbon dioxide extraction		–	[99–102]
				[103]
	Supercritical water extraction		–	[104]
	Accelerated solvent extraction		–	[105]
	Microwave-assisted extraction		–	[87,106]
	Solid–liquid extraction		–	[105,107]
	Automatic mode for direct continuous extraction	GC	–	[108]
	–	Direct acetylation	–	[109,110]
		GC-headspace analysis Space analysis	–	
	–	Higher order derivative spectrometry	–	[111]
	Benzene–water extraction	GC	–	[28]
	Microwave-assisted extraction	–	–	[87]
	0.1 M sodium hydroxide extraction	HPLC	13–6 μg ng	[112]
Methoxy groups	–	Zeisel hydriodic acid digestion	–	[113,114]

GC, Gas chromatography; *HPLC*, high-performance liquid chromatography; *MS*, mass spectrometry.

Nonylphenols

Lubecki [116] has delivered a simple and efficient method for the determination of 4-nonylphenols in marine sediments and soils. These are a group of xenobiotics of great environmental concern owing to their endocrine-disrupting properties; they are recognised as persistent priority pollutants by the Water Framework Directive. The analytical procedure involves ultrasonic extraction followed by two-step solid-phase extraction clean-up and gas chromatography (GC)–mass spectrometry (MS) analysis, without a derivatisation step. The method was validated and demonstrated to be suitable for marine sediments rich in organic matter from a eutrophic, contaminated area; 4-nonylphenol recovery rates were above 90%. This method was applied to the analysis of 4-nonylphenols in recent sediments from the Gulf of Gdańsk (Southern Baltic Sea). Sediments from this area are moderately contaminated with 4-nonylphenols (0–1 cm layer: 1–42 ng g d.w.$^{-1}$; 1–5 cm layer: 2–61 ng g d.w.$^{-1}$).

To summarise, a simple cost-effective reliable analytical method has been developed for the determination of 4-nonylphenols in marine sediments rich in organic matter from eutrophic and contaminated areas. The analytical procedure combines ultrasonic extraction followed by two-step solid-phase extraction clean-up and GC–MS analysis, without the necessity of derivatisation of 4-nonylphenols. The extraction procedure does not require special expensive equipment. The analytical method turned out to be suitable for marine sediments, with recovery rates of 4-nonylphenols exceeding 90% for both raw and preextracted sediment samples spiked with a standard. This method has been successfully applied to the analysis of 4-nonylphenols in recent sediments from the Gulf of Gdańsk. Results of this work show that sediments from the Gulf of Gdańsk are moderately contaminated with 4-nonylphenols.

Taking into account the different enantiomers, a total of 550 4-nonylphenol isomers are theoretically possible. It is assumed that technical 4-nonylphenol is a mixture of about 100 isomers, and its GC/MS chromatogram consists of a cluster of peaks. Despite the lack of derivatisation, chromatographic separation of 4-nonylphenols was satisfactory. The 4-nonylphenol identification was based on retention times, the presence of target ions and their relative abundance. The total concentration of the sum of 4-nonylphenols was quantified in this study, since at present, isomer-specific analysis of these compounds is not possible using conventional GC methods. To compare 4-nonylphenol composition patterns in the studied matrices, a cluster of 4-nonylphenol chromatographic peaks was divided into 12 fractions. The sum of peak areas of four g = fragmentation ions (i.e. 107, 121, 135 and 149 *m/z*) within a retention time window was used for quantifying the 4-nonylphenols. The 4-nonylphenol molecules are strongly fragmented under electron impact conditions in an MS detector, and therefore molecular ion

(220 *m/z*) is almost missing. It is interesting to note that the mass spectra of some 4-nonylphenol isomers vary significantly; this is mainly due to the degree of branching at position C1 in the nonyl chain [117]. Calibration curves were plotted on the basis of solutions of different concentrations of the nonylphenol technical mixture dissolved in isooctane. 4-*n*-nonylphenol with the aromatic ring label with C^{13} isotope containing a straight (not branched) nonyl chain was used as a surrogate internal standard (quantification ion – 113 *m/z*) to correct for possible analyte losses during the analysis.

In the absence of a certified reference material for 4-nonylphenols, recovery experiments were crucial in the validation procedure of this method. The 4-nonylphenol recovery was determined using both external and internal standard methods, each sample (~3 to ~4 g d.w.) being spiked with 2.5 μg of the 4-nonylphenol technical mixture. Good linearity of the detector signal ($\gamma^2 = 0.9998$) was observed for standard solutions of 4-nonylphenols in the 0.25–5.0 $\mu g\ mL^{-1}$ range. The internal standard (1 μg) was added to each sample prior to extraction. Relative response factors based on comparison of the GC/MS chromatogram peak areas of the 4-nonylphenol technical mixture and the internal standard were determined on each day of measurements. Procedural blanks were regularly analysed, and all results were corrected for blank values. Blank values contained quantifiable amounts of 4-nonylphenols but were satisfactorily low. The limit of detection, calculated as the sum of mean value and three times the standard deviation of the blank, for the 4-nonylphenol standard solution was ~0.05 $\mu g\ mL^{-1}$ and ~4 $ng\ g^{-1}$ dry weight for a sediment sample if the weight of sample extracted was ~3 g d.w. The precision of the 4-nonylphenol analyses by GC/MS was valuated from the relative standard deviations (RSD), which ranged from 2.5% to 5.5%.

Sirvent et al. [112] have described a simple and efficient method based on high-performance liquid chromatography (HPLC) for the determination of pollutant levels of phenols in soils containing high levels of organic matter. Alkaline extraction with 0.1 M sodium hydroxide of the phenolic compounds from soil with a high organic content is followed by their concentration in a highly cross-linked polystyrene-divinylbenzene sorbent (Bakerbond SDB-) and analysis by high-performance chromatography with ultraviolet (UV) detection. Detection limits ranged from 13 to 64 $\mu g\ kg^{-1}$. The method was applied to the analysis of spiked soils and to the evaluation of the stability of the analytes in these soils.

Organic acids and ketones

Supercritical fluid chromatography [112] has been used to determine organic acids and ketones in soils and rhubarb.

Solid samples were collected at depths of 0–2.5 and 2.5–5 cm from 10 cm sections along 100 cm transects from rhubarb plants at four locations in Denmark, and from seven layers in a soil profile to 80 cm depth at one location. Oxalate was extracted from the soil with 0.2 M phosphate at pH2 by reciprocal shaking for 23 hours and then determined by fast capillary zone electrophoresis method with 300 mM KH_2PO_4 electrolyte adjusted to pH7, developed and tested to analyse high-ionic strength. Rhubarb increased the oxalate content in soil under the leaves slightly. The average content of oxalate in the upper 0–5 cm soil was 444 μmol kg^{-1} at the Kaldred site, and 111–333 μmol kg^{-1} at the three other locations. In the soil profile the content of oxalate decreased from 500 μmol kg^{-1} in 0–5 cm depth to 110 μmol kg^{-1} at 75–80 cm depth. No significant seasonal changes in oxalate contents were observed, while an annual variation of 100 μmol kg^{-1} could be observed at 0–24 cm depth. During plant decay in autumn a slight increase in oxalate content was observed at 30 cm soil depth. In conclusion the role of oxalate in weathering and metal transport appears to be limited in soils under rhubarb.

Methoxy groups

The determination of the methoxy group (OCH_3) content of soils is important in studies concerned with the degree of humification of soil organic matter. Methods used to determine methoxy groups in soils have generally been based on volumetric modifications of the classical Zeisel method, which is a complicated and tedious procedure requiring specialised apparatus.

Magabhaes and Chalk [113] used a combination of Zeisel hydriodic acid digestion and GC of the produced iodomethane to determine methoxy groups in soil.

This was used to determine the content of methoxy in samples of soils from Eastern Australia. A wide range in the methoxy group concentration was observed, which accompanies the wide range in organic concentrations measured. Positive correlation ($P < .001$) was found between the methoxy groups and organic C contents of the soils.

15.3 Halogen-containing compounds in soil

The determination of these compounds in soil is reviewed in Table 15.3.

More recent methods are discussed in more detail later.

Chlorinated organic compounds

Yang and Her [131] used the principle of solid-phase microextraction combined with Fourier transform infrared (FT-IR) sensing to determine chlorinated organic compounds in soil.

The sensing device of this method was based on an IR hollow waveguide, the inner surface of which was coated with a hydrophobic film. Vaporised

TABLE 15.3 Reviews of early published literature for the determination of halogen compounds in soil.

Compound	Extraction procedure	Method of analysis	LD	Reference
Chloroaliphatic hydrocarbons	*n*-Pentane extraction	GLC		[114]
Chloroaliphatic				
Chloroaliphatic hydrocarbon	Hexane extraction	GLC	–	[118]
	–	Curie point pyrolysis GLC	–	[119]
	Methanol extraction	Purge and trap GL MS		[30]
	Grab sampling of soil	–	–	[120]
	Determination of distribution coefficient	–	–	[121]
Chloroaromatic hydrocarbons	Accelerated solvent extractions	–	–	[122–125]
Chloroaromatic compounds	Supercritical fluid extracts	–	–	[123,124,126]
	Solid-phase microextraction	–	–	[125,127–139]
	Methanol extraction	Purge and trap	–	[30]
	Lead space solid-phase microextraction	Fourier transform IR	–	[131]
Chlorophenols	Subcritical water extraction	–	–	[136–138]
	Soxhlet extraction	–	–	[132,133]
	Microwave extraction	–	–	[134]
	Ultrasonic extraction	–	–	[134]
	Accelerated solvent	GC–MS	–	[135]

	Acetone extraction hexane extraction	GC–MS	0.1 ppb	[140]
	Solvent extraction	Dilution – GC as methyl ethers	–	[141]
	Soxhlet extraction	–	–	[132,133]
	Microwave-assisted extraction	–	–	[134]
	Ultrasonic-assisted extraction	–	–	[134]
	Accelerated solvent extraction	–	–	[135]
	Subcritical hot water extraction	–	–	[136–138]
	Water extraction	–	–	[139,140]
Polychlorobiphenyls	Misc.	–	–	[107,142]
	Microwave extraction	–	–	[143–145]
	Subcritical water extraction	–	–	[146,147]
	Subcritical extraction	–	–	[48,57,76,108,146,148–155]
	Comparison of solvent for supercritical extraction	–	–	[146]
	Various	GC–MS	–	[74,145,156–159]
	–	Interference by sulphur by GC method	–	[160,161]
	–	Effect of air drying of soils prior to GC	–	[160,161]

(Continued)

TABLE 15.3 (Continued)

Compound	Extraction procedure	Method of analysis	LD	Reference
	Separation of PCBs and chlorinated insecticides	GC–MS	–	[162,163]
	Method of extraction or Soxhlet extraction	Enzyme-based immunoassay	9–12 μg kg^{-1}	[164–167]
	–	Adsorption of PCB into days	–	[149,167]
	–	Photo-activated luminescence analysis and room temperature phosphorescence analysis	–	[168]
	Cryotrapping	GC	0.23 PM	[169]
Polychlorodibenzo-dioxins and polychlorodibenzofurans	–	uv spectroscopy	–	[170–191]
Polychlorodibenzo-*p*-dioxins and polychlorodibenzofurans	General discussion and method extraction from soil	–	–	[170–191]
Polychlorodibenzo-*p*-dioxin and polychlorodibenzofurans	Extraction methods	–	–	[137,191–199]
	Supercritical fluid extraction	–	–	[151,193]
	Accelerated solvent extraction	–	–	[194–198]
	Soxhlet extraction	–	–	[195]

GC, Gas chromatography; *IR*, infrared; *MS*, mass spectrometry.

chlorinated aromatic compounds from soils were trapped onto the hydrophobic film of the hollow waveguide sampler following detection by the FT-IR spectrometry. The extraction process in this method was, in principle, similar to headspace solid-phase microextraction. Means of increasing the speed of transfer of the vaporised organic species to the sampler were also studied. Results indicated that, with a negative pressure on the end of the sample, the speed of transfer increased significantly. Vapour pressure of the analytes is used as an indication in order to test the limitations of this method in the analysis or organic compounds in soils. Results showed that analytes with vapour pressures lower than 1600 Pa could be deducted quantitatively. Typical R^2 values for the regression on the concentration and IR signals were around 0.99, and typical detection limits were in the range of hundreds of parts per billion.

Chlorinated aliphatic hydrocarbons

Neumayr [118] carried out soil atmosphere studies using capillary GC and electron-capture and flame-ionisation sequential detection and used this as a means of pinpointing zones of soil and groundwater contamination. Methods have been described for determining chlorinated aliphatic hydrocarbons in soil and chemical waste disposal samples. The latter method involves a simple hexane extraction and temperature-programmed gas chromatographic analysis using electron-capture detection and high-resolution glass capillary columns.

Chlorobenzoic acid

Chlorobenzoic acids are relevant environmental pollutants originating mainly from microbial degradation of polychlorobiphenyls. Their presence in sites historically contaminated by polychlorobiphenyls can lead to a substantial negative effect on further polychlorobiphenyl degradation because of their toxicity for polychlorobiphenyl-degrading bacteria. The majority of polychlorobiphenyl-degrading bacteria are not able to degrade chlorobenzoic acids, and therefore chlorobenzoic acids are accumulated in the contaminated sites. Only a few of the bacterial strains are able to degrade biphenyl and benzoates isolated from contaminated sites.

Kresinova et al. [200] prepared extracts by using accelerated solvent extraction. Methylated chlorobenzoic acids were separated by GC using a system with two different columns DB 55 and DB 200. A series coupled (tandem) arrangement and detected by electron impact MS was used in this procedure.

A clean-up with gel chromatography was carried out to remove soil interfering matrix compounds as well as a major proportion of polychlorobiphenyls. The limit of quantification ranged between 1 and 10 ng g^{-1} of individual chlorobenzoic acids in the soil.

A rapid derivatisation method has been developed by Kresinova et al. [200] used GC–MS to determine polychlorobiphenyl-contaminated soils for 15 isomers of mono-, di-, tri-, tetra- and penta-chlorobenzoic acids in chlorobenzoic acid mixtures. Two derivatisation agents (diazomethane and methyl chloroformate) and various conditions were evaluated in terms of efficiency. The optimum method was diazomethane and 1% methanol running 1 hour at 5°C for derivatisation of extracts of soils and river sediments from polychlorobiphenyl-contaminated sites.

Kresinova et al. [200] compared two types of derivatisation agents (diazomethane and methyl chloroformate) and various derivatisation protocols for chlorobenzoic acid methylation; the advantages of diazomethane were documented. Optimised conditions with diazomethane were used for the determination of chlorobenzoic acid in extracts of polychlorobiphenyl-contaminated soil from South Bohemia and river sediment from Slovakia. The whole analytical process involved accelerated solvent extraction, a purification step using gel permeation chromatography, derivatisation and tandem column GC/MS analytical extension. An indirect method of the evaluation of diazomethane derivatisation was successfully employed to monitor the reaction recovery, and this method was verified using commercially available standards. The presence of bacterial polychlorobiphenyl 'dead-end' metabolites in real soil was confirmed, where this fact suggests a possible reason for polychlorobiphenyl persistence in soil. The sum of chlorobenzoic acid was about 1% in comparison with the sum of polychlorobiphenyls (302.7 $\mu g\ g^{-1}$). The analytical protocol represents a complementary method for the determination of polychlorobiphenyls during bioremediation or biodegradation studies.

Perfluorooctane sulphonyl fluoride

These substances are widely applied to the production of water repellents, fire extinguishers and surface coating of paper and textiles in the 1960s because of their perfect surface activity and stability.

This group of compounds has been introduced into the environment in recent years. Because of their lipophilic characteristics they have been detected in human and animal tissue.

In May 2001 the Stockholm Convention on persistent organic compounds was brought into effect. In 2009 perfluorooctane sulphonate and its precursor perfluorooctanoic acid were listed.

Meng et al. [201] determined perfluorooctane sulphenyl fluoride and developed a method using LC with UV and fluorescence detection.

In this study a new method was developed by derivatising perfluorooctane sulphenyl fluoride with 1-naphthol to form 1-naphthylperfluorooctanesulphonate that allowed rapid qualitative and quantitative analysis using LC–UV and LC–FLD. The derivatising product was confirmed from the analyses by proton nuclear magnetic resonance (NMR) and quadrupole-time-of-flight MS. The LC–FLD method demonstrated good linearity in the 1-naphthylperfluorooctanesulphonatelate concentration range from 20 μg L^{-1} to 20 ng L^{-1} with correlation coefficient better than 0.00, and an instrumental detection limit of 1.5 pg μL^{-1}.

15.4 Nitrogen-containing compounds in soil

The determination of these compounds is reviewed in Table 15.4.

More recently developed methods are discussed in more detail later.

Nitro compounds

A surface-enhanced Raman spectroscopy [221] method has been used to determine nitro compounds in soil. Polymers attached to fibre optics [159] have been used to determine down to 5 ppb of 2,4-dinitrotoluene in soil.

Emery et al. [208] studied the binding of trinitrotoluene to soil using ^{2}H magic angle spinning (MAS) NMR.

Trinitrotoluene has been determined in soil using a field-portable continuous-flow immunosensor. Results agreed with those obtained by HPLC [209,210].

Nitrogen-containing explosives [214] and trinitrotoluene [215] have been determined in soil by GC with thermionic NP detection and reverse-phase HPLC. Wornhoudt et al. [216] used tunable IR laser detection to study the pyrolysis products of explosives in soil.

Polycyclic aromatic nitrogen heterocyclic

Koci et al. [222] have discussed the extraction of these compounds from spiked soil samples. Extraction recovery was tested of 10 selected polycyclic aromatic nitrogen heterocycles quinoline, 2-methylquinoline, 6-methylquinoline, 8-methylquinoline, acridine benzo[*h*]quinoline, phenantridine, indole, 2-methylindole and carbazole from spiked soil samples. Four different extraction techniques, pressurised solvent extraction, supercritical fluid extraction, Soxhlet warm extraction and standard Soxhlet extraction, were applied and compared. The RP-HPLC technique with a silica-based octadecyl stationary phase was used for recovery and determination of individual determination PANHs. Supercritical fluid extraction has been found to be the most effective method for the extraction of selected PANHs from soil, and pressurised solvent extraction and Soxhlet warm extraction methods offered similar results with slightly lower extraction recoveries; on the contrary, a standard Soxhlet extraction is a time-consuming method with a low recovery or target analytes.

TABLE 15.4 Determination of nitrogen-containing compounds.

Compounds	Extraction method	Method of analysis	LD	Reference
Aromatic amines	–	Supercritical fluid chromatography	–	[202–206]
	–	Hydrogen derivative spectrophotometry	–	[111]
Nitro compounds	–	Surface-enhanced Raman spectroscopy attached to fibre optics	5 ppb	[159]
	–	Determination	–	[207]
	–	^{2}H magic angle spinning	–	[208]
		NMR		
	–	Continuous flow immune sensor	–	[209,210]
Nitro aromatic compounds	Various extraction procedure	Study of effect of calcium hydroxide	–	[211]
Tricyclazole and tetracycline	–	APLC	–	[212,213]
Trinitrotoluene		GC with thermionic	–	[214,215]
		NP detector reverse phase		
		HPLC		
Trinitrotoluene	–	Ultrared laser detection	–	[216]
N-oxides	12 L benzene extraction	GC–MS and GC	–	[217]
Nitro amines	–	Toxicity of formation	–	[218]
Ethylenediamine tetraacetic acid	–	Reverse-phase HPLC	–	[219]
Acetonitrile	–	Purge and trap chromatography	–	[30]
		Liquid chromatography	0.3–1.5 pg per injection	[220]
		GC–MS		

GC, Gas chromatography; *HPLC*, high-performance liquid chromatography; *MS*, mass spectrometry; *NMR*, nuclear magnetic resonance.

Hydrazines

Smolenkov et al. [223] have described an ion chromatographic method for the investigation of unsymmetrical hydrazine degradation in soils.

Growth regulators

Zhang et al. [224] used GC–MS to study the dissipations of the plant growth regulator hexanoic acid 2-(diethylamino)ethyl ester through biological processes in soil. For this purpose a single step was used to extract this compound with dichloromethane from the aqueous–acetone extracts of soil. Average recoveries in soil were between 85% and 104% at both spiking levels 0.01 and 0.1 mg kg^{-1}.

Smolenkov [223] investigated the use of ion chromatography as a tool in the investigation of unsymmetrical hydrazine degradation in soil.

Miscellaneous

Krone et al. [206] used capillary column GC with nitrogen-specific detection and GC–MS to determine nitrogen-containing aromatics originating from creosote oil in solvent extracts of sediments taken in Eagle Harbour, Puget Sound, and in contaminated soils. Organic sediment extracts and the commercial creosote oil were fractionated by silica alumina column chromatography. No nitrogen-containing aromatics were detected in sediments from a pristine reference area.

15.5 Sulphur and phosphorus–containing compounds in soil

Ingram et al. [226] applied static secondary ion MS to determine down to 70 pg m^{-2} of tributyl phosphate in soil surfaces.

David and Seiber [227] compared the efficiencies of various extraction techniques, including supercritical fluid [226], high-pressure solvent and Soxhlet extraction, for the extraction of organophosphorus hydraulic fluids from soil. High-pressure solvent extraction at temperatures up to 200°C and pressure up to 17 MPa was the favoured technique.

Sulphur compounds dimethyl disulphide

With the rapid development of the methods for protecting vegetables, the prevention of soil-borne disease and parasitic nematodes in food production is becoming a major issue. Disinfecting the soil using fumigants is the most effective measure to control this problem. In the past, methylbromide was one of the most common fumigants used for controlling soil-borne disease, parasitic nematodes and weeds. However, due to its ozone depletion potential, methylbromide has been largely phased out under the 1987 Montreal Protocol and was banned worldwide in 2015 (except for the necessary use exemption).

Dimethyl disulphide is a volatile sulphur compound that is a perfect alternative for methylbromide due to its high activity for soil-borne pests and absence of the ozone-depleting potential. It is an insecticide used to control nematodes, soil-borne pathogens and weeds in iconic crops such as strawberries, cucumbers and tomatoes.

15.6 Miscellaneous organic soil

Earlier methods for the determination of mestranol, flame retardants and polystyrene are summarised in Table 15.5.

Humic and fulvic acid

Weber and Wilson [242] used anion and cation exchange resins to isolate fulvic and humic acids from soil and water.

Ion-selective electrodes have been used to determine the stability constants for the complexation of copper II ions with soil fulvic acids [243]. Two cases of binding sites were found with conditional stability constants of about 1×10^6 and 8×10^3.

Saar and Weber [244] compared methods based on spectrofluorimetry and ion-selective electrode potentiometry determining the complexes formed between fulvic acid and heavy metal ions. The fluorescence properties of two fulvic acids, one derived from the soil and the other from river water, were studied. The maximum emission intensity occurred at 445–450 nm upon excitation at 350 nm, and the intensity varied with pH, reaching a maximum at pH5.0 and decreasing rapidly as the pH dropped below 4. Neither oxygen nor electrolyte concentration affected the fluorescence of the fulvic acid derived from the soil. Complexes of fulvic acid with copper, lead, cobalt, nickel and manganese were examined, and it was found that bound copper II ions quench fulvic acid fluorescence. Ion-selective electrode potentiometry was used to demonstrate the close relationship between fluorescence quenching and fulvic acid complexation of cupric ions. It is suggested that fluorescence and ion-selective electrode analysis may not be measuring the same complexation phenomenon in the cases of nickel and cobalt complexes with fulvic acid.

Wilson et al. [245] carried out a compositional and solid-state NMR spectroscopic study of humic and fulvic acid and fractions present in soil organic matter. The ^{13}C NMR study utilised cross polarisation-MAS (CP-MAS) with spin counting. The elemental and functional group analyses provided input

TABLE 15.5 Determination of miscellaneous organic compounds in soil.

Compounds	Extraction method	Methods of analysis	LD	Reference
Mestranol ethinylestradiol	Extraction of acid medium – florisil linearity	GC, TLC	0.1 ppm	[228]
Trifluoroacetic acid	Methanolic sulphuric acid extraction	Derivatisation to methyl ester of – trifluoroacetic-headspace GC	0.20 ng g^{-1}	[229–232]
Brominated flame retardant, poly-brominated biphenyls	–	–	–	[233–235]
Polystyrene		Curie point flash evaporation pyrolysis	–	[236]
Persistent organic contaminants	–	Industrial hygiene gas detection tubes	–	[237,238]
	Adsorption–desorption of contaminants on soil	Electro-dynamic thermogravimetric acid	–	[239]
	–	Laser Raman spectroscopy		[240]
Organochlorine, contaminants	Ultrasonic extraction with acetone petroleum ether 1:1	GC and GC–MS		[241]

GC, Gas chromatography; *MS*, mass spectrometry.

for a series of analytical constraints calculations that yield an absolute upper limit for the amount of aromatic carbon, and estimates for both aromatic and noncarboxyl aliphatic carbon in each sample. Spin counting experiments demonstrate that less than 50% of the carbon in three of the fractions is observed in the NMR experiment, and even after correction for different relaxation, the amounts or aromatic and noncarboxyl aliphatic carbon determined by ^{13}C CP-MAS NMR are dissimilar to those obtained by calculation. Unambiguous evidence is presented for the predominance of aliphatic carboxyl groups in one of the fulvic acid fractions.

15.7 Volatile organic compounds in soil

Different methods, such as headspace analysis and purge and trap analysis, have been employed in the analysis of mixture of volatile organic compounds (VOCs) in soils [246–251].

The determination of VOCs in soil is reviewed in Table 15.6. More recent methods are now discussed.

A combination of headspace sampling and liquid-phase microextraction (LPME) has been successfully developed to solve sensitivity problems in attenuated total reflection (ATR) IR determination of volatile compounds [264]. The headspace sampling facilitates the selective extraction of the target volatile analytes from the sample matrix, while the LPME allows their preconcentration prior to IR analysis. The direct determination of extracted analytes in the acceptor solvent provides high preconcentration factors of the order of 200 with a reduced consumption of organic solvents and a minimum generation of wastes. The qualitative and quantitative capability of the proposed approach was evaluated on the basis of two different examples: (1) screening of benzene, toluene and xylene compounds in soil samples and (2) quantitative determination of toluene in cosmetic nail products.

Yang and Her [246] developed a rapid method for the determination of semivolatile compounds in contaminated soil samples by coupling solid-phase microextraction with attenuated total reflectance (ATR)-FT-IR spectroscopy. A trapezoidal internal reflection element was mounted horizontally in a flow cell with the inlet port connected to a temperature-controlled glass extraction chamber. Soil samples were placed inside the glass tube and heated to the desired temperature. Vaporised semivolatile compounds were carried out by a stream of nitrogen gas to the ATR/IR flow cell. To increase the trapping efficiency the ATR crystal was coated with hydrophobic polyisobutylene polymer that acted as the SPME phase. The method proved to be very sensitive in the detection of semivolatile compounds in soils. The relationship between various parameters affecting chemical quantitation, such as the film thickness, gas flow rate and water contents, was also studied. Three different compounds, 1-chloronapththalene, nitrobenzene and

TABLE 15.6 Determination of volatile organic compounds in soil.

Compounds	Extraction method	Method of analysis	LD	Reference
Volatile organic compounds (aromatic and chlorinated compounds)	Direct vapour or solvent extraction with methane	(Octanol–water)	–	[120,123,252–260]
Volatile organic compounds	Study of effect of headspace conditions	–	–	[258]
	Automated static headspace method	–	–	[257]
	Per evaporation method as alternative to headspace analysis	–	1 ng g^{-1}	[259]
Volatile organic compounds trichloroethylene benzene, toluene, chloroform methylene chloride	Comparison of purge and trap analysis (1) methane of immersion of soil and (2) hot soil vent extraction	Hot solved extraction most effective		[247]
Volatile organic compounds	Thermal vaporisation trapping tenax GC	GC–MS	–	[178,249]
	–	Multidimensional GC–MS, GC–IR		[251]
	–	Pyrolysis GC		[261,262]
	Inert gas purging	Purge and membrane	Low ng g^{-1}	[145,258,260,263]
Halogenated hydrocarbons benzene toluene, xylenes	Gas purging			
Volatile organic compounds trifluoroacetic acid	Extraction with methanolic sulphuric acid	Derivatisation to methyl to methyl ester headspace	–	[229]

GC, Gas chromatography; *IR*, infrared; *MS*, mass spectrometry.

2-chlorotoluene, were used to investigate the feasibility of this method in the analysis of organic compounds in sand and soil. Results indicated a linear relationship between concentration and IR signals can be obtained for the three analytes. The detection limit of this method was in the range of 200–300 ppb.

Yang and Chen [265] constructed a reflection–absorption IR-sensing device for the detection of semivolatile aromatic compounds in soils.

Hiatt [266] shows that the role of internal standards in their interaction with soils has an impact on the accuracy of volatile organics determinations. The workers describe an IR-sensing device for the examination of chlorinated aromatic compounds, namely in soils, namely from glass vial modified for use in the analysis of soil samples by conventional FT-IR spectroscopy. In this sampling device an aluminium plate coated with a hydrophobic film was placed on top of the cap of the sample vial to absorb the analytes that evaporated from the soil matrix. After this absorption process was complete, the cap was placed in an FT-IR spectrometer, and the absorbed analytes were detected in the reflection–absorption mode. To accelerate the rate of evaporation of the analytes, the soil samples were heated to various temperatures. Other factors, such as the moisture content, sampling time, thickness of the hydrophobic film and the volatilities and concentrations of the analytes, were also examined to optimise the analytical conditions. The results indicated that the time required to reach equilibrium conditions was short, and evaporated/absorption could be achieved within 10 minutes. With a water content of 10% (v/w) or less the intensities of the analytical signals were increased greatly when compared with those of dry samples; when the water content was above 10% (v/w), these intensities decreased partially as a result of the heating efficiency. After examining the compounds that had different vapour pressures, the analytical results indicated that this method was applicable to the examination of compounds that had vapour pressures below 1.0 Torr. Using the optimal conditions determined in this study, the detection limits for semivolatile aromatic compounds were lower than 100 ng g^{-1}, and the regression coefficients of the standard curves for compounds that had a vapour pressure lower than 1.0 Torr were larger than 0.99 in the concentration range of 1–100 μg g^{-1}.

15.8 Mixtures of organic compounds in soil

Understandably, the analyst is not concerned with the determination of a single organic compound in soil, but with soil extractions containing more than one component.

This is illustrated in Table 15.7 that lists mixtures of various compounds that have been found in soil extracts.

TABLE 15.7 Analysis of mixtures of organic compounds in soil.

Mixtures of Compounds	Extraction method	Method of analysis	LD	Reference
Kepone and DDT permethrin	–	GC	–	[118,267]
Phenolic residue	–	Multidimensional	–	[251]
		IR and MS		
	–	Direct acetylation – GC and direct acetylation – MS	–	[110]
Phenols and cresols	–	Direct acetylation – headspace analysis	–	[109]
Anthropogenic compounds, polycyclic aromatic compounds, haloorganics, aliphatic hydrocarbons and hetero-aromatic hydrocarbons	–	Flesh evaporation	–	[31]
Miscellaneous	Cryofocusing	Cryogenic GC		[225,268–272]
Volatile organic compounds	Purge and trap analysis methanol extraction and hot solvent extraction	–	–	[247]
Nitro compounds atrazine	–	HPLC–MS	–	[161,273]
Insecticide mixtures	Supercritical extraction fluid extraction	–	–	[274]
Polycyclic aromatic hydrocarbons, phenols organochlorine insecticide	Microwave-assisted extraction	–	–	

(Continued)

TABLE 15.7 (Continued)

Mixtures of Compounds	Extraction method	Method of analysis	LD	Reference
Mixtures of volatile organic compounds	–	HPLC–MS	–	[275]
Miscellaneous nitramines nitroaromatic compounds	Vacuum distillation	AC/MS	–	[276]
	Headspace membrane extraction	GC	–	[277,278]
Volatile organic compounds	–	Sampling devices	–	[276,279]
Miscellaneous organic compounds	–	Comparison of methods based on GC, MS, TLC	–	[225,269–274, 280–283]
Volatile organic compounds, polychlorobiphenyls polyaromatic pesticides and polychloro dibenzo-*p*-dioxins	–	Sample preparation	–	[282]
Miscellaneous organic compounds	–	Pyrolysis GC	–	[216,240,284,285]
Aliphatic and aromatic carboxylic acid substituted phenols, benzene diols, benzene triols, phenolic acids and phenolic amides		Pyrolysis GC		[240]
Explosives	–	Pyrolysis GC with turnable	–	[216]
		IR laser detectors		
Nitro compounds	Purge and trap concentration	Purge and trap GC	–	[286,287]
Misc. organics	Soxhlet extraction	–	–	

GC, Gas chromatography; *HPLC*, high-performance liquid chromatography; *IR*, infrared; *MS*, mass spectrometry.

References

[1] US EPA, Methods for Chemical Analysis of Water and Wastes EPA 600/14-79/020, US Environmental Protection Agency, Washington, DC, 1970.

[2] WRCB, Leaking Underground Fuel Tank, (LUFT) Field Manual, State Water Resources Control Board, Sacramento, CA, 1998.

[3] ASTM, Annual Book of ASTM Standards, Volume 11. 02, ASTM, West Conshohocken, PA.

[4] V. Camel, A. Tambut, M. Caude, Chromatography 642 (1993) 263.

[5] S. Laing, D.C. Tilotta, Anal. Chem. 70 (1998) 616.

[6] V. Janda, K.D. Bartle, A.A. Clifford, Chromatography 642 (1993) 283.

[7] S.B. Hawthorne, Anal. Chem. 62 (1990) 633A.

[8] S.E. Eckert-Tolitta, S.B. Hawthorne, D.J. Miller, Fuel 72 (1993) 1015.

[9] S.B. Hawthorne, D.J. Miller, K.M. Hegvik, Chromatogr. Sci. 31 (1993) 26.

[10] S.B. Hawthorne, K.M. Hegvik, Y. Yang, D.J. Miller, Fuel 73 (1994) 1876.

[11] Y. Yang, S.B. Hawthorne, D.J. Miller, J. Chromatogr. 699 (1995) 265.

[12] K.H. Schafer, P.R. Griffiths, Anal. Chem. 55 (1983) 1939.

[13] K. Jinno, M. Saito, Anal. Sci. 7 (1991) 361.

[14] T.J. Jenkins, M. Kaplan, M.R. Smimmonds, G. Davidson, M.A. Healey, M. Poliakoff, Analyst 116 (1991) 1305.

[15] D.L. Heglund, D.C. Tilotta, D.J. Miller, S.B. Hawthorne, Anal. Chem. 66 (1994) 3543.

[16] M.D. Burford, S.B. Hawthorne, D.J. Miller, J. Chromatogr. 685 (1994) 95.

[17] X. Lou, H. Janssen, C.A. Cramers, J. Chromatogr. 750 (1996) 215.

[18] G. Morel, O. Samhan, P. Literathy, M. Al-Hashash, L. Moulin, T. Saeed, et al., Fresenius J. Anal. Chem. 339 (1991) 699.

[19] G. Hazel, F. Buchholz, I.D. Aggarwal, G. Nau, K.I. Ewing, Appl. Spectrosci. 51 (1997) 984.

[20] U.D. Roe, M.J. Lacy, J.D. Stuart, G.A. Robbins, Anal. Chem. 61 (1989) 2584.

[21] J.L. Parr, G. Walters, M. Hoffman, Sampling and analysis of soils for gasoline range organics, in: P.T. Kostecki, E.J. Calabrese (Eds.), Hydrocarbon Contaminated Soils and Groundwater: Analysis, Fate, Environmental and Public Health Effects, Remediation, vol. 1, Lewis, Ann Arbor, MI, 1991, pp. 105–132.

[22] G. Xie, M.J. Barceelona, J. Fang, Anal. Chem. 71 (1999) 1899.

[23] M. Remmler, F.–D. Kponke, U. Stottmeister, Thermochem. Acta 263 (1995) 101.

[24] D.M. White, H. Luong, R.L. Irvine, J. Cold Reg. Engl. 21 (1998) 1.

[25] P. Peuron, S. Daugherty, J. Contam. Soils 2 (1997) 449.

[26] D.W. Ostendorf, L.E. Leach, E.S. Hinlein, Y. Xie, Ground Water Monit. Rem. 11 (1991) 107.

[27] G.A. Robbins, R.D. Bristol, V.D. Roe, Ground Water Monit. Rem. 9 (1989) 87.

[28] F.W. Karâsek, G.H. Charbonneau, G.J. Revel, H.Y. Tong, Anal. Chem. 59 (1987) 1027.

[29] M.J. Geerdink, C. Erklens, J.C. Dan, J. Frank, K.C.A.M. Luyben, Anal. Chim. Acta 315 (1995) 159.

[30] P.E. Kester, Analysis of Volatile Organic Compounds in Soils by Purge and Trap Gas Chromatography, Tekmar Company, Cincinnati, OH, 1987. 45222–1856.

[31] J.W. de Leeuw, E.W.B. de Leers, P.J.W. Damsté, Schuyl, Anal. Chem. 58 (1986) 1852.

[32] S. Jones, Sampling and analysis of soils for gasoline range organics I, II in: P.T. Kostecki, E. J. Calabrese (Eds.), Hydrocarbon Contaminated Soils and Groundwater Analysis, Fate Environmental and Public Health Effects, Remediation, vol. 1, Lewis, Ann Arbor, MI, 1993.

[33] A.M. Ilias, J. Janger, Sampling and analysis of soils for gasoline range organics, in: P.T. Kostecki, E.J. Calabrese (Eds.), Hydrocarbon Contaminated Soils and Groundwater: Analysis Fate, Environmental and Public Health Effects. Remediation, vol. 1, 3, Lewis, Ann Arbor, MI, 1993, p. 147.

[34] K.E. Karp, Ground Water Monit. Rem. 13 (1993) 101.
[35] M. Picer, V. Hocenski, Water Res. 28 (1994) 619.
[36] T.G. Greco, R.L. Grob, Environ. Sci. Health A 25 (1990) 185.
[37] A. Eschenboch, M. Kaesner, R. Bierl, G. Schater, B. Makro, Chemosphere 28 (1994) 683.
[38] J. Schawrtz, D. Slater, T.V. Larson, W.E. Pierson, J.Q. Keoning, Am. Rev. Respir. Dis. 147 (1993) 826.
[39] M.G. Zemanek, S.J. Pollard, S.L. Kenefick, S.E. Hrudey, Air Waste Manage. Assoc. 47 (1997) 1250.
[40] D. Schuetzle, T.E. Jensen, T.E. Bell, Environ. Inst. 11 (1985) 169.
[41] D. Schueszle, W.O. Siegle, T.E. Jensen, M.A. Dearth, E.W. Kaiser, R. Gorse, et al., Environ. Health Prospect. 102 (1994) 3.
[42] R.N. Westerholm, T.E. Alsberg, A.B. Frommelin, M.E. Strandell, U. Rannug, L. Winquist, et al., Environ. Sci. Technol. 22 (1988) 925.
[43] T. Alsberg, U. Stenberg, R. Westerholm, M. Strandell, U. Rannug, A. Sundcvall, et al., Environ. Sci. Technol. 19 (1985) 43.
[44] S.R. Wild, K.S. Waterhouse, S.P. McGrath, K.C. Jones, Environ. Sci. Technol. 24 (1990) 1706.
[45] M. Trapido, Environ. Pollut. 105 (1999) 67.
[46] N.J. Hayden, T.C. Voice, R.B. Wallace, J. Contam. Hydrol. 25 (1977) 271.
[47] P. Fine, E.R. Graber, B. Yaro, Soil Technol. 10 (1997) 133.
[48] M.G. Zemanek, S.J.T. Pollard, S.L. Kenefick, S.E. Hrudy, Environ. Pollut. 98 (1997) 239.
[49] J.C. Chuanang, M.A. Pollard, Y.L. Chou, R.G. Menton, N.K. Wilson, Sci. Total Environ. 224 (1998) 189.
[50] L.C. Marr, T.W. Kirchstetter, R.A. Harley, A.H. Miguel, S.V. Hering, S.K. Hammond, Environ. Sci. Technol. 33 (1999) 3091.
[51] O.H.J. Szolan, H. Rose, R. Brannan, A.P. Lochner, Anal. Chem. 74 (2012) 2397.
[52] G. Codina, M.T. Vaguero, L. Comellas, F. Broto Puig, J. Chromatogr. 672 (1994) 21.
[53] S. Lundstedt, B. Van, P. Haglund, M. Tysklind, L. Oberg, Chromatography 883 (2000) 151.
[54] J. Ramos, J.L. Vreule, U.A.T. Brinkman, J. Chromatogr. 891 (2000) 275.
[55] M. Saim, J.R. Dean, M.D. Abdullah, Z. Zakaria, Anal. Chem. 70 (1998) 420.
[56] S.B. Hawthorne, S. Trembley, C.I. Moniot, C.R. Grolonski, D.T. Miller, J. Chromatogr. 886 (2000) 237.
[57] J.J. Langenfield, S.B. Hawthorne, D.J. Miller, J. Rawliszyn, Anal. Chem. 65 (1993) 338.
[58] M.D. Burford, S.B. Hawthorne, D.J. Miller, Anal. Chem. 65 (1993) 1497.
[59] S. Reindt, F. Hoffler, Anal. Chem. 66 (1994) 1808.
[60] J.J. Barnabas, J.R. Dean, W.R. Tomlinson, S.P. Owen, Anal. Chem. 68 (1996) 2064.
[61] M.T. Tena, M.D. Luque de Castro, M. Valcarcel, Anal. Chem. 68 (1996) 2386.
[62] Environmental Protection Agency, SW-846, Test Methods for Evaluating Solid Wastes, Method 3545 US, third ed., Update 111 US GPO, Washington, DC, 1995.
[63] A. Meyer, W. Kleibömer, J. Chromatogr. 657 (1993) 327.
[64] J. Dankers, M. Groenboom, L.H.A. Schlotis, C. Van de Heiden, Chromatography 641 (1993) 357.
[65] H.B. Lee, T.E. Peart, R.L. Hang-You, R.D. Gere, Chromatography 653 (1993) 83.
[66] F. Guo, Q.X. Li, J.P. Alcantara-Licudine, Anal. Chem. 71 (1999) 1309.
[67] J. Hollender, B. Koch, C. Lutemann, W. Dott, Int. J. Environ. Anal. Chem. 83 (2003) 21.
[68] A. Dreyer, M. Radke, J. Environ. Anal. Chem. 85 (2005) 423.
[69] A. Rabbat, T.L. Liu, B.M. Abraham, Anal. Chem. 64 (1992) 1477.
[70] R. Hartman, Int. Anal. Chem. 62 (1996) 161.

[71] M.J. Dale, Ac Jones, S.J.T. Pollard, P.R.R. Longridge-Smith, A.G. Rowley, Environ. Sci. Technol. 27 (1993) 1693.
[72] R.P. Rodgers, Ac Reilly, W.B. Whitten, J.M. Ramsey, Anal. Chem. 72 (2000) 5040.
[73] G.A. Lopez, G.E. Blanco, A.J.I. Garcia, A. Sanz-Medel, Chromatographia 33 (1992) 225.
[74] S.M. Hankin, P. John, A.W. Simpson, G.P. Smith, Anal. Chem. 68 (1996) 3235.
[75] L. Nondek, M. Kuzilek, S. Krupicka, Chromatographia 37 (1993) 381.
[76] S.H. Lain, J. Huelton, Windruch, Int. J. Environ. Anal. Chem. 63 (1996) 245.
[77] R.J. Van de Nesse, G.J.M. Hoogland, J. Moel, J.M. De, C. Gooljee, U.A.T. Brinkman, et al., J. Chromatogr. 552 (1991) 613.
[78] P.J.A. Fowlie, T.L. Bulman, Anal. Chem. 58 (1986) 721.
[79] R. Maybury, Laboratory Manual for Pesticide Residue Analysis in Agricultural Products, Food Production and Inspection Branch, Agriculture, Canada, revised, 1984.
[80] B.K. Afghan, R.J. Wilson, Method for Determination of Polynuclear Aromatic Hydrocarbons in Environmental Samples HPLC-Multi-Detection System, Environment Canada Manuscript 20-AMD-3-81-BKA, 1981.
[81] P.P. McDonald, R.E. Almond, J.P. Mapes, S.B. Freidman, J. AOAC Int. 77 (1994) 466.
[82] D. Brűggemann, R. Freitag, J. Chromatogr. 717 (1995) 309.
[83] J.K. Lai, S.V. Filseth, C.M. Saowski, F. Morgan, Int. J. Environ. Chem. 40 (1990) 99.
[84] J. Bublitz, A. Christopherson, W. Schade, Fresenius J. Anal. Chem. 355 (1996) 684.
[85] Y. Shimizu, H.M. Lillsetrou, Water Sci. Technol. 23 (1991) 427.
[86] M. Niedere, Environ. Sci. Pollut. Res. Int. 5 (1998) 209.
[87] V. Lopez-Avila, R. Young, W.F. Beckett, Anal. Chem. 66 (1994) 1097.
[88] Medina-Vera, J. Appl. Pyrolysis 36 (1996) 27.
[89] M. Sparrevik, H. Jonassen, Soil Environ. 5 (1995) 537.
[90] R.S. Brown, J.H.T. Luang, J.D.H. Szolar, Anal. Chem. 68 (1996) 287.
[91] C. Luterman, W. Dott, J. Hollander, J. Chromatogr. 811 (1998) 151.
[92] E. Lichtfouse, H. Budzinski, P. Garrigues, T.I. Eglington, Org. Geochem. 26 (1997) 353.
[93] J.H. Martin, A.M. Siebert, R.C. Loehr, J. Environ. Chem. 117 (1991) 291.
[94] US EPA, Substances Control Act, Environmental Protection Agency, Washington, DC, 1979.
[95] P. Rigon, S. Saini, S.J. Setford, Int. J. Environ. Anal. Chem. 84 (2004) 979.
[96] J.C. Risdon, J.T. Pollard, K.J. Brassington, J.N. McEwen, G.I. Paton, K.T. Semple, et al., Anal. Chem. 80 (2008) 1090.
[97] T.L. Terleak, S.O. Agbo, A. Barendregt, J.L.M. Hermes, Anal. Chem. 40 (2006) 1307.
[98] A. Longbehn, H. Steinhart, High Resolut. Chromatogr. 17 (1994) 203.
[99] R. Koltainan, Atmos. Ecosyst. 29 (1995) 69.
[100] S.B. Hawthorne, D.J. Miller, Anal. Chem. 66 (1994) 4005.
[101] T.S. Reighard, S.V. Olesik, Anal. Chem. 68 (1996) 3612.
[102] J.E. Futter, P. Wall, J. Planar Chromatogr. 6 (1993) 372.
[103] M.P. Lrompart, R.A. Lorezo, R. Cela, J.R. Pare, Analyst 122 (1997) 133.
[104] S.B. Hawthorne, Y. Yang, D.J. Miller, Anal. Chem. 66 (1994) 2912.
[105] C.S. Hottenstein, S.W. Jourdan, M.C. Hayes, F.M. Rubio, D.P. Herzog, T.S. Lawruk, Environ. Sci. Technol. 29 (1995) 2754.
[106] R. Dean, A. Santamaria-Rekondo, E. Ludkin, Anal. Commun. 33 (1996) 413.
[107] D.E. Kimbrough, R. Chin, J. Wakakuwa, Analyst 119 (1994) 1283.
[108] M.A. Crespin, M. Gallego, M. Valcarcel, Anal. Chem. 71 (1999) 2687.
[109] M.P. Llopart-Vizoso, R.A. Lorenzo-Ferreira, R. Cela-Torrijos, J. High Resolut. Chromatogr. 19 (1996) 207.
[110] T.G. Danis, T.A. Albanis, Toxicol. Environ. Chem. 53 (1996) 9.

[111] G. Talsky, Int. J. Environ. Anal. Chem. 14 (1983) 81.
[112] G. Sirvent, J.M. Sanchez, M. Hidalgo, V. Salvado, Int. J. Environ. Anal. Chem. 89 (2009) 293.
[113] A.M.T. Magabhaes, P.M. Chalk, Analyst 111 (1986) 77.
[114] A.A. Deetman, P. Demenlemeester, M. Garcia, Anal. Chim. Acta 82 (1976) 1.
[115] B.W. Strobel, T. Kristense, H.C. Hansen, Int. J. Environ. Anal. Chem. 84 (2004) 909.
[116] L. Lubecki, Int. J. Environ. Anal. Chem. 94 (2014) 1360.
[117] J.R. Dean, Anal. Commun. 33 (1996) 191.
[118] V. Neumayr, in: G. Milde, R. Leschber (Eds.), Soil and Groundwater Protection, Gustav Fischer Verlag, Stuttgart, 1986, pp. 65–84.
[119] B.E.W. Leeuw, L. De Leer, Sinringh, Anal. Chem. 58 (1986) 1852.
[120] H.B. Kerfoot, Environ. Sci. Technol. 21 (1987) 1022.
[121] M. Mehran, R. Olsen, B.M. Rector, Groundwater 25 (1987) 275.
[122] L. Wennrich, D. Popp, M. Molder, Anal. Chem. 72 (2000) 546.
[123] S. Bowlat, S.B. Hawthorne, J. Chromatogr. A 703 (1995).
[124] R. Duester, N. Lubahn, C. Friedrich, W. Kleiböhomer, J. Chromatogr. 85 (1997) 227.
[125] S.B. Hawthorne, C.B. Grabanski, J.K. Hageman, D.J. Miller, J. Chromatogr. A 814 (1998) 151.
[126] J. Fisher, A.M.J. Scarlett, A.D. Stott, Environ. Sci. Technol. 31 (1997) 1120.
[127] C.L. Arthur, J. Pawliszyn, Anal. Chem. 62 (1990) 2145.
[128] D. Lough, S. Moltagh, J. Pawliszyn, Anal. Chem. 64 (1992) 1187.
[129] Z. Zhang, M.J. Yang, J. Pawliszyn, Anal. Chem. 66 (1994) 844A.
[130] F.J. Santos, O. Jauregui, F.T. Flinto, Galcerar, J. Chromatogr. 823 (1998) 249.
[131] J. Yang, I.E. Her, Anal. Chem. 72 (2000) 878.
[132] M.H. Tavendale, A.L. Wilkins, A.G. Langdon, K.L. Mackie, T. Stutridge, P.N. McFarlane, Environ. Sci. Technol. 29 (1995) 1407.
[133] J. Passivirta, H. Hakaal, J. Knuutinen, T. Otallinene, J. Särkkä, L. Welling, et al., Chemosphere 21 (1990) 1355.
[134] M.C. Alonso, D. Puig, I. Silgoner, M. Grasserbauer, D. Barcelo, J. Chromatogr. Anal. 823 (1998) 231.
[135] F. Höfler, J. Ezzell, B. Ritcher, Labor Prax. 4 (1995) 58.
[136] F.J. Santos, F.J.O. Jáuregui, P.J. Pinto, M.T. Galceran, J. Chromatogr. A 823 (1998) 249.
[137] Y. Yang, M. Belghazi, A. Lagadec, D.J. Miller, S.B. Hawthorne, Chromatogr. A 810 (1998) 149.
[138] K.J. Hageman, L. Mazeas, C.B. Grabansky, D.J. Miller, S.B. Hawthorne, J. Chromatogr. A 68 (1996) 3892.
[139] L. Wennrich, M. Moder, Anal. Chem. 22 (2000) 546.
[140] V. Lopez-Avila, P. Hirata, S. Kronska, M. Flanagan, J.H. Taylor, S.C. Hern, Anal. Chem. 57 (1985) 2897.
[141] L. Renberg, Anal. Chem. 46 (1974) 459.
[142] D.E. Kinbrough, R. Chin, J. Wakakuwa, Analyst 119 (1994) 1277.
[143] V. Lopez-Avila, J. Benedicto, C. Charon, et al., Environ. Sci. Technol. 29 (1996) 271.
[144] V. Lopez-Avila, J. Benedicto, C. Charon, R. Young, V.F. Becert, Environ. Sci. Technol. 29 (1995) 2709.
[145] R. Kostainen, T. Kotiaho, I. Matilla, T. Manskki, M. OJala, R.A. Ketola, Anal. Chem. 70 (1996) 3028.
[146] S.B. Hawthorne, J.J. Lagenfeld, D.J. Miller, M.D. Burford, Anal. Chem. 64 (1992) 1614.
[147] R. Yang, S. Bawadt, S.B. Hawthorne, Anal. Chem. 62 (1995) 4571.

[148] H. Hellman, Deutsch Gewasserkundlich Mitt 29 (1985) 111.
[149] B.O. Brady, C.C. Kao, K.M. Doolen, et al., Indus. Eng. Chem. 26 (1987) 261.
[150] S.B. Hawthorne, J.J. Lagenfeld, D.I. Miller, J. Rawliszyn, Anal. Chem. 65 (1993) 338.
[151] B. Von Bavel, M. Jaremo, L. Karlsson, G. Lindstorm, Anal. Chem. 68 (1996) 1279.
[152] Y. Yang, S. Bawadt, S.B. Hawthorne, D. Miller, J. Am. Chem. 67 (1995) 4571.
[153] L. Anlexandrou, J. Pawliszyn, Anal. Chem. 61 (1989) 2270.
[154] A. Glausch, G.P. Blanch, V. Schruig, J. Chromatogr. A 723 (1996) 399.
[155] B. Hawthorne, C.B. Grabinski, K.J. Hageman, P.J. Miller, J. Chromatogr. A 814 (1998) 155.
[156] D. Robbat, L. Tyng-Umi, B.M. Abraham, Anal. Chem. 64 (1992) 358.
[157] H. Budzinski, M. Letellier, S. Thompson, K. Le Menach, P. Carrigner Fresenius, J. Anal. Chem. 367 (2000) 165.
[158] E. Benicka, R. Novochovsti, J. Hronzek, I. Krupsih, D. Soudre, J. DeZeeun, J. High Resolut. Chromatogr. 19 (1996) 95.
[159] K.J. Albert, M.L. Myrick, S.B. Brown, D.L. James, F.D. Milanovich, D.R. Walt, Environ. Sci. Technol. 35 (2002) 193.
[160] G. Jensen, L. Renberg, L. Reutergard, Anal. Chem. 49 (1977) 316.
[161] Southwest Water Laboratory, Southwest Water Laboratory Method No. SP 8/71, Sediment Extraction Procedures, Athens, GA, 1977.
[162] J. Teichman, A. Beverue, J.W. Hylin, J. Chromatogr. 151 (1978) 155.
[163] A.L. Alford Stevens, J.W. Eichelberger, W.L. Budde, Environ. Sci. Technol. 22 (1988) 304.
[164] J.C. Johnson, J.M. Van Emon, Anal. Chem. 68 (1996) 162.
[165] A.J. Schuetz, M.G. Weller, R. Niessner, Fresenius J. Anal. Chem. 363 (1999) 777.
[166] S. Pullen, G. Haiber, H.H. Schöeler, R. Hock, Int. J. Environ. Anal. Chem. 65 (1996) 127.
[167] R.E. Alcock, C.J. Halsall, C.A. Harris, A.E. Johnston, W.W. Lead, G. Sanders, et al., Environ. Sci. Technol. 28 (1994) 1838.
[168] T. Vo-Dihn, A. Pal, T. Pal, Anal. Chem. 66 (1994) 1264.
[169] R.A. Kerwin, P.M. Crill, F.W. Tabot, Anal. Chem. 68 (1996) 899.
[170] R.E. Tucker, A.L. Young, A.P. Gray, Human and Environmental Risks of Chlorinated Dioxins and Related Compounds, Plenum, New York, 1993.
[171] R.D. Kimbrough (Ed.), Halogenated Biphenyls, Terphenyls, Naphthalenes, Dibenzodioxins and Related Products, Elsevier, Amsterdam, 1980.
[172] O. Hutzinyer, R. Frei, E. Merian, R. Pocchiara, Chlorinated Dioxins and Related Compounds. Impact on the Environment, Pergamon, New York, 1982.
[173] W.J. Nicolson, A.J. Moore, Health of Halogenated Aromatic Hydrocarbon, New York Academy of Science, New York, 1979.
[174] D.H.K. Lee, H.L. Falk, Environmental Health Perspectives Experimental Issue No. 5, US Department of Health, Education and Welfare, Publication No. (NIH), 1973, pp. 74–218.
[175] J.R. Huff, J.A. Moore, D.R. Saracci, L. Tomatis, Environ. Health Perspect. 36 (1980) 221.
[176] C. Rappe, H.R. Buser, H.P. Bosshardt, in: W.J. Nicholson, A.J. Moore (Eds.), Health Effects of Halogenated Aromatic Hydrocarbons, New York Academy of Science, New York, 1979, pp. 1–18.
[177] M.P. Esposito, T.O. Tiernan, F.E. Dryden, Dioxins, in: US EPA Report No. EPA-600/2-80-1997, US EPA, Washington, DC, 1980.
[178] K. Olie, P.L. Vermeullen, O. Hutzinger, Chemosphere 6 (1977) 455.
[179] B. Ahling, A. Lindskog, B. Jansson, G. Sundstorm, Chemosphere 8 (1977) 461.
[180] H.R. Basor, H.R. Bosshardt, C. Rapper, R. Lindahl, Chemosphere 7 (1978) 419.
[181] D.G. Crosby, A.S. Wong, Chemosphere 5 (1976) 327.

[182] L.L. Lamparski, R.H. Stehl, R.L. Johnson, Environ. Sci. Technol. 14 (1980) 196.
[183] F.E. McConell, in: R.D. Kinbrough (Ed.), Halogenated Biphenyls, Terphenyls, Naphthalenes Dibenzodioxins and Related Products, Elsevier, Amsterdam, The Netherlands, 1980, pp. 109–150.
[184] A.J. Goldstein, in: R.D. Kimbrough (Ed.), Halogenated Biphenyls, Terphenyls, Naphthalenes, Dibenzodioxins and Related Products, Elsevier, Amsterdam, The Netherlands, 1980, pp. 151–190.
[185] A. Di Domenico, G. Vivcano, G. Zapponi, in: O. Hutzinger, R.W. Rei, E. Marian, F. Pocchiari, (Eds.), Chlorinated Dioxins and Related Compounds – Impact on the Environment, Pergamon, New York, pp. 105–114.
[186] C.T. Ward, F. Matsumura, Arch. Environ. Contam. Toxicol. 7 (1978) 349.
[187] A.L. Young, in: R.E. Tucker, A.L. Young, A.P. Gray (Eds.), Human and Environmental Risks of Chlorinated Dioxins and Related Compounds, Plenum, New York, 1983, pp. 173–190.
[188] M.H. Bickel, S. Muhlback, in: O. Hutzinger, R.W. Frei, E. Marian, F. Pocchiaria (Eds.), *Chlorinated Dioxins and Related Compounds – Impact on the Environment*, Pergamon, New York, 1982, pp. 303–306.
[189] G.M. Decad, L.S. Birnbaum, S. Matthews, in: O. Hutzinger, R.W. Frei, E. Marian, F. Pocchiari (Eds.), Chlorinated Dioxins and Related Compounds – Impact on the Environment, New York, 1982, pp. 307–315.
[190] A.R. Isensee, Ecol. Bull. 27 (1978) 255.
[191] Y. Masuda, H. Kuroki, in: R.D. Kimsbrough (Ed.), Halogenated Biphenyls, Terphenyls, Naphthalenes, Dibenzodioxins and Related Compounds, Elsevier, Amsterdam, 1980, pp. 561–659.
[192] R.W. Walters, A. Guiseppi-Elie, Environ. Sci. Technol. 22 (1988) 819.
[193] F.I. Onuska, K.A. Terr, J. High Resolut. Chromatogr. 12 (1989) 357.
[194] B.E. Richter, J.L. Ezzell, D.E. Knowles, F. Hoefler, A.K.R. Muttulat, M. Scheutwinkel, et al., Chemosphere 34 (1997) 975.
[195] L.O. Kjeller, S.E. Kupl, B. Jonsseikon, C. Rappe, Toxicol. Environ. Chem. 39 (1993) 1.
[196] R. Hengstmann, R. Haman, Weber, A. Keltrun, Fresenius J. Anal. Chem. 335 (1989) 982.
[197] H. Otaka, M. Shinomya, T. Amagi, Int. J. Environ. Anal. Chem. 85 (2005) 503.
[198] H. Otaka, M. Shinomiya, T. Amagi, Int. J. Environ. Anal. Chem. 85 (2005) 515.
[199] Z. Vasilic, N. Peris, M. Wilken, V. Drevenkar, Int. J. Environ. Anal. Chem. 84 (2004) 1093.
[200] Z. Kresinova, L. Hostaena, J. Medkova, M. Cvanarova, T. Stella, G. Cajhami, Int. J. Environ. Anal. Chem. 94 (2014) 822.
[201] X. Meng, Z. Tang, Q. Li, Z. Cai, Int. J. Environ. Anal. Chem. 94 (2014) 1388.
[202] T.S. Oostdyk, R.L. Grob, J.L. Snyder, M.E. McNally, Anal. Chem. 65 (1993) 596.
[203] R.J. Peters, J.A.D. van Renesse Von Duivenbode, Fresenius J. Anal. Chem. 348 (1994) 249.
[204] W.C. Brumley, C.R. Brownrigg, G.M. Brilis, J. Chromatogr. 558 (1991) 223.
[205] A. Kido, R. Shinohara, S. Eto, M. Koga, T. Hori, Jpn. J. Water Pollut. Res. 2 (1979) 245.
[206] C.A. Krone, D.W. Burrows, D.W. Brown, P.A. Robisch, A.J. Friedman, D.C. Malins, Environ. Sci. Technol. 20 (1986) 1144.
[207] C.L. Grant, T.F. Jenkins, K.F. Myers, E.F. McCormick, Environ. Sci. Technol. 14 (1995) 1865.
[208] A.P. Emery, S.N. Chesler, W.A. MacCrehan, J. Chromatogr. 606 (1992) 221.
[209] I. Surugui, J. Svitel, L. Ye, K. Haupt, B. Danielsson, Anal. Chem. 73 (2012) 4388.

[210] P.R. Gauger, D.D. Holt, C.H. Patterson, P.T. Charles, L. Shriver-Lake, A.W. Kusterbeck, J. Hazard. Mater. 83 (2001) 51.
[211] M. Emmrick, Int. J. Environ. Anal. Chem. 83 (2003) 769.
[212] T. Tsukiokai, Analyst 113 (1988) 193.
[213] H. De Geus, B.N. Zegers, H. Lingeman, U.A.T. Lingeman, Int. J. Environ. Anal. Chem. 56 (1994) 119.
[214] A.D. Hewitt, T.F. Jenkins, T.A. Ranney, Field Anal. Chem. Technol. 5 (2001) 228.
[215] M.E. Walsh, T.F. Jenkins, P.G. Thorne, J. Energy J. 13 (1995) 357.
[216] J. Wornhoudt, J.H. Shorter, J.B. McManus, P.L. Kebabian, M.S. Zahniser, C.E. Davis, et al., Appl. Optometry 35 (1996) 3992.
[217] A. Kido, R. Shinoharg, S. Eto, M. Kugen, T. Hori, Jpn. J. Water Pollut. Res. 2 (1979) 245.
[218] A.J. Mills, M. Alexander, J. Environ. Qual. 5 (1976) 437.
[219] B. Nowak, F.G. Kari, S.U. Hilger, Y.L. Sigg, Anal. Chem. 68 (1996) 56.
[220] R. Svabenski, M. Oravec, A. Simek, Int. J. Environ. Anal. Chem. 89 (2009) 167.
[221] J.M. Sylvia, J.I.A. Janni, J.D. Klein, K.M. Spencer, Anal. Chem. 72 (2000) 5834.
[222] K. Koci, H. Petraska, A. Simek, E. Varadova, A. Syslova, Int. J. Environ. Anal. Chem. 87 (2007) 111.
[223] A.D. Smolenkov, P.P. Krechetov, A.A. Bendryshev, O.A. Shpigun, M.M. Martynova, Int. J. Environ. Anal. Chem. 85 (2005) 1089.
[224] H. Zhang, L. Xie, P. Xu, S. Jiang, Int. J. Environ. Anal. Chem. 88 (2008) 561.
[225] Capillary Column Use in Purge and Trap Gas Chromatography II: Use of the Model 1000 Capillary Interface, Tex/Data B021684, Tekmar Company, Cincinnati, OH.
[226] J.C. Ingram, G.S. Groenwald, A.D. Applehans, A.D. Dahl, J.E. Delmore, Anal. Chem. 68 (1996) 309.
[227] M.J. David, J.N. Seiber, Anal. Chem. 68 (1996) 3038.
[228] I. Okuno, A. Higgins, Bull. Environ. Contam. Toxicol. 18 (1977) 248.
[229] T.M. Cahill, J.A. Benesch, M.A. Gustin, J.E. Zimmerman, J.N. Seiber, Anal. Chem. 71 (1999) 4465.
[230] T.J. Wellington, W.F. Shnieder, D.R. Worsop, O.J. Neilson, J. Schested, W.J. Debruyn, et al., Environ. Sci. Technol. 28 (1994) 320A.
[231] D.J. Bowdem, S.L. Clegg, P. Brimblecombe, Chemosphere 32 (1996) 405.
[232] V.R. Kotamrthi, J.M. Rodriquez, M.K.W. Ko, T.K. Tromp, N.D. Sze, M. Prather, J. Geophys. Res. 103 (1998) 5747.
[233] L.W. Jacobs, S.-F. Chou, J.M. Teidje, J. Agric. Food Chem. 24 (1976) 1198.
[234] A. Norstrom, K. Andersson, C. Rappe, Chemosphere 4 (1976) 255.
[235] C.U.U. Ma, C.K. Boyne, Anal. Chem. 65 (1993) 772.
[236] J.W. de Leeuw, E.W.Z. Leer, J.S. Sinninghe Damste, Anal. Chem. 58 (1986) 1852.
[237] D.E. Wells, F. Hess, Technol. Instrum. Anal. Chem. 21 (2000) 73.
[238] A. Di Domenco, E. De Filip, F. Ferri, N. Iacovell, R. Miniero, E.S. di Tella, et al., Microchem. J. 46 (1992) 48.
[239] L. Tognotti, M. Fkytzani-Stephanopaulos, A.F. Sarofim, H. Kopsinis, M. Stankides, Environ. Sci. Technol. 25 (1991) 104.
[240] H.R. Schulton, C. Sorge, Eur. J. Soil Sci. 46 (1995) 567.
[241] L. Ling, D. Muir, X. Tong Wang, Xiao Ba Xu, Int. J. Environ. Anal. Chem. 85 (2005) 89.
[242] J.H. Weber, S.A. Wilson, Water Res. 9 (1975) 1079.
[243] W.T. Bresnahan, C.L. Grant, J.H. Weber, Anal. Chem. 50 (1978) 1675.
[244] R.A. Saar, J.H. Weber, Anal. Chem. 52 (1980) 2095.

[245] M.A. Wilson, A.M. Vassalo, E.P. Peruc, J.H. Reuter, Anal. Chem. 59 (1987) 551.
[246] Y. Yang, J.-W. Her, Anal. Chem. 71 (1999) 4690.
[247] M.D.F. Askari, M.P. Maskarine, S.M. Smith, P.M. Beam, C.C. Travis, Anal. Chem. 64 (1996) 3431.
[248] A.C. Roche, C.T. Miller, Fresenius J. Anal. Chem. 232 (1993) 19.
[249] A.P. Bianchi, M.S. Varney, J. Phillips, J. Chromatogr. 542 (1991) 413.
[250] Y. Yokouchi, M. Sano, J. Chromatogr. 555 (1991) 297.
[251] A.K. Krock, C.L. Wilkins, J. Chromatogr. 726 (1996) 167.
[252] A.B. Hewitt, Environ. Sci. Technol. 32 (1998) 143.
[253] A.D. Hewitt, P.H. Miyares, D.C. Leggett, T.F. Jenkins, Environ. Sci. Technol. 26 (1992) 1932.
[254] M.R. Milana, A. Maggio, M. Denaro, R. Feliciani, L. Gramiccioni, J. Chromatogr. 552 (1991) 205.
[255] A. Maggio, M.R. Milana, M. Denar, R. Feliciani, L. Gramiccioni, J. High Resolut. Chromatogr. 14 (1991) 618.
[256] S.G. Pavostathis, G.N. Mathavan, Environ. Technol. 13 (1992) 23.
[257] J.D. Stuart, M.E. Miller, M.L. Williams-Burnett, J. Soil Contam. 6 (1997) 439.
[258] K. Kawata, A. Tanabe, S. Saito, M. Sakai, A. Yasuhara, Bull. Environ. Contam. Toxicol. 58 (1997) 893.
[259] E. Papefstathion, M.D. Lugue, D. Castro, J. Chromatogr. 779 (1997) 352.
[260] K.J. James, M.A. Stack, J. High Resolut. Chromatogr. 19 (1996) 515.
[261] M.E. Barrio, J.L. Liberia, L. Comellas, F. Broto-Puig, J. Chromatogr. 716 (1996) 131.
[262] J. Drodz, J. Novák, J. Chromatogr. 165 (1979) 141.
[263] J. Gan, S. Papiernik, S.R. Yates, J. Agric. Food Chem. 46 (1998) 986.
[264] A. Gonzalevez, G. Garrigues, S. Armenta, M. De la Guardia, Anal. Chem. 82 (2010) 3045.
[265] Y. Yang, W.C. Chen, Int. J. Environ. Anal. Chem. 84 (2004) 1045.
[266] M.A. Hiatt, Int. J. Environ. Anal. Chem. 90 (2010) 591.
[267] R.P. Gambrell, C.N. Reddy, V. Collar, J. Water Pollut. Control Fed. 56 (1984) 174.
[268] J.W. Pankow, J. High Resolut. Chromatogr. Chromatogr. Commun. 9 (1986) 18.
[269] J.W. Pankow, M.W. Rosen, J. High Resolut. Chromatogr. Chromatogr. Commun. 7 (1984) 504.
[270] A.R. Trussel, J.G. Moncur, F.-Y. Lieu, Y.C. Leong, J. High Resolut. Chromatogr. Chromatogr. Commun. 6 (1983) 292.
[271] N. Kirschen, Am. Lab. 16 (1984) 12.
[272] R.G. Westendorf, Performance of a third generation cryofocusing trap for purge and trap gas chromatography, in: Paper Presented at the 39th Pittsburgh Conference, February 1988, New Orleans, LA, Tekmar Company, Cincinnati, OH, 1988.
[273] S. Papillion, W. Haerdi, S. Chiron, Environ. Sci. Technol. 30 (1996) 1822.
[274] J.L. Snyder, R.L. Grob, M.E. McNally, T.S. Oostydk, Anal. Chem. 64 (1992) 1940.
[275] R.J. James, M.A. Stack, J. High Resolut. Chromatogr. Commun. 19 (1996) 515.
[276] M.H. Hiat, Anal. Chem. 67 (1995) 4044.
[277] M.J. Young, J. Pawliszya, J. Microcolumn Sep. 8 (1996) 89.
[278] T.E. Barber, W.G. Fisher, E.N. Wanter, Environ. Sci. Technol. 29 (1995) 1576.
[279] T.L. Likela, K.B. Olsen, S.S. Teel, D.C. Lanigan, Environ. Sci. Technol. 30 (1996) 3441.
[280] M.D.F. Askari, H.P. Maskarinec, S.M. Smith, et al., Anal. Chem. 68 (1996) 3431.
[281] R.L. Grob, Chromatography Analysis of the Environment, Dekker, New York, 1975. 744 pp.

[282] J. Namiesnik, B. Zygmunt, M. Biziuk, et al., Environ. Stud. 5 (1996) 5.

[283] Tecator Soxtec System HT6. A.B. Tecator, Hoganas, Sweden, Tecator Ltd, Bristol, 1996.

[284] M.E. Barrio, J. Liberia, L. Li Comellas, F. Broto-Puig, J. Chromatogr. A 719 (1996) 131.

[285] M. Schnitzer, H.B. Schutler, Adv. Agron. 55 (1995) 167.

[286] United States Environmental Protection Agency, Volatile Organic Compound in Water by Purge and Trap Capillary Column Gas Chromatography-Mass Spectrometry, Method 524.2, 1986.

[287] United States Environmental Protection Agency, Volatile Organic Compounds in Water by Purge and Trap Capillary Column Gas Chromatography with Photoionization and Electrolytic Detectors in Series, Method 502.2 September, 1986.

Further reading

B.M. Abraham, T.Y. Liu, A. Robbat, Hazard. Waste Hazard. 10 (1993) 461.

J. Chiarenzelli, R. Sendator, G. Arnold, et al., Chemosphere 33 (1966) 899.

M.S. Crouch, Chem. Eng. Prog. 86 (1990) 41.

A. Di Domenico, F. Merli, L. Bonforti, I. Camoni, A. Muccio, F. Taggi, et al., Anal. Chem. 51 (1979) 735.

J.L. Ezzell, B.E. Ruchter, W.D. Felix, S.R. Black, J.E. Meikel, LC-GC Int. 13 (1995) 390.

T.M. Fahing, M.E. Paulatic, D.M. Johnson, Anal. Chem. 65 (1993) 1462.

R. Fanco, P.R. Griffiths, Anal. Chem., Rome 82 (1992) 235.

G. Garlucci, L. Airoldi, R. Fanelli, J. Chromatogr. 287 (1984) 425.

F.E. Garner, M.T. Homsher, J.G. Parsons, IASTM STP 925 (1986) 132.

R.T. Hudak, J.M. Melby, J.W. Stave, Proceedings of the 87th Annual Meeting of the Air and Waste Management Association, Paper No. 94, Cincinnati, OH, 19–24 June 1994.

P. Kurán, L. Soják, J. Chromatogr. A 733 (1996) 119.

M. Llompart, K. Li, M.F. Fingas, Microcolumn 11 (1999) 397.

R.F. Lopshire, J.T. Watson, C.G. Enke, Toxicol. Ind. Health 12 (1996) 375.

S. Meyer, S. Cartellieri, H. Steinhart, Anal. Chem. 71 (1999) 4023.

S. Morlaes-Munzo, J.C. Luque-Garcia, M.D. De Castro, Anal. Chem. 74 (2002) 4213.

B.E. Richter, B.A. Jones, J.L. Ezzell, N.L. Porter, N. Avdalovic, C. Pohl, Anal. Chem. 68 (1996) 1033.

S.E. Schwazbach, Private Communication.

L.M. Smith, D.I. Stalling, J.L. Jonson, Anal. Chem. 56 (1984) 1830.

H.Y. Tong, S.J. Monson, M.L. Gross, L.Q. Huang, Anal. Chem. 63 (1991) 2697.

T.K. Tromp, M.K.W. Ko, J.M. Rodriguez, N.D. Sze, Nature 376 (1995) 327.

T.C. Voice, B. Kolb, Environ. Sci. Technol. 27 (1993) 70.

I. Voznakova, J. Podehradska, M. Kohlickova, Chemosphere 33 (1996) 285.

Chapter 16

Determination of insecticides and herbicides in soil

Chapter Outline

16.1 Determination of chlorine containing insecticides and herbicides in soil 303
16.2 Determination of triazine herbicides in soil 307
16.3 Determination of phenoxy acetic acid herbicides in soil 307
16.4 Determination of carbamate type of insecticides in soil 307
16.5 Substituted urea-type herbicides in soils 307
16.6 Determination of imidazolinone herbicides in soils 307
16.7 Determination of organophosphorus-type herbicides in soil 316
16.8 Miscellaneous insecticides in soil 318
16.9 Review of earlier work on the determination of miscellaneous insecticides in soil 332
16.10 Determination of fungicides 332
References 332
Further reading 339

16.1 Determination of chlorine containing insecticides and herbicides in soil

Earlier work on the determination of organochlorine insecticides and pesticides is tabulated in Table 16.1.

More recently Vega et al. [47] have discussed the application of microwave-assisted micellar extraction combined with solid-phase microwave-assisted solid-phase extraction (SPE) and high-performance liquid chromatography (HPLC) with ultraviolet (UV) detection for the determination of organochlorine pesticides in mud samples. This method allows detection limits to be reduced with respect to microwave-assisted micellar extraction and also enables target organochlorine pesticides to be determined in complex matrices due to the clean-up procedure. Several variables affecting the microwave-assisted micellar extraction process were introduced and optimised. A nonionic detergent polyoxyethylene bound on ether polydimethylsiloxane–divinylbenzene fibre on 60 μm polydimethylsiloxane fibre was used for this approach. Optimum conditions provide satisfactory

Determination of Toxic Organic Chemicals In Natural Waters, Sediments and Soils.
DOI: https://doi.org/10.1016/B978-0-12-815856-2.00016-3

TABLE 16.1 Review of earlier published method for the determination of organochlorine insecticides and herbicides in soil.

Organic compounds	Method of extraction from soil	Method of analysis	LD	Reference
Chlorinated insecticides DDT, toxaphene and chlordane	Supercritical fluid extraction with CO_2	GL or GC–MS	–	[1–5]
Dichlorvos endrin, diendrin	Subcritical CO_2 with 3% methanol	GL or GC–MS	–	[2,3,6–12]
Diendrin aldehyde, *p,p′* DDT, mirex and decachlorobiphenyl				
Organochlorine and organophosphorus insecticides	Extraction, clean-up	GL or GC–MS	–	
Dieldrin and aldrin	Extraction studies, ultrasonic extraction			[13,14]
Polycyclic aromatic hydrocarbons	Pressurised liquid extraction	Determination of polycyclic aromatic hydrocarbons in sediments	–	[15,16]
	Hexane extraction	GC/MS		[17–19]
	–	Real-time aerosols MS	–	[20]
	–	Capillary GC	–	[21]
	Sequence solvent extraction	GC	–	[4]
Chlorinated insecticides	Microwave-assisted extraction	–	–	[22]
	Solvent extraction with florisil clean-up	–	–	[23]

Organochlorine insecticides, polychlorinated benzenes, polybrominated biphenyls and polybrominated diphenyl ethers	Various extraction methods	–	–	[24]
Alachlor	Methanol extraction on C_{18} cartridge	–	–	[25,26]
HC11 insecticides	Comparison of extraction methods	–	–	[27]
Metolachlor and alachlor	Miscellaneous	ELISA method	–	[28]
DDT, polychlorinated biphenyls	Temperature–pressure phase diagram for carbon dioxide		–	[29]
Organochlorine insecticides	Solid-phase carbon trap supercritical fluid extraction	Liquid chromatography	–	[30,31]
Polychlorobiphenyls and polychlorodibenzofurans				
Chlorpyrifos and acephate	–	Dissipation studies on soils	–	[32]
Chlorinated insecticides	Carbopack B columns and acetone–toluene extraction (II)	GC	–	[33,34]
Polychlorobiphenyls and chlorinated insecticides, DDT, dieldrin and heptachlor	Separation of these two on alumina columns	GC–MS	–	[35,36]
Miscellaneous organochlorine insecticides	Comparison of extraction solvents	–	–	[21,37]
DDT, dieldrins	Hexane–isopropanol, hexane, acetone, hexane–iso propanol acetone	GC	–	[38]

(*Continued*)

TABLE 16.1 (Continued)

Organic compounds	Method of extraction from soil	Method of analysis	LD	Reference
DDT, dieldrin, endrin, methoxychlor	Florisil column extraction	GC	–	[39]
Dieldrin	–	Reaction with BF_3 then GC	–	[40]
BHC isomers	Light petroleum extraction	GC, TLC	–	[41]
DDT	–	GC	–	[30]
Miscellaneous	–	GC	0.0VOS 0.08 ppm	[33,42]
	Acetone–toluene 1:1 extraction	GC	–	[34]
Ketone, DDT permethrin	–	Study of DDT breakdown	–	[34]
Miscellaneous	Acetone extraction	GC	–	[43]
	Miscellaneous	GC–MS	–	[35]
Dieldrin heptachlor, lindane,	Miscellaneous	GC		[36]
Lindane	–	GC–MS	–	[44]
Miscellaneous	Miscellaneous	GC–MS	–	[45]
Organochlorine insecticides	Microwave-assisted extraction	GC	–	[46]
DDT	–	GLC	–	[30]

precision (relative standard deviation less than 10%), good recoveries (70.78%–117.70%) and detection limits ranging between 28 and 136 ng g^{-1} for the pesticides studied. This method was successfully applied to the determination of target organochlorine pesticides in several kinds of mud samples of different physiochemical characteristics. Microwave-assisted micellar extraction solid-phase microextraction method was also validated and applied to a certified reference material.

The samples were analysed using HPLC-UV. The chromatogram obtained for the mixture of pesticides (0.8 μg g^{-1} for 4,4′-DDD, 4,4′-DDT, 2,4′-DDT and 4,4′-DDE, and 1.6 μg g^{-1} for aldrin and dieldrin) extracted from a spiked Hoya Pozuelo mud sample using microwave-assisted micellar extraction solid-phase microextraction procedure and wall separated peaks were obtained for each of these compounds. It can be seen that the mobile phased used (methanol:water, 84:16 v/v) allows a good separation of analytes and a short analysis time.

16.2 Determination of triazine herbicides in soil

The determination of these compounds is reviewed in Table 16.2.

16.3 Determination of phenoxy acetic acid herbicides in soil

Earlier work on the determination and phenoxy acetic acid herbicide is tabulated in Table 16.3.

16.4 Determination of carbamate type of insecticides in soil

Recently Cohen and Wheals [18] used gas chromatographic methods to determine then carbamate and urea type of insecticides in soils.

Earlier methods are reviewed in Table 16.4.

16.5 Substituted urea-type herbicides in soils

Earlier work on the determination of substituted urea herbicides in soil (see Table 16.5).

16.6 Determination of imidazolinone herbicides in soils

Earlier work on the determination of imidazolinone herbicides in soil is tabulated in Table 16.6.

Recently a method for the determination using imidazolines or using HPLC has been described [11,12,37]. This method utilises a combined soil column extraction and off-line-phase extraction for sample preparation. Analysis was by liquid chromatography–electrospray mass spectrometry

TABLE 16.2 Review of earlier published works for the determination of triazine herbicides in soil.

Compounds	Method of extraction	Analysis	LD	Reference
Terbutylazine and metabolites	Subcritical water extraction	–	–	[48]
Atrazine and cyanazine	Supercritical fluid extraction with methanol-modified carbon dioxide	Supercritical fluid chromatography	–	[49]
Triazine metabolites	Automated solid-phase extraction with methanol in water 4:1 v/v extraction octadecyl resin	–	0.1 μg kg^{-1}	[50]
Atrazine, linuron, metribuzin, simazine, triallate and phorate	Methanol extraction	Specific GC	4 μg kg^{-1}	[51,52]
3-Amino 1,2,4-triazole herbicides	Water leaching	Spectroscopy study of fate of in soil in presence of clay minerals montmorillonite	–	[53]
Atrazine, simazine, atratone, prometryn, desmetryn, megalo protamine, chlormequat and diquat	–	Triazines determined by isotachophoresis	–	[54]
Trifluralin, linuron, fluorochloridone, atrazine, alachlor, metolachlor and pendimethalin	Methylene chloride or ethyl acetate extraction	GC with NP detection		[55,56]
Atrazine, simazine, terbuthylazine and molinate	Solid-phase microwave-assisted extraction with methanol	GC–MS	1–10 ng g^{-1}	[57]
Atrazine, cyanazine, diethylatrazine and metolachlor	Supercritical fluid chromatography	GC and HPLC	–	[58]

Atrazine, simazine, linuron, metribuzin, triallate and phorate	Methanol extraction	GC met electron-capture detector	–	[51–54]
Atrazine	Hexane–acetone extraction	Isotope dilution MS	01–1 ppm	[44]
Atrazine	Acetone–hexane	GC–MS	–	[44]
Cyanazine	–	HPLC		[59]
Atrazine, simazine, propazine, diisopropulatrazine, deethylatrazine, hydroxyl and atrazine	Dynamic microscale solid-phase extraction with toluene	Liquid chromatography	0.02–0.04	[17,60,61]
Cyanazine	–	HPLC with microbore column with assay detection	–	[59]
Atrazine and metabolites	Cyclohexane solid-phase extraction	HPLC with photodiode array detection	–	[62–64]
Triazine	Methanol, C_{18} solid-phase extraction	Gradient C_{18}, HPLC with UV detection	ppb range	[65]
Tributyl azine and	Hot acetone then – absorption on cation exchange cartridges	HPLC with photodiodes	–	[66–68]
Atrazine and metabolites	Microwave-assisted and solid-phase extraction	HPLC with UV detection	–	[3,61,69]

TABLE 16.3 Review of published methods for the determination of phenoxy acetic acid herbicides in soil.

Chemical compound	Method of extraction from soil	Method of analysis	LD	Reference
Phenoxy acetic acid–type herbicide	Effect of types of soil on extraction efficiency in organic solvents	–	–	[15,20,70,71]
MCPA (4-chloro-2-methyl phenoxy acetic acid) and metabolites	Ether–acetone–pentane–hexane 2:1:1:1 extraction	Pentafluorobenzyl derivative	–	[15,72]
2:4 Dichloro phenoxy acetic acid, 2:4:5 trichloro phenoxy acetic acid	–	Derivatised to methyl or 2-chloro ethyl ester – GC	–	[73]
4-Chloro-2-methyl phenoxy acetic acid 2,4 chlorophenoxy acetic acid herbicides	Extraction with dichloromethane	Esterification with 2,4,4,5,6 pentafluorobenzyl bromide then ether extraction – GC	0.5 ng kg^{-1}	[16]

TABLE 16.4 Review of early published literature for the determination of carbamate type herbicides in soil.

Chemical compound	Method of extraction from soil	Method of analysis	LD	Reference
Urea and carbamate herbicides	Chloroform or hexane, acetone (8:1)	Separation of TLC plate followed by GC/MS	1–5 ng kg^{-1}	[18]
M–S Butyl phenyl methyl (phenylthio) carbamate	Dichloromethane chloroform or acetonitrile extraction, florisil cleaning	GC	0.05 ng	[19]
Methomyl	Extraction with dichloromethane and florisil clean-up	GC	0.05 ng kg^{-1}	[75]
Carbaryl and carbofuran	–	Conversion to pentafluoro esters, GC	–	[76]
Oxamyl	Extraction with dichloromethane or acetone florisil clean-up	GC	10 ng kg^{-1}	[77]
Carbosulphan, carbofuran	Extraction with hexane-2-propanol or methanol buffer	GC with N-specific detector	–	[78]
Carbofuran	Extraction with methanol–water (80:20)	GC with N-specific detector	Submicrogram	[79]
Aldicarb and metabolites *N*-methyl carbamate, benfuracarb, carbofuran	Extraction with chloroform	GC/MS	1 ng kg^{-1}	[80]
	Extraction with methylene chloride silica clean-up	HPLC	–	[81–84]

(*Continued*)

TABLE 16.4 (Continued)

Chemical compound	Method of extraction from soil	Method of analysis	LD	Reference
Oxamyl	–	Reactions with copper titration with ethylenediaminetetraacetic acid of excess copper	–	[85,86]
Chlorinated phenoxy acid and ester-type herbicides	–	Isotope dilution GC–MS	ng kg^{-1}	[87]
Chlorinated phenoxy acid and ester-type herbicides	–	Liquid chromatography combined with particle beam MS and UV spectrometry	ng kg^{-1}	[88]
Chlorinated and 2,4 dinitrophenoxy acid herbicides	–	Liquid chromatography particle beam MS	–	[18,19,75–81, 89–92]

TABLE 16.5 Review of earlier published literature for the determination of substituted urea herbicides in soil.

Chemical compound	Method of extraction from soil	Method of analysis	LD	Reference
Diuron and linuron	Supercritical fluid extraction	–	–	[93–97]
Triasulphuron	–	Magnetic particle–based chemiluminescent immunoassay	–	[98,99]
Linuron and metabolites	–	Study of presence of free anilines and other metabolites	–	[100,101]
				[102–107]
Sulphonylurea type	Supercritical fluid extraction	Supercritical fluid extraction coupled to supercritical fluid	–	[108]
	Supercritical fluid extraction with methanol-modified carbon dioxide	HPLC with UV detection or GC of the dimethyl derivatives of sulphonylurea herbicides	–	[109,110]
Substituted urea herbicides	–	GC	1–50 ng kg^{-1}	[18]
	Extraction with acetone	Conversion to anilines – GC	–	[111]
Linuron and metabolites isoproturon, dichlorprop, hexaflumuron, diuron, monuron, chlorocruorin and diuron	–	HPLC	–	[112–118]
Diuron	Extraction with methanol	HPLC or GC	–	[118,119]

(*Continued*)

TABLE 16.5 (Continued)

Chemical compound	Method of extraction from soil	Method of analysis	LD	Reference
Phenyl urea herbicides	–	GC or hydrolysis of urons to corresponding anilines GC calorimetric determination of chromophore	–	[119–126]
Chlorobromuron, diuron, monolinuron, linuron and chloroxuron	–	Pyrolysis liquid chromatography of the phenyl isocyanate derivative using electron-capture detector	–	[119–125,127]
Urea herbicides	–	HPLC or GC	–	[128–132]
	–	Alkylation then GC	–	[102,120,133]
Chlorobromuron, diuron, linuron, monolinuron, monuron, metobromuron	Extraction with methanol	HPLC with UV detection	–	[126]
Substituted ureas and aniline breakdown products	–	Presence of free anilines in soil, derived by decomposition of substituted ureas	–	[103–107,111–114, 122,124]
Chlorobromuron and metabolites	Soxhlet extraction with ethyl acetate derivatives	GLC and TLC of acetyl	–	[134]
Substituted urea types	–	Immunosorbents coupled with liquid chromatography–MS	–	[135]

TABLE 16.6 Review of earlier published literature for the determination of imidazolinone herbicides in soil.

Chemical compound	Method of extraction from soil	Method of analysis	LD	Reference
Imidazolinone	–	Studies of this relatively new class of herbicides	–	[136–143]
Imidazolinone	Carbon dioxide supercritical fluid	Carbon dioxide supercritical fluid chromatography with UV detection	–	[144,145]
Imazapyr, m-imizametha	Extraction with methanol	HPLC – electrospray MS	0.1–0.05 ng pg^{-1}	[146–149]
p-Imazamethabenz, imazamethabenz methyl	Ammonium carbonate microwave-assisted extraction			
Imazapyr	Extraction with methanol	HPLC with UV detection	–	[150]

(MS) in selected ion monitoring mode. Several different extractants were evaluated for the purpose of soil column extraction optimisation. The system that best optimises the extractability of imidazolines from the soil was found to be the mixture of methanol-ammonium carbonate (0.1 M, 50:50 v/v). The total recovery of each imidazolines from soil at each of the two levels investigated ranged from 87% to 95%. Under three selected ion monitoring conditions, the limit of detection (S/N = 3) was found to be 0.1–0.05 ng g^{-1} in soil samples.

Examples of this type of herbicide are imazapyr, *m*-imazamethabenz, *p*-imazamethabenz, *m,p*-imazamethabenzmethyl, imazethapyr and imazaquin. Imazapyr has been determined at the μg kg^{-1} level in 0.1 M ammonium acetate extracts of soil by microwave-assisted extraction using electron capture–negative chemical ionisation MS [149]. HPLC with UV detection at 250 nm has been used to determine imazapyr in methanol extracts of soil [150].

16.7 Determination of organophosphorus-type herbicides in soil

Earlier work on the determination of organophosphorus insecticides in soil is tabulated in Table 16.7.

More recent work is reviewed below.

De Pasquale et al. [100] used soil phase microextraction to determine organophosphorus insecticides adsorption in water and soil matrices. These workers showed that solid-phase microextraction coupled with gas chromatography (GC) enables rapid and simple analysis of organophosphorus pesticides in a range of complex matrices. Investigations were made into the extraction efficiencies from water of six organophosphorus insecticides (methamidophos, omethoate, dimethoate, parathion methyl, malathion and parathion ethyl). These showed a wide range of polarities. Three solid-phase microextraction fibres coated with different stationary phases, polydimethylsiloxane, polyacrylate and carbowax–divinylbenzene were investigated. Water was spiked with the pesticides at concentrations from 1 to 0.1 μg mL^{-1}, and these solutions used for optimisation of the procedure. The carbowax–divinylbenzene fibre, with a 65 μm coating, gave the best performance. The optimised experimental conditions were sample volume 10 mL at 20°C equilibration time 16 minutes, pH5 and presence of 10% w/v sodium chloride. Solid-phase microextraction analyses were performed in solutions obtained by equilibrating aqueous pesticide solutions with six certified soils with various physicochemical characteristics. Solid-phase microextraction data were also assessed by comparison with analyses performed by using conventional SPE. Results indicate the suitability of solid-phase microextraction. Recoveries were in the range 88%–104%.

TABLE 16.7 Review of earlier published literature for the determination of organophosphorus insecticides in soil.

Chemical compound	Method of extraction from soil	Method of analysis	LD	Reference
Methamidophos, omethoate, dimethoate, parathion methyl, malathion and parathion	Solid-phase microextraction	GC	–	[100]
Organophosphorus insecticide	–	Study of adsorption behaviour of organophosphorus or soil and peat	–	[101]
Organophosphorus insecticides	Soxhlet extraction with acetone-*n*-hexane	GC with nitrogen specific detector	95–220 g kg^{-1}	[151]
Trichlorophon	Solvent extraction	GC	50 ng kg^{-1}	[151–154]
Fenophos	–	GC	–	[155]
		Literature review of extraction, clean-up and analysis		[156]

Rotich et al. [101] studied the adsorption behaviour of organophosphorus insecticides in peat and soil samples and their degradation in aqueous solutions at different temperatures and pH values.

Sakellarides et al. [157] studied the photodegradation of selected organophosphorus insecticides under sunlight in different natural waters and soils.

In this work, photodegradation of four organophosphorus insecticides (ethyl parathion, methyl parathion, fenitrothion, fenthion) in difference natural soils was studied under sunlight. The original of the waters was from the region of Ioannina (underground, lake and river water) and from Preveza (sea water) in Northwestern Greece. The soils used had different percentages of organic matter (0.9%–3.5%), and their characterisation was sandy clay, clay and sandy loam. The photodegradation kinetics of these insecticides was followed by GC–MS. The half-lives of the organophosphorus insecticides vary from 0.4 to 35.4 days in natural waters and from 3.4 to 21.3 days in soils. The humic substances and the other components of these environmental matrices seem to influence the degradation kinetics. The use of GC–MS allowed the identification of some important photodegradation by-products such as fenthion, sulphone, fenthion sulphoxide, fenoxon, 4-*O*-methylthio-3,5-dimethyl phenol, *O*,*O*,*O*-triethyl phosphorothioate, paraoxon, 4-nitrophenol and aminoparathion.

Koleli et al. [158] studied the measurement and adsorption of methamidophos in clay loam and sandy loan. Batch sorption experiments showed that the soil texture and methamidophos concentration play a major role in the sorption and migration behaviour of methamidophos.

Methamidophos sorbs onto the clay soil more strongly than onto the sandy loam soil. The equilibrium isotherms for the sorption of methamidophos onto the sandy loam and clay soils were nonlinear and were best described by the Freundlich equation. The results of column experiments indicate that the recovery of methamidophos during desorption was incomplete due to either partially irreversible sorption to high-energy surface sites or strongly rate-limited desorption. Methamidophos was more readily leached out from the sandy loam soil column as consistent with the batch isotherm date.

Organophosphorus insecticides including diazinon, ronnel, parathion ethyl, methiadiathion and trichlorovinphos have been extracted from soil by subcritical carbon dioxide containing 3% methyl alcohol. At a pressure of 35.5 MPa and 50°C, recoveries of 85% were obtained [7].

16.8 Miscellaneous insecticides in soil

Earlier work on the determination of miscellaneous insecticides in soil is tabulated in Table 16.8.

More recent work is described below.

St-Amand and Girard [170] have described a procedure for the determination of acephate and its degradation product methamidophos in soil by

TABLE 16.8 Review of earlier works for the determination of miscellaneous insecticides in soils.

Chemical compound	Method of extraction from soil	Method of analysis	LD	Reference
Triazine	Extraction with methyl acetate	Study of feasibility simultaneous filtration plus liquid chromatography microsystem with GC/MS	–	[159]
Chloropropham, metribuzin, diazinon, dimethoate, fluazinam, aclonifen, azinphos-methyl	–	Study of retention process in organophosphorus insecticides in soil and vegetables	–	[160]
Pyrethroids and mirex	Ultrasonic extraction. Extraction with hexane–acetone, hexane–dichloromethane and isooctane–acetone	Use of GC/MS in study of determination of pyrethroids in soil	–	[161,162]
Flurazon-butyl	Extraction with methanol-hydrochloric acid dichloromethane	Spectroscopy at 225 and 270 nm	–	[163]
Paraquat and diquat	–	Isotachoelectophosoresis, liquid chromatography particle beam MS and UV spectroscopy	–	[54]
Chlorinated phenoxy acetic acid	–		–	[164,165]
Chlormequat	–	Capillary isotachophoresis	–	[54]
Methyl parathion	–	Electroanalytical method	–	[166]
Pyraflufen-ethyl	–	HPLC	–	[167]
Thifensulphuron-methyl	–	Polarography	–	[168]

(Continued)

TABLE 16.8 (Continued)

Chemical compound	Method of extraction from soil	Method of analysis	LD	Reference
Thiazopyr	–	Differential pulse polarography	–	[169]
Methamidophos	Solid-phase extraction	GC/MS	–	[170]
Imazapyr	–	Study of effects of two formulations on the persistence of imazapyr on soils	–	[171]
Artemisinin	Supercritical fluid extraction	HPLC	13 ng kg^{-1}	[172]
Trifluralin	–	Study of photodegradation and determination in soils	–	[173]
Imazapyr	–	Study of degradation and metabolism of imazapyr in soils	–	[174]
Pyrethroids and permethrin	–	GC with negative ion	–	[162]
Cyfluthrin and cypermethrin	–	Chemical ionisation MS		
ʎ-Cyhalothrin, deltamethrin, fenvalerate and mirex	–	Study of adsorption on soil		
Potato alkaloids	–	Study of determination of potato glycol alkaloids in soil	–	[175]
Hexazinone	–	Study of influence of the saturating cation, the sorbent herbicides ratio on the release properties of organoclay-based formulations	–	[176]

Parathion	–	Electroanalytical procedures, square voltammetry	0.4 L mg L^{-1}	[177]
Thifensulphurea, thifensulphuron-methyl	–	Polarographic method	1.05×10^{-7} M	[168]
Pyraflufen-ethyl	Ultrasonic extraction with acetone–water (80:20)	HPLC with UV detector and ion-trap MS	1.6 ng	[167]
Acephate and degradation products methamidophos	Solid-phase extraction with methylene chloride	Degradation studies using studies using GC/MS	–	[169,170]
Various herbicides and insecticides	Sonic-assisted extraction with ethyl acetate	Capillary GC electron-capture detection	0.01 ng g^{-1}	[178]
Imidacloprid	–	Study of factors influencing adsorption and dissipation on soil	–	[179]
Myclobutanil	–	GC/ion-trap MS	0.6 ng kg^{-1}	[180]
Paraquat	Extraction with toluene	GC	–	[181]
Paraquat	–	Continuous flow spectrometry	ng m^{-1} range	[182–184]
Paraquat, trifluralin and diphenamid	–	GC	–	[185]
Paraquat and diquat	Extraction with dichloromethane	GC	–	[186]
Paraquat and diquat	–	Catalytic dehydrogenation then GC	–	[74,187–189]
Paraquat	–	Enzyme-linked immunoassay	0.2 ng kg^{-1}	[190]
Acarol	–	GC of C_{14} herbicide	–	[191,192]
Picloram	Extraction with diethyl ether	Pyrolysis-electron capture GC	–	[87]
Dicamba	Amino propyl weak ion exchange and C_{18} exchange	HPLC	–	[193]

(Continued)

TABLE 16.8 (Continued)

Chemical compound	Method of extraction from soil	Method of analysis	LD	Reference
Bromacil, lenticil and terbacil	Water extraction then chloroform extraction	GC with NP detection	20 ng kg^{-1}	[194]
Bromacil, lenticil and terbacil	Miscellaneous	Miscellaneous	–	[99,195–203]
Fluazifop-butyl, fluazifop	Extraction with methanol, hydrochloric acid dichloromethane	Liquid chromatography	–	[204,205]
Fluazifop-butyl, pydriyloxy	–	GC	–	[206]
Phenoxy propionate	–	–	–	
γ-Fluazifon	–	–	–	
Fluazifop-butyl	Phenyl and cyano bound silica gel solid-phase extraction	Ion pair HPLC	–	[207]
Diclofop methyl, diclofop	–	GC	–	[208,209]
Diclofop methyl, diclofop	Extraction with methanol:water: methyl acetate:acetic acid (40:40:19:1)	Conversion to pentafluorobenzyl bromide derivative: GC	–	[210]
Frenock	Steam distillation toluene extraction	Mass fragmentography of 1-benzyl-3-polytriazine	–	[211]
Glyphosate	–	GC/MS fluorimetric detection with postcolumn oxidation then derivatisation	–	[212]
Cyperquat	–	GC/MS	–	[213]

Norflurazon	Extraction with methanol	C_{18} with fluorescent detection	–	[214]
Propanil	–	IR and GC/MS	–	[215]
Sencor	–	GC	–	[216]
Trifluralin and benefin	–	Electron-capture GC	50 ng	[217]
Miscellaneous herbicides	–	GC/MS	–	[218–220]
Miscellaneous herbicides	–	HPLC	–	[221]
Miscellaneous herbicides	–	GC/MS	–	[222]
Miscellaneous herbicides	–	TLC	–	[223]
Miscellaneous herbicides	–	Enzyme-linked immunoassay	–	[224]
Isomethiozin	–	Differential pulse polarography	–	[225]
Trichlorophon	Solvent extraction	GC	0.0002 ppm	[154]
Bromoxynil	Solvent extraction	Perfluoro acetylation then GC/MS	–	[217–226]
Toxaphene	–	Electron capture–negative spectrometry then HPLC and GC	–	[227]
Dimethoate	–	Spectrophotometric flotation reaction with molybdate and methylene blue and spectrometric finish	–	[228]
Bentazone	–	HPLC with photodiode array detection	–	[229]
Dichlorobenil	Distillation	HPLC	–	[230]
Chlorpyrifos	Supercritical fluid extraction and subcritical water extraction	–	–	[231]

(*Continued*)

TABLE 16.8 (Continued)

Chemical compound	Method of extraction from soil	Method of analysis	LD	Reference
Flumetol	Carbon dioxide supercritical fluid extraction	–	–	[232–235]
Hexazinone and metabolites	–	GC	–	[234]
Chlorpyrifos	–	Enzyme immunoassay	–	[235]
Metoxyl	–	Study of chiral separation		[236]
Imugen, *N*-formyl-*N*-dichlorophenol trichloro acetaldehyde	–	GC	–	[237]
Dicamba	–	GC/MS	–	[87]
Chlorophenoxy acetic acid	Supercritical fluid extraction with methanol-modified carbon	–	–	[238]
Diclofop and diclofop extraction with methanol water:ethyl methyl	Electron-capture GC acetate:acetic acid (40:40:19:1) and back extraction with 5% sodium chloride, 5% sodium carbonate	–	–	[209,239–241]
Diclofop methyl	Methylation with diazomethane	GC	–	[242]
Dacthal	Reaction with 0.4% hydrochloric acid, acetone, diazopropane to convert dacthal to ester	GC	–	[243–249]
Dacthal and metabolites	Supercritical carbon dioxide extraction then hot water extraction	GC	–	[247–250]

SPE followed by GC–MS. Both of these compounds are highly polar organophosphorus pesticides and are therefore highly soluble in water, which leads to difficulties when traditional methods of extraction, such as liquid–liquid extractions, are used. SPE is a relatively new, highly versatile method, which has proven successful in many cases that were considered problematic in the past. In this study, several adsorbents (polymeric and silica based) and parameters are considered and modified to obtain maximum recovery. Maximum recoveries for acephate and methamidophos were found to be 90%–95%, respectively, with Oasis HLB cartridges and methylene chloride as the elution solvent. In order to establish applicability and reliability, the matrix effect of several real water and solid compost and soil samples was evaluated. A 20%–30% diminution of recovery is noted for some samples with a complex matrix containing a high amount of dissolved organic matter.

Syversen and Haarstad [160] carried out a laboratory study to examine the retention process of pesticides through vegetated buffer zones compared to bare soil. Soil columns with low biological activity and vegetation columns with normal biological activity were tested. Pesticides frequently used in vegetable production (namely aclonifen, azinphos-methyl, chloropropham, diazinon, dimethoate, fluazinam, iprodione, linuron, metalaxyl, metamitron, metribuzin and propachlor) equal to 1/50 to 1/5 part of recommended doses, and nutrients equal to 1, 5 and 20 mg nitrogen L^{-1}, were added. The pesticide retention was more than 60% for all pesticides, except dimethoate, with retention of about 30% in columns with low microbial activity. Biological transformation and plant uptake were important for removal of nitrogen and organic matter. Nitrogen retention was high (over 90%) in vegetation columns. Plant uptake and phosphorus content in soil were important for phosphorus retention.

Tadeo et al. [178] described methods for the determination of various herbicides and insecticides belonging to different groups. In soil the method was based on the sonication-assisted extraction in small columns of pesticides using ethyl acetate. All pesticides were determined by capillary GC with electron-capture detection and their identity was confirmed by GC–MS.

Ali and Baugh [161,162] have carried out detailed studies of the determination of pyrethroids in soil. In one study [161], ultrasonic extraction was used to develop a suitable solvent system for the analysis of synthetic pyrethroid pesticides and mirex on soil. The analysis was carried out by GC with negative ion chemical ionisation MS. In the initial experiments, accurately weighed soil samples were spiked with a mixture of standard solution pyrethroids and mirex and shaken for 24 hours to ensure homogeneity, then extracted with solvent. The extracts were evaporated to dryness before the volumetric internal standard was added.

The binary solvents used in this study were various mixtures of hexane: acetone, hexane:dichloromethane, isooctane:acetone, isooctane:dichloromethane, representing different classes of polarity. The recoveries of all pyrethroids and mirex were satisfactory over three solvent systems: hexane:acetone: hexane dichloromethane and isooctane:acetone, but results of isooctane: dichloromethane produced low recoveries. The average recovery increased with the extraction time, but the increase was not statistically significant.

Ali and Baugh [162] also carried out studies on the determination of seven pyrethroids in soils. These were permethrin, cyfluthrin, cypermethrin, λ-cyhalothrin, deltamethrin and fenvalerate and mirex. These were determined in soils possessing a range of organic content (1.15%–2.46%).

Solutions (in deionised water, pH 6.5–7.5) of the samples were shaken using a mechanical shaker for 24 hours. The suspensions were centrifuged and aliquots of clear supernatant were passed through a C_{18} column. The eluates were concentrated to dryness before a volumetric standard was added. The analytes were determined by GC with negative ion chemical ionisation MS either in SIR or SCN mode. Sorption isotherm parameters (n and k) were calculated according to the Freundlich equation. The values n are around unity. Permethrin and cyfluthrin were the least sorbed pyrethroids, with k values less than 2. The effect of the pH on sorption was examined also (at pH values, 2, 4, 6 and 9). Sorption behaviour on different soils and silica was also examined. Desorption studies were conducted on the same pyrethroid solutions. After sorption, the supernatant was replaced with similar volume of deionised water. Desorption was achieved by removing all the supernatant from the centrifuged samples and then replacing it with deionised water. This equilibration process was repeated five times. Each time the suspension was centrifuged concentrated and analysed using GC/MS analysis. The residual amount of pyrethroid in the soil was calculated as the difference between the initial amount and the desorbed amount (mass balance).

Jessing et al. [172] developed a method of extraction of artemisinin in sand, clayey and humic soil samples by supercritical fluid extraction and determination by HPLC. Optimal supercritical fluid extraction conditions were reached using ethanol as modifier at a flow of 0.5 mL min^{-1} and a total extraction time of 20 minutes. The HPLC method had linearity up to greater than 535 mg kg^{-1} for all types, limit of detection was 13 μg kg^{-1} soil and limit of quantification was 43 μg kg^{-1} soil. Recovery for soil samples spiked with artemisinin 1 hour before extraction was determined to be 70%–80%. No matrix effect was observed. The method enabled quantification of artemisinin in three common soil types and was applied for determination of degradation kinetics of artemisinin in spiked soils. Degradation kinetics consisted of an initial fast degradation followed by a slower one. The slower reaction could be fitted by first-order kinetics resulting in rate constants of 0.05, 0.084 and 0.32 per day in sandy, clayey and humic soils, respectively. Both the rate of the fast and slow reaction appeared to increase with soil organic matter content.

Simoes et al. [177] described a fast and simple electroanalysed procedure for the determination of methyl parathion in a solution extracted from a typical Brazilian soil using square wave voltammetry and glassy carbon electrode. The effects of pH, scan rate and surface poisoning were studied in order to establish the optimum conditions for the electroanalysis of methyl parathion. It was observed that the substances commonly present in the soil solution modify the voltammograms, which improves the current values and displaces the peak potential to a less negative value. This was attributed to the more alkaline pH caused by dissolved organic matter, mineral colloids and other substances in the soil solution. The best response was obtained in neutral or in slightly acidic solutions. In such conditions the limits of detection were $0.32\ mg\ L^{-1}$ in pure water and $0.36\ mg\ L^{-1}$ in the soil-extracted solution.

Inam et al. [168] have described a polarograph determination of thifensulphuron-methyl herbicide in soil. The differential pulse polarographic procedure was based on a highly sensitive peak formed due to the reduction of thifensulphuron-methyl on a dropping mercury electrode over the pH range 1.00–10.00 in Britton–Robinson buffer. The polarographic reduction exhibits only a single peak in the pH range pH 3.0 and pH $\leq$ 6.0 and pH 10.0 located at potential values of −1.010, −1.350 and −1.610 V SCE, respectively. The single peak appeared as a maximum of pH 3.0 (−1.010 V) was well resolved and was investigated for analytical use. This peak showed quantitative increments with the additions of standard thifensuphfuron-methyl solution under the optimal conditions, and the cathodic peak current was linearity proportional to the thifensulphuron-methyl concentration in the range of 2×10^{-7}–5×10^{-5} M. The limit of detection and limit of quantification were obtained as 1.05×10^{-7} and 3.50×10^{-7} M, respectively, according to the relation $k \times SD/b$ (where $k = 3$ for limit of detection, $k = -10$ for limit quantification, SD is standard deviation of the blank, and b is the slope of the calibration curve). The method was applied to pesticide formulation and the average percentage recovery was in agreement with that obtained by the spectrophotometric comparison method, namely 97.82% and 106.6%, respectively. The method was extended to determination of thifensulphuron-methyl in spiked soil showing a good reproducibility and accuracy with a relative standard deviation of 4.55 and a relative error of +2.80.

Wang et al. [167] have described a method to analyse pyraflufen-ethyl residues by HPLC. The UV detector was used for routine analysis, and the ion-trap MS was used to confirm the identity of the compound. The residue levels of the pesticides and its dissipation rate in apples and soil in an apple orchard in Beijing were also studied. Primary secondary amine and octadecyl (C_{18}) SPE cartridges were used for the determination of pyraflufen-ethyl residues in applies and soil, respectively. The limit of detection was estimated to be 1.6 ng, and the limit of quantification of pyraflufen-ethyl in the samples was $0.01\ mg\ kg^{-1}$. Average recoveries were between 90.1% and

102.1% at three spiking levels of 0.01, 0.1 and 1 mg kg^{-1}, and relative standard deviations were less than 10% throughout the whole recovery test. A primary secondary amine column was found to provide effective clean-up for apple extract in the determination of pyraflufen-ethyl, and C_{18} could remove the greatest number of sample matrix interference in soil. A dissipation study showed that the half-life obtained for pyraflufen-ethyl in soil was approximately 11.89 days at 1.5 times of the recommended dosage, and no pyraflufen-ethyl residues were detected in apples in harvest.

In this method, soil samples (30 g) passed through a 2 mm sieve and were extracted by ultrasonic extraction with a mixture of acetone–water (80:20, v/v, 2 × 60 mL). The combined extracts were filtered and then concentrated under vacuum with a rotary evaporator at a bath temperature of 50°C until the final volume reached about 10 mL. The resultant mixture was dehydrated by passing through anhydrous magnesium sulphate and eluted with acetone. The eluate was then concentrated under vacuum at 40°C to dryness with a rotary evaporator. The residue of the extracts was redissolved with 3 mL acetonitrile–water (30:70, v/v) and centrifuged for 5 minutes at 5000 rpm for purification by C_{18} cartridges. The C_{18} cartridges were connected to a Visiprep 12-port SPE manifold and conditioned with acetonitrile (5 mL), followed by distilled water (5 mL). The extract (2 mL) was loaded onto the cartridges and passed through at a flow rate of one to two drops per second. The cartridge was washed with acetonitrile–water (3 mL, 50:50, v/v) and then dried with air. The column was eluted with acetonitrile (3 mL), and the elute was dried under a gentle stream of nitrogen. The residue was redissolved in acetonitrile (1 mL) and filtered through a 0.45 μm filter before HPLC-UV determination.

Wang et al. [174] also studied the degradation of metabolites of imazapyr in soils both under aerobic and anaerobic conditions. Lika and Tsiropoulos [251] studied the behaviour of residues in soil after it has been treated with microencapsulate and emulsified formulations.

Wang et al. [171] conducted a laboratory experiment to study the effects of two different formulations (25% Arsenel SL and 5.0% Arsenal G) and doses soil on the persistence was determined of imazapyr in four soils of Zhejiang province, southern eastern China. Based on the first-order kinetic equation, the calculated half-lives of imazapyr in the range 22.0–35.7 days in four soils were 30.9 days (highest) and 24.1 days (lowest). The highest mean half-lives were observed in coastal saline soil, pH 8.78. An increase in soil pH tended to lead to higher persistence of imazapyr in soil. The difference between the mean half-lives, corresponding to 0.5 was not significant, which showed that the different initial application rates had little impact upon degradation of imazapyr. In contrast, a greater impact of the different formulation type upon persistence of imazapyr was observed. Higher persistence was observed with the granular formulation ($t_{1/2}$ = 28.1 days) compared with the liquid formulation ($t_{1/2}$ = 26.2 days) for the lower dose,

TABLE 16.9 Review of earlier publications on miscellaneous insecticides and herbicides in soil.

Chemical compound	Method of extraction from soil	Method of analysis	LD	Reference
2,4D, dicamba, 3,6-dichloropicolinic acid, dichlorprop, picloram, 2,4,5-fenoprop, 2,3-TBA, bromoxynil and ioxynil	Extraction with saturated calcium hydroxide then ethylation with iodomethane and tetrabutyl ammonium hydrogen sulphate liquid–liquid extraction on macroreticular resin column	Electron-capture GC	0.01–0.05 ng g^{-1}	[252]
DDT, ketone and permethrin	–	Study of effect of time on recovery of insecticides analysis by GC	–	[253]
4,4′-DDT, 4,4-DDD, 4,4′-DDE, 2,4′ -DDT, γ-GHCG and ʆ-GHCG, metaphos, phosphamidon and phosalone	–	TLC, GLC	0.5–5 ng kg^{-1}	[254]
2,4D, dicamba, mecoprop	Extraction with acidified acetone then methylation with diazomethane	–	–	[255]
MCPA, MCPB, 2,4D	Extraction with dilute sulphuric acid and diethyl ether – chloroform – acetic acid	–	–	[256]
2,4,5T, 2,4D, dichloroprop, dicamba				
Picloram, 3,6-dichloropicolinic acid	Extraction with saturated calcium hydroxide	–	–	[257,258]

(Continued)

TABLE 16.9 (Continued)

Chemical compound	Method of extraction from soil	Method of analysis	LD	Reference
2,4D	–	Macroreticular resin XAD–2 is an efficient absorber of 2,4D	–	[259,260]
Various herbicides	–	Hydroxybenzo, nitrile herbicides are insufficiently volatile for GC methyl esters using diazomethane on boron – trichloride methanol or iodomethane or suitable derivatisation reagents for GC	–	[255,260–278]
Carbamates, substituted ureas and triazines	–	TLC	–	[279]
Atrazine, barban, diuron, linuron, monuron, simazine, trifluralin, bromoxynil, dalapon, dicamba, MCDB, mecoprop and dichloram	Extraction with chloroform or	TLC	–	[280]
Dichlorvos, diazinon, ronnel, parathion ethyl, methidathion, tetrachlorvinphos, endrin, endrin aldehyde, *p,p′* DDT, mirex and decachlorobiphenyl	Supercritical fluid chromatography with carbon dioxide 3%	Supercritical fluid chromatography	–	[281]

TABLE 16.10 Earlier works on the methods for the determination of fungicides in soil.

Chemical compound	Method of extraction from soil	Method of analysis	LD	Reference
Fenpropimorph and metabolites	Extraction with acetone water	GC with NP detection	–	[282]
Fenoxaprop, fenoxaprop-ethyl	Solvent extraction – florisil clean-up	HPLC with UV detection	–	[283,284]
Dichloro, 1,4-napthaquinone	–	Spectrophotometry	–	[285]
Cyprodinil	–	NMR spectroscopy	–	[286]
Fenhaxamid	–	Study of effect of soil microflora	–	[287]
2,6-Dichloroacetanilide and metabolites	–	Identification of metabolites	–	[288]
Benomyl	–	Study of degradation products in soil	–	[289]
Furaloxy and metalaxyl	Soxhlet extraction with acetone	GC–NP detector	–	[290]
Hexadecylpyridinium cation metalaxyl	–	Study of adsorption on clays	–	[207,213,256,291–299]

which was statistically significant, and an identical trend existed in the higher dose. Three major metabolites were separated by preparative TLC. On the basis of their spectral (IR, LC–MS and ^{1}H NMR), the structure of each compound was deduced and their formation pathway was also discussed.

Artemisinin, a bioactive compound in *Artemisia annua* (sweet wormwood), is used as an active ingredient in drugs against malaria. Cultivation of *A. annua* in field studies implies high amounts of artemisinin produced and potential high losses to soil with impact to vulnerable organisms in soil and leaching to the aquatic environment.

16.9 Review of earlier work on the determination of miscellaneous insecticides in soil

Earlier work on the analysis of multimixtures of insecticides and herbicides in soils is discussed in Table 16.9.

Obviously, soil samples contain not a single insecticide or herbicide but a mixture of these.

16.10 Determination of fungicides

Earlier work on the determination of fungicides in soils is tabulated in Table 16.10.

References

[1] J.M. Wong, Q. Xhi, B.D. Hammock, J.N. Seiber, J. Agric. Food Chem. 39 (1991) 183.
[2] J.L. Snyder, R.I. Grob, M.F. McNally, T.S. Oostdyke, Anal. Chem. 64 (1992) 1940.
[3] J.L. Snyder, R.L. Grob, M.E. McNally, T.S. Oostdyke, J. Chromatogr. Sci. 31 (1993) 183.
[4] J.M. Wong, B.D. Hammock, J.N. Seiber, J. Agric. Food Chem. 39 (1992) 1802.
[5] W.C. Bramley, E. Lateira, V. Kelliher, A. Marcus, D.E. Knowles, J. Liquid Chromatogr. 21 (1998) 1199.
[6] K.E. Novikova, U.S. Hur, Khim Obstich 18 (1972) 562.
[7] M.M. Schanz, S. Bawadt, B.A. Benner, S.A. Wise, S.B. Hawthorne, J. Chromatogr. A 816 (1998) 213.
[8] B.O. Brady, C.C. Kan, E.M. Dooley, Chem. Rev. 26 (1987) 261.
[9] B.K. Cooke, N.M. Western, Analyst 105 (1980) 490.
[10] E.S. Goodwin, R. Goulden, J.G. Reynolds, Analyst 86 (1961) 697.
[11] Stringer, J.A. Pickard, C.H. Lyons, Pest. Sci. 5 (1974) 587.
[12] P.A. Mills, B.A. LaVerna, J.A. Burke, Int. J. Am. Assoc. Off. Anal. Chem. 55 (1972) 39.
[13] R.E. Johnson, R.I. Starr, J. Agric. Food Sci. 20 (1972) 48.
[14] M. Chiba, H.V. Morely, J. Agric. Food Chem. 16 (1968) 916.
[15] M.A. Sattar, J. Passiovirta, Anal. Chem. 51 (1979) 598.
[16] S.H. Waliszewski, G. Szymczynski, Fresenius Z. Anal. Chem. 322 (1985) 510.
[17] W. Boonjob, Y. Yu, M. Hir, M.A. Segundo, J. Wang, V. Cerol, Anal. Chem. 82 (2010) 3052.

[18] I.C. Cohen, B.B. Wheals, J. Chromatogr. 43 (1969) 233.
[19] W.E. Westlake, I. Monika, F.A. Gunther, Bull. Environ. Contam. Toxicol. 8 (1972) 109.
[20] N.H. Gutermann, D.J. Lika, Int. J. Am. Assoc. Off. Anal. Chem. 47 (1964) 353.
[21] J.M. Miller, J. Singh, Bull. Environ. Contam. Toxicol. 16 (1976) 483.
[22] V. Lopez-Avila, R. Young, W.F. Beckert, Anal. Chem. 66 (1994) 1097.
[23] D.E. Kimbrough, R. Chin, J. Wakakuwa, Analyst 119 (1994) 1283.
[24] K. Hartonen, S. Bawadt, S.R. Hawthorne, M.L. Reikkola, J. Chromatogr. 774 (1997) 229.
[25] A.P. Schwab, P.S. Michael, L.S. Sonon, K.B. Hoddinon, T.A. O'Shay (Eds.), Application of Agricultural Analysis on Environmental Studies, ASTM, 1993. **ASTM STP 1162**.
[26] M.A.E. Chirnside, W.K. Ritter, in: K.B. Hoddinot, T.A. O'Shay (Eds.), Application of Agricultural Analysis in Environmental Studies, ASTM, West Conshohocken, PA, 1993, pp. 92–97. ASTM STP 1162.
[27] H. Steinwandter, Fresenius J. Anal. Chem. 327 (1987) 309.
[28] P. Casino, S.M. Marois, R. Puchade, A. Aguiera, Environ. Sci. Technol. 35 (2001) 4111.
[29] B.O. Brady, C.C. Kao, K.M. Dolley, et al., Chem. Res. 26 (1987) 261.
[30] P.H. Gooding, H.G. Philip, H.S. Tawnk, Bull. Environ. Contam. Toxicol. 7 (1972) 288.
[31] B. Von Bavel, M. Jarimo, L. Karlsson, G. Lindstrom, Anal. Chem. 68 (1996) 1279.
[32] L.K. Chai, N. Mohd-Tahir, H.C. Branum-Hansen, Int. J. Environ. Anal. Chem. 88 (2008) 549.
[33] M. Suzuki, Y. Yaommoto, Y. Wanatabe, Nippon Nogei Kagaku Kaishi 47 (1973) 1.
[34] F. Mangani, G. Crescentini, F. Bruaer, Anal. Chem. 53 (1981) 1627.
[35] M. Suzuki, M. Marimoto, J. High Resolut. Chromatogr. 9 (1986) 692.
[36] J. Teichman, A. Bevenu, J.W. Hylin, J. Chromatogr. 151 (1978) 155.
[37] M. Chiba, H.V. Morley, J. Agric. Food Chem. 16 (1968) 911.
[38] K.H. Denberb, Bull. Environ. Contam. Toxicol. 5 (1970) 379.
[39] R.J. Hesselberg, J.L. Johnson, Bull. Environ. Contam. Toxicol. 7 (1972) 115.
[40] D.W. Woodham, C.D. Loftis, C.W. Collier, J. Agric. Food. Chem. 20 (1972) 163.
[41] H. Mahel'ova, M. Sackmanereva, A. Szokolav, J. Kovac, J. Chromatogr. 89 (1974) 177.
[42] M. Suzuki, Y. Yaomotor, Y. Wanatabe, Environ. Sci. Technol. 11 (1977) 1109.
[43] R.B. Grambrell, G.N. Reddy, V. Collard, G. Green, W.H. Patrick, J. Water Pollut. Control 56 (1984) 174.
[44] V. Lopez-Avila, P. Hirata, S. Kooka, M. Flanagan, J.H. Taylor, S.C. Hern, Anal. Chem. 57 (1985) 2797.
[45] S. Pyle, A.B. Marcus, J. Mass. Spectrom. 32 (1997) 897.
[46] M. Barriade-Pereira, M.J. Gonsalez-Castro, S. Muniategui-Lorenzo, P. Lopez-Mahia, D. Proda-Rodregues, E. Fernandez-Fernandez, Int. J. Environ. Anal. Chem. 85 (2005) 325.
[47] D. Vega, Z. Sosa, J.J. Santona-Rodrigues, Int. J. Environ. Anal. Chem. 88 (2008) 185.
[48] Di Corcia, G.D. D'Ascenzo, M. Nazzari, S. Marchere, R. Samperi, Anal. Chem. 71 (1999) 2157.
[49] T.R. Steinheimer, R.C. Pfeffer, K.D. Scoggins, Anal. Chem. 66 (1994) 645.
[50] M.S. Mills, E.M. Thurman, Anal. Chem. 64 (1992) 1985.
[51] E.G. Cotterill, Analyst 104 (1979) 878.
[52] T.H. Byast, E.G. Cottrill, R.J. Hane, Methods for the Analysis of Herbicide Residues, second ed. Technical Report, Agricultural Research Council, Weed Research Organisation UK, No. 15, 1997.
[53] J.D. Russel, M.I. Cruz, J.L. White, J. Agric. Food Chem. 16 (1968) 21.
[54] Z. Stransky, J. Chromatogr. 320 (1985) 219.
[55] Miellet, Ann. Falsif. Expert Chim. 80 (1988) 467.

[56] C. Sanchez-Brunete, L. Martinez, J.L. Tadeo, J. Agric. Food Chem. 42 (1994) 2210.

[57] F. Hernandez, Beltran, F.J. Lopez, J.V. Gasper, Anal. Chem. 72 (2000) 2313.

[58] S. Pappiloud, W. Haerdi, Chromatographia 38 (1994) 514.

[59] F. Sanches-Rasoro, G.C. Dios, J. Chromatogr. 446 (1988) 426.

[60] E.N. Papadakis, E. Papdopoulou-Mourkidon, Int. J. Environ. Anal. Chem. 46 (2006) 573.

[61] S. Fingler, S. Stipicovic, V. Drevenkar, Int. J. Environ. Anal. Chem. 89 (2009) 89.

[62] Y. Xu, W. Lorenz, M. Bahadir, F. Korte, Z. Fresenus, J. Anal. Chem. 325 (1986) 377.

[63] Y. Xu, S. Pappiloud, W. Haerdi, S. Chiro, D. Bardelo, Environ. Sci. Technol. 30 (1986) 1822.

[64] T.R. Steinheimet, J. Agric. Food Chem. 41 (1993) 488.

[65] X. Qiao, R. Durnog, H. Hummel, Medel FAc Haandbound Risksunniv Gent 56 (1991) 949.

[66] R. Shiwes, F.X. Maidl, G. Fishbeck, J. Lschy, A. Von Gleissenhall, Suss, J. Chromatogr. 641 (1993) 89.

[67] R. Schives, S. Whilst, J. Lepschy, F. Von Gleissenhall, B. Asuss Hock, G. Fishbeck, Anal. Lett. 27 (1994) 487.

[68] J.L. Martinez-Vidal, M. Martinez-Galera, Ann. Chim. (Rome) 84 (1994) 177.

[69] R.M. Johnson, F. Hutomann, J. Liq. Chromatogr. 15 (1992) 2941.

[70] G. Yip, J. Chromatogr. Sci. 13 (1975) 225.

[71] R.P. Upchruch, P.D. Mason, Weeds 10 (1962) 9.

[72] F.K. Kawawara, Anal. Chem. B40 (1968) 1069.

[73] L. Renberg, Anal. Chem. 36 (1974) 459.

[74] Z. Niewola, J.P. Benner, H. Swaine, Analyst 111 (1986) 399.

[75] R.G. Reeves, D.W. Woodham, J. Agric. Food Chem. 22 (1974) 76.

[76] J.A. Caburn, B.D. Ripley, A.S.Y. Chau, Int. J. Am. Assoc. Off. Anal. Chem. 59 (1976) 188.

[77] R.H. Bromilow, Analyst, London 101 (1976) 982.

[78] B.C. Leppert, J.C. Markle, R.C. Helt, G.H. Fugie, J. Agric. Food Chem. 31 (1983) 220.

[79] A. Bilikova, A. Kuthan, Vodni Hosposdarstvi Ser. B 33 (1983) 215.

[80] K.K.H. Fung, Pest. Sci. 7 (1976) 571.

[81] B.D. McGarvey, Chromatography 642 (1993) 89.

[82] H.S. Okamoto, D. Wijekoon, C.E. Esperaza, J.C. Change, S.L. Park, ASTM, STP, 1075 Water Test Quality Assurance, vol. 3, ASTM, West Conshohocken, PA, 1992.

[83] L. Lin, W.J.J. Cooper, J. Chromatogr. 390 (1987) 285.

[84] J.R. Huff, J.A. Moore, L. Tomatis, Environ. Health Prospect. 36 (1980) 221.

[85] J.P. Singhal, S.U. Khan, O.P. Bansal, Analyst 103 (1978) 872.

[86] J.P. Singhal, S.U. Khan, O.P. Bansal, J. Agric. Food Chem. 25 (1979) 377.

[87] V. Lopez-Avila, P. Hirta, S. Kraske, J.H. Taylor, J. Agric. Food Chem. 34 (1986) 530.

[88] A. Di Corcia, M. Marchetti, Anal. Chem. 63 (1991) 819.

[89] A. Li, R.J. Magee, B.D. James, Anal. Chim. Acta 255 (1991) 167.

[90] I.S. Kim, F.I. Sasinos, R.D. Stephens, J. Want, M.A. Brown, Anal. Chem. 63 (1991) 819.

[91] E.A. Rochette, J.B. Hill, H.H. Hill Jnr, Talanta 40 (1993) 147.

[92] S.Y. Chau, K. Terry, Int. J. Am. Assoc. Off. Anal. Chem. 58 (1975) 1294.

[93] R.J. Bushway, T.S. Fan, B.F.S. Young, L.R. Paradise, L.,B. Perkins, J. Agric. Food Chem. 42 (1994) 1138.

[94] A. Roda, P. Rauch, P. Ferri, S. Girotti, S. Ghini, G. Carrea, et al., Anal. Chim. Acta 294 (1994) 35.

[95] M. Leistra, A. Dekken, A.M.M. Vanderberg, Water Air Soil Pollut. 23 (1984) 155.

[96] J.F. Lawrence, R.W. Frei, Anal. Chem. 44 (1972) 2046.
[97] P.M.E. McNally, J.R. Wheeler, J. Chromatogr. 447 (1988) 53.
[98] J.M.A. Schlaeppi, A. Kessler, W.E. Enery, J. Agric. Food Chem. 49 (1994) 1914.
[99] A. Benevue, J.H. Ogata, J. Chromatogr. 46 (1970) 110.
[100] C. De Pasquale, A. Jones, A. Charlton, G. Alonzo, Int. J. Environ. Anal. Chem. 85 (2005) 1101.
[101] H.K. Rotich, Z. Shan, Y. Zhao, F. Li, Int. J. Environ. Anal. Chem. 84 (2004) 289.
[102] H. Bieser, K. Grolinmund, J. Assoc. Off. Anal. Chem. 57 (1974) 1294.
[103] E. Grossbard, J.A.P. Marsh, Pest. Sci. 5 (1974) 609.
[104] T.J. Sheets, J. Agric. Food Chem. 12 (1964) 30.
[105] R. Bartha, J. Agric. Food Chem. 19 (1971) 385.
[106] H. Geissbul, in: P.C. Kearney, D.D. Kaufman (Eds.), Degradation of Herbicides, Marcel Dekker, New York, 1971, p. 79.
[107] W.D. Burge, Soil Biol. Biochem. 4 (1972) 379.
[108] P.M.E. McNally, J.R. Wheeler, J. Chromatogr. 435 (1988) 63.
[109] P. Khafferbach, P.J. Holland, J. Agric. Food Chem. 41 (1993) 396.
[110] P. Klafferbach, P.J. Holland, Mass Spectrom. 22 (1993) 565.
[111] J.D. Caverley, R.C. Denney, Analyst 103 (1978) 368.
[112] G. Henze, A. Meyer, J. Hauser, Fresenius J. Anal. Chem. 346 (1993) 761.
[113] A. Khoshab, R. Teasdale, J. Chromatogr. 660 (1994) 195.
[114] A. Liegeois, Y. Dehon, B. Brabant, D. Portetelle, A. Copin, Sci. Total Environ. 113 (1992) 17.
[115] E. Smith, K.A. Hand, J. Chromatogr. 107 (1975) 407.
[116] J.J. Kirkland, J. Chromatogr. Sci. 7 (1969) 7.
[117] A.J. Sidwell, J.H.A. Ruzicka, Analyst 101 (1976) 111.
[118] F.G. Cotterill, Analyst 105 (1980) 987.
[119] C.E. McKone, J. Chromatogr. 44 (1969) 60.
[120] S.U. Kahn, R. Greenhalgh, W.P. Cockrane, Bull. Environ. Contam. Toxicol. 13 (1975) 602.
[121] A. Dabrouchais, Int. J. Am. Assoc. Off. Anal. Chem. 56 (1973) 831.
[122] J.H. Onley, G. Yip, Int. J. Am. Assoc. Off. Anal. Chem. 54 (1971) 1366.
[123] J.J. Kirkland, Anal. Chem. 34 (1962) 428.
[124] H. Lokke, Pest. Sci. 5 (1962) 749.
[125] H.O. Freistead, Int. J. Am. Assoc. Off. Anal. Chem. 57 (1974) 221.
[126] D.S. Farrington, R.G. Hopkins, J.H.A. Ruzicka, Analyst 102 (1977) 377.
[127] A. Spengler, B. Hamroll, J. Chromatogr. 49 (1970) 1070.
[128] J.F. Lawrence, Int. J. Am. Assoc. Off. Anal. Chem. 59 (1976) 1066.
[129] T.H. Byast, J. Chromatogr. 134 (1977) 216.
[130] J. Pribyl, F. Herzel, J. Chromatogr. 166 (1978) 272.
[131] H.J. Jarczyk, Pfanzenschutz-Machr Bayer 28 (1975) 334.
[132] J.F. Lawrence, G.J. Laver, J. Agric. Food Chem. 23 (1975) 1106.
[133] A. Buchert, H. Lokke, J. Chromatogr. 115 (1975) 682.
[134] A. Katz, R.F. Strusz, Bull. Environ. Contam. Toxicol. 3 (1968) 258.
[135] V. Pichon, L. Chen, N. Durand, J. Chromatogr. 725 (1996) 107.
[136] L.D. Shaner, S.L. O'Connor, The Imidazolinone Herbicides, CRS Press, Boca Raton, FL, 1991.
[137] D.L. Shaner, P.C. Anderson, M.A. Stidham, Plant Physiol. 76 (1984) 545.
[138] P.C. Anderson, K.A. Hibert, Weed Sci. 33 (1985) 479.
[139] G. Wehtje, R. Dickens, J.A. Wilcut, B.F. Hajek, Weed Sci. 35 (1987) 858.

[140] R.N. Stoughaard, P.J. Shea, A.R. Martin, Weed Sci. 38 (1990) 67.
[141] M. Che, M.M. Loux, S.J. Trina, T. Logan, J. Environ. Qual. 21 (1992) 698.
[142] M.M. Loux, R.A. Liebl, R.W. Slife, Weed Sci. 37 (1989) 712.
[143] K.A. Renner, W.F. Meggit, D. Penner, Weed Sci. 36 (1988) 78.
[144] K.N. Reddy, M.A. Locke, Weed Sci. 42 (1994) 249.
[145] K.A. O'Bryan, B.J. Brecke, D.G. Shilling, C.L. Colvin, Weed Technol. 8 (1994) 203.
[146] V. Heber, J. Sieber, H.G. Holting, H.J. Vetten, R. Krenzig, M. Bahadire, Ferensice J. Anal. Chem. 360 (1998) 739.
[147] A. Lagana, G. Fago, A. Marino, Anal. Chem. 70 (1998) 121.
[148] A.J. Krynitsky, S.J. Stout, H. Nejad, T.C. Caolier, Int. J. Am. Assoc. Anal. Chem. 82 (1999) 956.
[149] S.J. Stout, A.R. Cunha, D.G. Aladice, Anal. Chem. 68 (1996) 653.
[150] W. Liu, A. Pusino, C. Gessa, Sci. Total Environ. 39 (1992) 123.
[151] J. Kjoholt, J. Chromatogr. 325 (1985) 231.
[152] S.B. Singh, G. Kulshrestha, J. Chromatogr. 637 (1993) 109.
[153] A. Robbat Jnr, C. Liu, T.-Y. Liu, J. Chromatogr. 625 (1992) 277.
[154] M. Devine, J. Agric. Food Chem. 21 (1973) 1095.
[155] S.U. Khan, H.A. Hamilton, E.J. Hague, Pest. Sci. 7 (1976) 553.
[156] K.F. Novikova, Zhur Uges Obstrich 18 (1972) 562.
[157] T.M. Sakellarides, M.G. Siskos, T.A. Albanis, Int. J. Environ. Anal. Chem. 83 (2003) 33.
[158] N. Koleli, C. Kantar, U. Cuvaler, H.H. Yilmaz, Int. J. Environ. Anal. Chem. 86 (2006) 1127.
[159] P.A. Jensen, B.J. Harder, B.W. Strobel, B. Svenmark, H.C. Hensen, Int. J. Environ. Anal. Chem. 87 (2007) 813.
[160] N. Syversen, K. Haarstad, Int. J. Environ. Anal. Chem. 85 (2005) 1175.
[161] M.A. Ali, P.J. Baugh, Int. J. Environ. Anal. Chem. 83 (2003) 909.
[162] M.A. Ali, P.J. Baugh, Int. J. Environ. Anal. Chem. 83 (2003) 923.
[163] M. Negri, A. Cignetti, J. Chromatogr. 387 (1987) 541.
[164] F.K. Kawahara, Environ. Sci. Technol. 5 (1971) 235.
[165] D. Corcia, M. Marchetti, Anal. Chem. 63 (1991) 819.
[166] A. Stransky, Int. J. Environ. Anal. Chem. 82 (2009) 95.
[167] H. Wang, J. Hu, H. Zhany, C. Chen, J.L. Chen, Int. J. Environ. Anal. Chem. 87 (2007) 99.
[168] R. Inam, T. Sariil, E.Z. Gulerman, N. Uncu, Int. J. Environ. Anal. Chem. 86 (2006) 1135.
[169] H. Mercan, R. Inam, Int. J. Environ. Anal. Chem. 88 (2008) 879.
[170] D. St-Amand, L. Girard, Int. J. Environ. Anal. Chem. 84 (2004) 739.
[171] X. Wang, H. Wang, D. Fan, Int. J. Environ. Anal. Chem. 85 (2005) 99.
[172] K.K. Jessing, T. Bowers, B.A. Strobel, B. Svensmark, H.C.B. Hansen, Int. J. Environ. Anal. Chem. 89 (2009) 1.
[173] A.D. Dimou, V.A. Sakkas, T.A. Albanis, Int. J. Environ. Anal. Chem. 84 (2004) 173.
[174] X. Wang, H. Wang, D. Fan, Int. J. Environ. Anal. Chem. 86 (2006) 541.
[175] L. Ramos, J.J. Vreule, U.A.T. Brinkman, Environ. Sci. Technol. 33 (1999) 3254.
[176] R. Celis, G. Facenda, M.C. Hermosin, J. Comejo, Int. J. Environ. Anal. Chem. 85 (2005) 1153.
[177] L.R. Simoes, R. Alves, de Toledo, J.R. Rodrigues, C.M. Petrovaz, Int. J. Environ. Anal. Chem. 89 (2009) 95.
[178] J.L. Tadeo, J. Castro, C. Sandez-Brunete, Int. J. Environ. Anal. Chem. 84 (2004) 29.
[179] L. Cox, M.C. Hermosin, J. Comejo, Int. J. Environ. Anal. Chem. 84 (2004) 95.
[180] X. Liu, F. Dong, X. Wang, Z.L. Zheng, Int. J. Environ. Anal. Chem. 89 (2009) 957.
[181] F. Herzel, J. Chromatogr. 193 (1980) 320.
[182] M. Agudo, A. Rios, M. Valcarcel, Anal. Chim. Acta 281 (1993) 103.

[183] A. Calderbank, O. Yuens, Analyst 90 (1965) 99.
[184] J.D. Pope, J.E. Benner, Int. J. Am. Assoc. Off. Anal. Chem. 57 (1974) 202.
[185] W.R. Payne, J.D. Pope, J.E. Benner, Agric. Food Chem. 22 (1974) 79.
[186] S.L. Khan, J. Agric. Food Chem. 22 (1974) 863.
[187] A. Niewola, S.T. Walsh, G.E. Davies, Int. J. Immunopharm. 5 (1983) 211.
[188] Z. Niewola, C. Hayward, B.A. Swymington, R.T. Robson, Clin. Chim. Acta 148 (1985) 149.
[189] G. Kohler, C. Milstein, Nature 256 (1975) 495.
[190] R.D. Cannizzara, T.E. Cullen, R.T. Murphy, J. Agric. Food Chem. 18 (1970) 728.
[191] S.D. Abbot, R.C. Hall, G.S. Giam, J. Chromatogr. 45 (1969) 317.
[192] R.C. Hall, G.S. Gian, M.G. Merkle, Anal. Chem. 42 (1970) 432.
[193] J.A. Kryszowska, G.F. Vance, J. Agric. Food Chem. 42 (1994) 1693.
[194] J.D. Caverly, R.C. Denney, Analyst 102 (1977) 576.
[195] W.B. Wheeler, N.P. Thompson, B.R. Ray, M. Wilcox, Weed Res. 19 (1971) 307.
[196] U.A. Jolliffe, B.E. Day, R.S. Jordan, J.D. Mann, J. Agric. Food Chem. 15 (1967) 174.
[197] N.G.W. Gutamann, D.J. List, Int. J. Am. Assoc. Off. Anal. Chem. 54 (1971) 975.
[198] H.L. Pease, J. Agric. Food Chem. 17 (1966) 121.
[199] H.L. Pease, J. Agric. Food. Chem. 14 (1966) 94.
[200] S. Gawronski, H. Skapsi, Zesz Nauk Akad Roln Warsz Ogrodnictwo 8 (1974) 59.
[201] D.J. Hamilton, J. Agric. Food Chem. 16 (1968) 152.
[202] F.G. Van Stryke, G.F. Zajecs, J. Chromatogr. 41 (1965) 125.
[203] H.J. Jarcjyk, Pflanzeneuschutz-Machr Bayer 37 (1975) 319.
[204] M. Negre, M. Gennari, A. Cignetti, J. Chromatogr. 387 (1987) 541.
[205] M. Patuni, C. Marucchini, M. Businelli, C. Vischetti, Pest. Sci. 21 (1987) 193.
[206] B.S. Clegg, J. Agric. Food Chem. 35 (1987) 269.
[207] M. Zanco, G. Pfeister, A. Kettrup, Fresenius J. Anal. Chem. 344 (1992) 39.
[208] K. Blaue, G. King, Handbook of Derivatives for Chromatography, Heyden, London, 1978.
[209] J.D. Gaynor, D.C. MacTavish, J. Agric. Food Chem. 29 (1981) 626.
[210] J.D. Gaynor, D.C. MacTavish, Analyst 107 (1982) 700.
[211] T. Tsukioka, S. Shimizu, T. Murakami, Analyst 110 (1985) 39.
[212] K.P. Spann, P.A. Hargreaves, Pest. Sci. 40 (1994) 41.
[213] S.U. Khan, K.S. Lee, J. Agric. Food Chem. 24 (1976) 684.
[214] W.T. Willan, T.C. Mueller, Int. J. Am. Assoc. Off. Anal. Chem. 77 (1994) 752.
[215] M. Oda, M. Yulimoto, O. Zesso, Kenkyu 20 (1975) 12.
[216] D. Prestel, I. Weisgerber, W. Klein, F. Korte, Chemosphere 5 (1976) 137.
[217] G.B. Downer, M. Hall, O.N.B. Mallen, J. Agric. Food Chem. 24 (1976) 1223.
[218] S. Stout, J. Ar DaChuna, M.M. Sarapour, Int. J. Am. Assoc. Off. Anal. Chem. 80 (1997) 426.
[219] A.E. Smith, Int. J. Environ. Anal. Chem. 46 (1992) 11.
[220] H.J. Turin, R.S. Brown, J. Environ. Qual. 22 (1993) 332.
[221] K. Michels, GIT Fachz Lab. 37 (1993) 28.
[222] C. Sanchez-Brunete, A.I. Garcia-Valcarcel, J.L. Tadeo, J. Chromatogr. A 675 (1994) 213.
[223] J. Kovac, J. Tekel, M. Kurucova, Z. Lebensm. Unters. Forsch. 184 (1987) 96.
[224] H.K.M. Bekheit, A.D. Lucas, F. Szurdoki, S.J. Gee, B.D. Hammock, J. Agric. Food Chem. 41 (1993) 2220.
[225] J.F.A. Valentin, R.B. Diez-Caballero, M.A.G. Altuna, Analyst 113 (1988) 629.
[226] C. Sanchez-Brunete, A.O. Carcia, J.L. −Valcarcel, Tadeo, Chromatographia 38 (1994) 624.
[227] F.I. Onuska, K.A. Terry, A. Seech, M. Antonic, J. Chromatogr. A 665 (1994) 125.
[228] J. Vanisha, K.N. Das Ramachandran, V.K. Gupta, Analyst 119 (1994) 1387.
[229] F. Sanchez, M.E. Rasero Baez, C.G. Dios, Sci. Total Environ. 123−24 (1992) 57.

[230] M. Smidt, R. Hamman, A. Kettrup, Int. J. Environ. Anal. Chem. 33 (1988) 1.
[231] M.M. Jimmez-Carmona, J.J. Manclu, A. Montoy, M.D. Luque, deCastro, J. Chromatogr. 785 (1997) 329.
[232] M.A. Locke, J. Agric. Food Chem. 41 (1993) 1081.
[233] Y. Lu, Y. Xue, W. Won, Int. J. Am. Assoc. Off. Anal. Chem. 75 (1992) 1100.
[234] J.C. Fenig, Can. J. Chem. 70 (1992) 1087.
[235] O. Novakova, Chromatographia 39 (1994) 62.
[236] H.R. Buser, D. Müller, T. Poiger, M.E. Balmer, Environ. Sci. Technol. 36 (2002) 221.
[237] N. Sotirion, F. Weisgerber, W. Klein, R. Korte, Chemosphere 1 (1976) 53.
[238] S.B. Hawthorne, D.J. Miller, D.E. Divens, D.C. White, Anal. Chem. 64 (1992) 405.
[239] R. Martens, Pest. Sci. 9 (1978) 127.
[240] A.E. Smith, J. Agric. Food Chem. 27 (1979) 1145.
[241] A.E. Smith, J. Agric. Food Chem. 25 (1977) 893.
[242] A.E. Smith, J. Chromatogr. 129 (1976) 209.
[243] C.R. Worthin (Ed.), The Pesticide Novel. A World Compendium of the Edition, British Crop Protection Council, London, 1983.
[244] H. Gershin, G.W. McClune, Contrib Bayr Thompson Institute 23 (1966) 291.
[245] J.H. Miller, D.E. Keeley, R.J. Thullen, C.H. Carter, Wood Sci. 26 (1976) 30.
[246] J.L. Ran, S. Nicosia, M.M. McChesney, J. Environ. Qual. 19 (1990) 715.
[247] K. Monahan, I.J. Tinsley, G.F. Shepherd, J.A. Field, J. Agric. Food Chem. 43 (1995) 2418.
[248] J.A. Field, K. Monahan, Anal. Chem. 67 (1995) 3357.
[249] A. Wettasingle, I.J. Tinsley, Bull. Environ. Contam. Toxicol. 50 (1993) 226.
[250] J.D. Field, K. Monohan, R. Reed, Anal. Chem. 70 (1982) 1956.
[251] D.T. Lika, N.G. Tsiropoulos, Int. J. Environ. Anal. Chem. 87 (2007) 927.
[252] E.G. Cottrill, Analyst, London 107 (1982) 76.
[253] R.P. Gambrell, C.N. Reddy, V. Callard, et al., Int. J. Water Pollut. Control Federation 56 (1984) 174.
[254] V.N. Kavetkii, L.I. Bublik, G.V. Fuzik, J. Anal. Chem. USSR 42 (1987) 1037.
[255] S.U. Khan, Int. J. Assoc. Off. Anal. Chem. 58 (1975) 1027.
[256] M. Solidadr Andrades, S. Rodriguez-Cruz, M.J. Sanchez-Martin, M. Shancez-Camasano, Int. J. Environ. Anal. Chem. 84 (2004) 133.
[257] T.M. Fahin, M.E. Paulatis, P.M. Johnson, M.E.P. MacNally, Anal. Chem. 65 (1993) 1462.
[258] C.E. McKone, E.G. Cotterill, Bull. Environ. Contam. Toxicol. b 11 (1974) 233.
[259] A.E. Smith, B.J. Hayden, J. Chromatogr. 171 (1979) 482.
[260] E.R. Johnson, T.V. Yu, M.I. Montgomery, Bull. Environ. Contam. Toxicol. 17 (1977) 369.
[261] S.F. Howard, G. Yip, Int. J. Am. Assoc. Off. Anal. Chem. 54 (1971) 970.
[262] L.E. St John, D.J. Lisk, J. Dairy Sci. 50 (1967) 582.
[263] W.H. Gutterman, D.J. Lisk, J. Agric. Food Chem. 11 (1963) 301.
[264] A.P. Thio, M.J. Kornét, S.I.H. Tan, D.H. Tomkins, Anal. Lett. 12 (1979) 1009.
[265] C.E. MrKone, R.J. Hance, J. Chromatogr. 69 (1972) 204.
[266] E.G. Cotterill, J. Chromatogr. 106 (1975) 409.
[267] W.H. Gutenmann, D.J. Lisk, Int. J. Am. Assoc. Off. Anal. Chem. 60 (1977) 1070.
[268] H. Agemian, A.S.Y. Chau, J. Assoc. Off. Anal. Chem. 60 (1977) 1070.
[269] S. Mierzwa, S. Witak, J. Chromatogr. 136 (1977) 105.
[270] E.G. Cotterill, J. Chromatogr. 171 (1979) 478.
[271] A.S.Y. Chau, K. Terry, Int. J. Am. Assoc. Off. Anal. Chem. 59 (1976) 633.
[272] M.A. Satar, M.L. Hattula, M. Lahtipera, J. Paasivirta, Chemosphere 11 (1977) 747.
[273] T.P. Garbrecht, Int. J. Am. Assoc. Off. Anal. Chem. 53 (1970) 70.

[274] C.A. Bache, L.E. St John, D.J. Lisk, Anal. Chem. 40 (1968) 1241.
[275] M. Garle, I. Petters, J. Chromatogr. 140 (1968) 165.
[276] O. Gyllenhaal, B. Naslund, P. Hartgig, J. Chromatogr. 156 (1978) 330.
[277] O. Gyllenaal, H. Ehrsson, J. Chromatogr. 107 (1975) 327.
[278] P. Hartvig, C. Fagerlund, J. Chromatogr. 140 (1977) 170.
[279] D.C. Abbott, P.J. Wagstaff, J. Chromatogr. 43 (1969) 361.
[280] A.E. Smith, A. Fitzpatrick, J. Chromatogr. 57 (1971) 303.
[281] P.C. Kearney, J.R. Plimmer, W.B. Wheller, A. Konston, Pest. Biochem. Physiol. 6 (1976) 229.
[282] H. Dieckmann, M. Stocmaier, R. Kreuzig, M. Bahadir, Fresenius J. Anal. Chem. 345 (1993) 784.
[283] L. Celi, M. Negre, M. Gennari, Pest. Sci. 38 (1993) 43.
[284] A.J. Stäb, M.J.M. Rozing, V. Van, W. Hattum, P. Cofino, U.A.T. Brinkman, J. Chromatogr. 609 (1992) 195.
[285] S.E. Burket, N.Y. Medvedeva, B.G. Ivanov, Anal. Khim. 24 (1969) 264.
[286] J. Dec, K. Haider, A. Benesi, V. Rangaswamy, A. Schäffer, U. Plüken, et al., J. Environ. Sci. Toxicol. 23 (1997) 1232.
[287] D. Borzi, C. Abbâte, F. Martin, L.F. Lahrent, N.E. Azalri, M. Gennari, Int. J. Environ. Anal. Chem. 78 (2007) 949.
[288] N.K. Van Alfen, T. Kosuge, J. Agric. Food Chem. 24 (1976) 584.
[289] R.P. Singh, M. Chiba, J. Chromatogr. 643 (1993) 249.
[290] D.I. Caverley, J. Unwin, Analyst 106 (1981) 387.
[291] M.A. Islam, V. Sakkas, T. Albanis, Int. J. Environ. Anal. Chem. 90 (2010) 357.
[292] D. Borzi, C. Abbate, F. Martin-Lineant, E.E. Ashori, Int. J. Environ. Anal. Chem. 84 (2004) 949.
[293] Y. Merdassa, J.F. Liu, N. Megarsa, H. Tezzenes, Int. J. Environ. Anal. Chem. 99 (2015) 225.
[294] H. Hong, Y. Yang, Q. Huang, X. Min, Int. J. Environ. Anal. Chem. 914 (2014) 619.
[295] S. Sharma, D. Singh, Int. J. Environ. Anal. Chem. 94 (2014).
[296] D.S. Kasyakov, N.V. Ul'yanovskii, S.A. Pokryshkin, D.E. Lakhmanov, O.A. Shpigun, J. Environ. Anal. Chem. 95 (2015) 1321.
[297] D. Han, D. Wang, P. Liu, W. Fang, D. Yan, Int. J. Environ. Anal. Chem. 96 (2016) 694.
[298] J. Cmelik, Z. Nainarovk, P. Rysanek, J. Environ. Anal. Chem. 95 (2015) 1090.
[299] I. Iokubauskaite, K. Aneleviciute, V. Hepane, A. Stepetiene, J. Slepetys, I. Lianuanskiene, et al., Int. J. Environ. Anal. Chem. 95 (2015) 508.

Further reading

D.C. Abbott, H. Egan, E.W. Hammond, J. Thomson, Analyst 89 (1964) 480.
R.J. Bushway, S.P. Perkins, S.A. Savage, B.S. Ferguson, Bull. Environ. Contam. Toxicol. 40 (1988) 647.
A. Dankwardt, B. Hock, GIT Fachz Lab. 37 (1993) 839.
A. Dankwadt, J. Seifert, B. Hock, Acta Hydrochim. Hydrobiol. 21 (1993) 110.
T.S. Lawruk, C.E. Lachman, S.W. Jourda, J.R. Fleeker, D.P. Herzog, F.M. Rubio, J. Agric. Food Chem. 41 (1993) 747.
V. Lopez-Avila, C. Charan, W.F. Beckett, Trends Anal. Chem. 13 (1994) 118.
M. Martinez-Galera, J.L. Martinez-Vidal, A. Garrido, Frenisch Anal. Lett. 27 (1994) 807.
Z. Stransky, J. Chromatogr. 320 (1983) 219.
C. Wittman, R.D. Schmidt, J. Agric. Food Chem. 42 (1994) 1041.

Chapter 17

Determination of organometallic compounds in soils

Chapter Outline

17.1 Organoarsenic compounds	341
17.2 Organolead compounds	345
17.3 Organotin compounds	345
17.4 Organomercury compounds	345
References	345
Further reading	347

17.1 Organoarsenic compounds

Earlier work on the determination of organoarsenic compounds in soils is reviewed in Table 17.1.

More recently, Barshick et al. [4] evaluated glow discharge mass chromatography–mass spectrometry for total element assays in soil [4]. Glow discharge mass spectrometry is of limited value for volatile elements such as arsenic or when the element is not an inorganic salt but is a volatile organometallic compound. A solid-phase microextraction fibre was shown to be an effective sampling medium for several organometallic compounds.

Dithiol derivatisation with solid-phase microextraction and gas chromatography–mass spectrometry has been used to determine organoarsenic compounds in soil [5].

Arsenic specks have been determined in soil using inductivity coupled plasma mass spectrometry coupled with secondary ion mass spectrometry and by ion exclusion chromatography coupled with plasma mass spectrometry [6].

Naidu et al. [7] showed that the separation of arsenic species from soil solutions could be performed in less than 5 minutes using capillary electrophoresis. The detection limit is 0.1–0.5 mg L^{-1}.

Determination of Toxic Organic Chemicals In Natural Waters, Sediments and Soils.
DOI: https://doi.org/10.1016/B978-0-12-815856-2.00017-5

TABLE 17.1 Review of early publications on the determination of organoarsenic compounds in soil.

Chemical compounds	Method of extraction from soil	Method of analysis	LD	References
Organoarsenic compounds	–	GC with hydride	–	[1]
		Generation with sodium borohydride		
Hydroxy dimethylarsenic	–	Conversion to iodo-dimethyl arsenic and GC	–	[2]
Monosodium methane arsenic acid	–	Study of metabolism of organoarsenic compounds in soil	–	[3]
Monosodium methane arsenic acid, arsenate and arsenile	–	TLC study of the effect of organoarsenic compounds on soil microbial growth	–	[3]
Organometallic compounds total arsenic	–	Glow discharge MS	–	[4]
Organoarsenic compounds	Solid-phase microextraction coupled with plasma MS	GC/MS	–	[5,6]
Organoarsenic species	–	Capillary electrophoresis	$0.1-0.5 \text{ mg L}^{-1}$	[7]
Volatile organoarsenic compounds	–	HPLC	–	[8]

TABLE 17.2 Review of early publications on the determination of organotin compounds.

Chemical compounds	Method of extraction from soil	Method of analysis	LD	References
Organotin compounds	Purge and trap or liquid–liquid extraction	Conversion to volatile hydrides then AAS, GC with flame photometric detection		[10–30]
Methyltin compounds	Purge and trap extraction	Conversion tin hydrides with sodium borohydride then purge and trap analysis – GC with mass spectrometric detection	–	[31]
Mono, di, tri and tetra-alkyltins compounds	–	GC coupled with helium microwave emission spectrometry	10–30 mg kg^{-1}	[32]
Butyl and phenyltin compounds	Supercritical fluid extraction of derivatisation products	–	–	[33]
Methyltin, butyltin and inorganic	Extraction with calcium chloride hydrochloric acid	Hydride generation AAS	0.6–2.2 ng kg^{-1}	[34]
Organotin compounds	Extraction with tributyl phosphate	AAS	–	[35,36]

TABLE 17.3 Review of early publications on the determination of organomercury compounds in soils.

Chemical compounds	Method of extraction from soil	Method of analysis	LD	References
Organomercury fungicides	–	Study of decomposition in soil	–	[37–39]
Organomercury compounds	–	Study of sorption and inactivation of organomercury compounds by soil microorganism	–	[40,41]
Methyl and ethyl mercury compounds	–	Spectrophotometric method using dithizone	–	[42–44]
Diphenyl and alkyl mercury compound	–	Spectrometric with dithizone	0.1–0.5 ng kg^{-1}	[45,46]
Organomercury compounds	–	Bomb digestion in the presence of mineral acids	–	[47]
Alkyl mercury compounds	–	GC	–	[48,49]
Organomercury compounds	–	AAS	–	[50]
Organomercury compounds	–	Soil digestion with potassium dichromate nitric acid perchloric acid sulphuric acid	–	[51]
Organomercury compounds	–	Digestion of soil with aqua regia and potassium permanganate	–	[52,53]
Organomercury compounds	–	Digestion of soil with sodium hydroxide–stannous chloride	–	[54]
Organomercury compounds	Extraction of organomercury compounds from soil with glutathione	–	–	[55–58]

17.2 Organolead compounds

Blais et al. [9] determined alkyl lead salts in soil. They demonstrated that previously published methods gave poor recoveries of lead and the formation of artefacts during the isolation and derivatisation procedures. An alternative procedure is described involving a series of selective extractions of tetra-alkylleads, ionic alkylleads and inorganic ionic lead salts from soils and street dust. Alkyllead salts were selectively extracted complexometrically from samples containing up to 1000 mg ionic lead per kg. The extracts were then butylated and analysed by gas chromatography–atomic absorption spectroscopy. Reextraction of the sample with methyl isobutyl ketone–dithizone permitted the recovery of ionic lead. In the sample tested, ethyllead salts were detected, but not methyllead salts. Concentrations of these analytes were significantly correlated with levels of extractable ionic lead, but with total lead.

17.3 Organotin compounds

Earlier work on the determination of organotin compounds is reviewed in Table 17.2.

More recently, Cal et al. [33] has described an in situ method for the determination of organotin compounds in soil.

17.4 Organomercury compounds

Earlier methods for the determination of organomercury compounds in soil are reviewed in Table 17.3.

References

[1] Y. Odanake, N. Tsuchly, S. Goto, Anal. Chem. 55 (1983) 929.
[2] C.J. Soderquist, D.G. Crosby, J.B. Bowers, Anal. Chem. 46 (1974) 155.
[3] D.W. Von, P.C. Dent Kearney, D.D. Kaufman, J. Agric. Food Chem. 16 (1968) 17.
[4] C.M. Barshick, S.-A. Barshick, P.F. Britt, D.H. Smith, Rapid Commun. Mass Spectrom. 10 (1996) 341.
[5] B. Szostek, J.H. Aldstadt, J. Chromatogr. A 807 (1998) 253.
[6] G. Koellensperger, J. Nurmi, S. Hanns, G. Stingeder, W.J. Fitz, W.W. Wenzel, J. Anal. At. Spectrom. 17 (2002) 1047.
[7] R. Naidu, J. Smith, R.G. McLaren, D.P. Stevens, M.E. Sumner, P.E. Jackson, J. Soil Sci. Soc. Am. 64 (2000), 122.
[8] P. Thomas, J.K. Finnie, J.G. Williams, Anal. At. Spectrom. 12 (1997) 1367.
[9] J.S. Blais, M.G. Memplasir, W.P. Marshall, Anal. Chem. 62 (1990) 111.
[10] V.F. Hodge, S.L. Snider, D. Goldberg, Anal. Chem. 51 (1979) 1256.
[11] M.O. Andrea, J.T. Byrd, Anal. Chim. Acta 156 (1984) 147.
[12] Y.K. Chau, P.T.S. Wong, G.A. Bengert, Anal. Chem. 54 (1982) 946.

[13] HMSO, Determination of Organic, Inorganic and Total Tin Compounds in Waters, London, 1992.
[14] P. Kenis, A. Zirino, Anal. Chim. Acta 149 (1983) 157.
[15] Y. Arakawa, O. Wadd, M. Wanatabe, Anal. Chem. 55 (1983) 1901.
[16] W. Langseth, Talanta 31 (1984) 975.
[17] T.-H. Yu, Y. Arakawa, J. Chromatogr. 258 (1983) 189.
[18] K.L. Jewett, F.E. Brinkman, J. Chromatogr. Sci. 19 (1981) 583.
[19] S. Tugrul, T.I. Balkas, E.G. Goldberg, Mar. Pollut. Bull. 14 (1983) 297.
[20] J.A. Jackson, W.R. Blair, F.E. Brinkman, W.P. Iverson, Environ. Sci. Technol. 16 (1982) 110.
[21] O.F.X. Donard, S. Rapsomanikis, J.H. Weber, Anal. Chem. 54 (1986) 772.
[22] C.L. Matthias, J.M. Bellama, G.J. Olson, F.E. Brickman, Environ. Sci. Technol. 20 (1986) 609.
[23] A. Woollins, W.R. Cullen, Analyst 109 (1984) 1527.
[24] C.J. Soderquist, D.G. Crosby, Anal. Chem. 50 (1978) 1435.
[25] Y.K. Chau, P.T.S. Wong, G.A. Bengert, Anal. Chem. 54 (1982) 246.
[26] H.A. Meinema, T. Burger-Wiersma, G. Hann Verhluis-de, E.G. Gevers, Environ. Sci. Technol. 12 (1978) 288.
[27] B. Zimmerli, H. Zimmermann, Fresenius Z. Anal. Chem. 304 (1980) 23.
[28] R.J. Maguire, Environ. Sci. Technol. 18 (1984) 291.
[29] R.J. Maguire, R.J. Tkacz, Y.K. Chau, et al., Chemosphere 15 (1986) 253.
[30] E. Möllhoff, Pflanzenschutz-Machr 30 (1977) 249.
[31] S.A. Sinex, A.Y. Cantillo, G.R. Helz, Anal. Chem. 52 (1980) 2342.
[32] R. Lobinski, W.M.R. Dirlex, F.C. Adams, Anal. Chem. 64 (1992) 159.
[33] Y. Cal, R. Aizaga, J.M. Bayona, Anal. Chem. 66 (1994) 1161.
[34] L. Randall, J.S. Hans, J.H. Ucher, Environ. Technol. Lett. 7 (1988) 471.
[35] H. Berling Gong, K. Matsumoto, Anal. Chem. 68 (1996) 2277.
[36] H. Li, B. Gong, K. Matsumoto, Anal. Chem. 68 (1976) 2277.
[37] J.R. Booer, Ann. Appl. Biol. 31 (1944) 34.
[38] A.E. Hitchcock, P.W. Zimmerman, Ann. N.Y. Acad. Sci. 65 (1957) 474.
[39] P.W. Zimmerman, W. Crocker, Contrib. Boyce Thompson Inst. 6 (1934) 167.
[40] H. Keissling, Svenski Papperstid 64 (1961) 689.
[41] W.C. Spanis, D.E. Munnercke, R.A. Solberg, Phytopathology 52 (1962) 455.
[42] Y. Kimura, V.L. Miller, Anal. Chem. 32 (1960) 420.
[43] Y. Kimura, V.L. Miller, Anal. Chem. 34 (1962) 325.
[44] Y. Kimura, V.L. Miller, J. Agric. Food Chem. 15 (1964) 253.
[45] D. Polley, V.L. Miller, Anal. Chem. 27 (1955) 1162.
[46] P.C. Leong, H.O. Ong, Anal. Chem. 43 (1971) 940.
[47] E.W. Breehauer, A.A. Moghissi, S.S. Snyder, N.W. Matthews, Anal. Chem. 46 (1974) 445.
[48] K.K.S. Pillay, C.C. Thomas, C.J.A. Sonde, C.M. Hyone, Anal. Chem. 43 (1971) 1419.
[49] J.E. Longbottom, R.C. Dressman, J.J. Litchenberg, J. Assoc. Off. Anal. Chem. 56 (1973) 1297.
[50] F.J. Lonngmhyr, J. Admondt, Anal. Chim. Acta 87 (1976) 483.
[51] C. Feldman, Anal. Chem. 46 (1974) 99.
[52] J.N. Bishop, L.A. Taylor, B.O. Nearby, Determination of Mercury in Environment, US Environmental Protection Agency, Cincinnati, OH, 1975, p. 120.

[53] Environmental Protection Agency, Methods for Chemical Analysis of Water and Wastes, US Environmental Protection Agency, Cincinnati, OH, 1974, p. 134.
[54] U. Umezaki, K. Iwamoto, Jpn. Anal. 201 (1971) 173.
[55] M. Adinarayana, U.S. Singh, T.S. Dwivedi, J. Chromatogr. 435 (1988) 210.
[56] R.A. Lucero, M.A. Otieno, L. May, G. Eng, Appl. Org. Chem. 6 (1992) 273.
[57] K. Matsunga, S. Takahashi, Anal. Chim. Acta 87 (1976) 487.
[58] F.J. Langmyhr, J. Aamodt, Anal. Chim. Acta 87 (1976) 483.

Further reading

D.H. Anderson, J.H. Evans, J.J. Murphy, W.W. White, Anal. Chem. 43 (1971) 1151.
W.B. Wright, M.L. Lee, G.M. Booth, J. High Resolut. Chromatogr. Chromatogr. Commun. 1 (1979) 189.

Chapter 18

Determination of organic compounds in sediments

Chapter Outline

18.1 Determination of concentration of typical organic compounds in sediment 349
18.2 Determination of hydrocarbons in sediments 349
18.3 Oxygen-containing compounds 349
18.4 Chlorine-containing compounds 351
18.5 Nitrogen-containing compounds 357
18.6 Phosphorus-containing compounds 358
18.7 Sulphur-containing compounds 358
18.8 Insecticides and herbicides 359
18.9 Miscellaneous organic compounds 362
References 366
Further reading 370

18.1 Determination of concentration of typical organic compounds in sediment

Typical concentrations of organic compounds in sediments are reviewed in Table 18.1.

18.2 Determination of hydrocarbons in sediments

Aliphatic hydrocarbons, see Table 18.2.

Polycyclic aromatic hydrocarbons, see Table 18.3.

18.3 Oxygen-containing compounds

Determination of oxygen-containing compounds in sediments is reviewed in Table 18.4.

Ambe and Hanya [45] have combined the Longwell and Maniece [46] methods using methylene blue with the infrared (IR) spectroscopic method of Sallee [47] to devise a method for the determination of alkylbenzene sulphonates. Methylene blue alkylbenzene sulphonate complexes give absorption peaks at 890 and 1010 cm^{-1}, the ratio of the heights being proportional to the ratio of the amount of sulphonate to the total amount of methylene blue sensitive substances in the complex.

Determination of Toxic Organic Chemicals In Natural Waters, Sediments and Soils.
DOI: https://doi.org/10.1016/B978-0-12-815856-2.00018-7

TABLE 18.1 Organic compounds in river, lake and marine sediments.

Compound	Type of sediment	Concentration (mg kg^{-1})	References
Aromatic hydrocarbons		0.001–3	[1]
1,3-Dihexachloro butadiene		0.05	[2,3]
1,3,5-Trihexachloro butadiene		0.25	[2,3]
1,2,4-Trihexachloro butadiene		0.07	[2,3]
1,2,3-Trihexachloro butadiene		0.10	[2,3]
1,2,3,5-Tetrahexachloro butadiene		0.01	[2,3]
1,2,3,4-Tetrahexachloro butadiene		0.27	[2,3]
Pentachlorobutadiene		0.15	[2,3]
Hexachlorobutadiene		1.2	[2,3]
1,3-Dichlorobenzene		0.031	[2]
1,4-Dichlorobenzene		0.081	[2]
1,3,5-Trichlorobenzene		0.004	[2]
1,2,4-Trichlorobenzene		0.020	[2]
1,2,3,5-Tetrachlorobenzene		0.004	[2]
1,2,4,5-Tetrachlorobenzene		<0.001	[2]
Chlorobenzene		0.003–0.07	[2]
Perchlorobenzene		0.004	[2]
Hexachlorobenzene		0.007	[2]
Diethylhexylphthalate		0.1–70.05	[4]
Dibutyl phthalate		<0.1–15.5	[4]
Alkylbenzene sulphonates		16.9–96.3	[5]
Methylene blue active substances		107–288	[5]
Nitrogen-containing aromatics		200–1200	[6,7]
Fluorescent whitening agents		0.25–1.35	[5]
Total organic carbon		2.4–65.6	[3,8]
		1.4–6.2	[9]
Total phosphorus		610–1870	[10]
Total sulphur		229	[11,12]

TABLE 18.2 Determination of aliphatic hydrocarbons in sediments.

Chemical compound	Method of extraction from sediment	Method of analysis	LD	Reference
Petroleum	Hydrocarbon	Fingerprinting of petroleum hydrocarbon	–	[13]

This method has been applied to bottom sediments and muds. The mud sample is centrifuged to separate the water, dried at the room temperature, ground and sieved. This residue is extracted for 1 hour at 80°C with methanol–benzene (1:1), the extraction is repeated twice, and the combined extracts are evaporated and the residue dissolved in water. Alkylbenzene sulphonates are then determined by IR spectroscopy.

18.4 Chlorine-containing compounds

Typical concentration of chlorine-containing compounds in sediments is reviewed in Table 18.5.

More recently Hollies et al. [89] have carried out an extensive study on the determination of chlorinated long chain ($C_{13}-C_{20}$) normal paraffins (Cerechlors) in river sediments. They considered liquid chromatography, gas chromatography (GC) and thin-layer chromatography. Chlorinated paraffins are separated from the sediment by Soxhlet extraction with petroleum ether. A concentrate of the extract is then cleaned up on an alumina column that adsorbs these compounds, allowing impurities to pass through. The chlorinated paraffins are then desorbed with toluene. Analysis of the extracts is carried out by thin-layer chromatography on silica. The plate is developed by covering with a second plate coated with alumina and heating face to face at 240°C. The alumina plate is then sprayed with silver nitrate to visualise the separated chloroparaffins as grey-black spots. Any chloroparaffins present in the extract are then identified by reference to the *R* values, which are approximately 0.74 and 0.80 for $C_{18}-C_{17}$ and $C_{20}-C_{30}$ chloroparaffins, respectively.

Chlorinated hydrocarbons that have been determined in extracts of river sediments by GC include higher chlorinated aromatic hydrocarbons and polychlorobiphenyls. Hawthorne [90] compared supercritical chlorophenols, nitrous oxide and carbon dioxide for the extraction of polychlorobiphenyls from sediments.

Murray and Riley [91,92] and Novak et al. [93] described a gas chromatographic method for the determination of trichloroethylene, tetrachloroethylene, chloroform and carbon tetrachloride in sediments. These sediments were separated and determined on glass column (4 m × 4 mm)

TABLE 18.3 Determination of polycyclic aromatic compounds in sediments.

Chemical compound	Method of extraction from sediment	Method of analysis	LD	References
Polycyclic aromatic hydrocarbons	–	Flame ionisation capillary GC	–	[14,15]
	–	HPLC with UV detection	0.1–1.0 ng kg^{-1}	[16]
	–	Supercritical fluid chromatography	–	[17]
	–	High resolution Shpol'skii spectrofluorimetry at 10K	–	[18]
	–	High-resolution spectrofluorimetry	–	[19]
	Solvent extraction	UV spectroscopy	–	[20]
	–	HPLC	–	[21]
	–	Review of application of GC/MS to sediment analysis	–	[22,23]
	–	Study of oil pollution of sedimentary GC	–	[24,25]
Polycyclic aromatic hydrocarbons in kerosene	–	Capillary GC	–	[26]
Polycyclic aromatic hydrocarbon	–	Fluorescence spectrometry	–	[27]

TABLE 18.4 Review of earlier published work for the determination of types of oxygen-containing compounds in sediments.

Chemical compound	Method of extraction from soil	Method of analysis	LD	References
Phenols	Extraction from sediments	–	–	[28,29]
4-Nonylphenol	Pressurised liquid extraction	–	–	[30]
4-Nonylphenol and nonylphenol mono and diethoxylates	Ultrasonic assisted extraction comparison of different solvents	HPLC with fluorescence detection, also GC/MS	–	[31]
Lignophenols	–	Reverse HPLC used for 14C dating	–	[32,33]
Carboxylic acids	Liquid–liquid extraction with various solvents	GLC	–	[34–38]
	–	Saponification and extraction	–	[39]
	–	Determination of carboxylic acids in sediment core	5 mol L^{-1}	[40]
Short-chain organic acids	–	Ion chromatography	–	[41]
Aldehydes and ketones	Extraction with 3×10^{-1} M 2,4-dinitrophenyl hydrazine in 60% acetonitrile to 40% water	–	0.009 ng M^{-3}	[42]
Phthalate ester: 2-ethylhexyl phthalate and di-*n*-butyl phthalate	–	HPLC	–	[43]
Linear alkylbenzene sulphonates	–	Methylene blue: IR spectroscopy	0.5 mg kg^{-1}	[44]

TABLE 18.5 Determination of halogen-containing compounds in sediments.

Chemical compound	Method of extraction from soil	Method of analysis	LD	References
Volatile haloparaffins	Removal in a closed-loop system collected on Poropak N, then elution with methanol	GC with electron-capture and photoionisation detection	7 ng L^{-1} or 1 ng g^{-1}	[48,49]
Volatile chloroparaffins	–	Chloroparaffins reduced to normal hydrocarbons, then with microcoulometric detection	–	[50]
Volatile chlorinated hydrocarbons	–	Review of application of chlorinated hydrocarbons in sediments	–	[51]
Volatile chlorinated hydrocarbons	–	Flame ionisation electron-capture and Coulson electrolytic detectors with GC	–	[52]
Volatile chlorinated hydrocarbons	–	GC/MS	–	[53,54]
Chloroform trichloroethylene chlorobenzene	Purge and trap analysis			[55]
Volatile chlorinated hydrocarbons		Column chromatography with microcoulometric detection	–	[56]
Nonvolatile monochlorodifluoromethane	–	General discussion of methodology	–	[57–59]
				[58,59]
				[60–63]
				[64–68]
				[69,70]

Polychlorinated biphenyls	–	Study degradation by aerobic bacteria	–	[59,71]
Polychlorinated biphenyls organochlorine pesticides and chlorobenzenes	–	GC with electron-capture detection	–	[72,73]
Polychlorinated biphenyls	–	GC/MS	–	[74]
Polychlorinated biphenyls	–	Pyrolytic desorption at 1000°C, then GC/MS	10 mg	[61]
Polychlorinated biphenyls	–	Thermal desorption GC/MS	–	[66]
Polychlorinated biphenyls	–	Segmentation tandem GC/MS	–	[71,75,76]
Polychlorinated dioxins	–	Electron-capture GC	–	[77–79]
Octachloro dibenzo-*p*-dioxin	–	Electron-capture GC	–	[70,80]
Polychlorinated dibenzo-*p*-dioxin and chlorinated dibenzo furan	–	GC and GC/MS	–	[81,82]
Polychlorinated dibenzo-*p*-dioxin	Pressurised extraction with water	Study of affected various extraction solvents	–	[73,83]
Polychlorinated dibenzo-*p*-dioxins polychlorinated	–	Miscellaneous studies on the determination of dibenzo-*p*-dioxins	mg kg^{-1}	[39,84–88]

packed with 3% of SE-52 on Chromosorb W (AW DMCS) (80–100 mesh) and operated at 35°C, with argon (30 mL min^{-1}) as carrier gas. An electron-capture detector was used, with argon–methane (9:1) as quench gas. Chlorinated hydrocarbons were stripped from water samples by passage of nitrogen and removed from solid samples by heating in a stream of nitrogen. In each case the compounds were transferred from the nitrogen to the carrier gas by trapping on a copper column (30 cm × 6 mm) packed with Chromosorb B (AW DMCS) (80–100 mesh) coated with 3% of SE-52 and cooled at −78°C, and subsequently sweeping on to the gas chromatographic column with a stream of argon. A limitation of this procedure was that compounds that boil above 100°C could not be determined.

Amin and Narang [49] stripped volatile haloparaffins from sediments and adsorbed the volatiles on Poropak N. The compounds were eluted with methanol, and the elute analysed for organic compounds by GC with electron-capture and photoionisation detection. A detection limit of 7 μg^{-1} for each photoionisation active substance and 1 ng g^{-1} for each electron capturing compound was achieved. Samples could be stored in methanol for up to 90 days without significant loss of the volatile compounds. Recoveries ranged from 71% (bromoform) to 111% fluorobenzene.

Chlorofluoromethane provided the highest recoveries, while methanol-modified carbon dioxide gas gave a 90% recovery of polychlorobiphenyls for sediments.

Herbert et al. [65] used microwave-assisted extraction, combined with headspace solid-phase microextraction and high-resolution GC with ion-trap tandem mass spectrometry (MS) to determine polychlorobiphenyls in sediments and soils. Optimisation of the headspace solid-phase microextraction was carried out for the most important parameters, such as extraction time, sample volume and temperature. The adopted methodology has reduced consumption of organic solvents and runtime analysis. Under the optimised conditions the method detection limit ranged from 0.6 to 1 ng g^{-1} when 5 g of sample was extracted, the precision on real samples ranged from 4% to 21% and the recovery from 69% to 104%. The proposed method, which included the analysis of a certified reference material in its validation procedure, can be extended to several other PCBs and used in the monitoring of soil or sediments for the presence of PCBs.

Kominar et al. [66] have described a method for the determination of polychlorobiphenyls in river sediments in which samples were extracted using ultrasonics into 1:1 *n*-hexane/acetone. The extract was partitioned with water and back extracted into benzene. Combined organic extracts were dried on sodium sulphate, reduced in volume and cleaned up by gel permeation chromatography and silica gel partitioning. Analysis of polychlorobiphenyls was carried out by GC with electron-capture detection.

Langenfeld et al. [17] studied the effect of temperature and pressure on supercritical fluid efficiencies of polychlorinated biphenyls in river sediments.

At a temperature of 50°C, raising the pressure from 350 to 650 in atmosphere had no beneficial effect on the recovery of polychlorinated biphenyls from sediments. Recovery was improved, however, as the extraction temperature was increased from 50°C to 200°C.

Langenfeld et al. [17] compared supercritical monochlorodifluoromethane, nitrogen dioxide and carbon dioxide for the extraction of polychlorobiphenyls from sediments. Monochlorodifluoromethane provided the highest recovery. Methanol-modified carbon dioxide provided a 90% recovery of polychlorobiphenyls from sediments.

Lea et al. [94] have described an in situ procedure for qualifying inorganic chlorine concentrations in environmental samples based on X-ray absorption near-edge structure (XANES) spectroscopy. Cl 1s XANES spectra reflect contributions from all chlorine species present in the sample, providing a definitive measure of total chlorine concentration in chemically heterogeneous samples. Spectral features near the Cl K-absorption edge provide detailed information about the bonding state of chlorine, whereas the absolute fluorescence intensity of the spectra is directly proportional to total chlorine concentration, allowing for simultaneous determination of chlorine speciation and concentration in soil, sediment and natural water samples. Absolute chlorine concentrations are obtained from chlorine 1s XANES spectra using a series of chlorine standards in a matrix of uniform bulk density. With the high sensitivity of synchrotron-based X-ray absorption spectroscopy, chlorine concentration can be reliably measured down to the 5–10 ppm range in solid and liquid samples. Referencing the characteristic near-edge features of chlorine in various model compounds, it was possible to distinguish between inorganic chloride Cl_{inorg} and organochlorine Cl_{org} and aromatic Cl_{org} with uncertainties in the range of ~6%. In addition, total organic and inorganic bromine concentrations in sediment samples are quantified using a combination of bromine 1s XANES and X-ray fluorescence (XRF) spectroscopy. Bromine concentration is detected down to ~1 ppm by XRF, and 1s XANES spectra. This allows quantification of the inorganic bromine and organic bromine fractions.

18.5 Nitrogen-containing compounds

Weber and Wolfe [88] have shown that aromatic diazo compounds in sediments were readily degraded by abiotic surface-mediated reaction. The exact nature of the reducing agent was not determined, but it appeared to be associated with the sediment. There was no apparent correlation between the rate of degradation and the measured reduction potential of the diazo dyes. The rate of degradation appeared to be controlled by the amount of partitioning on the sediment, with increasing partitioning inhibiting the reduction process. The experimental results were used to develop a model for the reduction process.

18.6 Phosphorus-containing compounds

Methods for the determination of organophosphorus compounds in sediments are reviewed in Table 18.6.

18.7 Sulphur-containing compounds

Methods for the determination of sulphur-containing compounds are reviewed next.

The analysis of organosulphur compounds has been greatly facilitated by the flame photometric detector [101]. Volatile compounds can be separated by a glass capillary chromatographic column and the effluent split to a flame ionisation detector and a flame photometric detector. The flame photometric detector response is proportional to the square of the concentration of the sulphate [102–105]. The selectivity and enhanced sensitivity of the flame photometric detector for sulphur permit the quantitation of organosulphur compounds at relatively low concentrations in complex organic mixtures. The flame ionisation detector trace allows the organosulphur compounds to be referenced to the more abundant aliphatic and/or polynuclear aromatic hydrocarbons.

Reliable flame photometric detector quantification of organosulphur compounds requires careful optimisation of the gas chromatographic parameters. Although the relative response of the flame photometric detector to various

TABLE 18.6 Review of earlier published methods for the determination of phosphorus-containing compounds in sediments.

Chemical compound	Method of extraction from soil	Method of analysis	LD	References
Trialkyl and triaryl phosphates		GC and GC/MS	–	[95–97]
Inositol phosphate	–	–	–	[98]
Adensosine-5-triphospate	Discussion of extraction procedures	–	–	[99]
Nucleotides	–	Liquid chromatography to separate then ICPAES or ELS MS/MS	–	[100]

sulphur compounds remains somewhat controversial [106], analysis of organosulphur compounds by flame photometric detector is now relatively straightforward.

Trehey et al. [107] used GC–MS to determine alkylbenzene sulphonates and dialkyltetralin sulphonates in sediments by this technique with a detection limit of 0.5 g kg^{-1}.

Shea and MacGreehan [108] determined hydrophilic thiols in sediment pore water using ion-pair chromatography coupled with an electrochemical detector. Down to 2p mole, absolute of these compounds could be determined including cysteine, monothioglycerol, glutathione, mercaptopyruvic acid, 3-mercaptopropionic acid and 2-mercaptopropionic acid. Dichloroethane extraction of the sediment, followed by elimination of elemental sulphur mercaptans, disulphide and dibenzothiophene on a copper column is followed by a gas chromatographic analysis with flame photometric detection of the organosulphur compounds. The detection limit is 1 ng as sulphur with a precision of ±10% [109].

18.8 Insecticides and herbicides

Earlier methods for the determination of insecticides and herbicides in sediments are reviewed in Table 18.7.

More recently Tang et al. [118] discussed the determination of atrazine and its deethylatrazine degradation product in sediment using GC ion-trap MS.

The isotope dilution technique was applied for the quantitative analyses of atrazine at parts-per-trillion levels. Water samples were preconcentrated by solid-phase extraction using a C_{18} cartridge while the sediment samples were extracted by sonication with methanol. The concentrated extracts were analysed by a GC/ion-trap MS operated in the MS/MS mode. The extraction recoveries for the analytes were better than 83% when 1 L of water or 10 g of sediment was analysed. The method detection limits were 0.75 and 0.13 ng g^{-1} for atrazine and deethylatrazine detected in water and sediment, respectively. The precisions of the method represented by the relative standard deviation were in the range of 3.2%–16.1%. The method was successfully applied to analyse surface water and sediment samples collected from Guanting reservoir in Beijing. Trace levels of atrazine at 35.9–217.3 ng L^{-1} and 2.4–8.4 ng g^{-1} were detected in the water and sediment samples, respectively. The levels of deethylatrazine were 5–20 times lower than those of atrazine.

Dragan et al. [130] have carried out a detailed study of the occurrence of polychlorinated biphenyls and organochlorine pesticides, such as DDT and analogues, hexachlorocyclohexane (HCH) isomers and hexachlorobenzene (HCB) in surface soils and sediments from Eastern Romania. A total of 39 soil samples from the forested zone, 8 soil samples from the municipal waste disposal site and 10 sediment samples from the Bahlui river along the Iassy

TABLE 18.7 Review of earlier methods for the determination of insecticides and herbicides in sediments.

Chemical compound	Method of extraction from sediments	Method of analysis	LD	References
Organophosphorus insecticides Dursban	Extraction with dichloromethane adsorption of silica	GC with electron-capture detection	0.01 mg kg^{-1}	[110]
Dursban	Solvent extraction	GC detection	0.01 mg kg^{-1}	[111]
Organophosphorus insecticides and metabolites	Soxhlet extraction with acetone–hexane	–	95–220 mg kg^{-1}	[112]
Organophosphorus insecticides parathion ethyl, methidathion and tetrachlorovinphos	Supercritical carbon dioxide extraction	–	–	[60]
Diazinon, ronnel, parathion, methidathion and tetrachlorvinphos	Comparison of supercritical extraction with sonication and Soxhlet extraction	Methyl alcohol–carbon dioxide	–	[113]
Organochlorine insecticides	Extraction with acetone, hexane: 1% ammonium hydroxide	Capillary GC	–	[114]
DDT and BHC, aldrin, polychlorinated biphenyls	–	Discussion of interference by organosulphur compounds in GC method	1–10 ppb	[115–119]
BHC isomers, DDE, DDT, hexachlorobenzene	Petroleum ether extraction in Soxhlet extractor	TLC and GC	–	[77,78,120–123]

TDE, aldrin, dieldrin, kepones, permethrin, β- and γ-BHC, DDT, *p,p′*-DDE, *p,p′*-TDE and *p,p′*-DD		Methods for the determination of miscellaneous insecticides	–	[39,86,124–130]
Organochlorine insecticides	–	Study of relationship pesticides in water and in sediment	–	[131]
Diazinon, dichlorvos, endrin, endrin aldehyde, decachloro biphenyls, *p,p′*-DDT and mirex	Supercritical with sonication and Soxhlet extraction	Comparison of extraction methods		[129]
Azine-type herbicides atrazine, triazine, phenyl and urea type	–	Immunoassay and MS	0.001–0.005 ng kg^{-1}	[115–118,132–136]
Atrazine, linuron	–	Study of adsorption on sediment	–	
Triazine herbicides and metabolites	Solid-phase extraction with 4:1 v/v methanol: water	Isotope dilution technique or GC/ion-trap MS	0.1 ng kg^{-1}	[117,118]
Carbamate-type herbicides methomyl	Extraction with dichloromethane	GC	0.5 mg kg^{-1}	[102]
Carbamate and urea herbicides	Extraction with acetone	Spectrophotometry and TLC		[119,137]
Triazine herbicides and metabolites	Automated solid-phase extraction with methanol; water (4:1 v/v)	–	0.1 ng kg^{-1}	[117,118]
Phenoxy acetic acid–type herbicides	–	Particle beam MS	ng kg^{-1}	[137,138]

city were analysed using accelerated solvent extraction and GC coupled to electron detection or MS. The low mean concentrations of organochlorine pesticides (11–31 and 22–84 ng g^{-1} for HCHs and DDTs, respectively) and polychlorobiphenyls (8–43 ng g^{-1}) in soil samples from the forested zone suggest that contamination at most of these sites occurred predominantly through atmosphere transport from zones where these compounds were used and subsequently through atmosphere deposition. Contrarily, soil samples collected in the vicinity of a waste disposal site near Iassy contained higher mean levels of polychlorobiphenyls (278 ng g^{-1}, range 34–1132 ng g^{-1}) than organochlorine (6 and 101 ng g^{-1}) of soil for HCHs and DDTs, respectively. The sediment samples collected along the Bahlui river throughout the Iassy city revealed higher mean levels of polychlorobiphenyls (59 ng g^{-1}, range 24–158 ng g^{-1}) compared with organochlorine pesticide levels (2 and 37 ng g^{-1}) of solid for HCHs and DDTs, respectively. Furthermore polychlorobiphenyl profiles and concentrations in the sediment samples varied considerably along the river due to a wide variety of sources, such as different industries and waste sites. Although their sources are difficult to evaluate, the presence of these compounds at most sites (especially at the waste disposal site) may constitute a potential health hazard.

18.9 Miscellaneous organic compounds

Earlier methods for the determination of miscellaneous organic compounds in sediments are reviewed in Table 18.8.

More recently Chou and Liu [146] determined total faecal sterols in wastewater sediments by GC–MS.

The method included direct saponification, solvent phase extraction, derivatisation with *N*-methyl-*N*-trimethyltrifluoroacetamide and catalyst, and separation by GC with and HP-50 + capillary column followed by qualitative and quantitative analyses by MS. Recoveries of nine sterols by this method were 78%–89%. The indicators of biopollution markers (coprostanone and coprostanol/epicoprostanol) in different sources of wastewater effluent were calculated as human 0.913 ± 0.251, pig 0.224 ± 0.135, cow 0.023 ± 0.001 and duck 0.007 ± 0.001; such indicators are feasible for distinguishing between different animal sources of faecal pollution in water.

Stroll and Giger [140] have described a reversed-phase high-performance liquid chromatographic method for the determination of detergent derived fluorescent whitening agent isomers in lake sediments.

Stereoisomers of two main laundry detergent fluorescent whitening agents of the diaminostilbene type 1 (DAS-1) and the distyrylbiphenyl type (DSBP), as well as total BLS were quantitated in sediments and water from Greifensee, a lake in Switzerland. The freeze-dried sediments were extracted in an ultrasonic bath using methanol with tetrabutylammonium hydrogen sulphate as an ion-pairing reagent. Aqueous samples were extracted with

TABLE 18.8 Review of earlier published methods for the determination of miscellaneous organic compounds in sediments.

Chemical compound	Method of extraction from soil	Method of analysis	LD	References
Humic and fulvic acids	Discussion of extraction methods	–	–	[138,139]
	–	Reverse phase	–	[140]
		Liquid chromatography		
Detergents alkylbenzene sulphonate	Extraction with dichloroethylene	IR spectroscopy of methylene blue complexes		[45,46]
				[47]
Alkylbenzene sulphonates	Extraction with methanol:benzene (1:1)	IR spectroscopy	–	[141]
Optical whiteners and alkylbenzene sulphonates	Extraction with methanol:benzene (1:1)	Spectrofluorimetry	0.2 mg kg^{-1}	[5]
Carbohydrates	–	Spectrophotometry at 485 nm after reaction with phenol and concentrated sulphuric acid	–	[8]
	–	Automated chromatography of alkaline tetrazolium blue	0.1 nm L^{-1}	[77,79,142,143]
Uronic acids aldoses	–	GC	–	[77,79,138–140,142–144]
Pharmaceuticals	–	HPLC/GC	–	[145,146]
Sterols	Reaction with potassium hydroxide and methanol isolated sterols	–	–	
Pesticide polychlorinated organic compounds triazine, chloroacetanilide and organophosphorus type	Fluid and microwave-assisted extraction	Details of multiresidue method for screening and analysis being GC/MS	–	[78,126,145,146]

(*Continued*)

TABLE 18.8 (Continued)

Chemical compound	Method of extraction from soil	Method of analysis	LD	References
Environmental protection agency (United States) priority pollutants	Extraction with dichloromethane	GC/MS	–	[1,39,69,147–152]
Separation of acidic, basic and neutral organic compounds	–	Silica modified ion exchange method followed by HPLC	–	[153,154]
Separation of basic and nonbasic azarenes	–	Amino cyanate-bonded silica HPLC		[153–156]
Hetro compounds		Combined polarity activity chromatography GC–MS		[156,157]
Organic compounds	–	GC	–	[157,158]
Miscellaneous organic compounds	–	Review of application	–	[16]
	–	Miscellaneous methods	–	[25,159–161]
				[20,28,39,162]
				[39,84]
				[93,163]
				[164,165]
				[111,115,138,166,167]
				[11,12,112]
				[168–170]
				[34,77]
				[79,124,125,171]
				[72,158]

C_{18} extraction disks, which were subsequently eluted by methanol with tetrabutylammonium hydrogen sulphate. Extracts from solid samples were analysed by reversed-phase high-performance liquid chromatography. Fluorescence detection was applied after postcolumn ultraviolet irradiation. Analytical reproducibility ranged from 1% to 12%. Relative standard deviation was 1–11 g kg^{-1} of dry matter. Recoveries ranged from 93% to 100% in solid samples. Concentrations of DAS-1 and DSBP ranged from 0.4 to 1.4 mg kg^{-1} of dry matter on top matter sediment layers. Concentrations of BLS were between 0.02 and 0.08 mg kg^{-1} of dry matter in sediment layers.

Himmelsbach et al. [143] determine pharmaceutical drug residues in particulate material in surface water. In this paper a gas chromatographic method with mass spectrometric detection is presented, which allows the determination of the particle-bound fraction of some pharmaceuticals commonly found in surface water. Determination limits are between 2 and 12 ng g^{-1} particles. Results from surface water samples indicate the possibility that less hydrophilic pharmaceuticals, such as mefenamic acid, are present as suspended particulate material, although the amounts are small in comparison with the concentrations found in the aqueous phase. Additional work will be necessary to evaluate the full importance of particle-bound pharmaceuticals with respect to transportation in the environment.

Chi et al. [172] determined synthetic musks in lake sediments using accelerated solvent extraction followed by GC–MS.

Synthetic substitutes for natural musks are widely distributed in the environment. They have been detected in waters, sludge, fish, shrimp, mussels and other aquatic animals, and even in human adipose tissue, blood and breast milk. In this study, Chi et al. [172] described a new extraction procedure, based on accelerated solvent extraction successfully coupled with GC–MS for the analysis of musks in sediment samples. With this method the limits of detection as low as 0.03–0.05 ng g^{-1} and the recovery rate of 86.0%–104% are achieved. When compared with Soxhlet and ultrasonic extractions, accelerated solvent extraction not only has the best extraction efficiency but also has advantage in extraction time and solvent consumption. Eight musks, including six polycyclic musks [Tonalide (ANYN), Galaxolide (HHCB), Phantolide (AHDI), Traseolide (ATII), Cashmeran (DPMI) and Celestolide (ADBI)] and two nitro musks [musk xylene (M) and musk tone (MK)], were evaluated in sediment samples collected from 15 selected locations of the Taihu lake, one of the largest freshwater lakes in China. The contents of synthetic musks in sediment samples range from 0.336 to 3.10 ng g^{-1} for Galaxolide, 0.184 to 1.21 ng g^{-1} for Tonalide, below detection limit to 0.349 ng g^{-1} for muskxylene ionalide, and below detection limit to 0.0786 ng g^{-1} for musk ketone. The contents of Cashmeran, DPMI and Celestolide are below detection limit. The results reflect the current status of fragrance compound pollution in this area and provide basic data for environmental policymaking.

Semenov et al. [173] have applied low-Z electron probe microanalysis with subsequent cluster analysis of the results to specify the chemistry of different groups of suspended particles. Results obtained show that all the particles may be divided into three groups: (1) mineral, (2) organic and (3) mixed. Contribution of these groups into the total composition of suspended sediments was evaluated. According to the abundance of the particle groups, two categories of the rivers have been distinguished. Both chemical and mineralogical differences in suspended particles between two categories of Baikal tributaries are discussed. They are mainly conditioned by natural landscape features and the land use on the watersheds. Gfrerer et al. [174] have described a multiresidue method for the screening and analysis of 66 common pesticides from hydrological samples, including sediment, suspended solids and water. The investigated pesticides belong to the following chemical classes: polychlorinated organic compounds, triazine and chloroacetanilide herbicides and organophosphorus insecticides and miscellaneous. The method includes fluidised-bed and microwave-assisted extraction for solid samples and solid-phase extraction on C_{18}^- cartridges for water samples, followed by a combined purification–separation step on adsorption chromatography using open silica gel columns. Two fractions were eluted separating the 66 analytes into the nonpolar and the more polar compounds. All analytes were identified and quantified by GC coupled with MS in selected ion-monitoring mode. The method was characterised by recovery experiments and statistical methods and finally applied to environmental river samples during a 1-year monitoring program. This method allowed the screening and measurement of contaminants in all parts of Liao-He and Yangtze rivers (Eastern China) at levels as low as 0.07 ng L^{-1} for HCB, with a precision better than 20%.

A wide range of compounds can be identified and quantified in sediments by this technique.

References

[1] B.T. Hargreave, G.A. Phillips, Environ. Pollut. 8 (1975) 193.
[2] F.I. Onuska, K.A. Terry, Anal. Chem. 57 (1985) 801.
[3] H.B. Lee, R. Hong-You, A.S. Chau, Analyst, London 111 (1986) 81.
[4] H.W. Swartz, G.S.M. Ansion, H.P.M. Vleit, L.J.W. Peereboom, U.A.T. Brinkman, Int. J. Environ. Anal. Chem. 6 (1979) 133.
[5] M. Uchiyama, Water Res. 13 (1979) 847.
[6] C.A. Krone, D.W. Burrows, P.A. Rbisch, A.J. Friedman, D.C. Halins, Environ. Sci. Technol. 20 (1986) 1144.
[7] G.L. Mills, L.G. Quinn, Chem. Geol. 25 (1979) 155.
[8] N.R. McQuaker, T. Fung, Anal. Chem. 47 (1975) 1434.
[9] R.A. Kerr, The Isolation and Partial Characterisation of Dissolved Organic Matter in Seawater, 1977.
[10] K.I. Aspila, H. Agemian, A.S.Y. Chau, Analyst 101 (1976) 187.
[11] D.H. Landers, M.P. David, M.J. Mitchell, Int. J. Environ. Anal. Chem. 14 (1983) 245.

[12] H. Budzinshi, M. Lebellier, S. Thompson, K. Le, P. Menoch, Garrigues, Fresenius J. Anal. Chem. 367 (2000) 163.
[13] D. Broman, A. Colmsjo, B. Ganning, Mar. Pollut. Bull. 18 (1987) 380.
[14] T. Tan, J. Chromatogr. 176 (1979) 210.
[15] R.W. Readman, J.W. Preston, F.C.R. Mantious, Mar. Pollut. Bull. 17 (1986) 208.
[16] A. Marcomini, A. Striso, B. Pavoni, Mar. Chem. 21 (1987) 15.
[17] J.J. Langenfeld, S.B. Hawthorne, D.I. Miller, J. Pawliszyn, Anal. Chem. 65 (1993) 338.
[18] A. Saber, J. Jarocz, M. Marin-Boer, J. Environ. Anal. Chem. 28 (1987) 171.
[19] P. Garriques, M. Emald, Chemosphere 16 (1987) 485.
[20] H.K. Lee, G.J. Wright, W.H. Swallow, J. Environ. Pollut. 49 (1988) 167.
[21] P.D. Vowles, R.F.C. Mantoura, Chemosphere 16 (1987) 109.
[22] L.H. Keith, P.H. Lin, M.P. Kilpatrick, Water Qual. Bull. 6 (1981) 34.
[23] H.F.R. Reijnders, D. Onderdelindin, M.G. Visser, B. Griepak, Water Res. 14 (1980) 1645.
[24] M. Blumer, J. Sass, Mar. Pollut. Bull. 3 (1972) 92.
[25] J.W. Farrington, J.G. Quinn, Estuary Coast Mar. Sci. 1 (1973) 71.
[26] P.D. Guiney, J.L. Sykora, G. Keleti, Environ. Toxicol. Chem. 6 (1987) 105.
[27] S.G. Wakeman, Environ. Sci. Technol. 11 (1977) 272.
[28] M. Goldberg, E.R. Weiner, Anal. Chim. Acta 112 (1980) 373.
[29] V. Lopez-Avila, R. Northcutt, J. Oustat, M. Wickham, Anal. Chem. 55 (1983) 881.
[30] W.H. Ding, J.C.H. Fann, J. Chromatogr. A. 856 (1999) 79.
[31] E.R. Ale, P. Babay, J. Magallanes, G. Polla, E. Gautier, Int. J. Environ. Anal. Chem. 89 (2009) 1005.
[32] J. Houn, Y. Huong, C. Brodsky, M.R. Alexandre, Anal. Chem. 82 (2010) 7119.
[33] R. Ishmatari, T. Hanya, Chromatography-mass spectrometric detection of organic compounds in a river water. In: Proceedings International Meeting ed. Technig. 6th, p. 1051.
[34] K.K. Carrol, in: G.V. Marinetti (Ed.), Lipid Chromatographic Analysis, 1, Academic Press, New York, 1976, pp. 174–212.
[35] R.W. Johnson, J.A. Calder, Geochim. Cosmochim. Acta 37 (1973) 264.
[36] S. Thompson, G. Eglington, Geochim. Cosmochim. Acta 42 (1978) 199.
[37] W. Van Hoevan, J.R. Maxwell, M. Calvin, Geochim. Cosmochim. Acta 33 (1969) 877.
[38] M. Nishimura, Geochim. Cosmochim. Acta 41 (1977) 1817.
[39] J.W. Farrington, J.G. Quinn, Geochim. Cosmochim. Acta 35 (1971) 735.
[40] Y.A. Mendoza, F.O. Gulacar, Z.L. Hu, A. Bucks, Int. J. Environ. Anal. Chem. 31 (1987) 107.
[41] R.S. Raman, P.K. Hopke, Int. J. Environ. Anal. Chem. 86 (2006) 767.
[42] J. Liggio, R. McLaren, Int. J. Environ. Anal. Chem. 83 (2003) 819.
[43] H.W. Schartz, G.J.M. Anzion, H.P.M. Vleit, Int. J. Environ. Anal. Chem. 6 (1979) 133.
[44] M.L. Trehy, W.E. Gledhill, R.G. Orth, Anal. Chem. 62 (1990) 258.
[45] Y. Ambe, T. Hanya, Jpn. Anal. 21 (1972) 252.
[46] N. Longwell, O. Maniece, Anal. Abstr. 22 (1955) 2244.
[47] E.M. Sallee, J.F. Fairing, R.W. Hess, R. House, P.M. Maxwell, F.W. Melpolder, et al., Anal. Chem. 28 (1956) 1822.
[48] Private Communication.
[49] T.A. Amin, R.S. Narang, Anal. Chem. 57 (1985) 648.
[50] V. Zirko, J. Assoc. Off. Anal. Chem. 57 (1974) 1253.
[51] J. Hassler, F. Rippa, Vyskummy Ustai, Veda a Vyskunm Praxi No. 50, 1977.
[52] W. Glaze, J.E. Henderson, J.E. Bell, A. Van Wheeler, J. Chromatogr. Sci. 11 (1972) 590.
[53] J. Teichman, A. Bevenue, J.W. Hylin, J. Chromatogr. A. 151 (1978) 155.
[54] J.H. Carey, J.H. Hart, Water Pollut. Res. J. Can. 21 (1986) 309.

[55] M.J. Charles, M.S. Simons, Anal. Chem. 59 (1987) 1217.
[56] V. Zirko, J. Chromatogr. 81 (1973) 152.
[57] D.F. Goerlitz, L.M. Law, J. Assoc. Off. Anal. Chem. 57 (1974) 176.
[58] M.A.T. Kerkhoff, A. De Vries, R.C.C. Wegman, A.W.M. Hofstee, Chemosphere 11 (1982) 165.
[59] J.F. Brown, D.L. Bedard, M.J. Brennan, J.C. Carnation, H. Feng, R.E. Wagner, Science 236 (1987) 709.
[60] A.L. Alford-Stevens, W.L. Budde, T.A. Bellar, Anal. Chem. 57 (1985) 2452.
[61] K.D. McMurtrey, N.J. Wildman, H. Tai, Bull. Environ. Contam. Toxicol. 31 (1983) 734.
[62] H.B. Lee, Y.A.S. Chau, Analyst 112 (1987) 37.
[63] S.R. Tobin, J.F. Ryan, B.K. Afghan, Water Res. 12 (1978) 783.
[64] H.B. Lee, J. Assoc. Off. Anal. Chem. 71 (1988) 803.
[65] B.P. Herbert, S. Morais, P. Palga, A. Alves, L. Santos, Int. J. Environ. Anal. Chem. 86 (2006) 391.
[66] A.J. Kominar, F.L. Onuska, K.A. Terry, J. High Resolut. Chromatogr. Chromatogr. Commun. 8 (1985) 585.
[67] E.A. Maris, E. Noroozian, R.R. Otten, J. High Resolut. Gas Chromatogr. Chromatogr. Commun. 11 (1988) 197.
[68] R.E. Eaganhouse, R.W. Gossett, Anal. Chem. 63 (1991) 2130.
[69] V. Lopez-Avila, J. Beneticto, C. Charos, Environ. Sci. Technol. 29 (1996) 271.
[70] R.A. Hites, Anal. Chem. 60 (1988) 657A.
[71] J.F. Quensen, J.M. Tiedje, S.A. Boyd, Science 242 (1988) 752.
[72] A.M. Muscalu, E.J. Reinal, S.N. Liss, T. Chen, Int. J. Environ. Anal. Chem. 90 (2010) 1.
[73] A.L. Bridic, J. Bos, S.J. Herzberg, J. St Petrol 59 (1973) 263.
[74] J.W. Eichelberger, L.E. Harris, W.J. Budde, Anal. Chem. 46 (1974) 227.
[75] D. Robbat, L. Tyng-Yumi, B.M. Abraham, Anal. Chem. 64 (1992) 358.
[76] B.A. Ruddy, D.T. Qudah, J.A. Aldstodt, H.A. Bootsma, Int. J. Environ. Anal. Chem. 88 (2008) 337.
[77] J. Janak, M. Sackmauereva, A. Szokolay, O. Pal'usova, J. Chromatogr. 91 (1974) 545.
[78] J. Janak, M. Sackmaureva, A. Szololay, A. Madric, Chem Zvesti 27 (1973) 128.
[79] P.A. Mills, J. Assoc. Off. Anal. Chem. 51 (1968) 29.
[80] W.A. Korfmacher, L.G. Rushing, D.M. Nestorick, Chemosphere 14 (1985) 841.
[81] W. Christman, W. Rotard, V. Shcinz, H. Bode, Chemosphere 15 (1986) 2077.
[82] F.W. Karasak, S.H. Charbornneau, G.J. Renel, H.Y. Tong, Anal. Chem. 59 (1987) 1027.
[83] H. Otaka, S. Ogawa, I. Minato, H. Utsugi, T. Amagai, Int. J. Environ. Anal. Chem. 86 (2006) 583.
[84] A. Thuren, Bull. Environ. Contam. Toxicol. 36 (1986) 33.
[85] J. Kjolholt, J. Chromatogr. A. 325 (1985) 231.
[86] L.M. Smith, D.L. Stalling, J.L. Johnson, Anal. Chem. 56 (1984) 1830.
[87] H.Y. Tong, D.E. Giblin, R.I. Lapp, Anal. Chem. 63 (1991) 1772.
[88] E.J. Weber, N.L. Wolfe, Environ. Toxicol. Chem. 6 (1987) 911.
[89] J.I. Hollies, P.F. Pinnington, A.J. Handley, Anal. Chem. 11 (1979) 201.
[90] S.B. Hawthorne, S.B. Langenfield, J.J. Miller, M.D. Burford, Anal. Chem. 64 (1992) 1614.
[91] A.J. Murray, J.P. Riley, Anal. Chim. Acta 65 (1973) 261.
[92] A.J. Murray, J.P. Riley, Nature 242 (1973) 37.
[93] J. Novak, J. Zluticky, V. Kubelka, J. Mostecky, J. Chromatogr. A. 76 (1973) 45.
[94] A.L. Lea, M.B. Hay, A. Lansirottz, W. Rao, S.C.B. Mymenti, Anal. Chem. 78 (2006) 5711.

[95] R.J. Ozretich, W.P. Schroeder, Anal. Chem. 58 (1986) 2041.
[96] S. Ishikawa, M. Taketomi, R. Shinohara, Water Res. 19 (1985) 119.
[97] J.F. Fairchild, T. Boyle, W.R. English, C. Rabeni, Water Air Soil Pollut. 36 (1987) 271.
[98] W.C. Weimer and D.E. Armstrong, Anal. Chim. Acta, 1977, **94,** 35
[99] S.R. Tobin, I.F. Ryan, B.F. Afghan, Water Res. (1978) 7813.
[100] H. De Brabender, N. Forsgard, L. Israelsson, J. Patterson, E. Rydin, M. Waldeba, et al., Anal. Chem. 80 (2008) 6689.
[101] S.S. Brody, J.E. Chaney, J. Gas Chromatogr. 4 (1966) 42.
[102] D.G. Greer, T.J. Bydalek, Environ. Sci. Technol. 7 (1973) 153.
[103] T. Suguyama, Y. Suzuki, T. Takeuchi, J. Chromatogr. 77 (1973) 309.
[104] A.P. Bentz, Anal. Chem. 48 (1976) 454A.
[105] M.E. Garza, J. Muth, Environ. Sci. Technol. 8 (1974) 249.
[106] C.H. Burnett, D.F. Adams, S.O. Farwll, J. Chromatogr. Sci. 16 (1978) 68.
[107] M.E. Trehey, W.E. Gledhill, R.G. Orth, Anal. Chem. 62 (1990) 171.
[108] D. Shea, W.A. MacGreehan, Anal. Chem. 60 (1988) 1449.
[109] T.S. Bates, R. Carpenter, Anal. Chem. 51 (1979) 551.
[110] J.R. Rice, H.J. Dishberger, J. Agric. Food Chem. 16 (1968) 867.
[111] M.E. Deusch, M.E. Westlake, F.A. Gunther, J. Agric. Food Chem. 18 (1970) 178.
[112] W.C. Weimer, D.E. Armstrong, Anal. Chim. Acta 94 (1977) 35.
[113] R.G. Grob, M.E. McNally, S. Oostyke, Anal. Chem. 64 (1992) 1940.
[114] R.C.C. Wegman, M.A.W. Hafster, Water Res. 16 (1982) 1265.
[115] R.D. Wauchope, R.S. Myers, J. Environ. Qual. 14 (1985) 132.
[116] I. Ferrer, M.C. Hennion, D. Barelo, Anal. Chem. 69 (1997) 450.
[117] M.S. Mills, E.M. Thurman, Anal. Chem. 64 (1992) 1985.
[118] W. Tang, G. Jiang, Z. Zlai, Int. J. Environ. Anal. Chem. 85 (2005) 1117.
[119] D. Spengler, A. Jumar, P. Planzenschutz, Nacht, Bayer 7 (1971) 151.
[120] M. Sackmaureva, O. Pal'usova, E. Hinchan, Vodni, Hospodarstvi 10 (1972) 267.
[121] M. Sacmaureva, P. Pal'usova, A. Szokolay, Water Res. 11 (1977) 537.
[122] H. Mahel'ova, M. Sackmaureva, A. Szokolou, J. Kovac, J. Chromatogr. 89 (1974) 177.
[123] A. Szokolay, J. Uhnak, A. Madaric, Chem. Zvestiz 25 (1971) 453.
[124] P.A. Mills, J. Assoc. Off. Anal. Chem. 42 (1959) 134.
[125] F. Frank, A.F. Armstrong, R.G. Baeleus, Pestic. Monit. 7 (1974) 165.
[126] J.H. Hamence, P.S. Hall, D.J. Coverly, Analyst 90 (1965) 649.
[127] T. Teichman, A. Bevenue, J.W. Hylin, J. Chromatogr. 59 (1985) 2797.
[128] R.P. Gambrell, C.N. Reddy, V. Collard, J. Water Pollut. Control Fed. 56 (1984) 174.
[129] J.L. Snyder, R. Grob, M.E. MacNally, Anal. Chem. 64 (1992) 1940.
[130] D. Dragan, J. Cuu-Man, A.C. Dirtu, R. Moncann, L.V. Vaeck, A. Covaci, Int. J. Environ. Anal. Chem. 86 (2006) 833.
[131] M.C. Godber, J. Sci. Total Environ. 24 (1987) 73.
[132] V. Pichon, L. Chen, M.-C. Hennion, Anal. Chem. 67 (1995) 2451.
[133] V. Pichon, L. Chen, N. Durand, J. Chromatogr. 725 (1996) 107.
[134] G.S. Rule, A. Mordehal, J. Henion, Anal. Chem. 66 (1994) 230.
[135] S.J. Shahtaheri, M. Katmeh, P. Kwasowski, D. Stevenson, J. Chromatogr. 697 (1995) 131.
[136] A. Marx, T. Giersch, B. Hick, Anal. Lett. 28 (1995) 267.
[137] I. Kim, F.T. Sasinos, R.D. Stephens, Anal. Chem. 63 (1991) 819.
[138] T. Klenke, M.W. Oskierski, K.G. Poll, B. Reichel, Gas U Wasserfach (Wasser, Abwasser) 127 (1986) 650.

[139] K. Hayase, J. Chromatogr. 295 (1984) 530.
[140] J.M.A. Stroll, W. Giger, Anal. Chem. 69 (1997) 2594.
[141] Y. Ambe, Environ. Sci. Technol. 7 (1973) 542.
[142] J.S. Walter, J.I. Hedges, Anal. Chem. 60 (1988) 988.
[143] M. Himmelsbach, W. Buchberger, H. Miesbauer, Int. J. Environ. Anal. Chem. 83 (2003) 481.
[144] K.K. Mopper, E.T. Regeus, Anal. Biochem. 45 (1972) 147.
[145] F. Dreier, A. Bucks, F.O. Gulakar, Geochim. Cosmochim. Acta 52 (1988) 1663.
[146] C.-C. Chou, Y.-P. Liu, Int. J. Environ. Anal. Chem. 84 (2004) 379.
[147] V. Lopez-Avila, R. Young, W.F. Beckert, Anal. Chem. 66 (1994) 1097.
[148] M. Ippolo, R. Alexander, R.I. Kagi, Org. Geochem. 18 (1992) 603.
[149] B. Bennett, J.B.F. Bowler, S. Larter, Anal. Chem. 68 (1996) 3697.
[150] T. Rezanka, J. Chromatogr. 627 (1992) 241.
[151] M. Yamamoto, G. Taguchi, K. Sasaki, Chem. Geol. 93 (1991) 193.
[152] J.B. Green, R.J. Hoff, P.W. Woodward, L.L. Stevens, Fuel 63 (1984) 1290.
[153] J.B. Green, J. Chromatogr. 358 (1986) 53.
[154] M. Li, S.R. Larter, D. Stoddart, Bjorøy, Anal. Chem. 64 (1992) 1337.
[155] S.J. Rowland, A.T. Revill, Chromatography of the petroleum industry, J. Chromatogr. Lib. 56 (1995) 127–141.
[156] H. Willisch, H. Clegg, B. Horsfield, Anal. Chem. 69 (1997) 4203.
[157] J. Faure, P. Viallet, P. Picat, La Tribune Dure Cabedes 28 (1973) 439.
[158] J. Slobodnik, J. Ramhalo, R.L.M. Baer, A.J. Louder, U.A.T. Brinkman, Chemosphere 41 (2000) 1469.
[159] W. Giger, C. Schanaffer, Anal. Chem. 50 (1978) 243.
[160] Y.L. Tan, J. Chromatogr. 176 (1979) 316.
[161] A. Bjorseth, J. Knutzen, J. She, Sci. Total Environ. 13 (1979) 71.
[162] J.W. De Leeuw, W.B.E. De Leer, S.S.J. Damste, P.J.W. Schuyl, Anal. Chem. 58 (1982) 1852.
[163] N.R. Quaker, T. Fung, Anal. Chem. 47 (1975) 1453.
[164] H.B. Lee, Y.D. Stokker, A.S.Y. Chau, J. AOAC 70 (1987) 1003.
[165] K. Schellenberg, C. Leuenberger, R.P. Schwartzebbach, Environ. Sci. Technol. 18 (1984) 652.
[166] D. Spengler, A. Jumar, Arch Pflanzenschutz 7 (1971) 151.
[167] R.G. Reeves, D.W. Woodham, J. Agric. Food Chem. 22 (1974) 76.
[168] W.J. Contello, R.P. Gambrell, Chemosphere 16 (1987) 1053.
[169] J.E. Stein, E. Han, E. Casillas, Mar. Environ. Res. 22 (1987) 123.
[170] M.J. Baedeckce, A. Nissenbaum, I.R. Kaplan, Geochim. Cosmochim. Acta 38 (1972) 1185.
[171] J.A. Burke, B. Malone, J. Assoc. Anal. Chem. 49 (1960) 1003.
[172] J.-S. Chi, R.-P. Yn, Q.-J. Song, L.-P. Wang, S.-F. Wu, Int. J. Environ. Anal. Chem. 91 (2011) 387.
[173] M.Y. Semenov, Z. Spolnick, L. Granina, R. Van Grieken, Int. J. Environ. Anal. Chem. 85 (2005) 377.
[174] G. Gfrerer, T. Wenzi, E. Lankmayr, Int. J. Environ. Anal. Chem. 83 (2003) 111.

Further reading

M. Ahnoff, B. Josefsson, Bull. Environ. Contam. Toxicol. 13 (1975) 159.
N. Alexandrou, J. Pawliszyn, Anal. Chem. 61 (1989) 2770.
J. Lawrence, F. Onuska, R. Wilkinson, B.K. Afghan, Chemosphere 15 (1986) 1085.

J. Minton, M. Adolfsson-Erici, T. Alsberg, Int. J. Environ. Anal. Chem. 91 (2011) 553.

S. Jensen, L. Renberg, L. Reutgard, Anal. Chem. 49 (1977) 316.

S. Jenson, A.G. Johns, M. Olsson, G. Otterlind, Ambio Spectrosc. Rep. 1 (1972) 71.

F.I. Onuska, K.A. Terry, J. High Resolut. Chromatogr. Chromatogr. Commun. 14 (1991) 829.

V. Raverdino, R. Holzer, J. Berset, Fresenius J. Anal. Chem. 354 (1996) 477.

Southeast Water Laboratory, Sediment Extraction Procedure, Method No. SP8/71, Athens, GA, 1977.

V. Taguchi, Chemosphere 15 (1986) 1147.

Chapter 19

Organometallic compounds in sediments

Chapter Outline

19.1 Organoarsenic compounds	374
19.2 Organolead compounds	374
19.3 Organotin compounds	376
19.4 Organomercury compounds	380
19.5 Organosilicon compounds	386
References	386

Some elements, such as arsenic, bismuth, selenium, mercury, lead and tin, can exist in solid matter in both inorganic and organic forms. The organic forms of the element may originate either by direct contamination of the solid by organic compounds of industrial origin or by naturally occurring biomethylation processes whereby inorganic forms of metals, such as mercury and selenium, are converted to the organic form, for example in fish tissues or in sediments. Suitable handling procedures both during sampling and analysis are necessary to avoid losses of metals in such instances.

Several sample digestion techniques, some involving a subsequent extraction with an organic solvent, have been described for the determination of organometallic compounds in sediments.

Organomercury compounds have been digested with sodium hydroxide prior to extraction with toluene and determination by cold vapour atomic absorption spectrometry, gas chromatography or high-performance liquid chromatography. Direct solvent extraction with benzene, toluene or chloroform separates organolead and organomercury compounds from sediments prior to analysis by gas chromatography or column chromatography.

Combinations of digestion reagents and sediment extraction have also been used. Thus methanolic hydrochloric acid or ethereal tropalone separate organotin compounds from sediments prior to gas chromatography analysis.

Pacheon-Argona et al. [1] have evaluated self-tuning single-mode microwave technology to perform the quantitative routine extraction of organometallic species from solid matrices of environmental interest. Species-specific isotope dilution analysis has been employed to better investigate the real

Determination of Toxic Organic Chemicals In Natural Waters, Sediments and Soils.
DOI: https://doi.org/10.1016/B978-0-12-815856-2.00019-9

influence of the microwave-assisted extractions on the final results. These workers discuss advantages of such methodology in comparison with other established microwave units for the routine speciation analysis of organomercury and organotin compounds. These advantages include a capability of using disposable glass vials, a self-tuning mode that provides an accurate control of the temperature and pressure inside the vials, and the possibility of performing automated sequence of extractions with low sample size. The results obtained in this work demonstrate that such technology provides a fast and reliable quantitative extraction of the organometallic species in a wide range of extraction conditions even when the multielemental (Sn and Hg) species-specific determination is carried out.

19.1 Organoarsenic compounds

Maher [2] has described a method for the determination of down to 10.01 mg kg^{-1} of organoarsenic compounds in marine sediments. In this procedure the organoarsenic compounds are separated from an extract of the sediment by ion-exchange chromatography; the isolated organoarsenic compounds are reduced to arsines with sodium borohydride and collected in a cold trap. Controlled evaporation of the arsine fractions and detection by atomic absorption spectrometry complete the analysis.

Odanaka et al. [3] have reported the application of gas chromatography with multiple ion detection after hydride generation with sodium borohydride to the determination of mono- and dimethyl arsenic compounds, trimethyl arsenic oxide and inorganic arsenic in soil and sediment. Recoveries in spiking experiments were 100%–102% (mono- and dimethyl arsenic compounds and inorganic arsenic) and 72% (trimethyl arsenic oxide).

19.2 Organolead compounds

Potter et al. [4] acidified a petroleum ether extract of the sediment with dilute nitric acid to pH 5–6. This aqueous extract was neutralised with aqueous sodium hydroxide prior to analysis by gas chromatography.

Potter et al. [4] showed that recovery of alkyllead salts obtained from sediments was 90% for Et_3PbCl, 75% for Et_2PbCl_2 and 40% for Me_2PbCl_2. Extraction of Et_2PbCl_2 added to sediment, containing no alkylleads, from a clean and polluted river, canal and road drainage grids gave recoveries of between 65% and 75%. The lowest detectable concentration of alkyllead salts gave a recovery of 2 mg kg^{-1} dry weight of sediment. Extraction of the lowest detectable concentration of tetraalkyllead was 0.02 mg kg^{-1} dry weight of sediment. By extracting samples of filtered water with one-tenth their volume of petroleum ether, tetraalkyllead could be detected down to concentrations of 0.002 mg L^{-1} in water.

Using these procedures, Potter et al. [4] found up to 100 mg kg^{-1} of tetraalkyllead in some samples of drainage grid sediment. No alkyllead compounds were detected in the filtered water from any of these sediment samples.

Chau et al. [5–8] in a series of papers issued between 1979 and 1984 discussed various methods for the determination of alkyllead compounds in sediments.

Chau [5] has described a simple and rapid extraction procedure for extracting the five tetraalkyllead compounds (Me_4Pb, Me_3EtPb, Me_2Et_2Pb, $MeEt_3Pb$ and Et_4Pb) from sediment. The extracted compounds are analysed in their authentic forms by a gas-chromatographic–atomic-absorption spectrometry at 217 nm. Other forms of inorganic and organic lead do not interfere. The detection limit for sediment (5 g) was 0.01 mg kg^{-1}. In this method the sediment is digested with EDTA and a hexane extract analysed by gas chromatography. Concentrations found in a sediment ranged from 8.3 (tetramethyl and methyltriethyl leads) to 12 mg kg^{-1} (dimethyl diethyl lead and tetraethyl lead). Recoveries in spiking experiments were between 81% and 84%.

Determination of the ionic forms of alkyllead compounds is difficult because of the incomplete extraction of the dimethyl and trimethyl species from sample matrices. A chelation extraction method followed by derivatisation to their butyl homologues has overcome all the previous difficulties to achieve quantitative extraction of the dimethyl and trimethyl leads from water samples at nanogram levels [6].

The application of a combination of gas chromatography and atomic absorption spectrometry to the determination of tetraalkyllead compounds has been studied by Chau et al. [7] and Segar [8]. In these methods the combination of gas chromatography and flame showed a detection limit of about 0.1 μg Pb. Chau et al. [6,7,9] have applied the silica furnace in the atomic absorption unit and have shown that the sensitivity limit for the detection of lead can be enhanced by several orders of magnitude. They applied the method to the determination of tetramethyllead in sediment systems.

The relative standard deviation was in the range of 10%–15% at the 5 ng level (as Pb). When the absorbances were plotted against lead concentrations, each of the five tetraalkyl compounds gave similar calibration curves; the response was linear up to at least 200 ng Pb, above which overlapping of the peaks occurred. If only one compound was present (e.g. tetramethyllead), the plot was linear up to at least 200 ng. For determination at the microgram level the flame atomic absorption spectrometric technique [7] is more suitable.

Chau et al. [9] have described the optimum conditions for extraction of alkyllead compounds from sediments originating in nonsaline and saline waters [8]. Analyses of some environmental samples revealed for the first time the occurrence of dialkyl- and trialkyllead in sediments in areas of lead contamination.

The various alkyllead species and lead (II) are isolated quantitatively by chelation extraction with sodium diethyldithiocarbamate, followed by *n*-butylation to their corresponding tetraalkyl forms, all of which can be determined by gas chromatograph using an atomic absorption detector. The method determines simultaneously the following species in one sample: tetraalkyllead (Me_4Pb, Me_3EtPb, Me_2Et_2Pb, $MeEt_3Pb$ and Et_4Pb) and ionic alkyllead (Me_2Pb^{2+}, $EtPb^{2+}$ and Me_3Pb^{2+}, Pb^{2+}). Detection limit expressed for lead was 15 μg kg^{-1} for sediment samples.

Reisinger et al. [10] used the gas-chromatographic–atomic adsorption technique to demonstrate that biomethylation of inorganic lead does not account for the presence of organolead compounds in sediments. Sulphide-induced chemical conversion of organic lead (IV) salts into alkyllead compounds is, however, possible.

Wong et al. [11] on the other hand, claim that the conversion of inorganic lead to tetramethyllead in river and marine sediments is purely a microorganism-induced biological process. These workers demonstrated that incubation of some lead-containing sediments generates tetramethyllead; also that Me_3Pb^+ are readily converted to tetramethyllead by microorganisms in lake water or nutrient medium, with or without the sediment, and that in the presence or the absence of light; that conversion of inorganic lead (such as lead nitrate or lead chloride) to tetramethyllead occurred on several occasions in the presence of certain sediments; and that the conversion is purely a biological process.

Detection limits achievable in the determination of organolead compounds in sediments range from 0.01 [4,5] to 0.00001 mg kg^{-1} [11].

19.3 Organotin compounds

Various workers have discussed the determination of alkyl and aryltin compounds in sediments.

Nemonic et al. [12] have carried out a critical evaluation of various extraction procedures for the speciation of butyltin compounds in sediments.

Ebdon et al. [13] have discussed a programme to improve the quality of analytical results in the environmental monitoring of organotin compounds. They discuss the evaluation of a sensitive, reliable and robust analytical method for the determination of tributyltin, with emphasis on the difficulties of determining it at the nanogram per litre levels at which it was usually encountered, more especially as other forms of tin frequently occurred together at similar levels. The preparation of a standard reference sample, for use in interlaboratory comparative determinations, under the aegis of the Bureau of Community Reference of the EU is described, and plans for subsequent distributions of blank, artificially spiked and genuinely affected sediments are sketched.

Three groups have independently reported on the methylation of both inorganic tin and organotin substrates by the mixed populations of microbial flora present in sediments collected from a Canadian fresh water lake [14], and from estuarine sites in San Francisco Bay and Chesapeake Bay [15–17]. Biogenesis of Me_4Sn was seen only to occur with additions of Me_3Sn^+ to incubated sediments [14,18], but redistribution reactions of intermediate methyltins to form Me_4Sn by nonbiological pathways must be noted as competitive events in such experiments [18,19]. The concentrations of tin compounds added to incubated sediments were consistent with values found in polluted sediments. The influence of other bioactive pollutant metals also commonly found in such sediments was not investigated.

A relatively recent sediment (period 1980–84) and a sediment from the late 19th century (1880–85) taken from Lake Zurich were investigated. No organotin compounds could be detected in the 1880 sediment, but a series of organotin compounds ranging from 280 $\mu g\ kg^{-1}$ tributyltin to 10 $mg\ kg^{-1}$ dichlorhexyltin were present in the recent sediment. The main components were again the butyltin compounds, indicating their frequent use, persistence and bioaccumulative power. $Cyhex_2Sn^{2+}$ and $Cyhex_3Sn^+$ were also identified, reflecting the use of the parent compound, trichlorhexyltin as a miticide in the region around Lake Zurich. The absence of organotin residues in the 1880 sediment is explained by the fact that technical use of these compounds started after 1936. The residue found in the recent sediment is considerably higher than those detected in surface sediment from the lower basin. As the sedimentation near the effluent of Lake Zurich in the shallow and oxygen-saturated water is dominated by processes leading to resuspension and oxidation of the fine, carbon-containing particles, the sediment taken at the deepest (and anoxic) part of the lake accumulates higher organotin residues and it is therefore more representative as the lake accumulates higher organotin residues and is therefore more representative of the overall situation in the lake sediment.

Unger et al. [20] have studied the sorption behaviour of tributyltin on estuarine sediments.

Rapsomkankis et al. [21] have studied the biological methylation of organotin sediments.

Gas chromatography predominates as the preferred method of analysis of organotin compounds.

Arakawa et al. [22] pointed out that methyltin compounds may be extracted from complex matrices and analysed by conventional gas chromatography. However, the procedure is lengthy, involving multiple steps where speciation may be altered and vessel adsorption effects may be large. Detection limits achievable with a flame ionisation detector are 10–100 μg.

Chau et al. [23] have pointed out that butylation of methyltin species before solvent extraction and the use of atomic adsorption spectrometry shortens the extraction procedure and reduces detection limits to about 0.1 ng.

In the method described by Hattori et al. [24] the sediment samples were extracted into methanolic hydrochloric acid, and then, following mixing with water and sodium chloride, the mixture was extracted with benzene. Following dehydration and concentration, the tin compounds were cleaned up on a silica gel column impregnated with hydrochloric acid and then hydrides generated using an ethanol solution of sodium borohydride [25]. The organotin hydrides were determined in sediment samples, and the detection limits was 0.02 mg kg^{-1}.

Mueller [26] has described a method for determining 0.5 μg kg^{-1} tributyltin in sediments in which tributyltin is first converted to tributyl methyltin and analysed using capillary gas chromatography with flame photometric detection, and also by gas chromatography–mass spectrometry (MS).

Mueller [27] has described a comprehensive method for determining traces of mono-, di- tri- and some tetrasubstituted organotin compounds in lake sediment. The ionic compounds are extracted from acidified sediment as chlorides using ethereal tropalone solution. The extracted organotin compounds are ethylated using a Grignard reagent (EtMgBr) and analysed by high-resolution gas chromatography with flame photometric and mass spectrometric detection. Ethylation using ethyl magnesium bromide was chosen for conversion of the various mono-, di- and tri- substituted organotin compounds in sediments into tetrasubstituted ones.

Ethylation was preferred over either methylation or alkylation using a larger alkyl group because methylation of tin (IV) and butyltin species seems to occur in the environment leading to methyltins mixed methyl alkyltins. Further methylation of these environmental metabolites in the derivatisation step would exclude the possibility of determining these conversion and degradation products. The ethylation reaction of these compounds leads to a series of tetrabutyltin compounds.

Furthermore, ethylation facilitates identification of organotin derivatives in the gas chromatogram. This is because the elution follows increasing degrees of substitution, which is not the case of hexylated products.

Lobinski et al. [28] speciated organotin compounds in sediment samples by capillary gas chromatography using helium microwave-induced plasma emission spectrometry as a detector. They used the procedure to determine mono-, di-, tri- and some tetralkylated tin compounds in sediments. The ionic tin compound were extracted as diethyldithiocarbamates into pentane then converted to pentyl magnesium bromide derivatives prior to gas chromatography. The absolute detection limit was 0.05 pg tin.

Gilmour et al. [29] determined picogram quantities of methyltins in sediments as their hydride derivatives (methylstannanes) using gas chromatography–quadrupole MS. Hydride derivatives were prepared by the addition of sodium borohydride in a closed, flow-through system. Borate buffer was added to the samples and hydrogen generated from the borohydride. This resulted in high purge efficiencies for mono-, di- and tri-methyltin. Selected ion monitoring with the mass spectrometer allowed detection limits of 3–5 pg as tin for methyltins. Detection limits for 5 g sediment samples were below picogram tin per gram levels with a standard deviation of 6%–8% depending on the methyltin species and the sample type.

Szpunar et al. [30] and Prange and Jensen [31] have determined organotin species by using gas chromatography with indirectly coupled plasma mass spectrometric detection. In one study, butyltin species were extracted from sediments and biomaterials in only 1–5 minutes by using microwave digestion [32]. Detection limits of ~50 fg (Sn), 100 fg (Pb) and 120 fg (Hg) were achieved.

Various other workers have discussed the application of gas chromatography or high-performance liquid to the determination of organotin compounds in nonsaline sediments [31,33–39].

Yang et al. [40] accomplished the speciation of organotin compound using reverse-phase liquid chromatography with inductively coupled plasma mass spectrometric detection. The separation was complete in 6 minutes, and detection limits were in the range of 2.8–16 pg of tin for various species.

High-performance liquid chromatography coupled with fluorescence detection [41,42] or ion-exchange high-performance liquid chromatography with detection by graphite furnace atomic absorption spectroscopy [43] proved to be sensitive methods, but they have limitations in separation power and ease of identification of unknown products.

Epler et al. [44] used laser-enhanced ionisation as a selective detector for the liquid chromatographic determination of alkyltin compounds in sediments. The analysis was performed on a 1-butanol extract of the sediment.

Other procedures that have been used in studies of the occurrence of organotin compounds in sediments include supercritical fluid chromatography [25,45], atomic absorption spectrometry [46], electrochemical methods [47], fluorimetric analysis [48] and X-ray absorption near-edge structure (XANES) spectrometry [49].

Takahashi et al. [49] have developed a direct method for the speciation of tin compounds in solid environmental samples by XANES spectroscopy. It was found that the method can provide the 'organic extent', the average number of organic ligands bound to tin, for environmental samples. For Sn XANES at the L_{111}, L_1 and K edges, systematic variations were found in the spectra for butyl-, phenyl- and methyl-substituted tin compounds depending on the organic extent. A quantitative relationship between the organic extent and the characteristics in the XANES spectra was determined based on the peak position, peak area ratio and peak width. The detection limit was better than $10\ \mu g\ g^{-1}$ tin when using the K edge, which is sensitive enough for some environmental samples, for example sediments, biological samples and antifouling paints; the sensitivity would be better if a more intense X-ray source, such as an undulator or wiggler, was used. This EXANE method is totally nondestructive, having the advantage that no complicated pretreatment procedures are needed, whereas such procedures are essential in conventional chromatographic analysis, and may cause experimental error by alteration of tin species and poor recovery during analyses. Although the XANES method only provides the average number of organic ligands, the direct speciation using XANES will be helpful for estimating roughly the ratio of organic and inorganic tin species, which can be used to study organotin transformation in sediment cores and the inspection of organotin compounds in antifouling paints. In particular, micro-XANES analysis based on this method is a promising tool in obtaining the distribution of organotin species in biological samples and specific phases in sediments.

Staniszewska et al. [50] have pointed out that the determination of organotin compounds in bottom sediments is a complex process that requires a number of analytical steps, that is sample collection, transport and storage; extraction of analytes from sediments; derivatisation; extract purification; enrichment; and the final chromatographic measurement. The whole process is time and labour consuming and is subject to securing a representative sample. These workers review the most frequently encountered problems, and the examples of possible analytical solutions are presented, which encompass the specific steps of speciation analysis of these toxic compounds.

Capillary gas chromatography with a helium microwave-induced plasma emission detector has been used to determine mono-, di- and tributyltin compounds.

Detection limits achievable for organotin compounds in nonsaline sediments are in the range of 3–4 ng L^{-1}.

19.4 Organomercury compounds

Earlier work on the determination of total mercury in river sediments includes that of Iskandor et al. [51]. Iskandor applied flameless atomic absorption to a sulphuric acid–nitric acid digest of the sample following reduction with potassium permanganate, potassium persulphate and stannous chloride. A detection limit of 1 μg kg^{-1} is claimed for this somewhat laborious method. Langmyhr and Aamodt [52] determined down to 0.1 μg L^{-1} of organomercury. Matsunaga and Takahasi [53], Craig and Morton [54] and the AOAC [55] also determined organic mercury in river sediments using cold vapour atomic absorption spectrometry.

A method [55,56] has been described for the determination of down to 2.5 μg kg^{-1} alkylmercury compounds and inorganic mercury in river sediments. This method uses steam distillation to separate methylmercury in the distillate and inorganic mercury in the residue. The methylmercury is then determined by flameless atomic absorption spectrometry and the inorganic mercury by the same technique after wet digestion with nitric and potassium permanganate. The well-known adsorptive properties of clays for alkylmercury compounds do not cause a problem in this method. The presence of humic acid in the sediment did not depress the recovery of alkylmercury compounds by more than 20%. In the presence of metallic sulphides in the sediment samples the recovery of alkylmercury compound decreased when more than 1 mg of sulphur was present in the distillate. The addition of 4 N hydrochloric acid, instead of 2 N hydrochloric acid before distillation completely, eliminated this effect giving a recovery of 90%–100%.

This excellent method was sufficiently sensitive to determine 0.02 mg kg^{-1} methylmercury and 9 mg kg^{-1} inorganic mercury in river sediment samples.

Jirka and Carter [57] described an automated determination of down to 0.1 mg kg^{-1} total mercury in river sediment samples with a precision of 0.13–0.21 μg Hg kg^{-1} and at the 1 mg Hg kg^{-1} level and with a standard deviations varying from 0.11 to 0.02 mg Hg kg^{-1} (i.e. relative standard deviations of 8.4%–12%). At the 17.2–32.3 mg Hg kg^{-1} level in sediments, recoveries in methylmercuric chloride spiking studies were between 85% and 125%. This method is based on the automated procedure of Alwady et al. [58] for the determination of total mercury in waters and wastewaters in which potassium persulphate and sulphuric acid were used to digest samples for analysis by the cold vapour technique. These workers proved that the use of potassium permanganate as an additional oxidising agent was unnecessary.

Aromatic organic compounds, such as benzene, which are not oxidised in the digestion, absorb at the same wavelength as mercury. This represents a positive interference in all cold vapour methods for the determination of mercury. For samples containing aromatics (i.e. those contaminated by some industrial wastes), blank analysis must be performed, and the blank results must be subtracted from the sample results. The blank analysis is accomplished by replacing the potassium persulphate reagent and the stannous chloride reagent with distilled water and reanalysing the sample.

Umezaki and Iwamoto [59] have reported that organic mercury can be reduced directly with stannous chloride in the presence of sodium hydroxide and copper (II). The determination of organic mercury can be simplified, particularly if the reagent used for back extraction does not interfere with the reduction of organic mercury. Matsunaga and Takahasi [53] found that back extraction with an ammoniacal glutathione solution was satisfactory. In this method, contamination only from the ammoniacal glutathione solution is expected. However, any inorganic mercury in this solution will be absorbed on the glass container walls with a half-life about 2 days (i.e. the blank value becomes effectively zero if the solution is left to stand for more than a week). This method for mercury in sediments does not distinguish between the different forms of organomercury. Down to 0.2 μg kg^{-1} mercury in sediments can be determined by this method with a standard deviation of 0.03 μg kg^{-1}. In this method a large weight sample (10–20 g) is extracted with hydrochloric acid for 2 days and organic mercury then extracted from the filtrate with benzene. Mercury is back extracted from the benzene with aqueous ammoniacal glutathione. This extract is then added to an aqueous solution containing sodium hydroxide, cupric sulphate and stannous chloride, and the elemental mercury released is swept off with nitrogen and, in a further concentration step, is collected on gold granules. Finally, the granules are heated at 500°C to release mercury that is determined by flameless atomic absorption spectrometry at 253.7 nm.

Workers at the Department of the Environment, United Kingdom [60] have described a procedure for the determination of methylmercury compounds in soils and sediments, which involves extraction with a carbon tetrachloride solution of dithizone, reduction to elemental mercury, then analysis by atomic absorption spectrometry.

Various other workers have discussed the application of atomic absorption spectrometry to the determination of organomercury residues in nonsaline sediments [9,52,54,60–69].

Bartlett et al. [70] and Longbottom et al. [71] observed unexpected behaviour of methylmercury-containing River Mersey sediments during storage. They experienced difficulty in obtaining consistent methylmercury values; supposedly identical samples analysed at intervals of a few days gave markedly different results. They followed the levels of methylmercury in selected sediments over a period, to determine if any change was occurring on storage. They found that the amounts of methylmercury observed in the stored sediments did not remain constant; initially there was a rise in the amount of methylmercury observed, and then, after about 10 days, the amount present began to decline to levels which in general only approximate those originally present. They observed this phenomenon in nearly all of the Mersey sediment samples they examined. It was noted that sediments sterilised, normally by autoclaving at approximately 120°C, did not produce methylmercury on incubation with organic mercury, suggesting a microbiological origin for the methylmercury. A control experiment was carried out in which identical samples were collected and homogenised. Some of the samples were sterilised by treatment with an approximate 4 wt.% solution of formaldehyde. Several samples of both sterilised and unsterilised sediments were analysed at intervals, and all of the samples were stored at ambient room temperature (18°C) in the laboratory. There is a difference in behaviour between the sterilised and unsterilised samples. Some of the samples were separately inoculated into various growth media to test for microbiological activity.

This work suggests that the application of laboratory-derived results directly to natural conditions could, in these cases, be misleading: analytical results for day 10, if extrapolated directly, might lead to the conclusion that natural methylmercury levels and rates of methylation are much greater than in fact they really are. Work in this area, with model or laboratory systems, needs to be interpreted with particular caution.

Bartlett et al. [70] used the method of Uthe et al. [72] for determining methylmercury. Sediment samples of 2–5 g were extracted with toluene after treatment with copper sulphate and an acidic solution of potassium bromide. Methylmercury was then back extracted into aqueous sodium thiosulphate. This was then treated with acidic potassium bromide and copper sulphate following which the methylmercury was extracted into pesticide grade benzene containing approximately 100 $\mu g\ L^{-1}$ of ethyl mercuric chloride as an internal standard. The extract was analysed by electron-capture gas chromatography. The detection limit was 1–2 $\mu g\ kg^{-1}$.

Rodriguez Martin-Doimeadios et al. [73] used speciated isotope dilution (SID) coupled with gas chromatography-undiluted coupled with plasma MS to determine and unravel the artificial formation of monomethylmercury in reference samples.

SID-MS is claimed to be an absolute method; however, it has been found to be affected by artefact monomethylmercury formation in sediments. The determination of chloromethylmercury in sediments was carried out by SID-MS after open-focused microwave extraction. The extracted mercury species were then ethylated and separated by capillary gas chromatography. Isotope ratios (peak area ratios at different masses) were measured by on-line ICP-MS detection of the capillary gas chromatographically separated compounds. Reproducibility of $^{202}Hg/^{201}Hg$ isotope ratio measurements was 0.60% for MeEtHg and 0.69% for Et_2Hg; and for $^{202}Hg/^{199}Hg$, 0.43% was determined. The absolute detection limits for capillary gas chromatography-ICPMD measurements were better than 26 fg for ^{202}Hg, 20 fg for ^{201}Hg, and 24 fg for ^{199}Hg. In the direct determination of monomethylmercury in sediment reference materials (CRM 580, IAEA 356 and IAEA 405), higher values than the certified were always found. Systematic experiments were carried out to localise the sources of the unintentional abiotic methylmercury formation during analysis. Different spiking and derivatisation procedures (either ethylation, propylation or derivatisation by Grignard reagents) were tested. In addition, isotopically enriched inorganic mercury was spiked. The amount of inorganic mercury initially present in the sample was found to be the critical factor that should be known and carefully controlled. A simple solvent extraction technique involving no critical steps was applied in order to reduce Hg^{2+} concentration when it is high. The method was applied to the determination of monomethylmercury in sediment reference material IAEA 405 with satisfactory results after organic solvent extraction. The limitations of applicability of the proposed method are evaluated as related to inorganic mercury, organic carbon and sulphur contents. The results obtained confirmed that available sediment reference materials are adequate to achieve traceable mercury speciation analysis and to detect potential sources of monomethylmercury artefact formation.

Jensen and Jernelou [74] reported that both mono- and dimethylmercury [CH_3Hg^+ and $(CH_3)_2Hg$] can be produced in lake sediments and in fish.

The gases evolved from incubated sediment samples were analysed by gas chromatography for monomethylmercury by conversion to methylmercury halide.

Cappon and Crispin-Smith [75] have described a method for the extraction, clean-up and gas-chromatographic determination of alkyl and aryl mercury compounds in sediments. The organomercury compounds are converted to their chloro-derivatives and solvent extracted. Inorganic mercury is then isolated as methylmercury upon reaction with tetramethyltin. The initial extract is subjected to a thiosulphate clean-up, and the organomercury species are isolated as their bromoderivatives. Total mercury recovery was in the range of 75%–90% and down to 1 $\mu g\ kg^{-1}$ of specific compounds can be determined.

Ealy et al. [76] determined methyl-, ethyl- and methoxyethylmercury compounds in sediments by leaching the sample with sodium iodide for 24 hours and then extracting the alkylmercury iodides into benzene. The iodides are then determined by gas chromatography of the benzene extract with electron-capture detection (^3H foil). Good separation of chromatographic peaks is obtained for the mercury compounds as chloride.

Andren and Harris [77] have reported a methylmercury concentration of 0.02–0.1 ng Hg g^{-1} in unpolluted sediments by using a gas chromatograph with an electron-capture detector.

A procedure has been described by Lee [78] in which 20 L of water containing alkylmercury compounds is concentrated with sulphydryl cotton fibre. The alkylmercury compounds adsorb on to the fibre quantitatively. The cotton fibre is then removed and extracted with a few millilitres of dilute hydrochloric acid/sodium chloride, that is the preconcentration stage in which a concentration factor of about 20,000 is achieved. The alkylmercury content of the extract is then determined by gas chromatography using an electron-capture detector. A detection limit of 0.04 ng L^{-1} alkylmercury is achieved in this procedure.

In a standard procedure [79] the sediment is wet oxidised with dilute sulphuric acid and nitric acids in an apparatus in which the vapour from the digestion is condensed into a reservoir from which it can be collected or returned to the digestion flask as required. The combined oxidised residue and condensate are diluted until the acid concentration is 1 N and nitrate is removed by addition of hydroxylammonium chloride with boiling. Fat is removed from the cooled solution with a carbon tetrachloride solution of dithizone. The extract is shaken with 0.1 N hydrochloric acid and sodium nitrite solution after treatment of the separated aqueous layer with hydroxylammonium chloride and then a solution of urea and finally EDTA solution are added. This prevents subsequent extraction of copper. The liquid is then extracted with a 0.10% solution of dithizone in carbon tetrachloride and mercury estimated in the extract spectrophotometrically at 485 nm.

Jirka and Carter [57] have described an automated determination of down to 0.1 $\mu g\,L^{-1}$ mercury in river sediment samples. This method is based on the automated procedure of El-Awady et al. [32] for the determination of total mercury in waters and wastewaters, in which potassium persulphate and sulphuric acid were used to digest samples for analysis by the cold vapour technique. These workers proved that the use of potassium permanganate as an additional oxidising agent was unnecessary.

There was no significant interference due to sulphide in the solutions containing 10 mg sulphide L^{-1}. However, a negative interference was observed for both organic and inorganic standards containing 100 mg sulphide L^{-1}, which is equivalent to 25,000 mg sulphide kg^{-1} in the sediment. This interference was overcome by ensuring that an excess of dichromate was present during the automated analysis.

This automated procedure was estimated to have a precision of 0.13–0.21 mg Hg kg^{-1} at the 1 mg Hg kg^{-1} level with standard deviations varying from 0.011 to 0.02 mg Hg kg^{-1}, that is relative standard deviations of 8.4%–12% at the 17.2–32.3 mg Hg kg^{-1} level in sediments. Recoveries in methylmercuric chloride spiking studies were between 85% and 125%.

Robert and Robenstein [80] carried out an indirect determination of Hg^{119} by preparing NMR spectra of methylmercury complexes, for example CH_3Hg^{2+} thiol ligands in sediment samples.

Feldman [62] digested solid samples with potassium dichromate, nitric acid, perchloric acid and sulphuric acid [62]. Bishop et al. [63] used aqua regia and potassium permanganate for digestion of organomercury compounds. Jacobs and Keeney [64] oxidised sediment samples using aqua regia, potassium permanganate and potassium persulphate [64]. The approved US Environmental Protection Agency digestion procedure requires aqua regia and potassium permanganate as oxidants [81].

These digestion procedures are slow and often hazardous because of the combination of strong oxidising agents and high temperatures. In some of the methods, mercuric sulphide is not adequately recovered. The oxidising reagents, especially the potassium permanganate, are commonly contaminated with mercury, which prevents accurate results at low concentrations.

In lakes and streams, mercury can collect in the bottom sediments, where it may remain for long periods of time. It is difficult to release the mercury from these matrices for analysis. Several investigators have liberated mercury from soil and sediment samples by the application of heat to the samples and the collection of the released mercury on gold surfaces. The mercury was then released from the gold by application of heat or by absorption in a solution containing oxidising agents [65,69].

Batti et al. [82] determined methylmercury in river sediments from industrial and mining areas.

Bretlhauer et al. [66] described a method in which samples were ignited in a high-pressure oxygen-filled bomb. After ignition the mercury was absorbed in a nitric acid solution. Pillay et al. [67] used a wet-ashing procedure with sulphuric acid and perchloric acid to digest samples. The released mercury was precipitated as the sulphide. The precipitate was then redigested using aqua regia.

As the concentration of organomercury compounds encountered in actual sediment samples can occur at levels as low as 0.00001–0.004 mg kg^{-1}, considerable experimental skill is needed to adapt methods such as preconcentration that will give results on actual samples that have a very low level of contamination.

19.5 Organosilicon compounds

Pellenberg [83] and others [10] analysed river sediment for silicone content by nitrous oxide–acetylene flame atomic absorption spectrophotometry. Pellenberg [83] showed that total carbon and total carbohydrates both correlate with silicone content, and the correlation between sedimentary silicone and presumed sewage material is good enough to suggest silicone as a totally synthetic, specific tracer for sewage in the aquatic environment.

Wanatabe et al. [84] have described a method for the separation and determination of siloxanes in sediment, using inductively coupled plasma emission spectrometry. A petroleum ether extract of the sediments is evaporated to dryness. The damp residue is dissolved in methyl isobutyl ketone and aspirated into the plasma. The detection limit is 0.01 mg kg^{-1}. Recoveries are about 50% with a coefficient of variation of about 11%.

Van der Post [85] has described a method for the determination of silanols in water, based on their ability to reduce nitrite or nitrate to ammonia at normal temperature. Individual silanols were identified by MS.

References

[1] J. Pacheon-Argona, P. Rodriguez, M. Valiente, D. Barclay, O.F.X. Donard, Int. J. Environ. Anal. Chem. 88 (2008) 923.
[2] W.A. Maher, Anal. Chim. Acta 126 (1981) 157.
[3] Y. Odanaka, W. Tsuchiya, O. Matono, S. Goto, Anal. Chem. 55 (1983) 929.
[4] H.R. Potter, A.W.P. Jarview, R.N. Markell, Water Pollut. Control 76 (1977) 123.
[5] Y.K. Chau, P.T.S. Wong, G.A. Bengert, P. Kramar, Anal. Chem. 51 (1979) 186.
[6] Y.K. Chau, P.T.S. Wong, P. Kramer, Anal. Chim. Acta 146 (1983) 211.
[7] Y.K. Chau, P.T.S. Wong, H. Saitoh, J. Chromatogr. Sci. 162 (1976) 14.
[8] D.A. Segar, Anal. Lett., London 7 (1974) 89.
[9] Y.K. Chau, P.T.S. Wong, G.A. Bengert, J.L. Dunn, Anal. Chem. 56 (1984) 271.
[10] K. Reisinger, M. Stoeppler, H.W. Nurnberg, Nature, London 291 (1981) 228.
[11] P.T.S. Wong, Y.K. Chan, P.L. Luxon, Nature, London 253 (1975) 26.
[12] Nemonic, R. Milacic, J. Sconcor, Int. J. Environ. Anal. Chem. 87 (2007) 615.

[13] L. Ebdon, S. Hill, B. Griepink, Environ. Technol. Lett. 9 (1988) 965.
[14] Y.K. Chau, Biological methylation of tin? Compounds in the aquatic environment, 3rd International Conference Organometal Coordinating Chemistry Germanium, Tin, Lead, University of Dortmund, West Germany, July 1980.
[15] L.E. Hallas (Ph.D. dissertation), University of Maryland, 1980.
[16] L.E. Hallas, J.C. Mearns, J.J. Cooney, Science 213 (1982) 1505.
[17] L.E. Hallas, J.J. Cooney, Appl. Environ. Microbiol. 41 (1981) 466.
[18] H.E. Guard, A.B. Cobet, W.M. Coleman, Science 213 (1981) 770.
[19] P.J. Craig, Environ, Technol. Lett. 1 (1980) 225.
[20] M.A. Unger, W.G. MacIntyre, R.J. Huggett, Environ. Toxicol. Chem. 7 (1988) 907.
[21] G. Rapsomkankis, O.F. Donand, J.H. Weber, Appl. Organomet. Chem. 1 (1987) 115.
[22] Y. Arakawa, O. Wadan, T.H. Yu, H. Iwai, J. Chromatogr. 216 (1981) 209.
[23] Y.K. Chau, P.T.S. Wong, G.A. Bengert, Anal. Chem. 54 (1982) 946.
[24] Y. Hattori, A. Kabeyashi, S. Takemoto, K. Takami, A. Sigimae, N. Nakamoto, J. Chromatogr. 315 (1984) 341.
[25] Y. Cai, R. Aizaga, J.M. Bayona, Anal. Chem. 66 (1994) 1161.
[26] M.D. Mueller, Fresenius Z. Anal. Chem. 317 (1984) 32.
[27] M.D. Mueller, Anal. Chem. 59 (1987) 617.
[28] R. Lobinski, W.M.R. Dirk, M. Ceulemans, F.C. Adams, Anal. Chem. 64 (1992) 159.
[29] C.C. Gilmour, J.A. Tuttle, J.C. Means, Anal. Chem. 58 (1986) 1848.
[30] J. Szpunar, V.O. Schmitt, R. Lobinski, J.L. Moned, J. Anal. At. Spectrosc. 11 (1996) 193.
[31] A. Prange, E. Jensen, J. Anal. At. Spectrosc. 10 (1995) 105.
[32] A.A. El-Awady, R.B. Miller, M.J. Carter, Anal. Chem. 48 (1976) 110.
[33] K.L. Lu, I.D. Pulford, H.J. Duncan, Anal. Chim. Acta 106 (1979) 139.
[34] A.M. Ure, M.I. Hernandez-Artiga, M.C. Mitchell, Anal. Chim. Acta 96 (1978) 37.
[35] B. Pederson, M. Willems, J.S. Jorgensen, Analyst 105 (1980) 119.
[36] E.L. Henn, Flame Atomic Absorption Analysis, ASTM STP 618, American Society of Testing Materials, Philadelphia, PA, pp. 54–64.
[37] J.M. Jones, Commun. Soil Sci. Plant Anal. 8 (1977) 340.
[38] R.L. Dahlquist, J.W. Knoll, Anal. Spectrus 32 (1978) 1.
[39] MAFF (Ministry of Agriculture, Fisheries and Food), Method 54: nickel, nitric perchloric acid solubles in soil, in: The Analysis of Agricultural Material, second ed., RB427, HMSO, London, 1979.
[40] H.J. Yang, S.J. Jiang, Y. Yong, C. Hwang, Anal. Chim. Acta 312 (1995) 141.
[41] W. Longsett Talanta, Talanta 31 (1984) 975.
[42] T.H. Yu, Y. Arakawa, J. Chromatogr. 258 (1983) 189.
[43] K.L. Jewitt, F. Brinkman, W.P. Everson, J. Chromatogr. Sci. 19 (1981) 583.
[44] K.S. Epler, T.C. O'Haver, G.C. Turk, W.A. MacCrehan, Anal. Chem. 60 (1988) 2062.
[45] S.A. Sinex, A.G. Cantillo, G.R. Helz, Anal. Chem. 52 (1980) 2342.
[46] M.D. Stephenson, D.R. Smith, Anal. Chem. 60 (1988) 696.
[47] P. Kenis, A. Zirino, Anal. Chim. Acta 149 (1983) 157.
[48] Y. Arakawa, D. Wade, M. Wantabe, Anal. Chem. 55 (1983) 1901.
[49] Y. Takahashi, N. Sakakibari, M. Nomura, Anal. Chem. 76 (2004) 4307.
[50] M. Staniszewska, B. Radke, J. Namiesnik, J. Boltatech, Int. J. Environ. Anal. Chem. 88 (2008) 747.
[51] J.K. Iskandor, J.K. Severs, L.W. Jakobs, D.R. Keenan, J.T. Gilmour, Analyst 97 (1972) 388.
[52] F.J. Langmyhr, J. Aamodt, Anal. Chim. Acta 87 (1976) 483.

[53] K. Matsunaga, T. Takahasi, Anal. Chim. Acta 87 (1976) 487.
[54] P.J. Craig, S.F. Morton, Nature 261 (1976) 125.
[55] Official Methods of Analysis of the AOAC, 22nd ed., 1976, p. 418.
[56] AOAC Official Methods of the Association of Official Analytical Chemists, 11th ed., 1979, p. 418.
[57] A.M. Jirka, J.M. Carter, Anal. Chem. 50 (1978) 91.
[58] A.A. Alwady, R.B. Miller, W.J. Carter, Anal. Chem. 48 (1976) 112.
[59] Y. Umezaki, K. Iwamoto, Jpn. Anal. 20 (1971) 173.
[60] Department of the Environment at National Water Council UK, Mercury in Waters, Eluents, Soils and Sediments, Additional Methods—22-AGENW, HMSOL, London, 1985.
[61] K. Irukayama, M. Fukiki, S. Tajima, S. Omori, Jpn. J. Public Health 19 (1972) 25.
[62] C. Feldman, Anal. Chem. 46 (1974) 1606.
[63] J.N. Bishop, L.A. Taylor, B.P. Neary, The Determination of Mercury in Environment Samples, Ministry of the Environment, Ontario, 1973.
[64] L.W. Jacobs, D.R. Keeney, Environ. Sci. Technol. 8 (1976) 267.
[65] P.C. Leong, H.P. Ong, Anal. Chem. 43 (1971) 940.
[66] T.W. Bretlhauer, A.N. Moghissi, S.S. Snyder, N.W. Matthews, Anal. Chem. 46 (1974) 445.
[67] K.K.S. Pillay, C.C. Thomas, C.J.A. Sonde, C.M. Hyone, Anal. Chem. 43 (1971) 1419.
[68] S. Jensen, M.P. Jernlov, Nature 223 (1969) 753.
[69] D.H. Anderson, J.H. Evans, J.J. Murphy, W.W. White, Anal. Chem. 43 (1971) 1511.
[70] P.D. Bartlett, P.J. Craig, S.F. Morton, Nature 267 (1977) 606.
[71] J.E. Longbottom, R.C. Dressman, J.J. Litchenberg, J. Assoc. Off. Anal. Chem. 56 (1973) 1297.
[72] J.F. Uthe, E.A.J. Armstrong, K.C. Tam, J. Assoc. Off. Anal. Chem. 54 (1972) 866.
[73] R.C. Rodriguez Martin Doimeadios, M. Monperros, E. Krupp, D. Amouroux, O.F.X. Donard, Anal. Chem. 75 (2003) 3202.
[74] S. Jensen, Jernelou, Nature 223 (1969) 753.
[75] C.J. Cappon, V. Crispin-Smith, J. Anal. Chem. 49 (1977) 365.
[76] J.A. Ealy, W.D. Schulz, J.A. Dean, Anal. Chim. Acta 64 (1973) 235.
[77] W. Andren, R.C. Harris, Nature 245 (1973) 256.
[78] Y.H. Lee, Int. J. Environ. Anal. Chem. 29 (1987) 263.
[79] Society for Analytical Chemistry, Report of Metallics, Impurities in Soils, Committee of the Society for Analytical Chemistry, 1965.
[80] J.M. Robert, D.L. Robenstein, Anal. Chem. 53 (1991) 2074.
[81] Environment Protection Agency, Methods for Chemical Analysis of Waters and Wastes, US Environment Protection Agency, Cincinnati, OH, 1974, pp. 134–138.
[82] R. Batti, R. Magnaval, E. Lonzala, Chemosphere 4 (1975) 13.
[83] R. Pellenberg, Mar. Pollut. Bull. 10 (1979) 267.
[84] N. Wanatabe, Y. Yasuda, K. Kato, Sci. Total Environ. 34 (1984) 169.
[85] P.C. Van der Post, Water Pollut. Control 77 (1978) 52.

Index

Note: Page numbers followed by "*t*" refer to tables.

A

Abate, 170–172
Abiotic surface-mediated reaction, 357
Absolute chlorine concentrations, 357
Absorption detectors, 220
Accelerated solvent extraction, 281
Acciothion, 170–172
Acetic acid, 37
Acetic anhydride, 86
Acetonitrile, 82, 229
Acid stannous chloride, 241–242
Acid-catalyst esterification reaction, 188
Acrylic acid, 37
Acynoester pyrethroid, 200
Adenosine triphosphate, 120
Adsorbed trialkyllead compounds, 257
AgGSTD1–1 enzyme, 200–201
AgGSTD1–6 enzyme, 200–201
Alcohols
  GC, 52
  spectrophotometric methods, 51–52
Aldehydes
  GC, 54–55
  miscellaneous, 55–56
  spectrophotometric methods, 55
  thin-layer chromatography, 55
Aldicarb, 195
Algal
  chlorophyll a, 222
  toxins and blooms, 226
Aliphatic amines
  GC, 103–104
  HPLC, 104
  miscellaneous, 104
Aliphatic fraction, 269–273
Aliphatic hydrocarbons, 1–17. *See also* Aromatic hydrocarbons
  continuous monitoring of oil slicks, 16–17
  fluorescence techniques, 11
  gas chromatography, 5–6
  gas stripping methods, 4–5
  head space analysis, 3–4
  high-performance liquid chromatography, 7
  infrared spectroscopy, 7–10
  metals in spillage oils, 16
  oil spillages, 13–16
  paper chromatography, 12
  sampling, 12–13
  in sediments, 351*t*
  thin-layer chromatography, 10–11
Alkali metal salt flame photometric detectors, 174
Alkali salts, 173–174
Alkyl esters, 187–188, 191
Alkyl phosphates, 119–120
Alkylbenzene sulphonates, 359
  surfactants, 135
Alkyllead salts, 345, 374
Alkylthiols, 124
Amberlite XAD-2 resins, 174, 225
Amberlite XAD-4 resins, 2, 78, 129
Ambersorb XE-34 cation-exchange resin, 50
American Petroleum Institute (API), 9
Ametryne, 192–193
Amides
  GC, 107
  HPLC, 107
  polarography, 107
Amidithion, 170–172
4-Aminiphenazone, 190
Amino acids
  GC, 106
  HPLC, 106
  miscellaneous, 106
4-Aminoantipyrine (4-APP), 47–48
Aminocarb, 195
Amiton, 170–172
Ammoniacal glutathione solution, 381–382
Ammonium cobaltothiocyanate spectrophotometric method, 130

Ammonium hydroxide, 27
Ammonium tetrathiocyanatocobaltate(II) complex, 132
Amperometric detector, 106
Anionic surface active agents, 133–136. *See also* Cationic surface active agents; Nonionic surface active agents
  gas chromatography, 133–134
  high-performance liquid chromatography, 134–135
  miscellaneous, 135–136
Anoxic waters, 222
Anthropogenic hormones, 227–228
Antibiotics, 228–229
API. *See* American Petroleum Institute (API)
Aromatic amines
  GC, 105
  HPLC, 105
  miscellaneous, 105–106
Aromatic fraction, 269–273
Aromatic hydrocarbons, 17–28. *See also* Aliphatic hydrocarbons
  column chromatography, 18
  fluorescence spectrometry, 26
  gas chromatography, 17–18, 22–24
  high-performance liquid chromatography, 24–25
  infrared spectrometry, 18–19
  miscellaneous, 26–28
  polycyclic aromatic hydrocarbons, 21–22
  spectrophotometric method, 18
  thin-layer chromatography, 25–26
  ultraviolet spectroscopy, 19–21
Aromatic organic compounds, 381
4-APP. *See* 4-Aminoantipyrine (4-APP)
Arsenic, 373
  specks, 341
Artemisia annua (sweet wormwood), 332
Artemisinin, 332
Artificial hydrocarbons, 2–3
Aryl phosphates, 119–120
Asphaltene, 2
Atomic absorption spectrometry, 375
  for nitriloacetic acid, 112
  nonionic surface active agents, 132–133
  organoarsenic compounds, 247
  organolead compounds, 245
  organomercury compounds, 241–243
  for phenols, 49
ATR. *See* Attenuated total reflection (ATR)
Atraton, 192–193
Atrazine, 192–193
Attenuated total reflection (ATR), 288
Aurlorite XA D-4 resin, 133
Autoanalyser spectrophotometric determination of nonionic surfactants, 131
Automated liquid-chromatographic–mass-spectrometric method, 86
Automatic flow-analysis procedure, 137
Automatic sampler, 165
Aviation fuels, 15
Azinphos-ethyl, 170–172
Azinphos-methyl, 170–172
Azothoate, 170–172
Azure A, 136

B

Baker's yeast cells, 194–195
Barium chloride–phosphomolybdic and spectrophotometric method, 130
Bensulide, 170–172
Benthiocarb, 195
Benzene, 381
Benzoic acid, 39–40
Benzophenone, 228
Benzylsuccinic acid, 20
5β-chlorestane-β-ol, 224–225
β-nicotinamideadenine dinucleotide [β-(NAD)$^+$], 38
Bio monitors for PAH determination, 28
Biological transformation, 325
Biomethylation, 376
Bis-(4-chlorophenyl) disulphide (DDS), 168
1,1-Bis-(4-chlorophenyl)ethylene (DMC ethylene), 168
Bismuth, 373
Bitumoids, 12
British Standard method for cationic detergents, 137
Bromine-containing compounds, 94–95
Bromophos, 170–172
Bromophos-ethyl, 170–172
Butonate carbophenothion, 170–172
Buturon, 182–183
Butyltin species, 235–236

C

$12C_{24}H_{14}$ polycyclic aromatic hydrocarbons, 27–28
Caesium bromide, 173–174

Capillary columns, 13
gas chromatography, 3, 17
GC–ECD method, 105
Capillary electrophoresis, 226, 341
for nitrophenols, 108
Capillary gas chromatography, 6, 68–69, 82, 257, 378, 380
Carbamate type, 195–199
enzymic assay, 198–199
gas chromatography, 195–196
high-performance liquid chromatography, 196–197
of insecticides in soil, 307, 311*t*
miscellaneous, 199
thin-layer chromatography, 197–198
Carbaryl, 195, 199
Carbine, 195
Carbofuran, 195, 198–199
Carbofuranphenol, 198
Carbohydrates
GC, 57
miscellaneous, 57–58
spectrophotometric method, 57
Carbon dioxide, 146, 263
Carbon disulphide, 125
Carbon tetrachloride, 10–11
Carbon-13 isotope (C^{13} isotope), 275–276
Carboxylic acids, 35–42, 256
fluorescence spectrometry, 40
GC, 36–37
high-performance liquid chromatography, 37–39
miscellaneous, 41–42
polarography, 41
potentiometry, 41
spectrophotometric methods, 39–40
thin-layer chromatography, 39
Catalyst, 362
Cation-exchange resins, preconcentration on, 50
Cationic surface active agents, 136–138. *See also* Anionic surface active agents; Nonionic surface active agents
gas chromatography, 136
high-performance liquid chromatography, 136
miscellaneous, 137–138
spectrophotometric methods, 136–138
titration methods, 136, 138
Cellulose, 186
Centrifugation, 27
Ch1-GSB, 222
Chain oils, 12–13
Channel thin-layer chromatography, 10
Charcoal filter, 75
Chelation extraction method, 375
Chemeatcher passive sampler, 238
Chlorfenvinphos, 170–172
Chlorinated aliphatic hydrocarbons, 281
Chlorinated aromatic compounds, 290
Chlorinated hydrocarbons, 66–67, 351
Chlorinated insecticides, 166–167
Chlorinated organic compounds, 277–281
Chlorine containing insecticides and herbicides in soil, 303–307
Chlorine-containing compounds, 351–357
4-(Chloro-2-methyl phenoxyl) butyric acid (MCPB), 186
4-Chloro-2-methylphenoxyacetic acid (MCPA), 186–188
4-(4-Chloro-2-methylphenoxyl)-butyric acid, 188
Chloro-*n*-paraffins, 71
Chloroacetanilide herbicides, 366
Chloroaniline, 109
Chloroaromatic compounds
GC, 82
liquid chromatography, 82
miscellaneous, 82
Chlorobenzo sulphonic acid, 125
Chlorobenzoic acid, 281–282
methylation, 282
Chlorobromuron, 184
Chlorofluoromethane, 356
Chloroform, 76–78
Chloromethanes, 74
Chlorooxuron, 182–184
Chlorophenols
GC, 84–86
high-performance liquid chromatography, 86–87
miscellaneous, 88
preconcentration, 88
thin-layer chromatography, 87
2-(2-Chlorophenyl)-2-(4-chlorophenyl)-1,1-dichlorothene (2,4'-DDE), 201
Chlorophyll, 220
chlorophylls a and b, 221
Chlorotoluron, 169, 182–184
Chlorphonium, 170–172
Chlorpyrifos, 173
Chromadistillation, 52
Classical Zeisel method, 277

CNP. *See* 2,4,4-Dichlorophenyl-4'-nitrophenylether (CNP)
Cobalamin, 226
Cold vapour atomic absorption spectrometry, 256, 373, 380
Coleman 50 mercury analyser system, 240
Colorimetric method, 48
Column chromatography
  for aromatic hydrocarbons determination, 18
  for nitriloacetic acid, 112
  for nitrophenols, 108
  nonionic surface active agents, 128
Commercial creosote oil, 285
Conducted chemiresistant sensors for GC detection, 121
Contaminants, 146
Continuous liquid–liquid extraction, 255
Continuous monitoring of oil slicks, 16–17
Continuous solvent extraction procedure, 137
Conventional solvent extraction, 259, 263
  procedures for organic compounds in soils, 260*t*
  from soil packed cartridges, 259
Coprostanol, 224–225
Coumaphos, 170–172
Coumithoate, 170–172
Coupled enzymic high-performance liquid chromatographic method, 256
CP-4010 purge-and-trap injector, 51
Cross polarisation-MAS (CP-MAS), 286–288
Crotoxyphos, 170–172
Cruformate, 182
Cryoconcentration, 175
Cyanophos, 170–172
Cyfluthrin, 200, 326
Cypermethrin, 326

D

DAS-1. *See* Diaminostilbene type 1 (DAS-1)
Dasanite, 170–172
2,4-DB. *See* 2-Dichlorophenoxyl-butyric acid (2,4-DB)
2,4'-DDE. *See* 2-(2-Chlorophenyl)-2-(4-chlorophenyl)-1,1-dichlorothene (2,4'-DDE)
DDS. *See* Bis-(4-chlorophenyl) disulphide (DDS)
2,4'-DDT. *See* 1,1,1-Trichloro-2(2-chlorophenyl)-2-(4-chlorphenyl)ethane (2,4'-DDT)
Degassing technique, 4–6
Deltamethrin, 326
Demephion, 170–172
Demephion-*O*, 170–172
Demephion-*S*, 170–172
Demeton, 170–172
Demeton-methyl, 170–172
Demeton-*O*, 170–172
Demeton-*O*-methyl, 170–172
Demeton-*S*, 170–172
Demeton-*S*-methyl, 170–172
Derivatisation, 85
Dialkyltetralin sulphonates, 359
Diaminostilbene type 1 (DAS-1), 362–365
Diazinon, 170–173
Diazomethane, 282
Dicamba, 190
1,2-Dichlorobenzene, 138
2-Dichlorophenoxyl-butyric acid (2,4-DB), 186
2,4,4-Dichlorophenyl-4'-nitrophenylether (CNP), 167
Dichlorvos, 166, 170–172
Diethyldithiocarbamates, 378
Differential pulse polarographic procedure, 327
Dimethyl disulphide, 286
Dimethyl sulphide, 124–125
Dimethyl sulphoxide, 25, 123–124
1,5-Dimethyl-1,5-diazoundecamethylene polymethobromide, 138
Dimethylarsinic acid, 248
Dimethylarsinite, 246
2,4-Dinitrophenylhydrazine, 38
2,4-Dinitrophenylhydrazones, 56
Dioxans, 53
Direct method, 190
Direct solvent extraction, 373
Direct thermal desorption methods, 155
Dispersive liquid–liquid microextraction, 45, 185
Dissolved humic compounds, 222
Distyrylbiphenyl type (DSBP), 362–365
Disulphides, 123
  gas chromatography, 123
  titration method, 123
Disulphoton, 170–172
Dithiocarbamate resins, 256
Dithiol derivatisation, 341
Diuron, 182–185
DMC ethylene. *See* 1,1-Bis-(4-chlorophenyl) ethylene (DMC ethylene)
Drop headspace cell, 155
DSBP. *See* Distyrylbiphenyl type (DSBP)

DTD–GC–MS. *See* Thermal desorption–gas chromatography–mass spectrometry (DTD–GC–MS)
Dual-flame ionisation detector (FID), 36, 104
Dual-response detection, 15
Dursban, 170–172
Dye–ionic surfactant complex, 137
Dynamic headspace techniques, 79

E

Edifenphos, 170–172
EDTA. *See* Ethylenediaminetetraacetic acid (EDTA)
Electrochemical methods, 180–181
Electrokinetic chromatography for esters, 54
Electron-capture
  detection, 281, 325
  detector, 66–67
  GC coupled with glass capillary columns, 162–165
  GC technique, 67, 195
Electron-impact-induced mass spectra, 127–128
Electrospray, 184
Environmental Protection Agency, 26
Enzyme-linked immunoassay, 49
Enzymic assay, 58, 198–199
Esters
  electrokinetic chromatography, 54
  GC, 53
  high-performance liquid chromatography, 53–54
Ethane diol, 125
Ethanol, 10–11
Ether polydimethylsiloxane–divinylbenzene fibre, 303–307
Ethion, 170–172
Ethyl ether, 259
Ethylation, 378
Ethylene thiourea, 124
Ethylenediaminetetraacetic acid (EDTA), 111
Ethyllead salts, 345
Ethynyloestradiol, 226
Extraction, 277–281
  procedures, 172–173
  solvents for concentration of chlorinated insecticides, 163*t*
  techniques, 263
    of organic compounds from soils, 263*t*
Extractive photometric method for phenol, 48

F

Fatty acids, 35–36
Fenchlorphos, 170–172
Fenitrothion, 181
Fensulthion, 182
Fenthion, 170–173
Fenuron, 182–183
Fenvalerate, 326
FID. *See* Dual-flame ionisation detector (FID)
Field desorption mass spectrometry, 128
Fingerprint, 15
Fingerprinting parameters, 2
Flame ionisation
  detector, 155, 174
  gas chromatography, 22–23, 119
  sequential detection, 281
Flame photometric
  detector, 166, 174, 358–359
    sulphur detector, 14
  system, 124
Flameless atomic absorption spectrometry, 380–381
FloraSil
  chromatography, 6, 8
  column clean-up procedure, 23–24
Flow injection method, 137
Flow microsensor, 55
Fluorescence
  detectors, 220
  spectroscopy, 221
    for aromatic hydrocarbons determination, 26
    for carboxylic acids, 40
    humic and fulvic acid, 224–225
  techniques, 11
    for phenols, 46
  whitening agents, 362–365
Fonofos, 170–172
Formaldehyde, 55
Fourier transform infrared (FT-IR), 277
Fourier transform infrared spectroscopy, 150
Freely dissolved pore water concentrations, 273
FT-IR. *See* Fourier transform infrared (FT-IR)
Fuel catechol, 27–28
Fulvic acid, 221–225, 286–288
  fluorescence spectroscopy, 224–225
  gel permeation chromatography, 223
  polarography, 223
  ultraviolet spectroscopy, 223–224
Fungicides, 331*t*, 332

G

Gas chromatography (GC), 22–24, 36, 67, 104, 141, 145, 162, 223, 275, 316, 351, 373–374
for acrylonitrile, 110
for alcohols, 52
for aldehydes, 54–55
for aliphatic amines, 103–104
for aliphatic hydrocarbons determination, 5–6
for amides, 107
for amino acids, 106
anionic surface active agents, 133–134
for aromatic hydrocarbons determination, 17–18
carbamate type, 195–196
for carbohydrates, 57
for carboxylic acids, 36–37
cationic surface active agents, 136
for chloroaromatic compounds, 82
for chlorophenols, 84–86
detection
conducted chemiresistant sensors for, 121
with supported copper–cuprous oxide island film, 121
surface acoustic wave sensors for, 121
determination of organic compounds, 149
for dioxans, 53
for esters, 53
extraction solvents for concentration of chlorinated insecticides, 163*t*
for glycols, 52
for halocarboxylic acids, 82
for haloforms, 77–78
ionisation–quadrupole mass spectrometer, 95
mercaptans and disulphides, 123
of multicomponent organic mixtures in natural water, 147*t*
for nitriloacetic acid, 112–113
for nitrophenols, 107–108
for nitrosamines, 110–111
nonionic surface active agents, 127–128
organoarsenic compounds, 247
organochlorine insecticides, 162–166
organolead compounds, 244–245
organomercury compounds, 239–241
organophosphorus insecticide, 173–176
for phenols, 42–44
for polychlorobiphenyls, 89
for saturated aliphatic chloro compounds, 66–74
separation, 256–257
techniques, 2, 307, 351–356
for identifying petroleum products, 14
for triarylphosphate esters, 119
triazine type, 191–193
for trinitrotoluene, 109
urea herbicides, 183–184
Gas chromatography with microcell electron detector (GC-μECD), 69
Gas chromatography-flame ionisation system, 21
Gas chromatography–atomic absorption spectroscopy, 345
Gas chromatography–mass spectrometry (GC-MS), 24, 51, 235–237, 255, 378
organochlorine insecticides, 167–168
organotin compounds, 235–237
for polychlorobiphenyls, 89–91
for polychlorodibenzo-*p*-dioxins and polychlorodibenzofurans, 83–84
for unsaturated chloroaliphatic compounds, 74
Gas flame ionisation chromatography, 104
Gas stripping methods, 4–5
Gas-chromatographic–atomic-absorption spectrometry, 375
Gas-liquid chromatographic–mass spectrometric approach, 22
Gas–liquid chromatography, 127
Gasoline recovery, 6
GC. *See* Gas chromatography (GC)
GC-AED, 51
GC-MS. *See* Gas chromatography–mass spectrometry (GC-MS)
GC-μECD. *See* Gas chromatography with microcell electron detector (GC-μECD)
GCXGC. *See* Two-dimensional GC (GCXGC)
Gel chromatography, 281
Gel filtration chromatography, 223
Gel permeation chromatography, 223
for haloforms, 81
Geosmin, 225–226
Glow discharge mass chromatography–mass spectrometry, 341
Glutathione-*S*-transferase (9ST), 200
Glycols
GC, 52
spectrometry, 52
thin-layer chromatography, 53

Graphite-furnace atomic absorption spectrometry, 248
Growth regulators, 285
GS-13005, 170–172

H

Halocarboxylic acids
GC, 82
isotope dilution mass spectrometry, 83
miscellaneous, 83
Haloforms, 75–81, 76*t*
GC, 77–78
gel-permeation chromatography, 81
headspace analysis, 79
high-performance liquid chromatography, 81
preconcentration, 81
purge and trap methods, 79–80
resin adsorption–gas chromatography, 80–81
spectrophotometric method, 81
Halogen-containing compounds. *See also* Nitrogen-containing compounds; Oxygen-containing compounds; Phosphorus containing compounds
in nonsaline waters, 66
bromine-containing compounds, 94–95
chloroaromatic compounds, 82
chlorophenols, 84–88
halocarboxylic acids, 82–83
haloforms, 75–81
miscellaneous, 93–94
polychlorobiphenyls, 88–93
polychlorodibenzo-*p*-dioxins and polychlorodibenzofurans, 83–84
polychloroterphenyls, 93
saturated aliphatic chloro compounds, 66–74
unsaturated chloroaliphatic compounds, 74–75
in sediments, 354*t*
in soil, 277–283, 278*t*
chlorinated aliphatic hydrocarbons, 281
chlorinated organic compounds, 277–281
chlorobenzoic acid, 281–282
perfluorooctane sulphonyl fluoride, 282–283
Halogenated aliphatic hydrocarbons, 78
HCB. *See* Hexachlorozbenzene (HCB)
HCH. *See* Hexachlorocyclohexane (HCH)
Headspace
analysis, 3–4
for haloforms, 79
for saturated aliphatic chloro compounds, 71–72
for unsaturated chloroaliphatic compounds, 75
gas chromatography, 17
gaschromatographic method, 83
single-drop microextraction, 69, 155
volume, 69
Helium, 14
discharge detector, 82
microwave-induced plasma emission detector, 380
plasma microwave emission spectrometry, 174
Herbicides, 359–362, 360*t*
chlorine containing insecticides and herbicides in soil, 303–307
imidazolinone herbicides in soils, 307–316
miscellaneous herbicides in soil, 329*t*
organophosphorus-type herbicides in soil, 316–318
phenoxy acetic acid herbicides in soil, 307
substituted urea-type herbicides in soils, 307
triazine herbicides in soil, 307
Hexachlorocyclohexane (HCH), 359–362
Hexachlorozbenzene (HCB), 359–362
Hexane 1,6-diamine, 104
High-performance liquid chromatography (HPLC), 105, 145, 255, 276, 303–307, 373
for aliphatic amines, 104
for aliphatic hydrocarbons determination, 7
for amides, 107
for amino acids, 106
anionic surface active agents, 134–135
for aromatic amines, 105
for aromatic hydrocarbons, 24–25
carbamate type, 196–197
for carboxylic acids, 37–39
cationic surface active agents, 136
for chlorophenols, 86–87
determination of organic compounds, 149–150
for esters, 53–54
for haloforms, 81
HPLC–electron-capture detector approach, 169
of organic mixtures in natural waters, 151*t*

High-performance liquid chromatography (HPLC) (*Continued*)
organochlorine insecticides, 168–169
organophosphorus insecticide, 176–177
for phenols, 44–45
plant pigments, 219–220
for polychlorobiphenyls, 91
for saturated aliphatic chloro compounds, 70–71
triazine type, 193–194
for unsaturated chloroaliphatic compounds, 75
urea herbicides, 184–186
High-performance liquid chromatography with fluorescence detection (HPLC–FLD), 129–130
High-pressure
oxygen-filled bomb, 386
solvent, 285
High-resolution capillary gas chromatograph, 256
High-resolution glass capillary columns, 281
High-resolution mass spectrometry coupled with selective ion monitoring (HRMS-SIM), 84
Highly cross-linked polystyrene-divinylbenzene sorbent, 276
HPLC. *See* High-performance liquid chromatography (HPLC)
HPLC–FLD. *See* High-performance liquid chromatography with fluorescence detection (HPLC–FLD)
HRMS-SIM. *See* High-resolution mass spectrometry coupled with selective ion monitoring (HRMS-SIM)
Humic acid, 221–225, 286–288
fluorescence spectroscopy, 224–225
gel permeation chromatography, 223
polarography, 223
ultraviolet spectroscopy, 223–224
Hydrazines, 109–110, 285
Hydrocarbons, 269–273, 270*t*
in nonsaline waters
aliphatic hydrocarbons, 1–17
aromatic hydrocarbons, 17–28
in sediments, 349
Hydrolysis, 119
kinetics of sulphonylurea herbicide, 186
Hydrophobic film, 277–281
3-Hydroxycarbofuran, 198
Hydroxydimethylarsine oxide, 247

I

IDLs. *See* Instrumental detection limits (IDLs)
Imidazolinone herbicides in soils, 307–316, 315*t*
In situ characterisation, 146
Indirect method, 190
Indirect potentiometric stripping analysis method, 112
Inductivity-coupled plasma mass spectrometry
organotin compounds, 238
Infrared spectroradiometry, 16–17
Infrared spectroscopy (IR spectroscopy), 349
for aliphatic hydrocarbons, 7–10
for aromatic hydrocarbons, 18–19
determination of organic compounds, 150–154
IR spectroscopy–gas chromatography, 2
Inorganic ionic lead salts, 345
Inorganic salt, 341
Inorganic tin species, 235
Inositol triphosphate, 120
Insecticides, 359–362, 360*t*
carbamate type of insecticides in soil, 307
chlorine containing insecticides and herbicides in soil, 303–307
miscellaneous insecticides in soil, 318–332, 319*t*, 329*t*
organophosphorus, 121
Instrumental detection limits (IDLs), 95
Integrated submersible sensor probe, 145
Ion chromatographic method, 285
Ion exchange technique for chlorophenols, 85
Ion-exchange chromatography
coupled with plasma mass spectrometry, 341
nonionic surface active agents, 129–130
organoarsenic compounds, 248
Ion-exchange high-performance liquid chromatography, 379
Ion-exclusion chromatography, 256
Ion-pairing reagent, 220
Ion-selective electrode potentiometry, 286
Ionic alkylleads, 345
Ionic chromatography, 51
IR spectroscopy. *See* Infrared spectroscopy (IR spectroscopy)
Isotope dilution, 359
GC–MS, 188, 193
mass spectrometry for chloroaromatic compounds, 83

J

JEOL Model, 225
  JCG-20K gas chromatograph
  JMA-2000 mass data analysis system, 225
  JMS-D 100 mass spectrometer, 225

K

Karmen–Guiffrida detector, 173–174
Kepone, 166
Kerosene, 13–15
3-Ketocarbofuran, 198
Ketones, 56, 276–277
Kieselguhr G, 220

L

Lactams
  enzymic assay, 58
  polarography, 58
  thin-layer chromatography, 58
λ-cyhalothrin, 326
Laser-enhanced ionisation, 379
Laser-excited resonance Raman spectroscopy, 46
Lauryl alcohol ethoxylates, 127–128
LC. *See* Liquid chromatography (LC)
Lead, 373
  chloride, 376
  nitrate, 376
Lectin glycoenzyme multilayer film–modified biosensor, 109
Lignins, 222
Lignosulphonic acids, 223
Limits of detection, 70
Linuron, 182–183, 185
Lipophilic characteristics, 282
Liquid chromatography (LC), 22, 71, 219, 351
  for chloroaniline, 109
  for chloroaromatic compounds, 82
  LC–electrospray MS, 307–316
Liquid extraction method, 3
Liquid-phase microextraction (LPME), 288
Liquid–liquid extraction
  solvents, 79
  techniques, 190–191
Low-boiling solvent, 259
LPME. *See* Liquid-phase microextraction (LPME)
Luminescence analysis, 120
Lyophilisation, 175

M

Macroreticular resins, 125
Macroreticular XAD-4 resin, 37, 41
Magic angle spinning (MAS), 283
Magnesium sulphate, 27
Malathion, 182
Malic acid, 40
MAS. *See* Magic angle spinning (MAS)
Mass spectrometry (MS), 104, 141, 166, 226, 275, 307–316, 341, 356, 378
  for unsaturated chloroaliphatic compounds, 75
MCPA. *See* 4-Chloro-2-methylphenoxyacetic acid (MCPA)
MCPB. *See* 4-(Chloro-2-methyl phenoxyl) butyric acid (MCPB)
Mercaptans, 123
  gas chromatography, 123
  titration method, 123
Mercury, 373
  Mercury-containing sample storage, 243–244
Mestranol, 226, 228
Metabenzthiazuron, 182–183
Metalkinate, 195
Metals in spillage oils, 16
Metaxuron, 182–183
Methacrylic acid, 40
Methamidophos, 318
Methanol-modified carbon dioxide, 357
Methanolic hydrochloric acid, 373, 378
Methanoyl, 195
Methiocarb, 195
Methocrotophos, 170–172
Method 501.1 for trihalomethane analysis, 80
Method 501.2 for trihalomethane analysis, 80
Methoxy groups, 277
1-Methoxyl-1-methyl-3-phenylurea herbicides, 183
Methyl chloroformate, 282
Methyl isobutyl ketone–dithizone, 345
Methyl orange, 136–137
Methyl parathion degradation, 181–182
Methyl *tert*-butyl ether, 51
3-Methyl-2-benzothiazolinone hydrazone, 47–48
Methylated chlorobenzoic acids, 281
Methylbenzylsuccinic acids, 20
Methylene blue, 138
Methylene dichloride, 259
Methylmercury compounds, 382
Metobromuron, 182–183, 185

Mexacarbate, 195
Microcoulometric detectors, 174
Microcystins, 226–227
Microextraction, 45
 of herbicides, 202–206, 203*t*
Micromolecular thin-layer chromatographic technique, 10–11
Microprocessor-controlled gas chromatograph–monitoring system, 78
Microsulphonation–gas chromatographic technique, 133–134
Microtrap, 154
Microwave-assisted extraction, 259–262, 316
 micellar extraction, 303–307
Mirex, 326
Mixed-mode electrokinetic capillary, 182
Modified EPA 8200 gas chromatography–mass spectrometry procedure, 6
Molecular emission cavity analysis, 49
Molecular extinction coefficients of cobalt thiocyanate complexes, 130
Monolinuron, 182–183, 185
Monomethylarsinate, 246
Monouron, 182–185
MS. *See* Mass spectrometry (MS)
M–*S*-Butyl phenyl methyl(phenylthio) carbonate, 195
Multiorganic compounds in nonsaline waters
 determination of organic compounds, 149–155
 preliminary extraction of organic compounds, 146–148
Multiresidue gas chromatographic method, 175
Multiresidue method, 141–142, 188, 366

N

N,N'-dimethylcarbamates, 197
N-acetyl trimethyl-ammonium bromide, 180–181
N-arylcarbamates, 197–198
n-hexyl derivatives, 236
N-methyl-*N*-trimethyltrifluoroacetamide, 362
N-Methylcarbamate, 197
N-Trifluoroacetyl methyl esters, 106
N-vinyl-2-pyrrolidone copolymer, preconcentration on, 50
1-Naphthol, 282–283
1-Naphthylperfluorooctanesulphonate, 282–283
Natural water
 GC of multicomponent organic mixtures in, 147*t*
 HPLC of organic mixtures in, 151*t*
Near-infrared method, 8
Neburon, 182–183
Negative-ion mass spectrometry, 72–73
Nemacur, 170–172
Neutron activation analysis
 organomercury compounds, 243
^{63}Ni electron-capture detector, 192
Nitriles, 110
Nitriloacetic acid, 111–115
 atomic absorption spectrometry, 112
 column chromatography, 112
 GC, 112–113
 polarography, 112–114
 spectrophotometric method, 114–115
Nitro compounds, 283
Nitrobenzene, 8–9
1-(4-Nitrobenzyl)-4–4(4-diethylaminophenylazo) pyridinium bromide, 135
Nitrogen-containing compounds, 357. *See also* Halogen-containing compounds; Oxygen-containing compounds; Phosphorus containing compounds
 in nonsaline waters
  aliphatic amines, 103–104
  amides, 107
  amino acids, 106
  aromatic amines, 105–106
  chloroaniline, 109
  EDTA, 111
  hydrazines, 109–110
  miscellaneous nitrogen compounds, 115
  nitriles, 110
  nitriloacetic acid, 111–115
  nitrophenols, 107–108
  nitrosamines, 110–111
  trinitrotoluene, 109
 in soil, 283–285, 284*t*
  growth regulators, 285
  hydrazines, 285
  miscellaneous, 285
  nitro compounds, 283
  polycyclic aromatic nitrogen heterocyclic, 283–284
Nitrophenols
 capillary electrophoresis, 108
 column chromatography, 108
 GC, 107–108
 spectrophotometric method, 108
 thin-layer chromatography, 108

Nitrosamines
GC, 110–111
nuclear magnetic resonance spectroscopy, 111
Nitrous oxide–acetylene flame atomic absorption spectrophotometry, 386
NMR. *See* Nuclear magnetic resonance (NMR)
Noncarboxyl aliphatic carbon, 286–288
Noninsecticidal compounds in soil
halogen-containing compounds in soil, 277–283
hydrocarbons, 269–273
miscellaneous organic soil, 286–288
mixtures of organic compounds in soil, 290–292
nitrogen-containing compounds in soil, 283–285
oxygen-containing compounds in soil, 273–277
sulphur and phosphorus–containing compounds in soil, 285–286
VOCs in soil, 288–290
Nonionic detergent polyoxyethylene, 303–307
Nonionic surface active agents, 127–133. *See also* Anionic surface active agents; Cationic surface active agents
atomic absorption spectrometry, 132–133
column chromatography, 128
gas chromatography, 127–128
ion-exchange chromatography, 129–130
miscellaneous, 133
spectrophotometric methods, 130–132
Nonpolar organic compounds, 255
Nonylphenols, 275–276
Nuclear magnetic resonance (NMR), 282–283
^{13}C NMR study, 286–288
spectroscopy, 121
for nitrosamines, 111

O

O-arylcarbamates, 197–198
Octanol–water partition coefficients, 273
Oestrogenic hormones, 227–228
Oil slicks, continuous monitoring of, 16–17
Oil spillages, 13–16
Online enrichment, 25
Online trace enrichment technique, 7
Oregon, 255
Organic acids, 255, 276–277
Organic compounds
in aqueous precipitation
carboxylic acids, 256
organolead compounds, 257
organomercury compounds, 256
organotin compounds, 256–257
pesticides, 256
phenols, 255
polycyclic aromatic hydrocarbons, 255
concentration in sediment, 349
determination, 149–155
GC, 149
HPLC, 149–150
infrared spectroscopy, 150–154
miscellaneous, 154–155
mixtures in soil, 290–292, 291*t*
in nonsaline waters
algal toxins and blooms, 226
anthropogenic, ostragenic and oestrogenic hormones, 227–228
antibiotics, 228–229
cobalamin, 226
geosmin, 225–226
humic and fulvic acid, 221–225
mestranol and ethynyloestradiol, 226
microcystins, 226–227
miscellaneous pollutants, 230
pharmaceuticals, 229–230
plant pigments, 219–221
preliminary extraction, 146–148
in sediments
chlorine-containing compounds, 351–357
concentration of typical organic compounds in sediment, 349
hydrocarbons in sediments, 349
insecticides and herbicides, 359–362
miscellaneous, 362–366, 363*t*
nitrogen-containing compounds, 357
oxygen-containing compounds, 349–351
phosphorus-containing compounds, 358
in river, lake and marine sediments, 350*t*
sulphur-containing compounds, 358–359
Organic sediment extracts, 285
Organic soil, miscellaneous, 286–288, 287*t*
humic and fulvic acid, 286–288
Organic solvents, 52, 146, 373

Organoantimony compounds, 248
Organoarsenic compounds, 246–248, 341, 342*t*, 374
  atomic absorption spectrometry, 247
  gas chromatography, 247
  ion-exchange chromatography, 248
  miscellaneous techniques, 248
  polarography, 246
Organochlorine insecticides, 176, 303, 304*t*
  GC, 162–166
  GC–MS, 167–168
  HPLC, 168–169
  mixtures of chlorinated insecticides and polychlorinated biphenyls, 166–167
  other techniques, 169
  preconcentration, 170
  thin-layer chromatography, 169
Organocopper compounds, 249
Organogermanium compounds, 249
Organolead compounds, 244–246, 257, 345, 374–376
  atomic absorption spectrometry, 245
  gas chromatography, 244–245
  polarography, 245
  preconcentration, 246
Organomercury compounds, 239–244, 256, 345, 373, 380–386
  atomic absorption spectrometry, 241–243
  gas chromatography, 239–241
  miscellaneous techniques, 243
  neutron activation analysis, 243
  in soils, 344*t*
  storage of mercury-containing samples, 243–244
Organometallic compounds
  in nonsaline waters
    organoantimony compounds, 248
    organoarsenic compounds, 246–248
    organocopper compounds, 249
    organogermanium compounds, 249
    organolead compounds, 244–246
    organomercury compounds, 239–244
    organoselenium compounds, 249
    organosilicon compounds, 249
    organotin compounds, 235–238
  in sediments
    organoarsenic compounds, 374
    organolead compounds, 374–376
    organomercury compounds, 380–386
    organosilicon compounds, 386
    organotin compounds, 376–380
  in soils
    organoarsenic compounds, 341
    organolead compounds, 345
    organomercury compounds, 345
    organotin compounds, 345
Organophosphonil pesticides, 182
Organophosphorus compounds, 121
Organophosphorus hydraulic fluids, 285
Organophosphorus insecticide, 170–182, 366
  electrochemical methods, 180–181
  extraction procedures, 172–173
  GC, 173–176
  HPLC, 176–177
  miscellaneous, 181–182
  and pesticides, 121
  spectrometric methods, 177–180
  thin-layer chromatography, 177, 178*t*
Organophosphorus-type herbicides in soil, 316–318, 317*t*
Organoselenium compounds, 249
Organosilicon compounds, 249, 386
Organotin compounds, 235–238, 256–257, 343*t*, 345, 376–380
  gas chromatography–mass spectrometry, 235–237
  inductivity-coupled plasma mass spectrometry, 238
  miscellaneous techniques, 238
Ormetoprim, 228
Orthogonal array design, 45
Osazone, 38
Oseltamivir, 230
Ostragenic hormones, 227–228
Oxygen-containing compounds. *See also* Halogen-containing compounds; Nitrogen-containing compounds; Phosphorus containing compounds
  in nonsaline waters
    alcohols, 51–52
    aldehydes, 54–56
    carbohydrates, 57–58
    carboxylic acids, 35–42
    dioxans, 53
    esters, 53–54
    glycols, 52–53
    ketones, 56
    lactams, 58
    methyl *tert*-butyl ether, 51
    phenolic acids, 50
    phenols, 42–50
  in sediments, 349–351, 353*t*
  in soil, 273–277, 274*t*

methoxy groups, 277
nonylphenols, 275–276
organic acids and ketones, 276–277
oxalates, 273–274
Oxytetracycline hydrochloride, 228

P

P-value approach, 174–175, 190–191
PAHs. *See* Polycyclic aromatic hydrocarbons (PAHs)
Paper chromatography, 12
Paper electrophoresis, 189–190
Paraffins, 2, 19–20
Paraoxon, 170–172
Parathion, 170–172
Partisil aluminosilicate, 186
PCBs. *See* Polychlorinated biphenyls (PCBs)
PDMS. *See* Polydimethylsiloxane (PDMS)
PEG 20M, 166–167
Pentachlorophenol, 88
Pentafluorobenzoyl chloride, 86
Pentane extraction technique, 68–69
Perfluorooctane sulphonyl fluoride, 282–283
Perhydropyrene, 10–11
Permethrin, 326
Pesticides, 256
organophosphorus, 121
survey, 206
Petroleum oils, 3–4
Petroleum products, 11
pH-change colorimetric assay, 200–201
Phaeophytin a, 220–221
Pharmaceuticals, 229–230
Phenolic acids, 50
Phenols, 42–50, 174, 255
atomic absorption spectroscopy, 49
GC, 42–44
high-performance liquid chromatography, 44–45
miscellaneous, 49
preconcentration
on cation-exchange resins, 50
on N-vinyl-2-pyrrolidone copolymer, 50
on silica gel–modified cellulose, 50
on 5-vinyl pyridine-divinylbenzene copolymer resin, 49
Raman spectroscopy, 46
spectrophotometric method, 47–48
thin-layer chromatography, 45–46
Phenoxy acetic acid–type herbicides, 186–191
miscellaneous, 190–191
paper electrophoresis, 189–190
in soil, 307, 310*t*
thin-layer chromatography, 189
Phenoxyalkanecarboxylic acids, 188
Phenoxyalkanoic acid herbicides, 187
Phenyltin compounds, 236–237
Phenylurea herbicides, 183
Pheo-GSB, 222
Phorate, 170–172
Phosphine, 120
Phosphorus containing compounds. *See also* Halogen-containing compounds; Nitrogen-containing compounds; Oxygen-containing compounds
in nonsaline waters
adenosine triphosphate, 120
alkyl and aryl phosphates, 119–120
inositol triphosphate, 120
organophosphorus compounds, 121
organophosphorus insecticides and pesticides, 121
phosphine, 120
plytase-hydrolysable phosphate, 120
in sediments, 358, 358*t*
in soil, 285–286
Photobromination process, 10–11
Photodegradation, 318
Photolysis, 181–182
Phthalates, 53
esters, 176
Plant pigments, 219–221
HPLC, 219–220
spectrophotometric and spectrofluorimetric methods, 220–221
thin-layer chromatography, 220
Plytase-hydrolysable phosphate, 120
Polar materials, 7
Polarography, 180
for amides, 107
for carboxylic acids, 41
humic and fulvic acid, 223
for lactams, 58
for nitriloacetic acid, 112–114
organoarsenic compounds, 246
organolead compounds, 245
for polychlorobiphenyls, 92
Poly-ethylated secondary alcohol ethoxylates, 129
Poly(oxyethylene)-nonylphenyl ethers, 128

Poly(oxyethylene)alkylphenyl ethers, 128
Polychlorinated biphenyls (PCBs), 95, 165–167, 176
Polychlorinated compounds, 73
Polychlorinated organic compounds, 366
Polychlorobiphenyl-contaminated soils, 282
Polychlorobiphenyl-degrading bacteria, 281
Polychlorobiphenyls, 88–93, 281, 356
  gas chromatography–mass spectrometry, 89–91
  GC, 89
  high-performance liquid chromatography, 91
  miscellaneous, 92–93
  polarography, 92
  thin-layer chromatography, 91–92
Polychlorodibenzo-*p*-dioxins, 83–84
  gas chromatography–mass spectrometry, 83–84
Polychlorodibenzofurans, 83–84
  gas chromatography–mass spectrometry, 83–84
Polychloroterphenyls, 93
Polycyclic aromatic compounds in sediments, 352*t*
Polycyclic aromatic hydrocarbons (PAHs), 21–22, 24, 26–27, 149, 255, 273
  of $C_{24}H_{14}$ isomer class, 27–28
Polycyclic aromatic nitrogen heterocyclic, 283–284
Polycyclic hydrocarbons, 8–9
Polycyclic musks, 365
Polydimethylsiloxane (PDMS), 141–142, 273
Polymeric nanofilm coated optical sensor, 20
Polyoxyethylene glycol 1550, 14
Polyoxyethylene-*n*-dodecyl analysis, 131
Polyurethane foam, 26–27, 92
Potassium chloride, 173–174
Potassium permanganate, 385
Potassium persulphate, 239
Potassium picrate active substances, concentration of, 131
Potentiometry, 41
  titration, 222
Preconcentration
  organochlorine insecticides, 170
  organolead compounds, 246
Pressurised liquid extraction, 259
Pressurised solvent extraction, 283
Primiphos-ethyl, 170–172
Primiphos-methyl, 170–172
Prometon, 192–193
Prometryne, 192–193
Propazine, 192–193
Propham, 195
Propoxur, 195
Prothoate, 170–172
Purge-and-trap
  analysis
    for haloforms, 79–80
    for saturated aliphatic chloro compounds, 72
    for unsaturated chloroaliphatic compounds, 75
  GC, 51
  GC–MS system, 53
PX-21 active carbon, 146
1-Pyrenyl diazomethane, 38
Pyrethroids, 199–201
Pyrolysis-gas chromatography–mass spectrometry with single-ion monitoring, 125
Pyxine Coralligera, 28

Q

Quantitative gas chromatographic analysis, 5–6

R

Rainwater, 255
Raman spectroscopy, 145
  laser-excited resonance, 46
  for phenols, 46
  surface-enhanced, 283
Reflection–absorption IR-sensing device, 290
Relative standard deviations (RSDs), 70, 142, 167, 228, 276
Resin adsorption method, 78
Resin adsorption–gas chromatography for haloforms, 80–81
Reversed-phase
  adsorption liquid chromatography, 128
  high-performance liquid chromatography, 220, 362–365
  liquid chromatography, 379
ρ-bromophenacyl esters, 256
Risk assessment, 269–273
RSDs. *See* Relative standard deviations (RSDs)
Rubidium sulphate, 173–174

S

s-triazines, 193
Saccharomyces cerevisiae, 194–195
Salithion, 170–172
Sample composition technique (SSC technique), 90–91
Sampling technique, 12–13
Saturated aliphatic chloro compounds
gas chromatography, 66–74
headspace analysis, 71–72
high-performance liquid chromatography, 70–71
miscellaneous, 73–74
negative-ion mass spectrometry, 72–73
purge and trap analysis, 72
thin-layer chromatography, 71
Saturated hydrocarbons, 19–20
Secbumeton, 192–193
Secondary ion mass spectrometry, 341
Sediments
concentration of typical organic compounds in, 349
hydrocarbons in, 349
organometallic compounds in
organoarsenic compounds, 374
organolead compounds, 374–376
organomercury compounds, 380–386
organosilicon compounds, 386
organotin compounds, 376–380
Segmental injection techniques, 137
Selenium, 373
Self-tuning single-mode microwave technology, 373–374
Semivolatile compounds, 288–290
Sephadex G100, 223
Shpol'skii effect, 26
SID. *See* Speciated isotope dilution (SID)
Siduron, 182–183
Signal losses in pesticide analysis, 168
Silica gel–modified cellulose, preconcentration on, 50
Siloxanes, 386
Simazine, 192–193
Single-drop microextraction technique, 69–70
Sodium borohydride, 374
reduction, 247
Sodium dodecylbenzene sulphonate, 135–136
Soil(s), 269
carbamate type of insecticides in, 307, 311*t*
chlorine containing insecticides and herbicides in, 303–307
halogen-containing compounds in, 277–283
hydrocarbons in, 270*t*
imidazolinone herbicides in, 307–316, 315*t*
miscellaneous insecticides in, 318–332, 319*t*
nitrogen-containing compounds in, 283–285
organophosphorus-type herbicides in, 316–318, 317*t*
oxygen-containing compounds in, 273–277
packed cartridges, 259
phenoxy acetic acid herbicides in, 307, 310*t*
soil-borne pathogens, 286
substituted urea-type herbicides in, 307, 313*t*
sulphur and phosphorus–containing compounds in, 285–286
texture, 318
triazine herbicides in, 307, 308*t*
VOCs in, 288–290
Solar photocatalysis, 51
Solid-phase carbon trap, 146
Solid-phase extraction (SPE), 303–307, 318–325
Solid-phase microextraction, 141, 263, 277
analyses, 316
fibre, 341
Solvent extraction
of hydrocarbons, 16
of phenols, 43
spectrophotometric methods, 130
Sonication, 263
sonication-assisted extraction, 325
Sorption coefficients, 273
Sorption isotherm parameters, 326
Soxhlet extraction, 263, 269–273, 285, 351
warm extraction, 283
SPE. *See* Solid-phase extraction (SPE)
Speciated isotope dilution (SID), 383
Species-specific isotope dilution analysis, 373–374
Spectrofluorimetry, 220–221, 286
Spectrometric methods, 177–180
Spectrophotometric methods
for alcohols, 51–52
for aldehydes, 55
for aromatic hydrocarbons, 18
for carbohydrates, 57
for carboxylic acids determination, 39–40
cationic surface active agents, 136–138
for glycols, 52

Spectrophotometric methods (*Continued*)
for haloforms, 81
for nitriloacetic acid, 114–115
for nitrophenols, 108
nonionic surface active agents, 130–132
for phenols, 47–48
plant pigments, 220–221
for trinitrotoluene, 109
Spillage oils, metals in, 16
Square-wave polarography, 180
Square-wave stripping voltammetry, 230
SSC technique. *See* Sample composition technique (SSC technique)
9ST. *See* Glutathione-*S*-transferase (9ST)
Standard Soxhlet extraction, 283
Stereoisomers, 362–365
Subcritical fluid extraction, 263–264
Subcritical water extraction, 263
Substituted urea-type herbicides in soils, 307, 313*t*
Sulphadimethoxine, 228
Sulphonylurea, 184
herbicides, 186
Sulphosulphuron, 186
Sulphur compounds dimethyl disulphide, 285–286
Sulphur species, 125
Sulphur-containing compounds
in nonsaline waters
alkylthiols, 124
chlorobenzo sulphonic acid, 125
dimethyl sulphoxide, 123–124
ethylene thiourea, 124
mercaptans and disulphides, 123
miscellaneous, 125
thiobenzamide, 125
in sediments, 358–359
in soil, 285–286
sulphur compounds dimethyl disulphide, 285–286
Supercritical fluid, 285
carbon dioxide, 146
chromatography, 276
extraction, 283
Supercritical monochlorodifluoromethane, 357
Surface acoustic wave sensors for GC detection, 121
Surface active agents in nonsaline waters
anionic surface active agents, 133–136
cationic surface active agents, 136–138
nonionic surface active agents, 127–133
Surface-enhanced Raman spectroscopy, 283
sweet wormwood. *See Artemisia annua* (sweet wormwood)
Synchrotron-based X-ray absorption spectroscopy, 357
Syringe and cartridge method, 12

T

Tannins, 222
Temperature programmed gas chromatograph, 6, 281
Terbutylazine, 192–193
Tertiary octylphenol ethoxylates, 127–128
Tetraalkylleads, 244, 345
Tetrachlorvinphos, 170–172
Tetrahydrothiophen, 125
Tetramethylbutyltin derivatives, 237
Texas banding, 269–273
Texas Risk Reduction Program, 269–273
Thermaerosol detector, 174
Thermal desorption–gas chromatography–mass spectrometry (DTD–GC–MS), 142
Thermionic detector. *See* Karmen–Guiffrida detector
Thermospray, 184
Thin-layer chromatography, 351
for aldehydes, 55
for aliphatic hydrocarbons, 10–11
for aromatic hydrocarbons, 25–26
carbamate type, 197–198
for carboxylic acids, 39
for chlorophenols, 87
for glycols, 53
for lactams, 58
for nitrophenols, 108
organochlorine insecticides, 169
organophosphorus insecticide, 177, 178*t*
for phenols, 45–46
phenoxyacetic acid–type herbicides, 189
plant pigments, 220
for polychlorobiphenyls, 91–92
for saturated aliphatic chloro compounds, 71
triazine type, 194
Thiobenzamide, 125
Thiophanate-methyl, 169
Time-of-flight mass spectrometry (TOF-MS), 149
Tin, 373
compounds, 237
Titration methods, 123, 136, 138

TOF-MS. *See* Time-of-flight mass spectrometry (TOF-MS)
Toluene, 146
Trapezoidal internal reflection element, 288–290
Trialkyllead species, 257
Trialkyltin compounds, 237
Triazine, 191–195, 366
GC, 191–193
herbicides in soil, 307, 308*t*
high-performance liquid chromatography, 193–194
miscellaneous, 194–195
thin-layer chromatography, 194
Tributyltin, 236, 376
Trichlorfon, 166
1,1,1-Trichloro-2(2-chlorophenyl)-2-(4-chlorphenyl)ethane (2,4'-DDT), 201
Trifluoroacetic acid, 83
Trihalomethanes, 81
Trinitrotoluene, 109, 283
Triphenyltin compounds, 237
Triton X-100, 131–132
Turbojet fuel, 13–14
Two-dimension (2D)
conductometric detection, 39
thin-layer chromatographic method, 22
Two-dimensional GC (GCXGC), 149

U

Ultrasonic extraction, 275
Ultraviolet (UV), 276, 303–307
detectors, 223
fluorescence detection, 255
irradiation, 165, 182
photochemical systems, 242
spectrophotometric method, 73
spectroscopy
for aromatic hydrocarbons, 19–21
humic and fulvic acid, 223–224
Ultraviolet–visible fluorescence spectroscopy (UV–Vis fluorescence spectroscopy), 145
Unidimensional thin-layer chromatographic separation technique, 25
Unsaturated chloroaliphatic compounds, 74–75
gas chromatography–mass spectrometry, 74
headspace analysis, 75
high-performance liquid chromatography, 75
mass spectrometry, 75
purge and trap analysis, 75
Uranin, 146
Urea herbicides, 182–186
gas chromatography, 183–184
high-performance liquid chromatography, 184–186
UV. *See* Ultraviolet (UV)
UV–Vis fluorescence spectroscopy. *See* Ultraviolet–visible fluorescence spectroscopy (UV–Vis fluorescence spectroscopy)

V

Vaporised semivolatile compounds, 288–290
Vapour pressure, 277–281
Velsicol VC-506, 170–172
Victoria Blue B, 135–136
5-Vinyl pyridine-divinylbenzene copolymer, 49
Volatile chlorinated hydrocarbons, 75
Volatile compounds, 358
Volatile organic compounds (VOCs), 141, 288
evolution in groundwaters, 142
in soil, 288–290
water contaminants with, 141–142
Volatile organometallic compound, 341

W

Water-only reversed phase liquid chromatography (WRPLC), 155
Water–air systems, 5
Weathering of nC_8–nC_{40} aliphatic hydrocarbons in soil, 269

X

X-ray absorption near-edge structure spectrometry (XANES spectrometry), 94, 357, 379
X-ray fluorescence spectroscopy (XRF spectroscopy), 94, 357
XAD-7 cation-exchange resin, 199

Kolibri 360 GmbH
Ferdinand-Jühlke-Straße 7
99095 Erfurt, Deutschland
produktsicherheit@kolibri360.de